AF843882

Pulse Doppler Radar

Pulse Doppler Radar

Principles, Technology, Applications

by

Clive Alabaster
Cranfield University

SciTech Publishing

Edison, NJ
scitechpub.com

SciTech PUBLISHING

Published by SciTech Publishing
911 Paverstone Drive, Suite B
Raleigh, NC 27615
(919) 847-2434, fax (919) 847-2568
scitechpub.com

Copyright © 2012 by SciTech Publishing, Edison, NJ. All rights reserved.

No part of this publication may be reproduced, stored in a retrieval system or transmitted in any form or by any means, electronic, mechanical, photocopying, recording, scanning or otherwise, except as permitted under Sections 107 or 108 of the 1976 United Stated Copyright Act, without either the prior written permission of the Publisher, or authorization through payment of the appropriate per-copy fee to the Copyright Clearance Center, 222 Rosewood Drive, Danvers, MA 01923, (978) 750-8400, fax (978) 646-8600, or on the web at copyright.com. Requests to the Publisher for permission should be addressed to SciTech Publishing, 911 Paverstone Drive, Suite B, Raleigh, NC 27615, (919) 847-2434, fax (919) 847-2568, or email editor@scitechpub.com.

The publisher and the author make no representations or warranties with respect to the accuracy or completeness of the contents of this work and specifically disclaim all warranties, including without limitation warranties of fitness for a particular purpose.

Editor: Dudley Kay
Production Manager: Robert Lawless
Typesetting: MPS Limited
Cover Design: Brent Beckley

This book is available at special quantity discounts to use as premiums and sales promotions, or for use in corporate training programs. For more information and quotes, please contact the publisher.

ISBN: 978-1-891121-98-2

Library of Congress Cataloging-in-Publication Data Applied for

Brief Contents

Contents

Preface

"So, how do you actually choose the precise values of pulse repetition frequency (PRF) to use in a medium PRF pulse Doppler radar?" This was the innocent, yet highly valid, question asked of one of my colleagues following his lecture on the subject which set me and my colleague, Evan Hughes, on a long line of inquiry several years ago. It was the issue that first fired our research interest in pulse Doppler radar. Years later, we have provided our own answer to this question, which has fostered an intrigue in all aspects of pulse Doppler radar. Our own research, consultancies, projects for industry, supervision of student projects and personal experience together with the contributions others have made in the field of pulse Doppler radar have culminated in the present book of the same title. So it seems fitting to start with the question that set it all off. That original question and many others like it are addressed in this book.

Actually, I had already been interested in pulse Doppler radar as a result of seven years' experience as a microwave engineer engaged in the design and development of airborne pulse Doppler radars. When I later became a lecturer to the military on radar techniques, I realized that pulse Doppler radars were used in air defense, missile seekers, naval systems, airborne early warning radars, and a whole host of other military radars as well. In fact, the range of applications was enormous and getting bigger all the time. Fourteen years ago, I joined Cranfield University as a lecturer in radar systems at what is now known as the Defence Academy of the United Kingdom in Shrivenham. So when the question of PRF selection was posed, it reignited my interest in all things pulse Doppler.

Given the plethora of pulse Doppler radars performing all manner of tasks, it came as rather a surprise to find so little of the real detail in the open literature. What was published was very informative but still left many questions unanswered. Discussions with other academics, government scientists, and engineers in the industry confirmed the importance of the subject but still left some issues unresolved. Yet, in spite of the growing relevance of pulse Doppler techniques, there has only been a gradual trickle of papers on the subject appearing in the open literature in recent years. Two conflicting perceptions seemed to be at large: first, that pulse Doppler radar has been fully developed; and second, a sense that certain aspects of pulse Doppler radar were something of a *black art*. The more I study the subject of pulse Doppler radar, the more I find myself dispelling both these perceptions as myths. Recent work has advanced the subject, and sound engineering principles have been brought to bear on the problems to demystify them. Heeding my own views, I contend that there is ample scope for future developments.

This book attempts to explain pulse Doppler radar from the basics through to the latest research. It is designed to appeal to engineers, academics, researchers, users, those involved in the technical aspects of specification and procurement and (postgraduate) students. It is written in an instructive style for the most part when discussing established techniques and the current state-of-the-art but occasionally adopts the style of a review when covering recent research that may not necessarily be implemented on current systems but that may influence the design of future systems. In this way, the material in this book provides the reader with an up-to-date view as well as a glimpse of what might lie ahead in the near future. It draws on much of the research effort conducted at Cranfield University and the broad lessons learned from some of our personal experiences and the consultancies and studies conducted for industry. Material is also referenced from standard texts, the early seminal papers in the subject, and recently published research papers.

The book is organized into three parts, each one building on the material of the previous sections. Part I (Chapters 1–8) covers the basic principles to lay sound foundations for the following parts of the book. It emphasizes classic processing techniques, especially the fast Fourier transform (FFT), and microwave engineering issues, antennas, and hardware. The second part of the book deals with the theory and techniques specific to pulse Doppler radar. This is subdivided into Part IIA (Chapters 9–10), which covers high PRF pulse Doppler, and Part IIB (Chapters 11–15), which covers medium PRF pulse Doppler. A major theme is that of PRF selection and optimization, other waveform design issues, and the problem of ghosting. While high and medium PRF pulse Doppler techniques have become synonymous with airborne fire control radars, they are used over a broad spectrum of airborne- and surface-based radar applications. For sure, Part II does emphasize the airborne radar case, but it does not neglect the surface-based radar. Finally, Part III (Chapters 16–19) presents a series of four case studies. Each of these case studies applies the material of Part II whilst also highlighting additional radar techniques (and, in some cases, non-radar considerations) specific to the application. Such is the prevalence of pulse Doppler radars today; the number of case studies that could have been considered is well into double figures. However, the four presented here suffice to illustrate the wide variety of pulse Doppler radar applications.

The book progresses the story of pulse Doppler radar through successive chapters and so is best read from start to finish. However, few of us use a textbook of this nature like this and tend to refer to specific topics. Hence, I hope to have presented the material in a manner that supports the reader who prefers to dip in and out of the various subjects whilst minimizing any repetition.

Finally, I would be pleased to hear from you, particularly if you have any comments, questions, or corrections.

Clive Alabaster
Shrivenham, UK
April 2012

Author Acknowledgements

First and foremost, I would like to acknowledge the important input of Evan Hughes, my friend and colleague at Cranfield University, with whom I have worked on pulse Doppler radar for many years. Evan's contribution to this subject area has been huge in recent years and is deserving of special mention. Furthermore, through our work together, he has supported, advised, and encouraged me in the writing of this book.

I would also like to acknowledge the contributions made by a succession of master's students whose projects Evan and I have supervised over recent years. We have been blessed with some very talented and willing "sorcerer's apprentices" who have done a great deal of work to advance the science of pulse Doppler radar.

I also owe a huge debt of thanks to my family, especially my wife, Lindsay, for their forbearance of my long hours of incarceration in the study while I have been writing; the regular supply of coffee and the hours spent proofreading are much appreciated.

My thanks also go to my employers, Cranfield University at Shrivenham, UK, for allowing me the time and space to work on this book.

I'd also like to thank anyone who has contributed to the subject matter of this book in any way—from the authors of the papers and books referenced here and the reviewers of my scripts and colleagues at Cranfield University to those with whom and for whom I have worked on various research and consultancy projects relating to pulse Doppler radar.

Finally, my thanks to those at SciTech Publishing who have made this project possible, for their patience, encouragement, and friendly help and advice. Dudley Kay as overall editor encouraged and supported the book in its proposal and development stages. Robert Lawless was invaluable as the book's production manager, keeping the details squared away for accuracy and consistency, while also keeping me on schedule. Finally, marketing director Brent Beckley contributed the very stunning cover design and a strong early interest in the book's success. It is quite gratifying for an author to work with a close-knit and responsive team of publishing professionals.

Publisher Acknowledgements

SciTech Publishing gratefully acknowledges the selfless time and labor from members of the radar community who appreciate and help ensure our publication of high quality radar books:

Dr. Theodore G. Kostis, *research scientist*, University of the Aegean, GR
Dr. Mark A. Richards, *professor*, Electrical and Computer Engineering Dept, Georgia Institute of Technology
John D. Sahr, *professor*, Dept of Electrical Engineering, University of Washington
Francois Le Chevalier, *VP*, Scientific Director, Thales Air Operations, FR
Christopher Baker, Ohio Research Scholar: Integrated Sensor Systems, Ohio State University
Patrick Dever, P.E., *fellow engineer*, Northrop Grumman Corporation

PART I

Basic Concepts

CHAPTER 1

Historical Justification for Pulse Doppler Radar

The very name of *radar* hints at one of its most basic properties; that of being able to determine the *range* of a target. The term radar was coined by an American officer in 1943 as an acronym for RAdio Detection And Ranging. So the ability of radar to detect the presence of a target and determine the range of that target is embedded into its own name. Range may be measured by timing the delay between the transmission of a pulse of radio frequency (RF) energy and the reception of its subsequent echo from the target since it is known that RF, as with all bands of the electromagnetic spectrum, propagates at a velocity equal to the speed of light. Hence, pulsed radar has become synonymous with radar ranging. This is still very true today, although other modulation methods are also exploited for the measurement of range. However, radar can offer more than just a measurement of range.

The ability of a radar to determine the angle to a target in both azimuth (the horizontal plane) and elevation (the vertical plane) is determined by the radiation pattern of its antenna. A typical radiation pattern of a directive antenna is illustrated in Figure 1-1 and indicates the variation of antenna gain over angles of azimuth and elevation. The width of the antenna main beam is determined by the ratio of the wavelength of the RF signal, λ, to the antenna dimension, D. A narrow beam is achieved if the antenna dimension is many times the wavelength of the RF signal it supports. Hence, by knowing the orientation of the radar beam at the time of target detection, the azimuth and elevation angles to the target can be deduced. Once azimuth, elevation, and range are established, a target can be located in three-dimensional space. For conventional radars, the angular resolution is set by the beamwidth of the receiving antenna. It is worth noting that the horizontal and vertical dimensions of a radar antenna need not necessarily be equal, resulting in unequal beamwidths, and therefore unequal angular resolutions, in the azimuth and elevation planes. Early radars that emerged just before and during the first half of World War 2 typically used metric wavelength signals. Thus, it became difficult to achieve fine angular resolution unless very large antennas were used. This was particularly troublesome in early airborne intercept (AI) radars, which operated on a wavelength, λ, of 1.5 meters (frequency = 200 MHz), and the aircraft that carried them were quite clearly restricted on the size of antenna they could accommodate. This resulted in wide main beams, which in turn gave rise to poor directionality.

Note that wavelength (λ) and frequency (f) are related by

$$\lambda = \frac{c}{f} \tag{1-1}$$

in which c is the speed of light ($= 3 \times 10^8$ m/s).

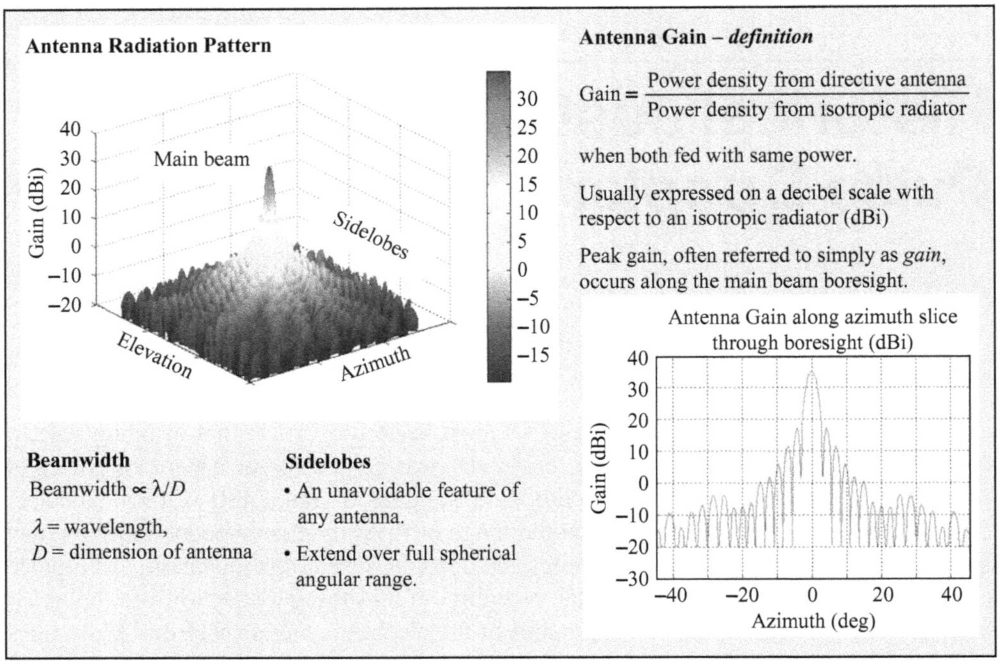

FIGURE 1-1 Radiation Pattern of Directive Antenna

Antenna limitations also lead to large antenna sidelobes, which would illuminate the ground even when the main beam was directed upward. If the main beam was directed downward, as it would be when searching for low-flying targets from a high-flying platform, the main beam, too, would illuminate the ground. The ground acts as a large extended target filling a wide angular view and extending in range from one's own altitude out to the horizon. A considerable echo from the ground known as *surface clutter* is therefore often present and may often overwhelm the echoes from smaller targets. Surface clutter is usually undesirable because it can be the limiting factor determining the detection of targets of interest. In fact, early AI radar was said to be limited to a range equal to its own altitude for this very reason. Clearly, some form of target discrimination is required to counter the clutter problem. The most obvious discriminant is *velocity*, since in most cases the targets of interest have a significantly higher velocity than the unwanted clutter return.

The advent of the magnetron in 1940 gave birth to high-power centimetric wave radar. Early in 1942, these were built into the AI Mk VIII, which operated on the 10 cm band (3 GHz). Its 71 cm diameter antenna boasted a beamwidth of 12° and was capable of detecting an enemy aircraft at a range of 4 miles (6.4km) [1]. Versions of AI Mk IX completed after the end of the war also included a scanning antenna and angle lock facilities. A useful derivative of the centimetric wave AI radars was the H2S radar system, also operating on 3 GHz and used as a navigation aid on British bombers from the start of 1943. H2S was underslung and aimed toward the ground to map the ground with high resolution. In this application, the ground and surface features and structures on the ground form the targets of interest. Clearly, the move into centimetric wave radar paved the way for a new generation of finer-resolution radars for navigation, tracking, and weapons-aiming purposes.

Another interesting development from World War 2 was the emergence of *chaff*. Chaff is a collection of thin metallic strips that, when deposited in quantity into the sky, form a large radar target. A chaff cloud is used as a radar countermeasure since the large target it represents is very confusing to an enemy radar operator or radar tracker. These strips are normally cut to half of the wavelength of the victim radar so that they act as half-wave dipoles and present a large radar cross section (RCS) to the radar. Chaff was independently developed in Britain, where it was originally known by the codename *Window,* and in Germany, where it was known as *Düppel*. Both sides refrained from using it since its use would immediately give the secret away to the other side, which could then copy it, eliminating its advantage since at the time no counter to it had been developed. Finally, the British used *Window* as part of Operation Gomorrah, the fire raids against Hamburg, in July 1943. The counter to chaff did not materialize until moving target indicator (MTI) radar and Doppler sensing radars emerged several years later, although chaff remains an effective radar countermeasure to this day.

Early radars had no means of measuring target velocity directly. A succession of detection points acquired over several scans or several seconds of data can be used to form a target track and can give an indication of target speed and direction. However, in many applications a more immediate measurement of target velocity is often desirable. Furthermore, by introducing some form of velocity discrimination, many of the problems associated with the earlier radars can be ameliorated. The *radial velocity* or *range rate* of a target can be measured by virtue of a measurement of its *Doppler shift*. The *Doppler shift* is the frequency difference between a transmitted signal and a received signal when the transmitter and receiver are moving with respect to each other. In the case of radar there is a two-way Doppler effect: first, the Doppler shift between the transmitter and target (which may be regarded as the receiver of the RF signal) on the outward path; and, second, the Doppler shift between the target (now seen as the emitter of the RF echo signal) and the receiver on the return path. The magnitude of the Doppler shift is proportional to the *radial* velocity between the radar and target, that is, the velocity difference between the radar and target resolved along the line of sight from the radar to the target. This is also known as the *range rate,* since it is the first derivative of the target range with respect to time. For typical velocities, the Doppler shift is a very small percentage change in the frequency of the RF carrier. Such a small frequency shift can be measured with reasonable resolution only by using coherent detection methods. (Some non-coherent methods have been used in the past by referencing the Doppler shifted return from a moving target to a large unshifted clutter echo, but these need not be considered any further here.) Doppler sensing radars place strict specifications on the stability and purity of the signals used and the preservation of phase coherency through the transmitter and in the reception of echoes in a receiver. The processing of returns is typically accomplished using fast Fourier transform (FFT) techniques, which must operate in near real time. The implementation of a Doppler sensing radar exploits recent advances in both hardware and software.

A pulse Doppler radar is the result of combining Doppler sensing techniques with pulsed radar operation. Such radars offer the powerful scope for direct measurements of both target range and velocity, even in the face of large clutter returns and even in the presence of chaff or other interference. To radar, the measurement of a time delay is tantamount to range, whereas the measurement of a (Doppler) frequency shift is tantamount to velocity. Thus, one must consider the design of suitable waveforms in both the time and frequency domains. However, there is a complex interplay between the waveform parameters, particularly when it comes to the selection of the radar pulse repetition frequency

(PRF). All too often, the requirements for the measurement of range clash with those for the measurement of velocity. The result is that multiple waveforms may be required which depend on the nature of the targets and clutter conditions. The study of pulse Doppler radar is thus inextricably bound up with issues of waveform design and their associated processing methods whilst always maintaining a watchful eye on the target scenario and radar environment.

Before tackling these waveform and processing issues, it is necessary to establish several fundamental radar concepts, which are covered in the ensuing chapters.

CHAPTER 2

Radar Detection Performance

Chapter Outline

2.1 NOISE-LIMITED RADAR RANGE EQUATION

2.1.1 Derivation of the Basic Radar Range Equation

Consider a transmitter transmitting a single pulse of radio frequency (RF) of peak power, P_{pk}, via an isotropic radiator. The radiation pattern of an isotropic radiator is a sphere and, at a distance, R, away from the transmitter the radiation has expanded to a sphere of radius, R, whose surface area is $4\pi R^2$. Since all the transmitted power must intersect this sphere, then

$$\text{Power density on surface of sphere} = \frac{P_{pk}}{4\pi R^2} \text{ watts/m}^2$$

If the isotropic radiator is replaced with a directive antenna of main beam boresight gain $= G$, then

$$\text{Power density on boresight} = \frac{P_{pk}G}{4\pi R^2} \text{ watts/m}^2$$

If a target of radar cross section (RCS) $= \sigma$ is on the antenna boresight at a distance R from the transmitter, then

$$\text{Power incident on target} = \frac{P_{pk}G\sigma}{4\pi R^2} \text{ watts}$$

From the definition of RCS, the target scatters this power equally in all directions; that is, it may be considered as an isotropic scatterer (see Section 2.5.1). The target therefore acts as a source of radiation of $\frac{P_{pk}G\sigma}{4\pi R^2}$ watts having a spherical scattering pattern. In truth, targets are not isotropic scatterers but the RCS of a target at any particular angle of illumination is defined in terms of an equivalently sized isotropic radiator that returns the same signal strength as obtained from the actual target. The power density back at the point of the original transmitter, a distance R away, is given by

$$\text{Power density back at transmitter} = \frac{P_{pk} G \sigma}{4\pi R^2 \cdot 4\pi R^2} = \frac{P_{pk} G \sigma}{(4\pi)^2 R^4} \text{ watts/m}^2$$

In a monostatic radar, the receiver is colocated with the transmitter. The effective antenna aperture, A_E, is a measure of the area of the receiving antenna that intercepts the returning signal. It is related to the actual area of the antenna, A, by

$$A_E = \eta A \tag{2-1}$$

in which η represents the antenna efficiency factor. A value of $\eta = 2/3$ is typical. This accounts for antenna mismatch losses and ohmic losses. The effective aperture, A_E, represents the area over which the antenna collects the returning signal with 100% efficiency. This is illustrated in Figure 2-1. In monostatic radars, the same antenna may be used for both transmission and reception. The effective aperture and gain of an antenna are related by

$$G = \frac{4\pi A_E}{\lambda^2} \tag{2-2}$$

The receiver is fed by an antenna having an effective aperture, A_E. Therefore,

$$\text{Received signal power, } S = \frac{P_{pk} G A_E \sigma}{(4\pi)^2 R^4} \text{ watts}$$

Rearranging this for range gives

$$R^4 = \frac{P_{pk} G A_E \sigma}{(4\pi)^2 S} \tag{2-3a}$$

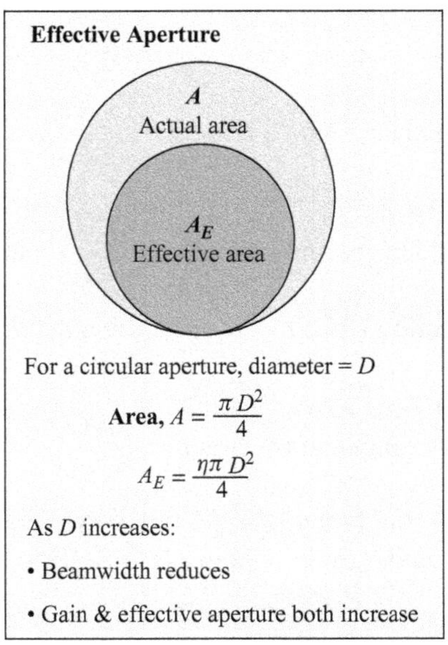

FIGURE 2-1 Effective Antenna Aperture

Substituting (2-2) into (2-3a) obtains

$$R^4 = \frac{P_{pk} A_E^2 \sigma}{(4\pi)\lambda^2 S} \tag{2-3b}$$

or

$$R^4 = \frac{P_{pk} G^2 \lambda^2 \sigma}{(4\pi)^3 S} \tag{2-3c}$$

Equations (2-3a, 2-3b, and 2-3c) represent the most basic form of the radar range equation. Maximum range, R_{max}, is achieved when the received signal power is the minimum necessary to meet the specified detection performance, S_{min}. In these circumstances the radar range equation can be recast as

$$R_{max} = \sqrt[4]{\frac{P_{pk} G A_E \sigma}{(4\pi)^2 S_{min}}} \tag{2-4}$$

It is worth noting that the antenna features twice within the basic radar range equation (2-4), since it is the last element in the transmission of a signal and the first element in the reception of the signal. This makes the antenna doubly important and adds weight to the old argument that *a well-designed RF system starts with a good antenna.* Noting that A_E and therefore G are both proportional to D^2, where D is the diameter of a circular antenna aperture, and that the radar range, R, is proportional to the fourth root of the product of GA_E, one obtains

$$R_{max} \propto D \tag{2-5}$$

Doubling the diameter of the antenna doubles the detection range, all other factors being equal, whereas if one was to double the detection range by increasing the transmitted power alone then a 16-fold increase in RF power would be required. Similarly, halving the antenna diameter halves the range, whereas halving the RF power reduces the range to 84% $\left(= \sqrt[4]{0.5}\right)$ of its original value. Radar range is relatively insensitive to transmitted power; such is the nature of the fourth root law of the radar range equation.

Other parameters indirectly affecting detection performance are also a function of antenna size, D, and are neglected from equation (2-5). Antenna beamwidth is inversely proportional to D, so for a fixed scan rate the beam dwell time on target reduces as D increases. This reduces the scope for pulse integration (Section 2.4) and hence the processing gain and, ultimately, the radar detection performance.

2.1.2 Losses

Equation (2-4) gives an overly optimistic estimate of the maximum detection range since atmospheric losses and other system losses must be considered.

The atmosphere imparts some attenuation and scattering to the passage of an RF signal propagating through it. These losses tend to increase with increasing frequency and with increasing atmospheric moisture content (rain, fog). Atmospheric losses are also temperature and pressure dependent and so vary with altitude. Let the two-way atmospheric loss factor be denoted by L_A.

A host of other loss mechanisms should also be considered:

- Plumbing losses, which are associated with the transmission line connecting the transmitter to the antenna (often waveguide where high power transmitters are used)
- Receiver losses, which are associated with the transmission line and components between the antenna and receiver input not accounted for as part of the receiver
- Radome losses

Note that antenna mismatch losses and efficiency have already been accounted for in the antenna gain term. Let the total system losses be denoted by L_{sys}. Including these losses into the radar range equation now results in:

$$R_{max} = \sqrt[4]{\frac{P_{pk} G A_E \sigma}{(4\pi)^2 S_{min} L_A L_{sys}}} \tag{2-6}$$

2.2 DETECTIONS IN NOISE

2.2.1 Threshold Detection

A target detection is declared when the detected voltage exceeds a suitable threshold level. Target returns compete with noise, interference, and clutter for detection. For most of the time one of these dominates over the other two and therefore sets the fundamental limit of any radar to detect weak target returns. All three exhibit statistical fluctuations, so the composite signal of target echo plus noise/interference/clutter also exhibits a statistical fluctuation. This results in a situation in which the target plus noise/interference/clutter signal may be sufficiently strong to exceed the detection threshold at one moment but too weak to exceed the detection threshold the next. The *probability of detection*, P_D, for any given target is the probability that the target plus noise/interference/clutter signal exceeds the detection threshold, and it is therefore a statistical quantity. A $P_D = 0$ represents a certainty that the target cannot be detected, whereas a $P_D = 1$ represents a certainty that the target can be detected. In practice, both these extremes are impossible; hence, there always exists a finite probability that a target may be detected, however small or distant it may be. Correspondingly, there is never a guarantee that a target can be detected no matter how large or close it may be. Furthermore, in the absence of any target, the noise/interference/clutter signal may be sufficiently strong to exceed the detection threshold on occasions and thereby trigger a false alarm. This gives rise to a *probability of false alarms*, P_{FA}, and therefore a *false alarm rate, FAR*.

The probability of false alarms, P_{FA}, is defined as the proportion of the time that noise exceeds the detection threshold, that is,

$$P_{FA} = \frac{\text{time noise exceeds threshold}}{\text{unit time}} \tag{2-7}$$

A more useful quantity to a radar operator is the false alarm rate, *FAR*, which is the number of false alarms incurred within a given interval of time. *FAR* is therefore a frequency. Its inverse, the false alarm time, T_{FA}, is therefore the mean time between false alarms. *FAR* is a function of P_{FA} and also of the bandwidth in which the noise is detected,

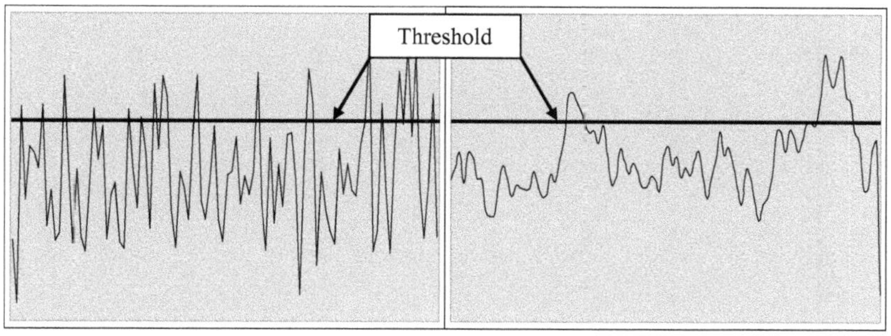

FIGURE 2-2 Wide-Band Noise (left) and Narrow-Band Noise (right)

the so-called *noise bandwidth*, B_N. These quantities are related by

$$FAR = P_{FA} B_N \tag{2-8}$$

The role of the noise bandwidth is illustrated in Figure 2-2, which depicts a burst of wide-band noise (left) and narrow-band noise (right). As can be seen, the wide-band noise is spikier and more threshold crossings (false alarms) of shorter duration are incurred compared with the narrow-band noise. The P_{FA} is identical in both cases, although, obviously, the *FAR* differs. Since the (noise) bandwidth of many radars is quite high, typically in the order of 100s of kHz or several MHz, the P_{FA} needs to be an exceedingly small quantity to keep *FAR* down to acceptable levels. Older radars typically required the P_{FA} to be in the order of 10^{-8} to 10^{-12} or lower. However, modern trackers have become very good at eliminating false tracks. A higher *FAR* is now acceptable since the tracker can apply a second thresholding process that eliminates many false targets.

Trackers relate a succession of target returns, typically over several scans, such that previous target detections are correlated to current ones and may be used to predict the position of future detection points. This is a useful feature to view the history of the target locations and judge, for example, its heading, speed, and likely intention. Trackers can also maintain tracks on targets even if the target detection is temporarily missed. The random nature of false alarms means that they are very unlikely to form tracks.

If all external sources of interference, clutter, and noise are excluded from the radar, thermal noise is still present on the receiver input. The voltage of thermal noise is described by a Gaussian probability density function (PDF) and has a uniform spectral power density, which means that the strength of the noise remains the same over all frequencies. This leads to thermal noise being dubbed "white" on account of its optical equivalent with visible light. A radar receiver processes signals using a filter whose bandwidth is typically quite narrow compared with its center frequency. The processing of white thermal noise using such a narrow-band filter results in a noise voltage whose envelope is described by a Rayleigh PDF [2]; this is plotted in Figure 2-3. The Rayleigh PDF of the envelope of the narrow-band noise indicates that the probability of obtaining a large noise voltage is low but is never zero. Particularly large noise voltages are very unlikely but not impossible. Therefore, there is always a finite possibility that the noise alone can exceed the threshold and trigger a false alarm. The probability of incurring a false alarm may be quantified by integrating the area under the Rayleigh curve describing the noise distribution that lies above the threshold, that is, from a lower limit of the threshold voltage to an upper limit of infinity.

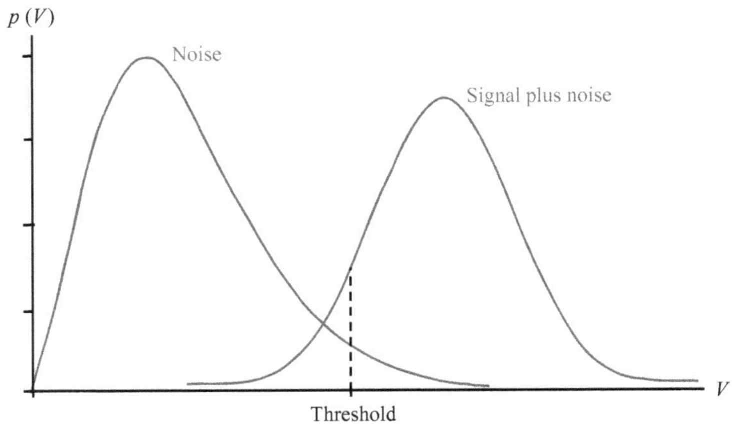

FIGURE 2-3 Noise and Signal-Plus-Noise Voltage Envelope Distributions

Target signals are also received in the presence of (thermal) noise. It is convenient for the time being to consider the case in which the target echo remains constant. This results in a signal-plus-noise voltage envelope adhering to a Rician PDF, which is also plotted in Figure 2-3. The peak of this distribution is set by the signal voltage arising from the target echo, and its spread is proportional to the variance in the noise voltage. It is desirable to have a situation in which the majority of this distribution lies above the threshold voltage. However, it is in the nature of the Rician PDF that the probability of obtaining small voltages is very low but is never zero. Again, very small signal-plus-noise voltages are exceedingly unlikely but not impossible. So there always exists a finite possibility that the signal comprising the target-response-plus-noise fails to exceed the detection threshold and results in a missed detection. A detection is declared when the signal-plus-noise voltage exceeds the threshold. The probability of detection can be quantified by integrating the area under the Rician curve describing the signal-plus-noise distribution that lies above the threshold, that is, from a lower limit of the threshold voltage to an upper limit of infinity. P_D and P_{FA} are determined from the integrated areas under these curves that, in many cases, are evaluated from graphical solutions or approximations. If the theory is extended to include the statistical fluctuation of realistic targets, the situation may grow more complex or simpler, depending on the statistical model of target fluctuation assumed. This is considered further in Section 2.5.4.

An example of a target detected in Gaussian noise is given in Figure 2-4. A target of 4 volts is combined with a set of Gaussian noise samples having a zero mean and a standard deviation of 1 volt that, for the Gaussian distribution, is the root mean square (RMS) noise voltage. The ratio of the signal voltage to the (RMS) noise voltage is the *voltage signal-to-noise ratio*, SNR_V, which equals 4.0 in this example. The left-hand plot shows the bipolar voltages, as may be observed prior to the detection stage, whereas the right-hand plot shows the unipolar voltage envelope, as would be observed on the output of a linear detector. It is also commonplace to refer to the ratio of the signal power to noise power as simply the *signal-to-noise ratio* (SNR); here it is understood and therefore unstated that SNR refers to a *power* ratio. Note also that $SNR \propto (SNR_V)^2$. In Figure 2-4, the combined target-plus-noise signal is apparent in sample 50 and has a voltage of 4.16 volts. A random noise spike of 3.0 volts is also seen in sample 7. Automated detection may be provided by comparing the received signal against a threshold. Clearly, a threshold is required that is exceeded by the target-plus-noise signal but is not exceeded by the noise

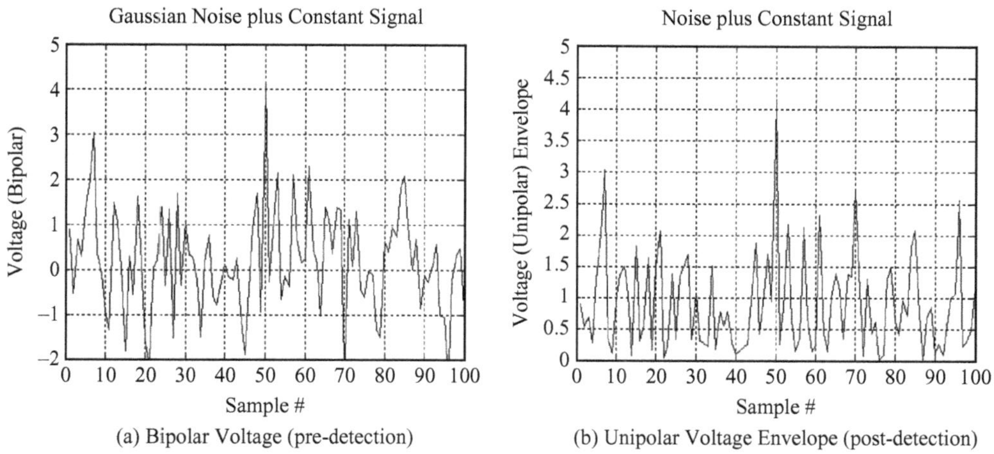

FIGURE 2-4 Gaussian Noise Plus Target

alone. In this case a threshold level of 3.5 volts would seem reasonable. If the threshold is reduced to 2.5 volts, weaker targets would be detected, but more false alarms, such as the noise spikes in sample numbers 7, 70 and 96, would be incurred. The number of false alarms is very sensitive to the threshold to noise ratio. If, on the other hand, the threshold is increased to 5 volts, there would be greater resilience against false alarms but also a risk of missing genuine target returns. Lower thresholds increase both P_D and P_{FA}, whereas higher thresholds reduce both P_D and P_{FA}. The better the ratio between the desired signal return from the target and the noise/interference/clutter level, the better a suitable threshold yielding a high P_D and a low P_{FA} may be selected. This is considered further in Section 2.5.5. The combination of P_D and P_{FA} denotes the *radar detection performance*.

2.2.2 Cumulative Probability of Detection

Some texts also refer to the *cumulative probability of detection*, P_{DC}. This is defined in various ways. Skolnik [2] defines P_{DC} as *the probability of detecting a target at least once over N scans*. Such a definition must be mindful of the fact that the P_D may differ from one scan to the next if the target range changes over N scans. Stralka and Fedarko [3] define P_{DC} in terms of a closing target as *the probability that the radar will detect a closing target at least once by the time the target has closed to a specified range*. These two definitions do have an equivalence that may be expressed thus: the cumulative probability of detection represents the probability that a target is detected at least once given several opportunities, perhaps over successive scans or simply in time. P_{DC} is a function of the P_D applicable at each individual opportunity to detect the target and accumulates over successive opportunities. Thus, P_{DC} is initially equal to P_D at the most distant range but increases with decreasing range more rapidly than does P_D. Alternatively, P_{DC} is initially equal to P_D on the first scan but is subsequently higher than P_D at any specific scan number.

The use of the cumulative probability of detection is perhaps less prevalent than its definition suggests. Many radars, particularly pulse Doppler radars, do not declare a target on its first detection but seek confirmation of the target on a second, third, or fourth scan before alerting the operator. The performance of tracking algorithms is dictated largely on how an algorithm handles missed detections rather than when a first detection is made.

2.3 MINIMUM DETECTABLE SIGNAL, S_{min}

2.3.1 Thermal Noise

If all sources of clutter, external noise, and interference could be excluded from the radar, thermal noise would still be present on the input to a radar receiver. Thermal noise is generated naturally in all conductors and semiconductors and ultimately sets the limit on the receiver sensitivity. It arises due to the random thermal agitation of electrons in conductors and gives rise to the flow of random currents and therefore random noise voltages across resistances. The thermal noise voltage has a Gaussian PDF. Under matched input conditions the thermal noise power, N_i, is given by

$$N_i = kTB_N \tag{2-9}$$

where k is Boltzmann's constant $= 1.38 \times 10^{-23}$ (J/K), T is temperature in degrees Kelvin (K), and B_N is noise bandwidth (Hz). By convention, a standard temperature, $T_0 = 290$ K, is taken for reference purposes. This gives rise to a spectral noise power density on the receiver input of

$$N_i = 4 \times 10^{-21} \text{ watts/Hz} \quad \text{or} \quad 4 \times 10^{-18} \text{ milliwatts/Hz} \quad \text{or} \quad -174 \text{ dBm/Hz}$$

This forms a fundamental lower limit on the thermal noise spectral power density.

2.3.2 Noise Figure

As will be seen in Section 8.4, no receiver is perfect. The receiver components contribute some additional noise onto the signal path and may impart losses that reduce the power of the target signal (losses do not affect the noise level since a level of at least kT_0B_N is always present). Both these effects degrade the signal-to-noise ratio as a target return progresses along the receiver chain. The added noise and loss effects are characterized by the *noise figure* or *noise factor*[1], F, of the receiver.

Noise figure is defined as:

The factor by which the noise on the output of an ideal device must be increased to give the same noise level out of the actual device.

This definition and the associated signal flow are depicted in Figure 2-5.

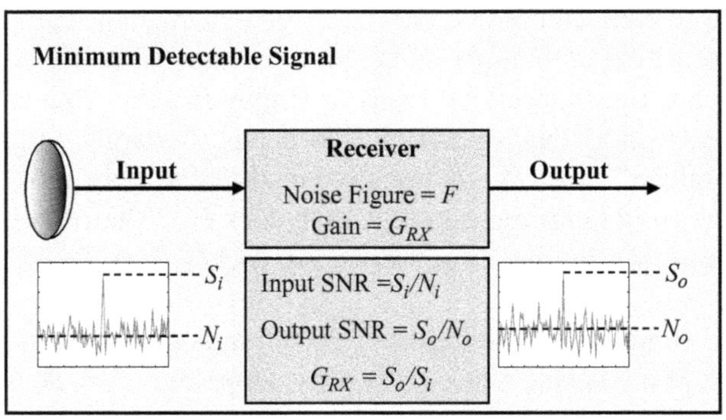

FIGURE 2-5 Minimum Detectable Signal

[1] Both terms are in common usage, however, the term *noise figure* is used here.

Let S_i = signal power on input, N_i = noise power on input, and S_o = signal power on output, N_o = noise power on output. Under ideal conditions, N_i is matched thermal noise only; that is,

$$N_i = kT_0B_N \tag{2-10}$$

The gain of the receiver,

$$G_{RX} = S_o/S_i \tag{2-11}$$

Therefore, the signal strength on the output is $S_o = G_{RX}S_i$. In the case of an ideal, noiseless receiver the noise on the output would be $G_{RX}N_i$. However, from the definition of noise figure, the actual noise on the output is

$$N_o = FG_{RX}N_i \tag{2-12}$$

Substituting (2-10) and (2-11) into (2-12) gives

$$N_o = kT_0B_NF(S_o/S_i) \tag{2-13}$$

Rearranging (2-13) for F gives

$$F = \frac{(SNR_i)}{(SNR_o)} \tag{2-14}$$

where

$$(SNR_i) = S_i/N_i \text{ denotes the input SNR}$$

$$(SNR_o) = S_o/N_o \text{ denotes the output SNR}$$

This indicates that the noise figure is the ratio of the input SNR to the output SNR, and this relationship is often taken as the definition of noise figure. This relationship stems from the earlier definition, which is based on the assumption that the input noise is matched thermal noise only. Equation (2-14) does not hold when external noise or noise-like interference is received by the radar, so caution must be exercised if using (2-14) as a definition of noise figure. However, equation (2-14) is a useful illustration of the fact that the noise figure of a receiver degrades the signal-to-noise ratio. The ideal receiver would have a noise figure, $F = 1$ (0 dB), and the input and output signal-to-noise ratios would be equal. A practical receiver has $F > 1 (> 0 \text{ dB})$ and has a lower output SNR than its input SNR.

2.3.3 Minimum Input Signal

Rearranging (2-13) for S_i gives

$$S_i = kT_0B_NF(S_o/N_o)$$

or

$$S_i = kT_0B_NF(SNR_o)$$

The minimum tolerable input signal is determined by the minimum output signal-to-noise ratio necessary to achieve the specified detection performance. Hence,

$$S_{min} = kT_0B_NF(SNR_o)_{min} \tag{2-15}$$

The expression for S_{min} from (2-15) can now be substituted back into the radar range equation (2-6) to give

$$R_{max} = \sqrt[4]{\frac{P_{pk} G A_E \sigma}{(4\pi)^2 k T_0 B_N F (SNR_o)_{min} L_A L_{sys}}} \tag{2-16}$$

2.4 PROCESSING GAIN VIA PULSE INTEGRATION

2.4.1 Integration in Early Radar

The radar range equation developed thus far gives the detection range for a single pulse of RF. In most cases, however, several pulses illuminate the target, and if their returns are integrated then the signal-to-noise ratio improves. This effect was apparent to early radar operators who observed cathode ray tube (CRT) displays. As echo pulses were received, each pulse modulated the glow on the face of the CRT display. A rapid succession of pulses returning at the radar PRF increased the intensity of the glow of the spot corresponding to the target detection on the display, and the afterglow of the target detection coupled with the persistence of human vision contributed to an integration effect that improved the contrast between targets and background noise and clutter. The integration effect improved the signal-to-noise ratio and made it easier for the operators to pick out real targets from the background noise/interference/clutter. Modern signal processing methods now perform pulse integration on a succession of returns, and, as with the older methods, they all improve the signal-to-noise ratio and so improve detection performance.

2.4.2 Coherent and Non-Coherent Integration

Pulse integration methods fall into two broad categories. *Non-coherent integration* adds pulse magnitudes of successive pulses with no regard to the preservation of their phases. This may be applied after a detection process in which phase is not preserved, such as an envelope detector. Non-coherent integration is therefore synonymous with *post detection integration*. This contrasts with *coherent detection,* in which the integration process forms the vector sum of successive echoes. This requires both magnitude and phase information to be preserved and so has become synonymous with *predetection integration.*

2.4.3 Quantifying the Benefits of Integration

Any practical integrator is predicated on the idea that the circuit must exhibit a degree of short-term memory to sum pulses over several successive receiving periods but also must exhibit some degree of loss of memory over the longer term so that target responses are not retained indefinitely once detected. The integrator presents a narrow-band filter operation. Digital signal processing (DSP) methods integrate received data over a suitable period in order to improve the signal-to-noise ratio.

An improvement in the signal-to-noise ratio is obtained because noise is uncorrelated from pulse to pulse, whereas target returns are highly correlated in the short-term. Gaussian noise decorrelates over a time scale of $1/B_N$, which is considerably shorter than one pulse repetition interval (PRI = 1/PRF). Therefore, noise samples received over successive receiving periods are completely uncorrelated. When the number of Gaussian noise samples is n and the RMS voltage of each noise sample is v_N, then $\sqrt{n}v_N$ is the resulting integrated RMS noise voltage. The noise power therefore increases by a factor of n.

Target signals do tend to vary in time, an effect known as *scintillation* (see Section 2.5.3). The timescale for which target returns remain correlated depends on the following:

- The rate at which the illumination and reception angles change
- The rate at which the target alters its orientation
- The distribution of scatterers on the target
- The radar frequency

Under ideal circumstances target returns remain correlated over several receiving periods. This is quite reasonable for slowly moving targets but may not be valid for rapidly moving or agile targets. Perfect correlation over a suitable integration period is assumed for now. If, in addition to this, a truly coherent train of pulses is transmitted and their coherency is preserved through the receiver, then all target returns are received having the same amplitude and phase. When the voltage of each target sample is v_S the resulting integrated signal voltage is nv_S. The signal power therefore increases by a factor of n^2.

Comparing the integrated signal and noise powers, one can see that the signal-to-noise ratio (powers) increases by a factor of n. This is the ideal case and demonstrates that the best processing gain achievable is the maximum integration improvement factor [2] given by

$$I_i(n) = n \tag{2-17}$$

It was reckoned that the early integration process of the CRT afterglow gave rise to an actual integration improvement factor of

$$I_i(n) = \sqrt{n} \tag{2-18}$$

Most practical integrators usually manage to achieve an integration improvement factor somewhere between $n^{0.5}$ and n^1. The actual processing gain achieved by virtue of integration depends on whether a coherent or non-coherent integrator is used, the exact nature of the integration process, and the number of pulses to be integrated.

Since most integrators fail to achieve the maximum improvement factor, $I_i(n) = n$, one may conveniently express the actual improvement as the product of n with an *integration efficiency*, $E_i(n)$[2]:

$$I_i(n) = nE_i(n) \tag{2-19}$$

Efficiencies of less than 100%, that is, $E_i(n) < 1$, give rise to the notion that there is an *integration loss*, $L_i(n)$[2] in which

$$L_i(n) = \frac{1}{E_i(n)} \tag{2-20}$$

This *loss* must be understood in the context that it is a loss with respect to the ideal case of 100% efficiency. $L_i(n)$ tends to increase ($E_i(n)$ decreases) with increasing n.

The benefits of pulse integration are demonstrated in Figure 2-6, in which a similar burst of data as that of Figure 2-4(b) is integrated 100 times. This amounts to an integration of the unipolar voltage envelope of the noise and target signals. Since this voltage can

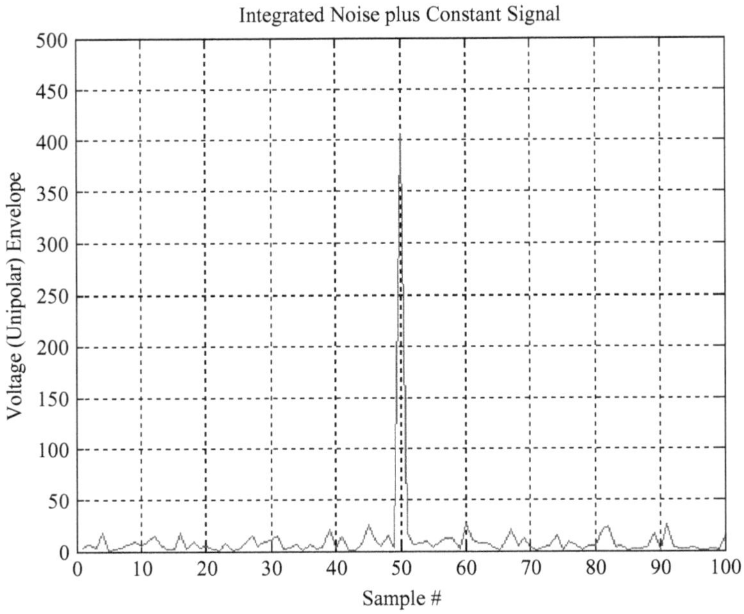

FIGURE 2-6 ■ SNR Improvements from Pulse Integration

be thought of as the output of a linear detector with no preservation of signal phase, the integration considered here is an example of non-coherent integration. Recall that the input signal resides in sample 50 and has an amplitude of 4.0 V, the input (bipolar) noise voltage has a Gaussian distribution with zero mean and standard deviation = 1.0 V (= RMS noise level). The original SNR_V for a single pulse is thus 4; as a result of the integration on the detected voltage envelope the signal has grown to 406 V, and the RMS value of the noise is 9.73 V, giving an improved $SNR_V = 41.7$.

Since the integration process improves the signal-to-noise ratio, a superior detection capability is obtained. This means that the signal-to-noise ratio required for each individual pulse may be somewhat lower than that demanded by the specification on the radar's detection capability because when n such pulses are integrated the signal-to-noise ratio is boosted back to the value originally required. It now becomes pertinent to distinguish between the signal-to-noise ratio of an individual pulse when no integration takes place and the signal-to-noise ratio of an individual pulse when integration does takes place.

Let the minimum acceptable signal-to-noise ratio required of an individual pulse for a given detection capability when no integration takes place be $(SNR_1)_{min}$ and let the minimum acceptable signal-to-noise ratio of an individual pulse for the same detection capability when integration of n such pulses does takes place be $(SNR_n)_{min}$.

Note the relationship:

$$I_i(n) = \frac{(SNR_1)_{min}}{(SNR_n)_{min}} \tag{2-21}$$

Therefore

$$E_i(n) = \frac{(SNR_1)_{min}}{n(SNR_n)_{min}} \tag{2-22}$$

As a result of the integration illustrated in Figure 2-6, $I_i(n) = 41.7/4.0 = 10.42_5$, $E_i(n) = 0.104$, and $L_i(n) = 9.59$. Note that this result is in close agreement with the predicted result of $I_i(n) \sim n^{0.5} = \sqrt{100} = 10$, as may be expected from non-coherent integration for a good input SNR. $I_i(n)$ is somewhat less than $n^{0.5}$ for very low input SNR, typically < 1 (0 dB).

Some older radars use RF detectors that produce an output current approximately proportional to the input voltage squared when the input signal is small. This small signal approximation means that the detector output is sensitive to input power. Such a detection process does not preserve the signal phase, degrades the SNR, and alters the statistics of the noise; a Gaussian noise voltage on the detector input results in an output noise signal adhering to an exponential distribution. This has profound implications on where the detection threshold must be set to control noise-generated false alarms. Modern systems digitize the baseband signals from a quadrature detector (see Section 8.4.1), which preserves the signal phase and hence the SNR, using analog-to-digital converters (ADCs). This entails a quantization of the received noise and signal voltages into discrete levels and their representation by a digital word. The ADC represents a process of linear "detection" (although signals may subsequently be squared and processed). The averaging of n noise samples reduces its RMS by $\sqrt{n}$, irrespective of their statistical distribution. When the integration is conducted on signals whose phase is preserved (e.g., at RF, before the detector or on the outputs of a quadrature detector), it leads to coherent integration, which can yield a considerable improvement in SNR. When the integration process is implemented on signals whose phase is not preserved (e.g., unipolar voltage envelopes from linear or square law detectors), it leads to non-coherent integration. Non-coherent integration can be very efficient when the input SNR_1 is reasonable, which is consistent with the integration of a few pulses (typically $n \leq 10$), but is not as efficient as coherent integration as n increases, since SNR_1 is very much smaller.

2.4.4 Integration Improvements in the Radar Range Equation

When the integration of n pulses takes place, the radar range equation can be written as [2]:

$$R_{max} = \sqrt[4]{\frac{P_{pk} G A_E \sigma}{(4\pi)^2 k T_0 B_N F (SNR_n)_{min} L_A L_{sys}}}$$

Substituting in (2-22) gives

$$R_{max} = \sqrt[4]{\frac{P_{pk} G A_E \sigma n E_i(n)}{(4\pi)^2 k T_0 B_N F (SNR_1)_{min} L_A L_{sys}}} \tag{2-23}$$

Substituting (2-19) into (2-23) gives

$$R_{max} = \sqrt[4]{\frac{P_{pk} G A_E \sigma I_i(n)}{(4\pi)^2 k T_0 B_N F (SNR_1)_{min} L_A L_{sys}}} \tag{2-24}$$

Equation (2-24) differs from (2-16) by including the $I_i(n)$ term in the nominator of the right-hand side. The integration improvement factor, $I_i(n)$, represents a *processing gain*.

Pulse Doppler radar processing entails the coherent integration of several pulses with a high degree of integration efficiency leading to a significant processing gain. The duration of this integration is *nPRI*, where *PRI* is the pulse repetition interval, and is known as a *coherent processing interval* (CPI). In Section 2.1.1 it was noted that a 16-fold increase in transmitted power would be required to double the detection range. A doubling of radar range can also be accomplished through the integration of 16 pulses with 100% efficiency, or 32 pulses with 50% efficiency, and is a far more practical strategy than making dramatic increases in RF power.

A casual glance at (2-24) might lead to the notion that the detection range is largely determined by the peak transmitted power. While this has some truth, it is perhaps rather misleading, since, as was shown above, a lack of power can be compensated for through pulse integration. The noise bandwidth term, B_N, is the bandwidth over which noise is admitted to the detection system. In Section 3.6 it will be shown that the best signal-to-noise ratio arises when the receiver bandwidth is matched to the transmitted waveform. A good approximation to this for a simple, unmodulated pulse is obtained when $B_N = 1/\tau$, where τ represents the transmitted pulse width. Making this substitution into (2-23) gives

$$R_{max} = \sqrt[4]{\frac{P_{pk}\tau G A_E \sigma n E_i(n)}{(4\pi)^2 k T_0 F (SNR_1)_{min} L_A L_{sys}}} \tag{2-25}$$

The product, $P_{pk}\tau$, in the nominator of the right-hand side of (2-25) is the transmitted *energy, E*, in each pulse.

$$E = P_{pk}\tau \tag{2-26}$$

Substituting (2-26) into (2-25) gives

$$R_{max} = \sqrt[4]{\frac{E G A_E \sigma n E_i(n)}{(4\pi)^2 k T_0 F (SNR_1)_{min} L_A L_{sys}}} \tag{2-27}$$

Furthermore, the product $EnE_i(n)$ in the nominator of the right-hand side of (2-27) is the total integrated energy over all the pulses in one CPI. Therefore, the *energy* in the transmitted waveform over a CPI—rather than the peak transmitted power of individual pulses—is the key parameter determining a radar's detection range.

These considerations present the radar system designer with an interesting dilemma regarding waveform design: is it better to use a low PRF, in which a modest number of long duration pulses illuminate the target, or a higher PRF, in which a large number of shorter pulses illuminate the target, assuming that both waveforms put the same energy on the target in each processing period? From the point of view of detection performance only, it would be better to use the waveform that entails the simplest processing because this can be accomplished with the greatest efficiency. This favors the low PRF/few pulses approach; in the limit, the best detection range would be obtained for a single pulse of very long duration. However, other considerations are likely to dictate the waveform design; low PRF waveforms are not practical for airborne pulse Doppler radar applications.

2.5 RADAR CROSS SECTION

2.5.1 Definition

The RCS of a target is its equivalent echoing area and has units of area, m^2. The formal definition of RCS is [2, p33]

> *The radar cross section of a target is the (fictional) area intercepting that amount of power which, when scattered equally in all directions, produces an echo at the radar equal to that from the target.*

This definition assumes isotropic scattering from a perfectly reflective body. Hence, it may be useful to imagine a hypothetical target such as metal sphere grown to a size until the echo from the metal sphere equals that from the actual target. When this point of equality is obtained, the cross sectional area of the sphere equals the radar cross section of the actual target. The RCS definition may be expressed mathematically as

$$\sigma = \frac{\text{power reflected towards source/unit solid angle}}{\text{incident power density}/4\pi} = \lim_{R\to\infty} 4\pi R^2 \left|\frac{E_r}{E_i}\right|^2 \qquad (2\text{-}28)$$

where E_r is the reflected field strength at the radar, and E_i is the incident field strength at the target.

2.5.2 Factors Determining RCS

Most targets of interest have a complex geometry and are composed of a wide variety of materials. These targets are a complicated array of scattering centers that reflect, diffract, and reradiate signals back to the observing radar. Some resonances may be associated with certain structural features of the target, or some resonances or cancellation effects may arise between multiple features. The target return is formed from the vector sum of all these individual components. The RCS of a target depends on the following:

- Target geometry (e.g., size, shape, orientation)
- Target composition
- Angle of illumination/reception
- Transmitted polarization
- Receiver polarization
- Range
- Frequency

Range may become a factor when target range is very short. Normally, one would expect the target to be in the far field of the transmitting antenna, in which case it may be assumed that the target is illuminated by a plane wave. There may be exceptions to this assumption for very short-range and short wavelength applications. As the echo embarks on the return path, the target is now seen as a source of RF, and the receiving antenna would normally be expected to be in the far field of the target. This assumption breaks down far more readily for large targets and short wavelengths. However, as can be seen by the definition of (2-28), RCS is defined as a far field limit since it is based on the limit as range becomes infinitely long.

Frequency can become significant for a variety of reasons. When the wavelength is considerably greater than the dimensions of a target, its RCS reduces very rapidly with increasing frequency. This is the so-called *Rayleigh region* in which $\sigma \propto 1/\lambda^4$. When the wavelength is considerably less than the target dimensions, its RCS is reasonably constant. This is known as the *optical region*. This is typical of microwave radar observing aircraft, vehicles, ships, and personnel. When the wavelength is comparable to the target dimensions the RCS fluctuates rapidly with frequency. This is known as the *resonance region*. Thus, target structures of differing size couple to the radar signal by differing degrees.

The RCS of a target is very sensitive to the angle of illumination; for complex targets considerable variation is incurred over angular increments of less than 1 degree. However, it is typical for a plot of RCS against illumination/reception angle to exhibit certain gross features that are attributable to the major features of the target. The RCS of a complex target is also characterized by a rapid variation over small angular increments. This is attributable to the fine structural detail of the target. It is clear that structural details of dimensions comparable to the wavelength have a significant effect on the target RCS. Millimetric wave radars, whose wavelengths are between 10 mm (i.e., 30 GHz) and 1 mm (i.e., 300 GHz), may well be sensitive to the quality of a welded joint on an armored vehicle target.

2.5.3 Scintillation

As was pointed out in Section 2.4.3, the RCS of a target can fluctuate quite considerably due to small changes in the factors given in the previous bulleted list. This leads to the idea that target RCS is also a statistical quantity and may be described by a PDF and that these fluctuations cause the RCS to decorrelate over a certain timescale. Various proposed models have described the PDF and decorrelation period of the RCS of certain target types. The variation in target RCS is known as *scintillation*. Mean values of target RCS are normally used in the radar range equation. Scintillation can cause targets to fade and therefore give rise to a loss in detection performance. This may be catered for by increasing the required value of $(SNR_1)_{min}$ by an extra margin of a few decibels (typically 3 to 10 dB), depending on the detection capability specification of the radar one is attempting to maintain and the decorrelation statistics and timescale assumed of the target. Even in spite of this, some slow-moving targets can fade into deep nulls for some considerable time. In these circumstances it becomes desirable to force the decorrelation of the target return over a much shorter timescale than might naturally happen. This can be achieved by varying the frequency of the radar since RCS is particularly sensitive to frequency. Under certain assumptions about the target, the RCSs of a target at two frequencies are completely uncorrelated if the two frequencies are spaced by the inverse of the time resolution of the radar. Generally, frequency agility over as wide a bandwidth as possible is desirable. Frequency agility is also used to force the rapid decorrelation of surface clutter (see Section 6.5.2). If the decorrelation of the targets of interest is minimal but the decorrelation of the clutter is more complete, then an improvement in the signal-to-clutter ratio (SCR) can be gained. Some radars now transmit each CPI on a different channel frequency to combat potential clutter and target fading problems.

2.5.4 The Swerling Models of RCS Fluctuation

The most well-known models of target fluctuation are the Swerling models [2]. Five cases are defined which describe the PDF of the RCS fluctuation and the timescale over which returns are correlated.

Case 1

- Assumes that the target consists of a random assembly of scatterers, all equally weighted
- Pulses within a scan correlate but are uncorrelated from scan to scan, that is, target aspect changes slowly
- Ignores the effect of beam shape on echo amplitude
- Model is appropriate to a ship

- Probability density function for the RCS is given by

$$P(\sigma) = \frac{1}{\sigma_0} \exp\left(-\frac{\sigma}{\sigma_0}\right) \tag{2-29}$$

where σ_0 is the mean cross section over all target fluctuations

Case 2

- Like case 1 but pulses are uncorrelated from pulse to pulse, that is, target aspect changes rapidly
- Model is appropriate to a ship detected using a pulse to pulse frequency agile radar
- Has same probability density function as case 1

Case 3

- Assumes that the target consists of one dominant reflector plus several small subreflectors
- Pulses within a scan correlate but are uncorrelated from scan to scan, that is, target aspect changes slowly
- Model is appropriate to a modern aircraft
- Probability density function for the RCS is given by

$$P(\sigma) = \frac{4\sigma}{\sigma_0^2} \exp\left(-\frac{2\sigma}{\sigma_0}\right) \tag{2-30}$$

Case 4

- As for case 3 but pulses are uncorrelated from pulse to pulse, that is, aspect changes rapidly
- Model is appropriate to a highly maneuvering aircraft or an aircraft detected using a pulse to pulse frequency agile radar
- Same probability density function as case 3

It is also helpful for the purposes of comparison to define a case 5 (sometimes also called case 0), which can be taken as a reference by representing returns that do not fluctuate.

2.5.5 SNR Requirements as a Result of Target Fluctuation

The *SNR* assuming additive Gaussian noise, for Swerling case 1 and 2 type targets, is given by a conveniently neat function of P_D and P_{FA}, namely [4],

$$SNR = \frac{\log P_{FA}}{\log P_D} - 1 \tag{2-31}$$

Closed-form solutions like (2-31) do not exist for other models of RCS variation or even case 5, the non-fluctuating target. P_D versus *SNR* from (2-31) is graphed in Figure 2-7 for values of $P_{FA} = 10^{-4}, 10^{-6}, 10^{-8}, 10^{-10}, 10^{-12}, 10^{-14}$, and 10^{-16} (lines running left to right, respectively). The figure shows that for good detection ($P_D = 0.9$ or 90%) and low false alarms ($P_{FA} = 10^{-10}$) a *SNR* of around 23.4 dB (= 218) is required. Reducing P_{FA} to 10^{-6}, as may be possible with modern trackers, leads to an *SNR* requirement of

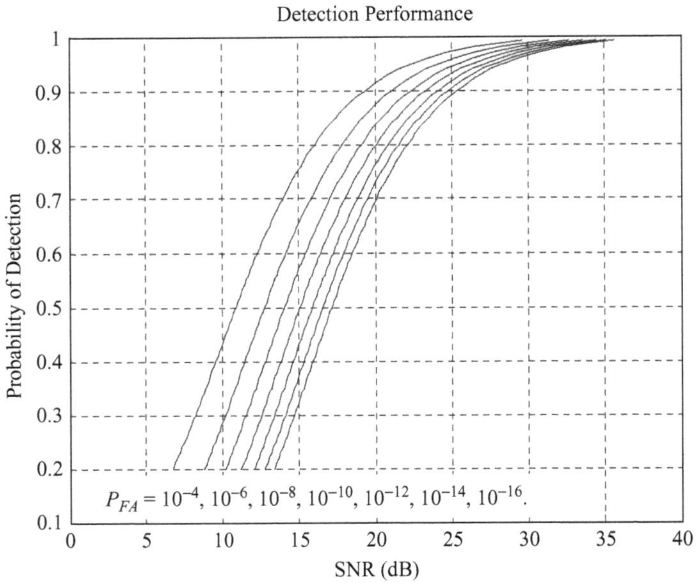

FIGURE 2-7 ■ P_D versus SNR and P_{FA} for Swerling Case 1 and 2 Type Targets

21.1 dB. The solutions for case 1 and 2 type targets given by equation (2-31) tend to lead to high SNR requirements. Graphical solutions (or numerical approximations) for cases 3, 4, and 5 tend to result in lower SNR requirements.

An alternative solution to (2-31) can be derived in which P_D is plotted as a function of P_{FA} for a fixed SNR. This gives an indication of the trade-off that can be made between P_D and P_{FA} and is known as a *receiver operating curve* (ROC). An example is given in Figure 2-8 for a Swerling case 1 or 2 type target with an SNR of 20 dB. Note that the horizontal scale of Figure 2-8 is plotted on a decibel scale for ease of viewing.

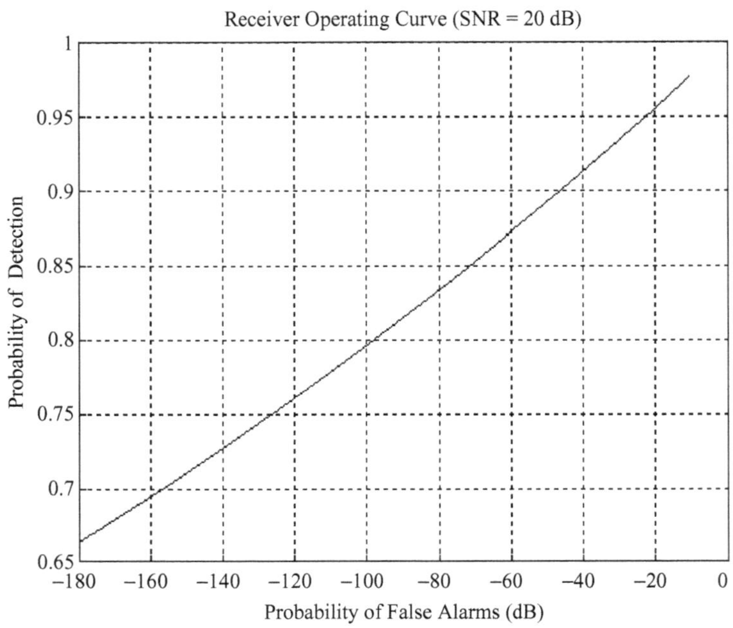

FIGURE 2-8 ■ Receiver Operating Curve

CHAPTER 3

Pulsed Radar

Chapter Outline

3.1 INTRODUCTION TO PULSE PARAMETERS

Pulsed radar entails the transmission of a succession of radio frequency (RF) pulses. As noted in Chapter 1, pulse modulation enables the range of a target to be measured. Pulsed modulation also permits a high-power transmitter to be used, which helps achieve a long detection range. High transmitted powers can be used if the sensitive receiver can be isolated from the transmitted signal during the pulse. This is easily achievable in pulsed radars since electronic switches can be thrown to isolate the receiver from any input signals during the transmitted pulse. Once the transmitted pulse has died down, the switches open the receiver path. Hence, when the transmitter is *on* the receiver is *off*, and when the transmitter is *off* the receiver is *on*. This alternation also means that no echo returns can be received during the transmitted pulse and consequently that the radar is blind to targets at certain ranges.

3.1.1 Pulse Repetition Frequency

The number of pulses transmitted per second is known as the radar *pulse repetition frequency* (PRF), often also denoted by f_r. The pulsed cycle of a radar is best illustrated using a timing diagram of the transmitter modulation, as in Figure 3-1. Figure 3-1 also depicts the *pulse repetition interval* (PRI), which is the inverse of the PRF ($PRI = 1/f_r = T_r$), and the transmitted pulse width, τ.

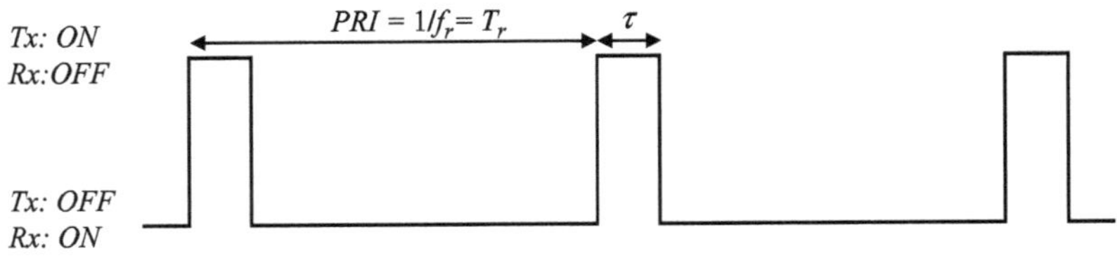

FIGURE 3-1 Pulse Timings

3.2 PEAK AND AVERAGE POWER

The *duty ratio* (or *duty cycle*) is defined as the ratio of the transmitter on time to the total periodic time, that is,

$$\textit{duty ratio} = \frac{\tau}{T_r} = \tau f_r \tag{3-1}$$

The average power in the signal can be derived by considering the preservation of energy. The energy in a pulse is given by the product of the peak pulsed power and the pulse width. The same energy is also the product of the average power and *PRI*. Hence,

$$E_T = P_{pk}\tau = P_{av}T_r$$

Rearranging, one obtains

$$P_{av} = P_{pk}\frac{\tau}{T_r} = P_{pk}\tau f_r = P_{pk} \times \textit{duty ratio} \tag{3-2}$$

The average power is the equivalent power level of a continuous wave (CW) signal.

3.3 PULSE DELAY RANGING

3.3.1 Range as a Measure of Time Delay

Originally, radars employed leading edge detection; that is, the time of a returning echo was given by the time of its rising edge. The tendency to refer to leading edge times is a legacy that has lived on long past the days of leading edge detection. If a returning echo having an arbitrary delay time, t, is received, the timing diagram now looks like that of Figure 3-2.

Range is a measure of distance. Since *distance* = *speed* × *time* and since the speed in this case is the velocity of light in free space, c ($c = 3 \times 10^8$ m/s), then the range, R, is given by

$$R = \frac{ct}{2} \tag{3-3}$$

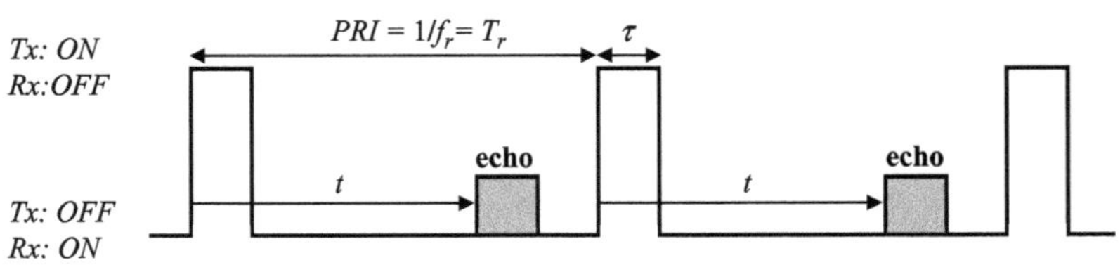

FIGURE 3-2 Pulse Timings with Target Echo

The division by 2 is included since the one-way range is required whereas the total distance is the two-way path. It is worth noting that each receiving period contains a repetition of the target echo on account of the periodicity of the waveform.

3.3.2 Minimum Range

Pulsed radar has a minimum range since the echo cannot be received until the receiver is on. Thus, the minimum time delay resulting in a detectable echo is $t = \tau$. In practice, the receiver does not become active until the transmitted power has died down to a safe level and the receiver protection switches are in their low loss state. This may take several tens of nanoseconds after the 50% point on the falling edge of the transmitted pulse. However, this delay is often assumed to be negligibly small compared with the transmitted pulse width, τ. The minimum range, R_{min}, is given by

$$R_{min} = \frac{c(\tau + T_{tr})}{2} \approx \frac{c\tau}{2} \tag{3-4}$$

in which T_{tr} is the transmit to receive safe switching delay time, and the approximation is based on the assumption that $T_{tr} \ll \tau$. The minimum range timing is illustrated in Figure 3-3.

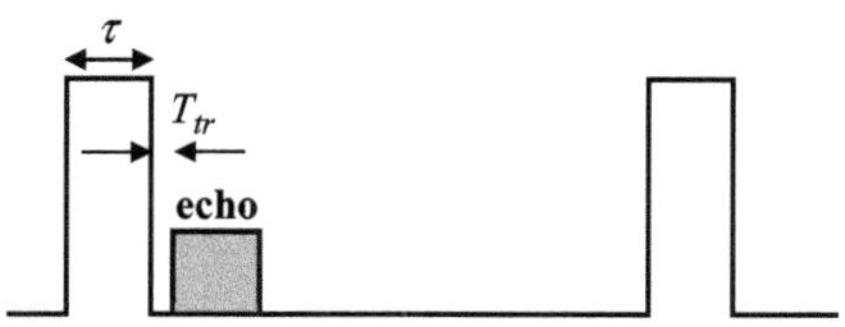

FIGURE 3-3 Minimum Range

Many radars no longer employ leading edge detection, in which case equation (3-4) is not strictly valid. For ranges shorter than R_{min} given by (3-4), the time delay is $t < \tau$. This means that part of the echo pulse falls within the first part of the receiving period and part coincides with the last part of the transmitted pulse. Whether such a target is detectable depends on whether sufficient echo energy falls within the receiving period. Therefore, at very short ranges, $t \ll \tau$, P_D is low but improves as the delay time increases, $t \rightarrow \tau$.

3.3.3 Eclipsing

At the end of the receiving period a similar situation is encountered in which the received pulse starts to overlap the next transmitted pulse. When the overlap is complete, none of the echo energy is detectable and the radar is blind to that target ($P_D = 0$). In fact, a whole series of blind ranges is possible for which the time delay is a multiple of *PRI*.

$$R_{blind} = \frac{cxT_r}{2} \tag{3-5}$$

where $x = 1, 2, 3, \ldots$

The loss of targets in these blind ranges is known as *eclipsing*. For a partial overlap between the echo and the receiver off period, part of the echo energy is lost and P_D degrades, as it does close to the minimum range. This situation is known as a *partial eclipse*. The variation in the received echo energy with time delay (range) is graphed in Figure 3-4.

3.3.4 Range Gating

To measure the delay time, a clock marks out regular increments of time throughout the receiving period. This has the effect of quantizing time into a series of *range gates* or *range*

FIGURE 3-4 Variation of Received Echo Energy Due to Eclipsing

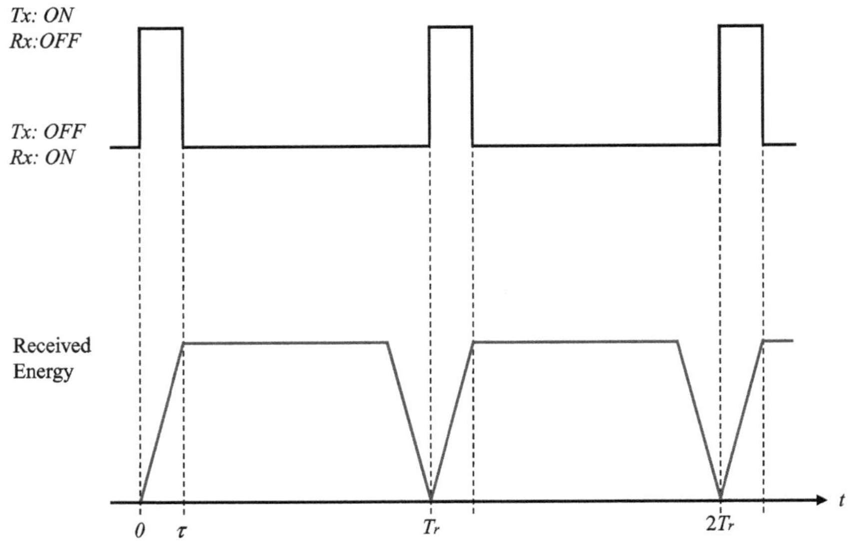

FIGURE 3-5 Range Gating

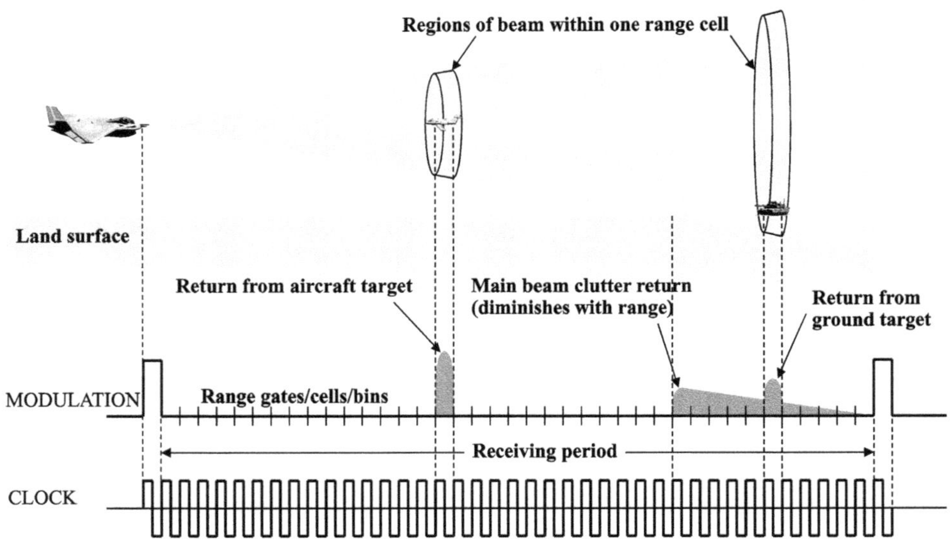

cells. For simple, unmodulated pulses (as here), the width of each range cell is normally set approximately equal to the transmitted pulse width. This is illustrated in Figure 3-5. Individual range gates isolate returns from a thin volume of space within the beam. Ground targets are detected in a range gate in which the thin volume of space intersects the ground.

3.4 LOW PRF RADAR

3.4.1 Range Ambiguity

If all targets result in an echo that falls within the first receiving period, $t < T_r$, then target range can be determined unambiguously. All target echoes are then known as *first trace echoes*. If the radar could detect a target at a sufficiently long range it may be possible that

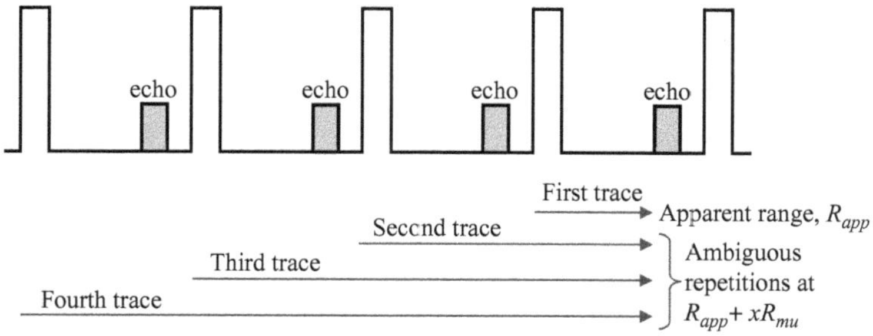

FIGURE 3-6 Range Ambiguity

an echo is received for which $t > T_r$. If this echo falls within the second receiving period it is known as a *second-trace echo.* Similarly, third-, fourth-, and fifth-trace echoes may be received for when $t > 2T_r$, $t > 3T_r$, $t > 4T_r$, respectively. Should second-trace echoes (or higher) be detectable, then all target ranges are ambiguous. The mere possibility that second-trace echoes could be detected places doubt on the indicated range of all targets. Thus, there is a potential problem of range ambiguity, as depicted in Figure 3-6.

This range ambiguity is avoided if all targets are first trace, $t < T_r$. In the limit, the maximum time delay admissible to avoid range ambiguity is $t = T_r$. Therefore, any PRF has an associated maximum unambiguous range, R_{mu}, given by

$$R_{mu} = \frac{cT_r}{2} = \frac{c}{2f_r} \tag{3-6}$$

This is also the limiting value for the first blind range.

The true range to the target could be its apparent range plus multiples of its maximum unambiguous range, that is,

$$R_{true} = R_{app} + xR_{mu} \tag{3-7}$$

where $x = 1, 2, 3, \ldots$

3.4.2 Definition of Low PRF

Clearly, the lower the PRF the longer the maximum unambiguous range. Low PRF radar is simply defined as having a PRF sufficiently low so as to avoid range ambiguities. Thus, when

$$R_{mu} > R_{max} \tag{3-8}$$

range is unambiguous and the PRF is low.

3.5 SPECTRUM OF PULSED RADAR

3.5.1 Theory of the Spectrum of a Pulsed Carrier

Pulsed signals are a form of amplitude modulation, and, in keeping with the theory of amplitude modulation, the spectrum of a pulsed carrier comprises the spectrum of the modulation signal that is reproduced in the upper and lower sidebands of the carrier. The modulating signal is a periodic rectangular function, as given in Figure 3-1. Being a periodic function, its spectrum is comprised of an infinite series of lines corresponding

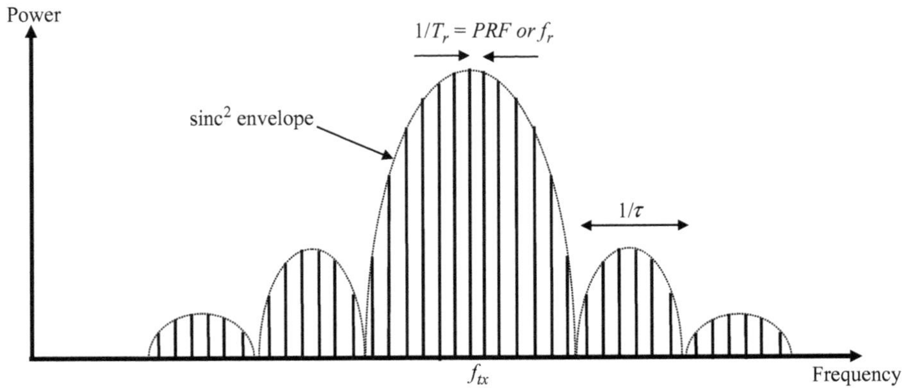

FIGURE 3-7 Power Spectrum of Pulsed Modulated Carrier

to the Fourier series of the modulating waveform. The Fourier series of the waveform of Figure 3-1, $f(t)$, is given by

$$f(t) = \frac{2V\tau}{T_r}\left[\frac{1}{2} + \sum_{m=1}^{\infty} \frac{\sin m\pi\tau/T_r}{m\pi\tau/T_r}\cos(m\omega_r t)\right] \tag{3-9}$$

where V is the peak voltage of the pulse, ω_r is the fundamental angular frequency of the pulsed waveform, m is the harmonic number, and t is time. It may be noted that the series $f(t)$ consists of an infinite number of harmonics of the fundamental frequency $(m\omega_r)$ (i.e., multiples of the PRF) and that the overall amplitudes of these spectral components follow a sinc function. There is also a zero frequency term whose amplitude is given by the product of the peak pulse amplitude and the duty ratio. The power spectrum of the pulsed modulated carrier is shown in Figure 3-7.

From Figure 3-7 one can see that the spectrum is characterized by a series of lines at harmonics of the PRF above and below the central transmitted carrier frequency, f_{tx}, and therefore the frequency spacing between any adjacent pair of lines equals the PRF. The amplitude (power) envelope of these lines is the classic sinc^2 function, which is characterized by a series of nulls: the first nulls above and below f_{tx} are at a frequency offset of $1/\tau$; subsequent nulls are offset at $1/\tau$ from each other. Thus, a narrow pulse width, τ, results in a broadband signal, whereas a wide pulse width results in a narrow-band spectrum. (Note that the null frequencies need not necessarily coincide with a harmonic of the PRF lines; in fact this occurs only if T_r is an integer multiple of τ.) For a true rectangular modulating waveform having infinitely fast pulse rise and fall times the spectrum becomes infinitely wide; however, this cannot be generated in practice. Figure 3-7 depicts the spectrum over a band out to only $\pm 3/\tau$ either side of f_{tx}. In truth, the sidelobe structure of the spectrum continues beyond this band, albeit at lower levels with increasing offset frequency.

3.5.2 Spectral Containment

In many cases it is necessary to confine the transmitted spectrum to avoid generating co-channel interference. Indeed, legislation in the high-frequency (HF) band (3–30 MHz) makes this compulsory. The bandwidth may be confined by transmitting non-rectangular pulses. This entails the transmission of shaped pulses that are time weighted. The weighting functions shape the rise and fall profiles of the pulses; for example, triangular-shaped pulses give rise to a Fourier series having a sinc^2 voltage envelope and therefore a sinc^4 power spectrum. Strictly, shaped pulses also have an infinite bandwidth, but their decay to levels below the ambient noise level takes place over a much narrower bandwidth than rectangular pulses. Pulse shaping tends also to reduce the energy in the pulse and degrade range resolution (see Section 3.7). Weighting functions are considered in more detail in Chapter 7. In practice, radar transmitters produce pulses having a finite rise and fall time, so the hardware limitations of the transmitter modulator impart some degree of pulse shaping.

3.5.3 Phase Spectrum and Phase Coherency

The transmitted signal also possesses a phase spectrum. The phase of all spectral components in the upper sideband of the central main lobe is 0°, whereas the phase of components in the upper first sidelobe is 180°. The phase of the spectral components continues to alternate between 0° and 180° within each successive sidelobe. The phase of spectral components in the lower sideband is 180° out from the corresponding spectral components in the upper sideband. The neat line spectrum of Figure 3-7 and its associated phase spectrum are obtained only for phase coherent signals. Some sources of radar waveforms such as magnetrons do not possess the necessary coherency to yield the neat line spectrum; such signals are not suitable for pulse Doppler radar applications.

3.6 MATCHED RECEPTION

The power spectrum of the transmitted signal shown in Figure 3-7 also represents the spectrum of the returning echo pulses, albeit at much lower amplitudes, assuming the absence of any distortions over the path to and from the target and due to the signal's interaction with the target. Since radar detection performance can be maximized for the maximum signal-to-noise ratio (SNR) [5], it becomes pertinent to consider the conditions giving rise to the best SNR in the reception/detection process. From Figure 3-7 it is obvious that returning signal components toward the center of the central main lobe of the spectrum contain the highest power, whereas (thermal) noise has a constant spectral power density. Hence, it seems reasonable to expect the best SNR to be achieved if the receiver employs some sort of filter that emphasizes the larger spectral components but attenuates noise at frequencies where little signal strength is present. The ideal filter shape is one that matches the envelope of the spectrum. This is known as the *matched* filter and gives rise to the notion of a *matched receiver* [5]. The matched receiver yields the optimum SNR. The mathematical basis for this result is presented in the following section.

3.6.1 Theory of Matched Reception

It is possible to consider the general case where signal and additive white Gaussian noise are fed into a linear filter, as shown in Figure 3-8.

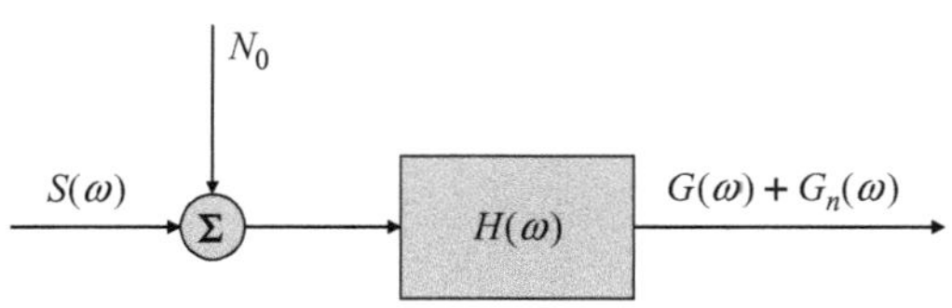

FIGURE 3-8 Signal Corrupted by Additive White Noise

The output signal in the frequency domain is

$$G(\omega) = S(\omega)H(\omega) \tag{3-10}$$

where $S(\omega)$ is the target return, N_0 is the peak noise voltage, $H(\omega)$ is the transfer function of the receiver filter, $G(\omega)$ is the target signal on the filter output, and $G_n(\omega)$ is the noise signal on the filter output.

The output signal in the time domain is the inverse transform

$$g(t) = \frac{1}{2\pi}\int_{-\infty}^{\infty} S(\omega)H(\omega)e^{j\omega t}d\omega \tag{3-11}$$

The output noise power spectral density is

$$G_n(\omega) = \frac{N_0}{2}|H(\omega)|^2 \tag{3-12}$$

The mean output noise power is then

$$\sigma_n^2 = \left(\frac{1}{2\pi}\right)^2 \int_{-\infty}^{\infty} \frac{N_0}{2}|H(\omega)|^2 d\omega \tag{3-13}$$

Integrating over all frequencies, the output signal-to-noise ratio is therefore

$$SNR = \frac{g^2(t)}{\sigma_n^2} = \frac{\left|\int_{-\infty}^{\infty} S(\omega)\,H(\omega)e^{j\omega t}d\omega\right|^2}{\frac{N_0}{2}\int_{-\infty}^{\infty} |H(\omega)|^2 d\omega} \tag{3-14}$$

One has to find $H(\omega)$ that maximizes this expression.

The Cauchy–Schwarz inequality states that the inner product has a maximum value when the two vectors are aligned:

$$\left|\int f(x)g(x)dx\right|^2 \leq \int |f(x)|^2 dx \int |g(x)|^2 dx \tag{3-15}$$

from which the SNR is a maximum when

$$\left|\int S(\omega)H(\omega)\,dx\right|^2 = \int |S(\omega)|^2 dx \int |H(\omega)|^2 dx \tag{3-16}$$

Thus, when

$$H(\omega) = kS^*(\omega)\,e^{-j\omega t'} \tag{3-17}$$

where k is an arbitrary scale factor, t' is an arbitrary phase shift factor (time delay), and $*$ represents the complex conjugate, the receiver output *SNR* is

$$SNR = \frac{k\left|\int_{-\infty}^{\infty} S(\omega)\, kS^*(\omega)\, e^{j\omega(t-t')}d\omega\right|^2}{\frac{N_0}{2}\int_{-\infty}^{\infty} |S(\omega)|^2 d\omega}$$

$$= \frac{\int_{-\infty}^{\infty} |S(\omega)|^2 d\omega}{\frac{N_0}{2}} = \frac{2E}{N_0} = \frac{2 \cdot signal\ energy}{Noise\ PSD} \tag{3-18}$$

where *PSD* stands for the power spectral density. This peak occurs at one specific instant, t' in this case. In other words, the SNR is a maximum when the receiver transfer function is equal to the complex conjugate of the signal spectrum, multiplied by an arbitrary scaling factor k, and shifted by an arbitrary phase $j\omega t'$. The complex conjugate spectrum equates to time reversal of the waveform.

3.6.2 Ideal and Practical Matched Filtering

A filter transfer function that is truly matched to the signal spectrum over an infinite bandwidth is very difficult to implement; however, a good practical approximation is a filter of bandwidth, B, given by

$$B = \frac{1}{\tau} \tag{3-19}$$

This filter covers the central half of the main lobe of a pulsed spectrum, as illustrated in Figure 3-9, and therefore filters off high-order harmonics of the PRF lines. As a result, the signal on the matched filter output is distorted and is no longer a series of rectangular pulses.

A rectangular pulse on the input of the matched filter and the filter output waveforms, $g(t)$, are illustrated in Figure 3-10 for the case of the ideally matched filter and the practical approximation to the matched filter case. If the filter is truly matched over an infinite

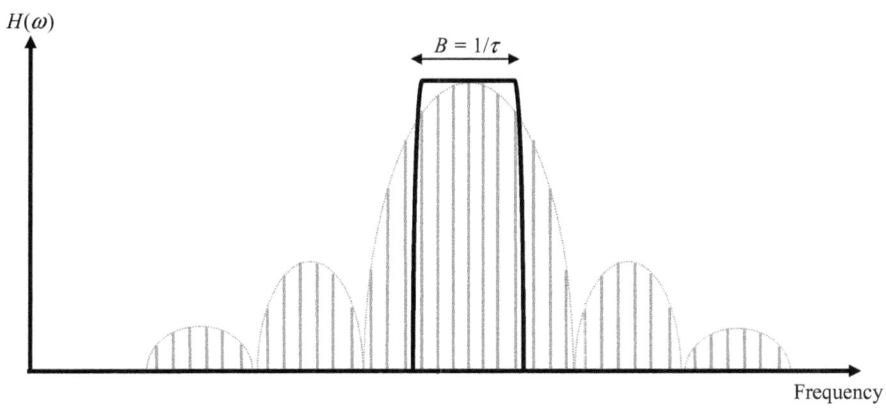

FIGURE 3-9 Approximation to a Matched Filter for Simple Pulsed Radar

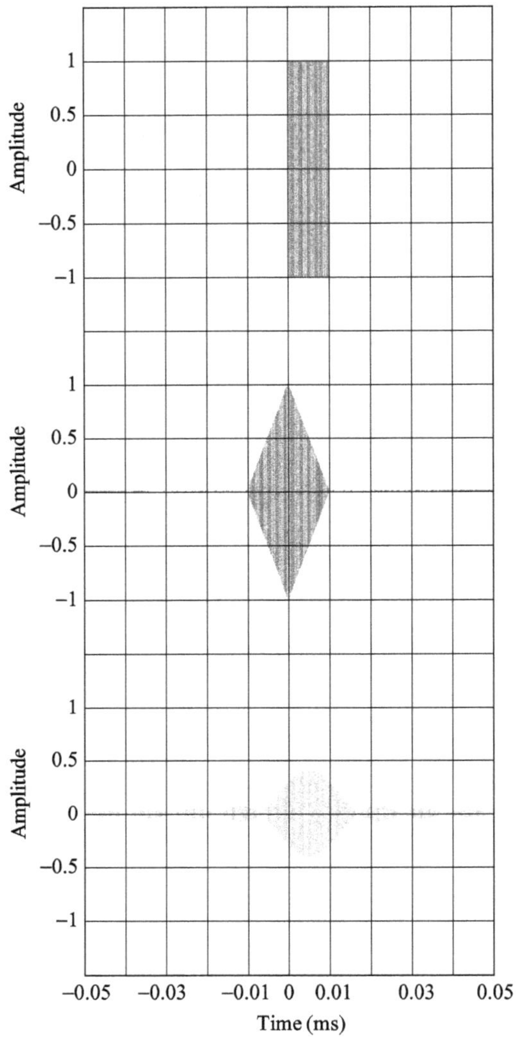

FIGURE 3-10 Matched Filter Response to a Simple Pulse: Upper Trace = rectangular input pulse, peak amplitude = 1, pulse width τ = 10 μs; Middle trace = ideal matched filter output with peak amplitude normalized to unity; Lower trace = output from practical band pass filter centered on carrier frequency and with bandwidth $B = 1/\tau$

bandwidth, the output waveform is triangular with a width twice that of the rectangular input pulse. This represents the integral of the input. Further theoretical analysis of matched filtering presented in Section 3.9.3 describes matched filter reception as a process of autocorrelation of the input signal. It is worth noting here that the autocorrelation duration is twice that of the input waveform and is consistent with the width of the triangular output waveform being twice that of the input pulse. If the voltage of the input pulse is A, then the peak output voltage from the ideal matched filter is $A^2\tau$, and all the energy in the input signal has been preserved [6]. If, however, the practical approximation to the matched filter having a bandwidth $B = 1/\tau$ is used, as illustrated in Figure 3-9, the output waveform is somewhat distorted but is still elongated compared with the input pulse. This waveform exhibits sidebands in time that is the classic ringing effect of a narrow-band filter being hit with a wideband pulse. An approximate rule of thumb is that the bandwidth, $B \approx 1/(1.3t_r)$, where t_r is the 10% to 90% rise time (or 90% to 10% fall time). The output waveform in this case reaches a considerably lower peak voltage compared with the output of the ideally matched filter, and it is clear that some energy has been lost; this corresponds to the energy in the high-order harmonics of PRF lines that were filtered away. The elongation of the pulse has ramifications on the range resolution of the radar (Section 3.7). It is also worth noting that for the simple pulses considered thus far the matched receiver bandwidth B defined in (3-19) is approximately equal to the receiver noise bandwidth, B_N, discussed in Section 2.2.1.

While the matched receiver response yields the best SNR and hence the best detection performance, it has not been universally used. Some short-range tracking radars use a much wider bandwidth, in the order of $5/\tau$ or $10/\tau$ in the interests of minimizing the distortion of the pulse. The distortion illustrated in the lower trace of Figure 3-10 shifts the rising and falling edges or energy centroid of the output with respect to the input pulse. This results in a range error that cannot be tolerated in highly accurate range tracking systems. This distortion is minimized when the detection is made in a much wider bandwidth, and although this results in a suboptimal SNR an adequate SNR may nevertheless be achieved over short ranges or for large targets. Notwithstanding this, it may be assumed that most radars employ matched receiver filtering, particularly long-range surveillance systems. Matched reception is assumed in the remainder of this book.

3.7 RANGE RESOLUTION

The range resolution of a radar refers to the minimum range difference between two or more targets for them to be perceived as separate, distinct targets. It may also be applied to the resolution of multiple scattering features distributed along the range axis of a single target, for example, the separation of the returns corresponding to the engines on an airliner when viewed from the side. Should two (or more) closely spaced targets be present, their echoes may overlap and they would be mistaken for a single extended target. For the targets to be resolved in range, their echoes must be separated in time by at least their own pulse width, τ. The resolution of two targets in range is illustrated in Figure 3-11. On the limit of resolution, the difference in the time of reception of two targets, $\Delta t = \tau$. This leads to the following expression for the range resolution, ΔR:

$$\Delta R = \frac{c\tau}{2} \tag{3-20}$$

Substituting the matched filter bandwidth condition of (3-19) gives

$$\Delta R = \frac{c}{2B} \tag{3-21}$$

From (3-19) to (3-21) one can see that the narrower the pulse the finer the resolution and the broader the bandwidth. In fact, the inverse relationship between range resolution and signal bandwidth is fundamental to radar and applies not just to pulsed radars but to any modulation imparted onto the carrier which permits the measurement of range.

3.7.1 Factors Degrading Range Resolution

In spite of the simplicity of the analysis, there are a number of reasons the range resolution achieved in practice may be somewhat coarser than that indicated by (3-21).

- Targets are often regarded as occupying a single point in space. This so-called *point target* model is used for convenience even though it is never true. Real-life targets actually occupy a volume and as such have a down-range dimension. Thus, the return

FIGURE 3-11 ■ Range Resolution

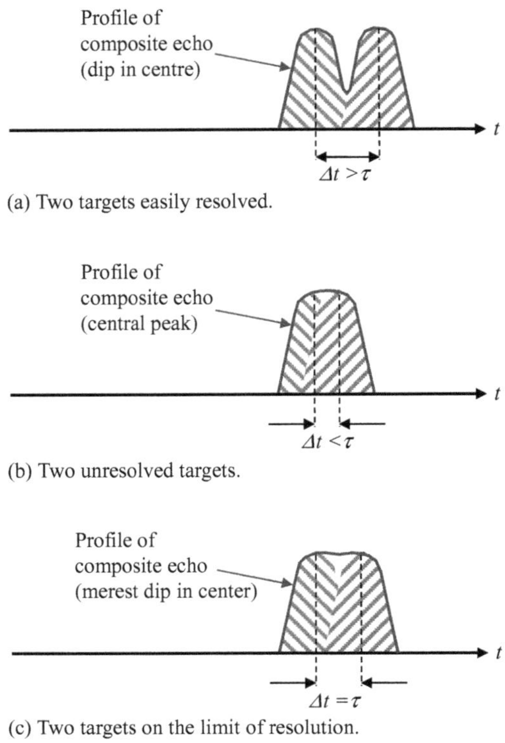

from the leading edge of the target and its trailing edge differ in range by the down-range extent of the target. The resulting echo is therefore elongated in range by the range extent of the target. This becomes an argument over semantics since one could also argue that if the range extent of the target exceeds ΔR, the echoes from the leading and trailing edges become resolvable; otherwise they may be regarded as two unresolved echoes. Nevertheless, the extension of the target return by its range extent is an effect that has to be borne in mind when resolving two such targets that are closely spaced. When ΔR is considerably less than the range extent of the target one obtains a range profile of the target. Such high-range resolution radars (HRRRs) are used for target recognition purposes.

- The distortion due to matched filter reception elongates the pulse width. While the peaks in the received waveforms arising from two targets on the limit of resolution remain τ seconds apart after passing through the matched filter, it may be rather more difficult to resolve them in practice.
- Shaped pulses result in a reduced bandwidth and hence a coarser level of range resolution.
- Resonances on the target cause its return to become smeared in time and hence range.

3.7.2 Terminology

Finally, it is worth noting that the terms *high resolution* and *low resolution* are in common usage. This is potentially confusing since *high range resolution* is taken to mean that ΔR is a small quantity and, conversely, that *low range resolution* is taken to mean that ΔR is a large quantity. This book uses the terms *fine resolution* and *coarse resolution*, respectively.

3.8 | RANGE ACCURACY

3.8.1 Range Cell Straddling

Target returns typically straddle two or more successive range cells. The receiving period is normally quantized into increments of time (range), known as range cells matched to the range resolution, ΔR, as described in Section 3.7. A target may be declared in two (or more) consecutive range cells occupied by its return. This may be perceived as two (or more) targets whereas in fact only one is present. This is an issue of range *accuracy* (i.e., the accuracy of the range to a single target declared by the radar) and should not be confused with range resolution, which refers to the minimum range difference between two or more targets for them to be perceived as separate targets. However, two nearby targets spaced beyond the limit of ΔR defined by (3-21) may merge together if they both straddle a common range cell. This situation is depicted in Figure 3-12.

Range cell straddling gives rise to a *straddling loss* since the full energy of the target return may not be present in one range cell as it is distributed over several successive range cells. Therefore, the detection performance varies across all the range cells occupied by the target and is never as good as if all the energy had been concentrated into just one range cell. However, it is possible to compare the target energies in each of the range cells to arrive at an accurate estimation of the energy centroid of the return and hence its true range. This is implemented in range tracking systems using a technique known as *split gate ranging*. This can yield a range accuracy of about 1/10th to 1/20th the range resolution, depending on the SNR.

3.8.2 Range Accuracy and the Importance of the SNR

The importance of the SNR has previously been stressed in the context of radar detection performance. It has also been noted that the matched bandwidth is a key parameter in establishing the optimum SNR. However, the importance of SNR goes well beyond the detection performance of a radar. In general, the accuracy with which any parameter of the target may be measured (e.g., angle, range, or velocity) is inversely proportional to the square root of the SNR [2]. In fact, *accuracy* $\approx$ *resolution*/$\sqrt{SNR}$. This also indicates how the SNR is fundamental in determining the quality of the information provided by the radar.

3.8.3 Accuracy and Resolution

The previous discussion has differentiated between *accuracy* and *resolution*; at least it has done so in the context of range measurement, although much the same could be said of other parameters such as angle and velocity. Clearly, issues of accuracy and resolution are closely bound up with one another. It is also reasonable to expect that as resolution improves so too does accuracy, although, strictly, this may not always be the case.

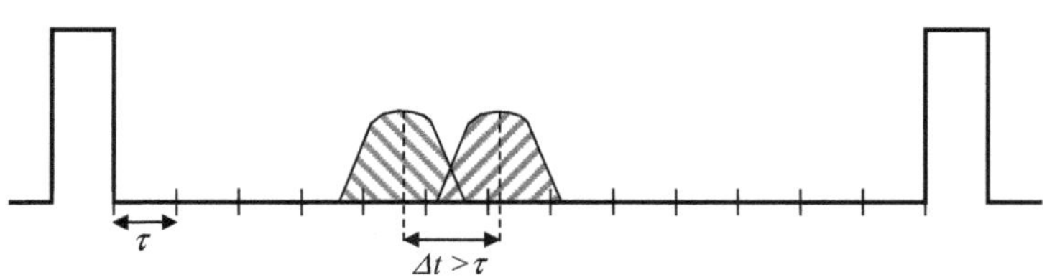

FIGURE 3-12 Range Cell Straddling

3.9 PULSE COMPRESSION

3.9.1 The Concepts of Pulse Compression

The analysis on range resolution (Section 3.7) made the point that the matched receiver bandwidth, B, and the receiver noise bandwidth, B_N, are approximately equal to each other for simple pulses, that is, pulses with no other modulations on them. This ties the range resolution, ΔR, to the detection performance of the radar and creates a dilemma when selecting a suitable transmitted pulse width. A wide transmitted pulse contains lots of energy, has a narrow (noise and signal) bandwidth, and leads to good detection performance but coarse range resolution whereas a narrow transmitted pulse contains little energy, has a wide (noise and signal) bandwidth, and leads to poor detection performance but fine range resolution. The solution to this dilemma is to employ pulse compression. This entails the transmission of a long pulse width (narrow, B_N) which is modulated over a very wide signal bandwidth (wide, B). The received pulses are processed in a matched filter that compresses the long input pulses into very much shorter pulses. The compressed pulse width $= 1/B$, where B is the bandwidth of the modulation employed. The signal bandwidth is now dominated by the modulation used, which can be taken as B. The ratio of the transmitted pulse width to the compressed pulse width is known as the compression ratio; values may range from single figures up to several thousand. This strategy now divorces the detection performance from the range resolution.

3.9.2 Pulse Compression Waveforms

The compression is brought about by the wide-band modulation of the transmitted waveform and its subsequent matched filter reception. Several frequency and phase modulation schemes are commonly used. These include linear frequency modulation (linear FM or LFM) also known as *chirp* because an audio signal with a linear FM would sound like a chirp, biphase modulation codes such as Barker codes and maximal length pseudo-random code sequences, and polyphase codes such as Frank codes, P1, P2, P3, and P4 codes. Other, less common modulations include non-linear FM, stepped frequency CW waveforms, and polyphase Barker codes, to name but a few [5, 6, 7]. When using the radar range equation to calculate the detection performance of a pulse compression radar, it is worth emphasizing the point made in Section 2.4.4 that the *noise* bandwidth is the inverse of the transmitted pulse width, that is,

$$B_N = \frac{1}{\tau} \tag{3-22}$$

In other words, the presence of the frequency or phase modulations on the transmitted pulse does not have any impact on the radar range or detection performance. The noise bandwidth, B_N, should not be confused with the signal bandwidth, B, in which $B > B_N$. Equations (3-19) and (3-20) are no longer valid for modulated pulses. However, when modulated pulses of bandwidth, B, are transmitted and they are received into a matched filter, equation (3-21) is still valid.

3.9.3 Theory of Pulse Compression

The mathematical basis for pulse compression is presented now [7]. The *correlation function* relating two variables $f_1(t)$ and $f_2(t)$ is given by

$$r_{xy}(t') = \lim_{T_0\to\infty} \frac{1}{T_0} \int_{-T_0/2}^{T_0/2} f_1(t) f_2(t+t')\,dt \tag{3-23}$$

where t' represents a time displacement, and T_0 is the correlation period. When the signals being compared are of limited duration, then it is appropriate to use the *finite* version of the correlation function given by

$$r_{xy}(t') = \int_{-\infty}^{+\infty} f_1(t) f_2(t+t')\,dt \tag{3-24}$$

If the two signals are in fact one and the same function, then one obtains the *autocorrelation function* (ACF). The ACF of a signal waveform is an average measure of its time domain properties. Formally, the ACF is defined as

$$r_{xx}(t') = \lim_{T_0\to\infty} \frac{1}{T_0} \int_{-T_0/2}^{T_0/2} f(t) f(t+t')\,dt \tag{3-25}$$

It is therefore equal to the average product of the signal $f(t)$ and a time-shifted version of itself and is a function of the time shift, t'. Strictly, the above definition applies to a continuous signal of infinite duration. If applied to a signal of limited duration (e.g., a signal pulse) it is more appropriate to use the finite ACF, which is defined as

$$r'_{xx}(t') = \int_{-\infty}^{+\infty} f(t) f(t+t')\,dt \tag{3-26}$$

With reference to Figure 3-8 and in the absence of noise, let us assume that a radar target returns a signal, $S(\omega)$, which is processed by a receiver having a transfer function, $H(\omega)$. The signal outputted from the receiver, $G(\omega)$, is given by

$$G(s) = S(\omega)H(\omega) \tag{3-27}$$

which may be represented in the time domain as the convolution of s and h:

$$g(t) = \int_{-\infty}^{+\infty} s(t')h(t-t')\,d\tau \tag{3-28}$$

and is very similar to the finite version of the cross-correlation function (3-24). The only significant difference between (3-24) and (3-28) is that the convolution integral (3-28), unlike the correlation integral (3-24), requires the reversal of one of the time functions.

If the receiver is *matched* to the received signal, then the receiver transfer function is the complex conjugate of the received signal on its input. i.e. if

$$H(\omega) = S^*(\omega) \tag{3-29}$$

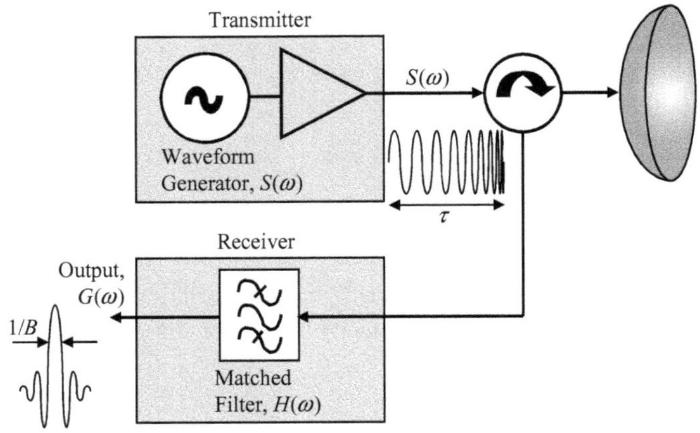

FIGURE 3-13 Linear FM Chirp Pulse Compression

then

$$G(\omega) = S(\omega)S^*(\omega) \tag{3-30}$$

The complex conjugate of a function amounts to a time-reversed version of the function in the time domain. Therefore

$$g(t) = \int_{-\infty}^{+\infty} s(t')s(t+t')\,d\tau \tag{3-31}$$

which is now the equivalent of the finite ACF of (3-26). Convolution of the matched filter's impulse response with the input signal is equivalent to autocorrelating the input. The upshot is that matched reception of a signal amounts to the autocorrelation of the signal relevant to a particular time shift.

3.9.4 Implementation of Pulse Compression

Figure 3-13 illustrates the waveforms and compression of a linear FM (chirp) modulated pulse. The waveform generator produces the modulated pulse of duration, τ, with LFM over the bandwidth, B. On reception, the matched filter compresses the received signal into a pulse width $= 1/B$.

3.9.5 Compression Ratio

The spectrum of an LFM chirp pulse is illustrated in Figure 3-14. This signal has been generated for a transmitted pulse length of 1 ms and peak amplitude of 1 and LFM chirp from 95 to 105 MHz (nominal carrier frequency = 100 MHz, B = 10 MHz). Plot (a) in the top left shows the spectrum of the signal. It is obvious from this plot that the spectrum approximates a rectangular function over the bandwidth of the chirp; that is, the spectral power density is near uniform over the chirp bandwidth. Plot (b) in the top right shows the linear FM sweep across the 1 millisecond pulse. The compression ratio is given by

$$CR = \frac{\text{transmitted pulse width}}{\text{compressed pulse width}} = \frac{\tau}{1/B} = \tau B \tag{3-32}$$

which, in this example, gives $CR = 1\text{ ms} \times 10\text{ MHz} = 10{,}000$.

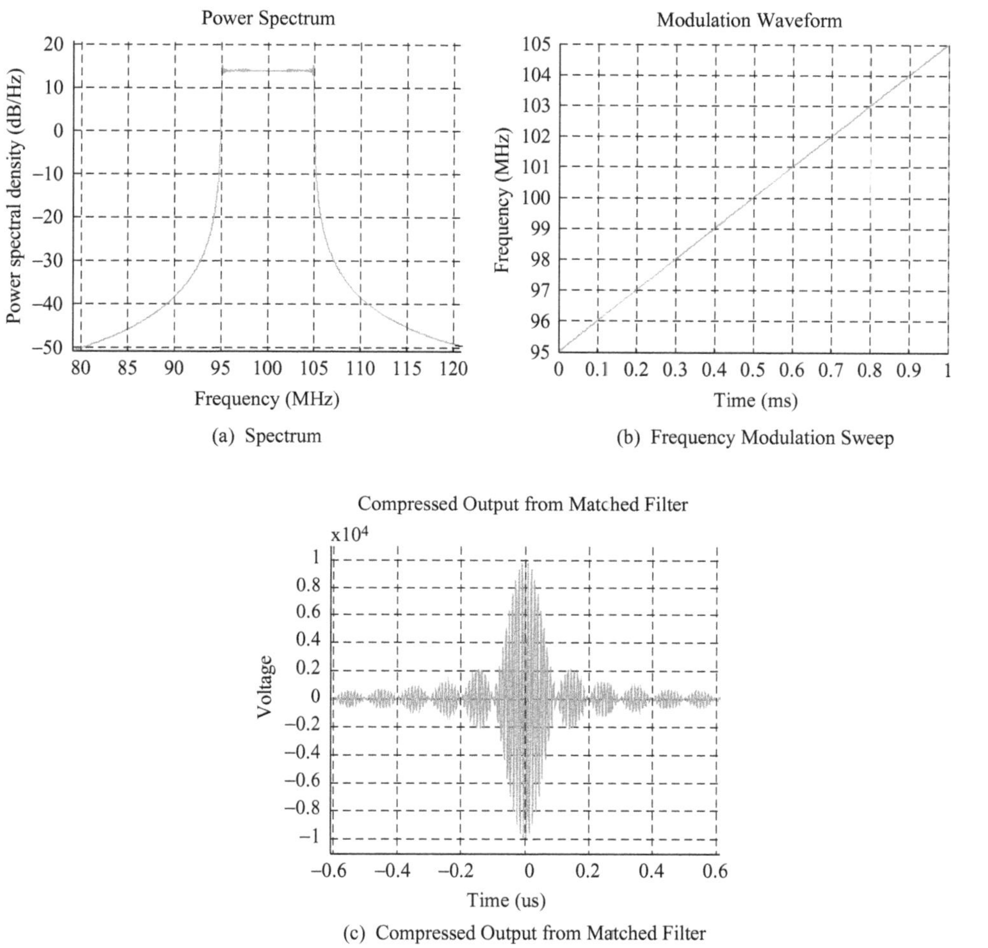

FIGURE 3-14 (a) Power Spectrum of Linear FM Chirp; (b) Frequency Modulation Sweep; (c) Compressed Output from Matched Filter

Plot (c) at the bottom center shows the compressed pulse available from the output of the matched filter. This illustrates how the rectangular function in the frequency domain has been transformed via an inverse Fourier transform into the classic sinc function envelope in the time domain. The carrier within this envelope runs at a constant frequency equal to the nominal center frequency of the linear FM chirp (i.e., 100 MHz in this example). The width of the main lobe of the compressed output is $1/B = 1/10\,\text{MHz} = 0.1\ \mu\text{s}$. Note also that the peak amplitude in the compressed output is 10,000 (this compares with a peak amplitude of the uncompressed signal of 1). Thus, the energy in the input signal has been preserved leading to an increase in the peak output amplitude by a factor in keeping with the compression ratio.

3.9.6 Range Sidelobes

One further characteristic of the compressed waveform of Figure 3-14c is the presence of sidelobes in the time (range) dimension known as *range sidelobes* or *time sidelobes*. These are undesirable features because the range sidelobes associated with strong target returns could give rise to false alarms at ranges just before and after the main target or could mask the presence of smaller targets at ranges just before or after the large target.

Since the spectrum of an LFM chirp is rectangular, the range sidelobes in the compressed waveform are 13 dB below the peak response. The range sidelobes may be minimized by introducing a weighting profile across the chirp bandwidth to taper the spectrum, much like the transmitter pulse shaping discussed in Section 3.5.2. This introduces a degree of mismatched filtering and results in a loss of SNR and resolution. Phase coding techniques of achieving pulse compression must use codes that minimize these range sidelobes. Range sidelobes lower than 30 dB below the peak are desirable and achieved with many waveforms [8].

3.9.7 Effects of Eclipsing on Pulse Compression

One further point of note is that eclipsing (Section 3.3.3) can be quite damaging to the performance of pulse compression systems [8]. Partially eclipsed returns cause the receiver to be blind to some of the returning pulse due to a clipping of the leading or trailing edge of the pulse by the receiver protection switches. Consequently, part of the echo energy is lost and the SNR and detection performance suffers. In addition to this, part of the modulation is also lost. This means that the full band of FM chirp waveforms is missing. The remaining signal partially correlates within the receiver but causes a loss in range resolution in sympathy with the loss of bandwidth. The sidelobe levels would be unaffected. In phase coded waveforms, the loss of bits from the phase code tends to cause increases in the range sidelobes; however, the bit duration is preserved in the uneclipsed portion, so there is no degradation in range resolution. Thus, it is also desirable to design codes that are robust to eclipsing.

3.9.8 Doppler Corruption

Finally, the presence of a Doppler shift on a pulse compression signal results in an imperfect correlation in the receiver and can, in extreme cases, corrupt phase codes. A Doppler shifted return imparts a frequency offset on the modulation. For linear FM this causes the returning FM to be offset by the Doppler shift. The receiver matched filter is now no longer perfectly matched to the returning waveform. As a consequence the peak in the correlation output is displaced in time, either earlier or later depending on the sign of the Doppler shift. This imparts a range error and is known as *range–Doppler coupling*. Furthermore, the magnitude of the correlation peak is not the maximum value expected from a non-Doppler shifted return, and there is a slight loss in detection performance and the range sidelobes are increased. In phase coded systems, a Doppler shift imparts a rate of change of phase. A large phase error can accrue over a long transmitted pulse or high Doppler shifts that may ultimately corrupt the phase coding. For any given phase coding there is a *critical velocity* beyond which the phase error that accrues across the pulse width corrupts the phase code [8]. The compressed waveform is also subject to similar losses in its correlation properties, as previously described for linear FM waveforms. Linear FM is more tolerant of Doppler corruption than phase coding modulation schemes because even the largest Doppler shifts tend to be a very small fraction of the chirp bandwidth. Linear FM tends to be favored in airborne radars where high target velocities are expected.

The effect that frequency (Doppler) shifts and time shifts have on radar waveforms will be considered in Chapter 5.

CHAPTER 4

Doppler Sensing Radar

Chapter Outline

4.1 THE DOPPLER EFFECT

4.1.1 The Doppler Shift

The phase shift of a radar signal over a distance, $2R$, where R is the one-way range to a target, is given by (ignoring any phase change upon reflection from the target):

$$\varphi = 2R\frac{2\pi}{\lambda} \tag{4-1}$$

If the range, R, is changing linearly with time, then there exists a rate of change of the phase of the returning echo. A constant rate of change in R arises due to a velocity difference between the radar and the target. The rate of change in phase amounts to a frequency shift whose magnitude and sign depend on the magnitude and sign of the relative velocity between the radar and target. This is known as the *Doppler effect,* and the corresponding frequency change is known as the *Doppler shift frequency* (or often just the *Doppler shift* or the *Doppler frequency*). The Doppler effect is exploited by Doppler radars for the discrimination of target returns of differing relative velocities and for the measurement of the relative velocity of a target. In its strictest sense, the Doppler effect is a relativistic effect; however, when the relative velocities between the radar and target are small compared with the speed of light, $c(= 3 \times 10^8$ m/s), then a non-relativistic treatment is a perfectly adequate approximation. In radar systems, the two-way Doppler effect needs to be considered. On the outward path, there exists a Doppler frequency between the signal transmitted by the radar and a hypothetical receiver carried by the target. Then on the return path, the target is seen as the source of a signal, and there is once again a Doppler frequency between the return off the target and the radar receiver. For monostatic radar (i.e., colocated transmitter and receiver) the outward and return paths are reciprocal courses of each other and the relative velocity is the same over both legs of the journey. This is not necessarily the case for bistatic or multi-static radars. This book considers the monostatic case for relative velocities $\ll c$.

4.1.2 The Doppler Shift for CW Radar

Taking the first-order time derivative of both sides of (4-1) leads to

$$\dot{\varphi} = 2\dot{R}\frac{2\pi}{\lambda}$$

or

$$\omega_d = 2V_r\frac{2\pi}{\lambda}$$

where $\dot{\varphi} = \frac{d\varphi}{dt} = \omega_d$ is the Doppler angular frequency and $\dot{R} = \frac{dR}{dt} = V_r$ is the relative velocity.

Since the substitution $\omega_d = 2\pi f_d$, where f_d is the Doppler frequency, can be made, after cancellation

$$f_d = \frac{2V_r}{\lambda} \tag{4-2}$$

4.1.3 The Doppler Shift for Pulsed Radar

It is also possible to derive (4-2) from the point of view of a pulsed radar in which the phase change is captured for each successive pulse in a series of coherent radar pulses. It is worthwhile considering this derivation since it establishes a key feature of pulse Doppler radar and its processing: that of a regular phase difference between successive pulses as the target moves at a constant velocity. Consider the situation depicted in Figure 4-1 in which a target is advancing toward a radar at a relative velocity, V_r. The target is shown frozen in time at the point at which pulse #1 reflects from the target. The target is also shown frozen in time at the point at which pulse #2 reflects from the target. In the time between pulse #1 and pulse #2 the target has advanced through a distance, d. Since the time between successive pulses is the pulse repetition interval (PRI) $= T_r(= 1/f_r)$, then recalling that velocity = distance/time:

$$V_r = \frac{d}{T_r} = df_r \tag{4-3}$$

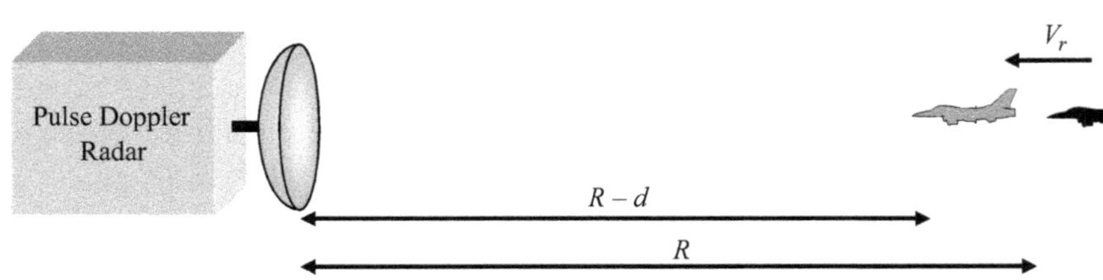

FIGURE 4-1 ■ The Doppler Effect in Pulsed Radars

The difference in phase between the two returning pulses is

$$\Delta\varphi = 2d\frac{2\pi}{\lambda} \tag{4-4}$$

This phase difference has been incurred over a time interval, $T_r = 1/f_r$. Hence, the rate of change of phase = phase difference / time difference. This rate of change of phase is the Doppler angular frequency, that is,

$$\omega_d = \frac{\Delta\varphi}{T_r} = \Delta\varphi f_r = \frac{2d2\pi f_r}{\lambda}$$

Recasting this in terms of the Doppler frequency

$$f_d = \frac{\omega_d}{2\pi} = \frac{2df_r}{\lambda} \tag{4-5}$$

Substituting (4-3) into the right-hand side of (4-5) gives $f_d = \frac{2V_r}{\lambda}$, which is equation (4-2).

4.1.4 Coherent Processing Interval

As long as the target motion continues in the same fashion, successive pulse returns continue to exhibit the same phase difference. This phase difference is a function of the wavelength, pulse repetition frequency (PRF), and relative velocity of the target. Since the wavelength and PRF are known to the radar system, the relative velocity may be measured by virtue of the phase difference between successive pulse returns. In practice, this phase difference is measured over several pulse returns over a time duration of a corresponding number of PRIs. This time duration is known as a *coherent processing interval* (CPI). During the CPI, it is necessary for the radar to transmit successive pulses that have a definite and fixed phase relationship with each other of $2\pi f_{tx}/f_r$ radians. If there were random phase differences between successive transmitted pulses, there would be similar random phase shifts superimposed onto their corresponding echoes and no measurement of the phase differences arising from the target motion would be possible. Similarly, there must be a definite and fixed phase relationship maintained on the pulse returns as they progress through the receiver. If the receiver were to impart random phase shifts onto the pulses, then by the same token no measurement of the phase differences arising from the target motion would be possible. Thus, it is incumbent on the radar to produce a coherent train of transmitted pulses over the CPI and for phase coherency to be maintained by the receiver over the CPI. This issue of phase coherency on transmit and receive is a necessary condition for the success of pulse Doppler radar; it will be considered again in Chapter 8.

4.1.5 Proportional Doppler Shift

The transmitted and received frequencies, f_{tx} and f_{rx}, respectively, are related by

$$f_{rx} = f_{tx} + f_d \tag{4-6}$$

The convention is that f_d is positive for closing targets ($f_{rx} > f_{tx}$), and f_d is negative for opening targets ($f_{rx} < f_{tx}$). The magnitude and sign of f_d is indicative of the magnitude and sign of the relative velocity difference between the target and radar. For typical target velocities, f_d represents a very small percentage of f_{tx}. While the magnitude of f_d increases at shorter wavelengths (or higher transmitter frequencies, f_{tx}), the percentage change in f_d is independent of f_{tx}. Since

$$f_{tx} = \frac{c}{\lambda} \tag{4-7}$$

the Doppler frequency may also be expressed as

$$f_d = \frac{2V_r f_{tx}}{c} \tag{4-8}$$

So the proportional Doppler shift is

$$\frac{f_d}{f_{tx}} = \frac{2V_r}{c} \tag{4-9}$$

For example, for $V_r = 1000$ m/s, the proportional Doppler shift is 6.67×10^{-6} or 0.000667%.

The Doppler shift given by equations (4-2) is inversely proportional to its wavelength, or from equation (4-8) it directly proportional to the carrier frequency, f_{tx}. Therefore, the higher the carrier frequency the greater the magnitude of the Doppler shift for a given target velocity. The use of higher carrier frequencies increases the sensitivity of the Doppler shift on target velocity. However, the proportional Doppler shift given by equation (4-9) is independent of the carrier frequency.

In making measurements of such small percentage changes it is necessary to compare the received frequency with the transmitted frequency in a coherent detection system. This can be accomplished routinely using modern phase locking synthesis techniques, superheterodyne receiver architectures, and coherent detection systems (Chapter 8).

4.2 | DOPPLER DISCRIMINATION

The Doppler shift is proportional to the *relative velocity* or the *radial velocity* of the target with respect to the radar. This is also known as the *range rate*. All these terms are in common usage, and all refer to the component of the velocity difference between the radar and the target resolved along the line of sight linking the radar and target. For radars at fixed locations the picture is fairly simple since the relative velocity corresponds to the component of the target's ground speed in the direction of the radar. In such cases, the ground return always appears at zero velocity and hence has a zero Doppler frequency. It therefore becomes possible to discriminate between the ground clutter return, unwanted slow-moving targets, and faster targets of interest by virtue of their differing Doppler shift frequencies. For many radar applications it is desirable to eliminate the ground clutter return and slow-moving ground-based targets. This can be accomplished through the use of a suitable high-pass filter in the demodulated Doppler band, which blocks zero Doppler returns and attenuates returns having a low Doppler frequency. This is sometimes known as the *clutter notch filter*. However, the clutter notch filter renders the radar blind to targets that may be of interest but are stationary or very slow moving. There is a delicate balance between the rejection of unwanted clutter and the possible rejection of targets of interest. This ultimately proved to be the counter to the chaff problem discussed in Chapter 1.

4.3 | PLATFORM MOTION COMPENSATION

4.3.1 Doppler Shift of Ground Return from Airborne Radar

The situation is rather more complex for radars that move with respect to the ground. This is applicable to airborne- or space-based systems and those mounted on moving surface platforms such as a ship or land vehicle. Platform motion over the ground generates a Doppler shift on the ground return that is dependent on the platform ground speed and the radar look angle with respect to its velocity vector. The situation for an airborne system is illustrated in Figure 4-2, in which the radar is detecting a ground return through a look angle of θ_{az} in azimuth off its velocity vector and a depression angle of θ_{el} in elevation. The relative velocity between the radar and the patch of ground along the radar direction of look is

$$V_r = V \cos\theta_{az} \cos\theta_{el} \tag{4-10}$$

where V is the platform velocity. This patch of ground therefore has a Doppler shift of

$$f_d = \frac{2V \cos\theta_{az} \cos\theta_{el}}{\lambda} \tag{4-11}$$

It must be borne in mind that the radar antenna scan angles and the platform pitch, roll, and yaw angles all contribute to θ_{az} and θ_{el}.

4.3.2 Ground Referenced Velocities

In many airborne radar applications it is commonplace to compute the Doppler shift from (4-11) along the orientation of the main beam boresight and provide a compensation for this Doppler shift onto all target returns [5]. This is called *platform motion compensation* (PMC). The PMC adjustment has to be made very quickly and updated very often since it must change as the antenna scans and as the platform rolls, pitches, and yaws. Applying PMC has the effect of offsetting one's own relative velocity along the main beam boresight such that the indicated velocity of all target detections in the main beam boresight are ground referenced rather than platform referenced. As a result, stationary ground clutter returns received via the main beam boresight appear at zero Doppler/zero velocity. The clutter notch filter would reject these returns.

Should a moving radar detect a moving target, the target returns exhibit a Doppler commensurate with the combination of both platform and target velocities resolved along

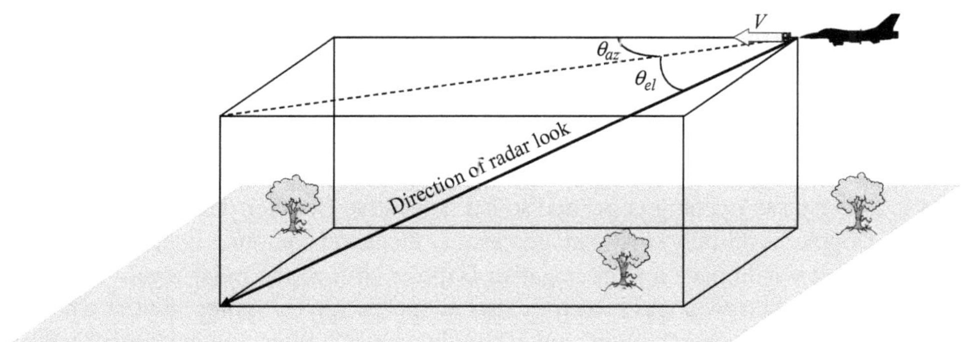

FIGURE 4-2 Relative Velocity of Ground from an Airborne Platform

the line of sight linking the radar and target. If PMC is applied on the radar then the Doppler shift from targets detected on its main beam boresight is commensurate with the component of the target ground speed resolved along the line of sight linking the radar and target.

4.4 DOPPLER BLINDNESS

In Section 4.2 the point was made that many radars employ a clutter notch filter to reject zero and low Doppler shifted returns. This is a useful and often necessary expedient to reject ground clutter but may lead to the rejection of certain targets of interest.

4.4.1 Air-to-Air Case of Blindness

In the air-to-air situation, certain targets are rejected by the clutter notch filter. The extent of target blindness depends on the bandwidth of the rejection notch. This bandwidth is set on the basis of providing rejection of main beam clutter and slow ground-moving targets (GMTs), see Chapter 6. If PMC is not applied, an airborne radar is blind to other airborne targets traveling at a near identical speed and heading. For a forward-facing airborne fire control radar, this could lead to blindness of a target in a tail-chase situation. Low-flying targets with (near-) zero ground speed (e.g., the return from the airframe of a helicopter in a hover) are also very difficult to detect because their return may be overwhelmed by the ground clutter return. If PMC is applied, targets of identical speed and heading may be detected when they are in the forward and rearward sectors, but those on a look angle that is perpendicular to the radar heading, or close to the perpendicular, would not be detected. Blind zones extend broadside to the radar track in such cases (but only for targets on the same course and heading, that is, flying parallel to the radar platform). Targets of (near-) zero ground speed would be rejected and thus blind.

4.4.2 Ground-to-Air Case of Blindness at the Zero Crossing Point

In general, the ground-to-air case can also be considered from Figure 4-2. If a radar is located on the ground and observes an airborne target at the angles θ_{az} and θ_{el} with respect to the velocity vector, V, of the target, as indicated in the diagram, the Doppler shift is also given by (4-11). In the ground-to-air case, a ground-based Doppler radar would be blind to certain airborne targets. As ever, targets of (near-) zero ground speed would be rejected by the clutter notch filter. An airborne target crossing the field of view of a ground-based radar is an interesting scenario. A plan view of this situation together with the Doppler history of the encounter is presented in Figure 4-3. The target is initially seen close to the head-on case and therefore has a large positive Doppler shift. As the target continues to approach the radar, the radar look angle opens out, and the Doppler shift reduces as the cosine of this angle. At the point where the target is moving at right angles to the bearing to the radar, its Doppler shift becomes zero. As the target continues past the zero crossing point it is seen as an opening target and so has a negative Doppler. The magnitude of this negative Doppler is initially small but increases as the target continues on its way. At some distant point it would have a large negative Doppler shift as the radar would see its full ground speed. A narrow angular sector exists about the zero crossing point at which the target Doppler falls within the stopband of the clutter notch filter, and the target cannot be

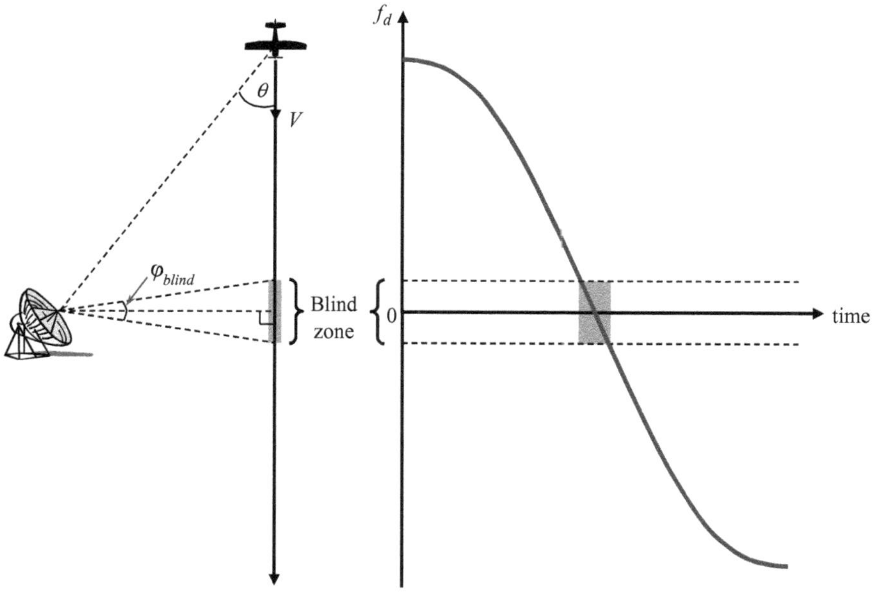

FIGURE 4-3 Crossing Targets

detected. The width of the blind angular zone around the zero crossing point depends on the rejection bandwidth of the clutter notch filter.

If the clutter notch filter has a cutoff frequency of f_{cc}, target visibility constrains the angle, θ, by

$$\frac{2V\cos\theta}{\lambda} > f_{cc} \tag{4-12}$$

In the limit, the target becomes blind when

$$\varphi_{blind} = 2\sin^{-1}\left(\frac{\lambda f_{cc}}{2V}\right) \tag{4-13}$$

Another scenario in which a ground-based Doppler radar would be blind to an airborne target arises should the target fly a circular path with the radar at the center of the circle. The target would sit continuously at the zero crossing point. This is perhaps an unlikely situation that would not persist for very long.

4.5 CONTINUOUS WAVE RADAR

4.5.1 Doppler Sensing CW Systems

A continuous wave (CW) radar transmits continuously and therefore receives the whole time. Since no modulation is imparted onto the transmitted signal, no measurement of range is possible. However, target motion causes a Doppler shift on the target return that affords a measurement of the relative velocity of the target. CW radar has therefore become synonymous with motion detection and velocity measurement applications. Two block diagrams of CW radars are presented in Figure 4-4. The block diagram of a single antenna system is given on the left-hand side, and that of a two antenna system is given on the right.

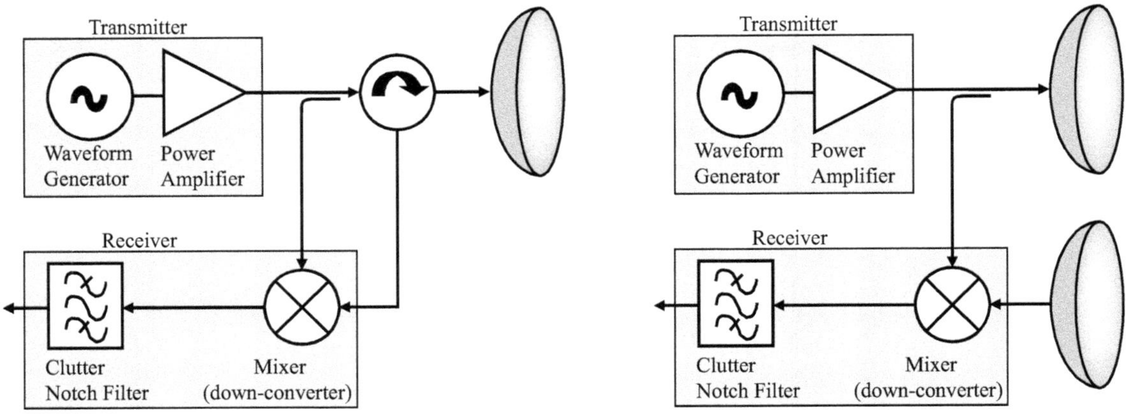

FIGURE 4-4 ■ CW Radar Block Diagrams: Single Antenna System (left); Double Antenna System (right)

Let the transmitted signal be given by $A_1 \sin(\omega_{tx} t)$, a small portion of which is coupled off as the local oscillator to the receiver mixer. The received signal is

$$A_2 \sin\left[(\omega_{rx} t) - 2\omega_{tx}\frac{R}{c}\right]$$

where $\omega_{rx} = \omega_{tx} \pm \omega_d$, and the $2\omega_{tx}\frac{R}{c}$ term represents the phase shift over the two-way journey to the target and back. Therefore, the received signal is

$$A_2 \sin\left[(\omega_{tx} \pm \omega_d)\, t - 2\omega_{tx}\frac{R}{c}\right]$$

The received signal is coherently mixed with the coupled portion of the transmitted signal in a multiplicative mixer. This results in a homodyne down-conversion of the received signal to a baseband output.

The mixer output is therefore

$$= A_1 \sin(\omega_{tx} t) A_2 \sin\left[(\omega_{tx} \pm \omega_d)\, t - 2\omega_{tx}\frac{R}{c}\right]$$
$$= \frac{A_1 A_2}{2} \cos\left[(2\omega_{tx} \pm \omega_d)\, t - 2\omega_{tx}\frac{R}{c}\right] + \frac{A_1 A_2}{2} \cos\left[(\pm\omega_d t) - 2\omega_{tx}\frac{R}{c}\right]$$

The first term runs at twice the transmitted frequency and is easily filtered off leaving just the component given by the second term, which runs at the Doppler frequency. Thus, the filtered baseband output is

$$= \frac{A_1 A_2}{2} \cos\left[(\pm\omega_d t) - 2\omega_{tx}\frac{R}{c}\right] \tag{4-14}$$

The peak amplitude is somewhat arbitrary.

Most modern systems use the more sensitive superheterodyne receivers in which incoming signals are down-converted to a baseband output in two or three mixing stages. This is described further in Chapter 8.

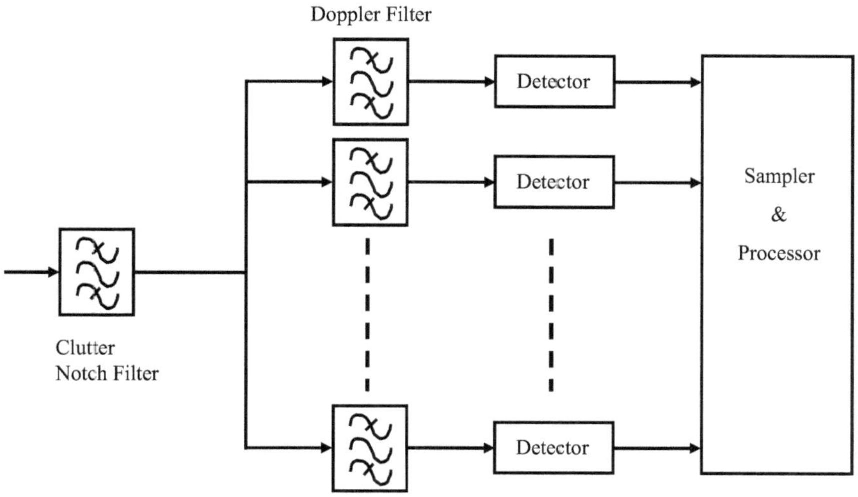

FIGURE 4-5 Doppler Filter Bank

4.5.2 Doppler/Velocity Resolution

Target velocity may be measured using a bank of narrow-band filters in the baseband signal, as shown in Figure 4-5. Each filter is tuned to pass a fine band of Doppler frequencies, and each successive filter is offset from neighboring ones so that, collectively, the bank of filters covers the expected spread of target Doppler frequencies. The output of each filter is detected and compared with a suitable threshold level to declare the presence of a target. The bandwidth of the filters establishes the Doppler resolution and hence the velocity resolution. The filter bandwidth is limited to the inverse of the integration time or the coherent processing interval:

$$\Delta f_d = \frac{1}{t_{int}} \tag{4-15}$$

in which Δf_d is the Doppler resolution, and t_{int} = CPI duration (integration time). Substituting (4-15) into (4-2) and rearranging gives a velocity resolution, ΔV_r, of

$$\Delta V_r = \frac{\lambda \Delta f_d}{2} = \frac{\lambda}{2t_{int}} \tag{4-16}$$

4.5.3 Doppler/Velocity Gating

The use of a filter bank to provide velocity resolution gives rise to the notion that the Doppler band may be subdivided into a series of contiguous Doppler resolution cells (or *bins*) as shown in Figure 4-6. Hence, the processing of baseband signals in this way

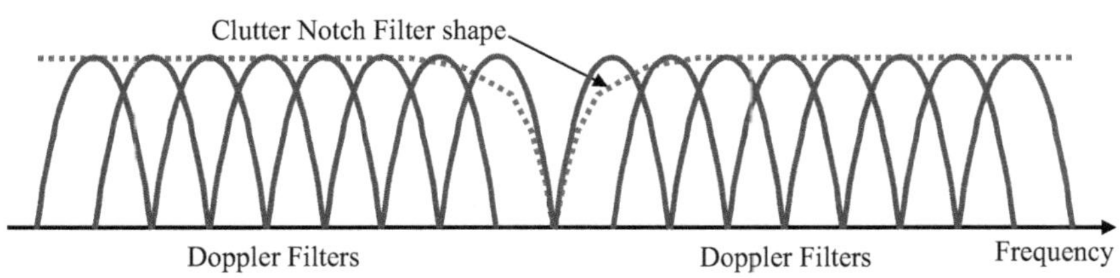

FIGURE 4-6 Doppler/Velocity Gating

gives rise to a set of quantized velocity cells in an equivalent manner in which range cells are used in the measurement of range. This subject will be covered in greater depth in Chapter 7.

4.5.4 Limitations of CW Radar

CW radars suffer from a few limitations. Since the transmitted signal is present the whole time, the receiver cannot be isolated from it using switches. The receiver is prone to being overloaded with the transmitted power. In the single antenna design, transmitter-to-receiver isolation is limited by the isolation of the duplexer and the return loss of the antenna feed structure. The isolation may be improved by using separate transmitting and receiving antennas, as used in the two antenna design. Here, the isolation is limited by the coupling between the two antennas. In spite of the improved isolation between the two antennas, one is still limited on the transmitter power that may be used. Transmitter leakage into the receiver has zero Doppler shift and so appears as a large clutter return. Even though this may be rejected by a clutter notch filter, its radio frequency (RF) power level may nevertheless saturate or even damage front-end receiver components.

The use of two antennas in a confined volume also constrains their size. Small antennas have a wide beamwidth and low gain resulting in poor angular resolution and poor detection performance. The low transmitted powers and low gain antennas tend to restrict CW radars to short-range applications. A reasonable range performance may be restored if long processing times or high processing gains are used, but this has an impact on the update rate.

A further shortcoming of CW radar is that target range is not directly available. However, range can be measured if a suitable frequency modulation (FM) or phase coding is imparted onto the waveform. FM ranging will be covered in more detail in the context of high PRF pulse Doppler waveforms in Chapter 10. When several targets are detected in the same antenna beam position, the only discriminant that may resolve them is Doppler. Multiple targets may be difficult to resolve from each other, and confusion could result unless some ranging capability is also introduced.

In spite of this, CW radars do have several important applications such as active missile seekers and radar altimeters. Furthermore, the use of very wideband FM, phase coding, or frequency-hopping modulations coupled with low peak powers and long processing intervals renders the waveforms very difficult to intercept. These waveforms are used in *low probability of intercept* (LPI) radars [7].

4.6 APPLICATION TO PULSED RADAR

Pulsed operation overcomes the limitations of CW radars. First, a high-power transmitter and single antenna system can be used because the receiver is isolated from the transmitted power using the receiver protection switches (see Section 3.1). Pulsed operation also enables range to be measured using the pulse delay method, as described in Section 3.3.

4.6.1 Baseband Waveform

In a pulsed system one obtains a similar baseband output as for the CW radar case except that the waveform described by (4-14) is obtained only for short durations corresponding to the returning pulses. The baseband waveform is shown in Figure 4-7.

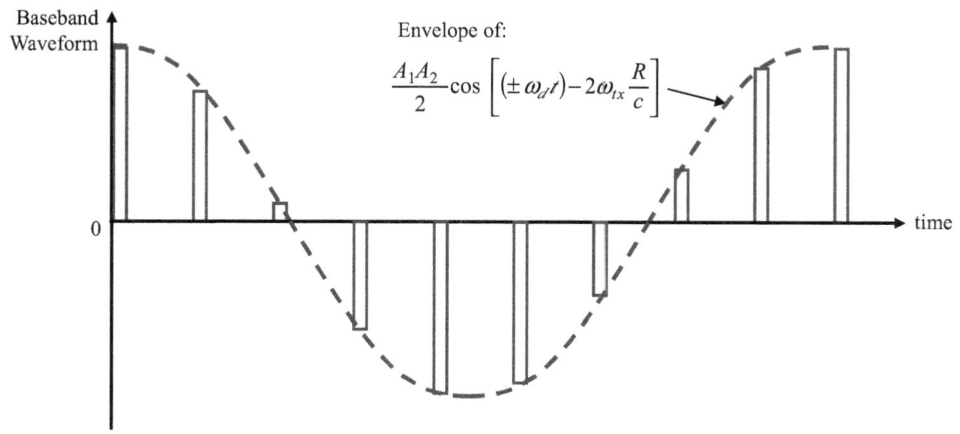

FIGURE 4-7 Baseband Output

The pulsed nature results in a train of returning pulses that are modulated in accordance with the Doppler frequency. Hence, these pulses represent a series of samples of the Doppler waveform, that is, samples of the waveform given by (4-14). These provide samples of the Doppler waveform at fixed intervals of phase. The sampling frequency is given by the radar PRF and the sample duration is given by one time resolution cell. The phase offset between successive samples is that given by (4-4) and remains constant so long as the transmitted frequency (or wavelength), PRF, and target velocity remain fixed. Since the radar has knowledge of its PRF and transmitted frequency, the phase offset between samples, $\Delta\phi$, and hence the relative velocity of the target can be computed from a succession of returns (samples). In the case of a static target, $\omega_d = 0$, and the CW baseband output would be a constant level. For pulsed systems, all pulses would have the same amplitude and $\Delta\phi = 0°$.

4.6.2 Doppler Ambiguity

Ambiguity arises if the sampling rate (i.e., the PRF) is less than twice the highest frequency being sampled. When a low PRF is used there may be inadequate sampling of the Doppler waveform. Therefore, low PRF radars result in unambiguous range measurement, but velocity ambiguities may arise.

CHAPTER 5

The Ambiguity Function

Chapter Outline

In Section 3.9.3, the point was made that a radar receiver performs a cross-correlation between the received signal and a reference of the transmitted waveform. Furthermore, the received signal may differ from the transmitted signal on account of being time delayed and Doppler shifted. The design of a matched receiver usually entails designing the receiver to match the transmitted waveform and so may no longer match the received signal on account of the time and frequency displacements of the received signal over the transmitted one. Matched reception is tantamount to autocorrelation of the received signal at a particular time shift. In radar, it is most convenient to think of the particular time shift associated with a given range cell. While the autocorrelation function handles shifts in time, it offers no provision for handling shifts in frequency. The *ambiguity function* describes the response of a matched receiver to a finite duration signal [7]. The use of the ambiguity function in ambiguity analysis considers the receiver to be matched to a signal received from a target at a given time delay (range cell) and frequency. The ambiguity function describes the matched receiver response as a function of any additional time delay and any additional Doppler frequency. The ambiguity function, $\chi(t', f_d)$, is often given as its magnitude squared, that is,

$$\left|\chi(t', f_d)\right|^2 = \left|\int_{-\infty}^{\infty} u(t)\, e^{2\pi j f_d t}\, u^*(t - t')\, dt\right|^2 \qquad (5\text{-}1)$$

where $u(t)$ is the complex envelope of the transmitted (and received) waveform, t' is the time delay, and f_d is the Doppler frequency shift. Note that while it is common to consider the square of magnitudes, the ambiguity function (5-1) resembles the finite autocorrelation function (3-26) but with the inclusion of the exponential term in Doppler frequency.

5.1 | AMBIGUITY DIAGRAMS

The ambiguity function is a function of two variables: t' and f_d. The three-dimensional plot of the ambiguity function, as a function of t' and f_d, is called an *ambiguity diagram*. The peak value of the ambiguity function occurs at the origin, $t' = 0$ and $f_d = 0$ and has a value $\chi(0,0) = (2E)^2$, where E is the total energy in the waveform. Note also that the total energy under the curved surface is also $= (2E)^2$ and is always constant, irrespective of the waveform.

Ambiguity diagrams are a convenient means of depicting the delay/Doppler response of a matched receiver. It is commonplace to plot the normalized amplitude $\chi(t', f_d)/(2E)^2$. The ideal ambiguity function would have a single peak at $t' = 0$ and $f_d = 0$ i.e. $\chi(0,0)$ and zero elsewhere. However, such an ideal is not attainable. In practice, the peak response has a finite width in Doppler frequency and time delay as determined by the frequency and time resolutions, respectively. In the Doppler dimension, the frequency resolution is given by the inverse of the processing interval ($\Delta f_d = 1/t_{int}$), and the time resolution is given by the inverse of the signal bandwidth ($\Delta t = 1/B$). Since the area under the curve is fixed by the energy, not all that energy can be contained within the peak at the origin, so the response is characterized by a series of sidelobes and nulls in both the range and velocity domains. The closest practical ambiguity diagram to the ideal is therefore a thin peak at $\chi(0,0)$ sitting on a low but constant pedestal and would resemble an upturned drawing pin. Note also that a cut of the ambiguity function at zero Doppler is the autocorrelation function of the transmitted modulation given by $u(t)$.

5.1.1 Ambiguity Diagram for a Simple Pulse

The ambiguity diagram of a simple pulse is given in Figure 5-1.

A cut through zero Doppler indicates the effect of a time shift and is presented in Figure 5-2 (left). This reveals a triangular function. A cut through zero time indicates the effect of Doppler frequency shift and is presented in Figure 5-2 (right). This reveals a sinc function.

5.1.2 Ambiguity Diagram for Linear Frequency Modulation Chirp

The ambiguity diagram of a linear frequency modulation (LFM) *chirp* pulse is illustrated in Figure 5-3. Its associated slice through zero Doppler and zero time is illustrated on the left- and right-hand sides of Figure 5-4, respectively. This shows a diagonal ridge in the plot. The zero Doppler slice shows a modified sinc function in which the sidelobes

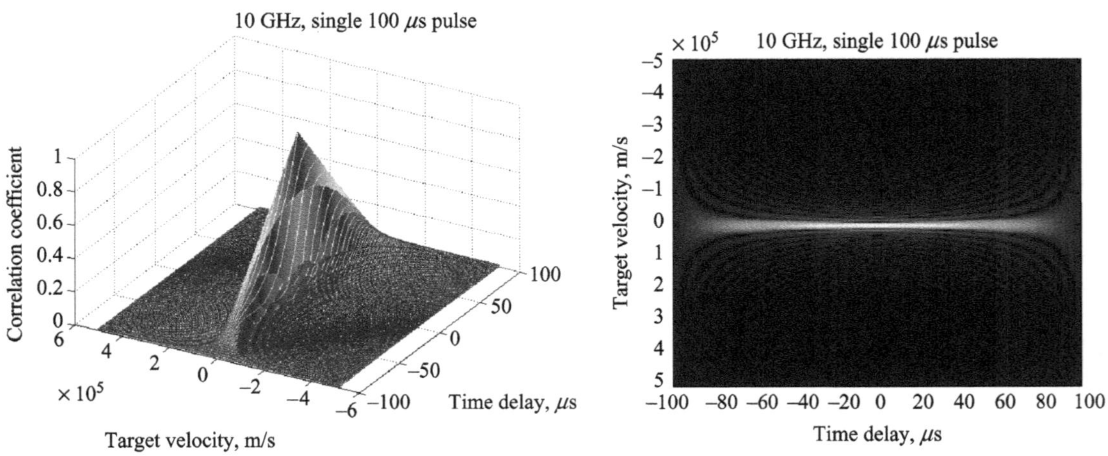

FIGURE 5-1 ■ Ambiguity Diagram of a Simple Pulse: 3-D View (left); Plan View (right)

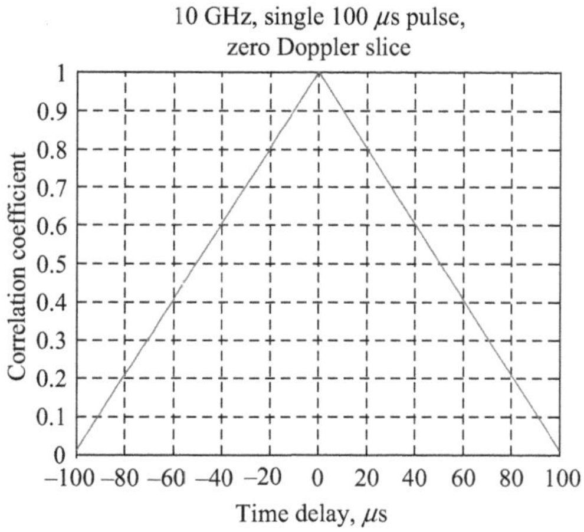

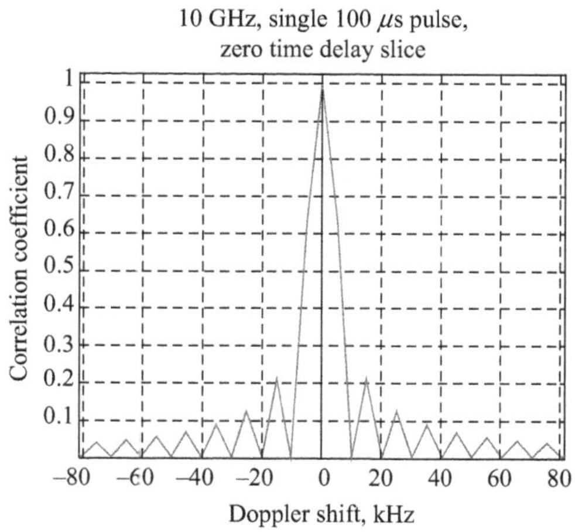

FIGURE 5-2 Ambiguity Diagram of Simple 100 μs Pulse, 10 GHz Carrier Frequency: Along Zero Doppler (left); Along Zero Time (right)

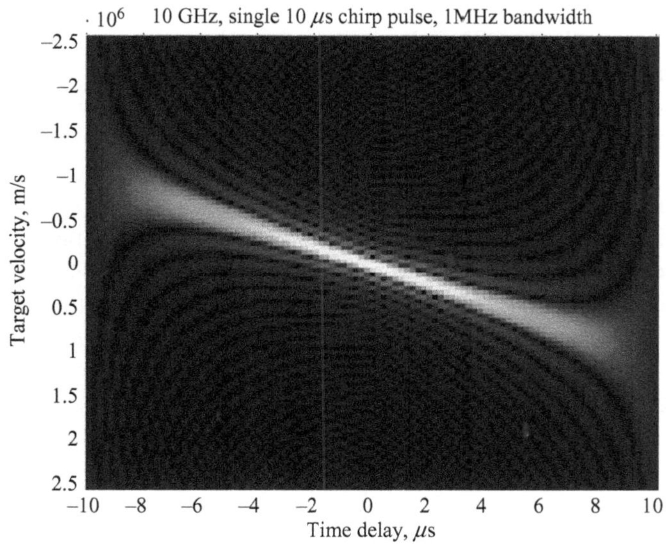

FIGURE 5-3 Ambiguity Diagram of Linear FM Pulse

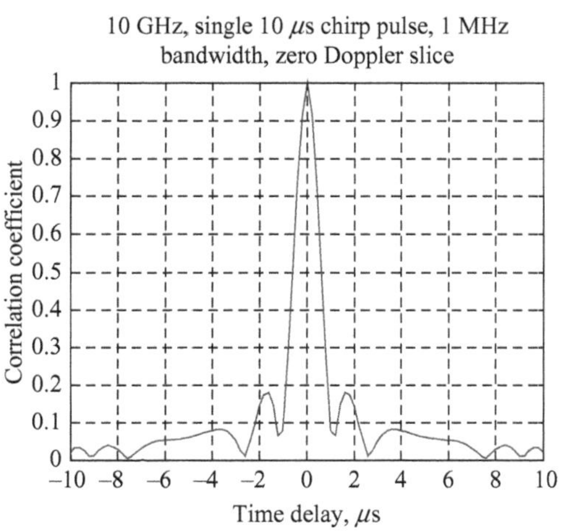

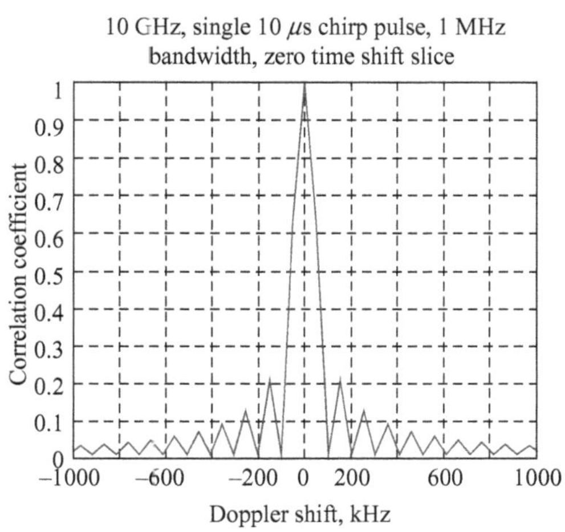

FIGURE 5-4 Ambiguity Diagram of a Linear FM Pulse, Pulse Width = 10 μs, FM Bandwidth = 1 MHz, Carrier Frequency = 10 GHz: Zero Doppler Slice (left); Zero Time Slice (right)

appear "stretched." These are the classic time sidelobes characteristic of FM chirp pulse compression.

If a Doppler shift is imparted on the linear FM modulation, the diagonal ridge in the ambiguity diagram becomes displaced in time (Section 3.9.8). The peak output now appears at a slightly displaced range, and the sidelobes in the zero Doppler and time slices become asymmetrical and distorted. The presence of the Doppler shift has corrupted the range measurement. This is known as *range–Doppler coupling*. In many practical cases, the magnitude of the Doppler shift is very much less that the FM chirp bandwidth and the range–Doppler coupling is minimal. Linear FM pulse compression is quite robust to range–Doppler coupling.

5.1.3 Ambiguity Diagram for a 13-Bit Barker Coded Biphase Modulated Pulse

The ambiguity diagram of a biphase coded pulse using the 13-bit Barker code is presented in Figure 5-5. Note that the correlation duration is given by $2N - 1 (= 25)$ bit times, where $N (= 13)$ is the number of bits in the code.

Figure 5-6 shows the zero Doppler slice (left) and zero time slice (right) of the ambiguity diagram for this waveform. The zero Doppler slice shows regular time sidelobes of amplitude $= 1/13$. The zero time slice shows a sinc-like structure with Doppler shift, although, as correlation falls at zero time, it rises either side due to strong diagonal components.

The presence of a Doppler shift introduces a rate of change of phase on the returning signal (Section 3.9.8). Large Doppler shifts can result in a large phase error accruing, particularly over a long duration phase code until eventually the phase coding becomes corrupted. For biphase coding one may imagine that the two phase states can be designated as 0° and 180°. If a phase change of more than 90° can accrue over the duration of the phase coded pulse, then the phase of the radio frequency (RF) transgresses the midway point between the two allowable phase states and the code could be inverted. This inversion would happen part way through the phase coded pulse and would result in only partial

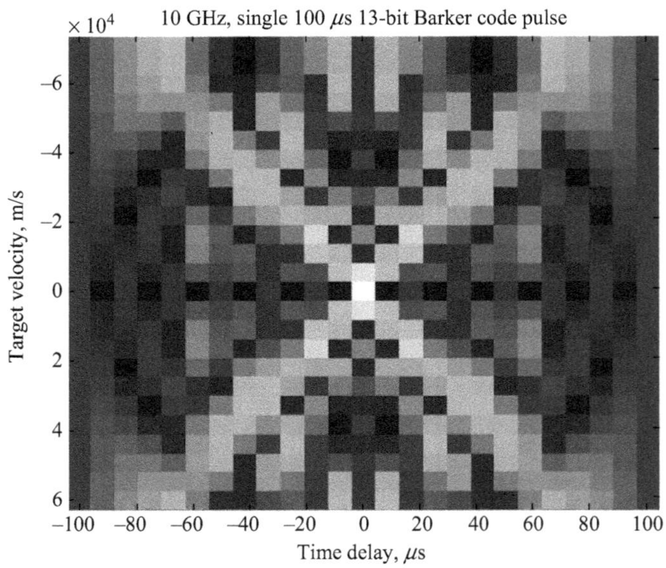

FIGURE 5-5 Ambiguity Diagram of 13-Bit Barker Code Biphase Modulated Pulse

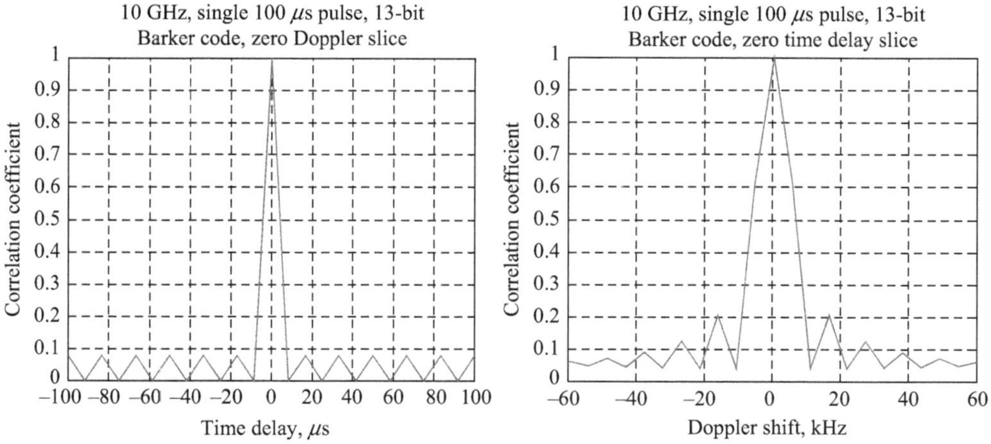

FIGURE 5-6 Ambiguity Diagram of 13-Bit Barker Code Biphase Modulated Pulse of Width = 100 μs, Carrier Frequency = 10 GHz: Zero Doppler Slice (left); Zero Time Slice (right)

correlation and large increases in the time sidelobes. Therefore, there is a limit to the permissible maximum Doppler shift and code duration. As a result, phase coded waveforms tend to be less Doppler tolerant than FM modulations. The radial velocity must therefore be constrained by

$$V_r < \frac{\lambda}{8\tau} \tag{5-2}$$

to avoid the code inversion of a biphase modulated pulse of transmitted pulse width, τ.

5.1.4 Ambiguity Diagram for Low Pulse Repetition Frequency Pulse Train

The ambiguity diagram of a low pulse repetition frequency (PRF) waveform is shown in Figure 5-7. The low PRF pulse train consists of a train of responses, each one looking

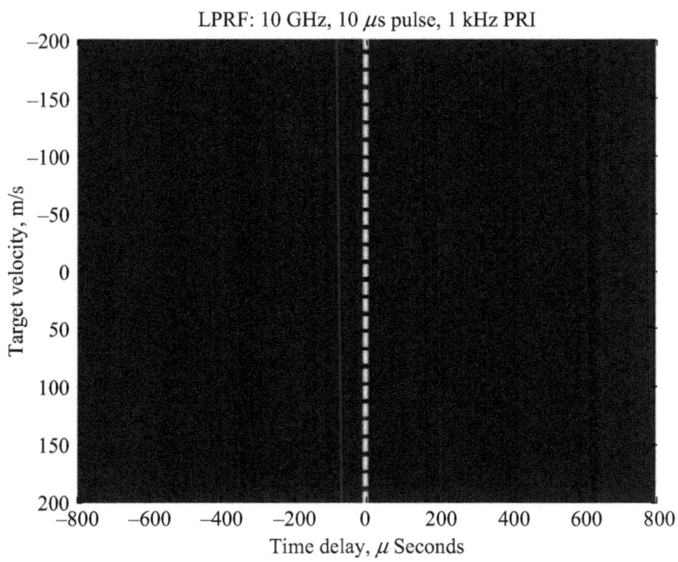

FIGURE 5-7 Ambiguity Diagram of Low PRF

like that of the simple, unmodulated pulse. If pulse compression is used in conjunction with a low PRF then each response in the ambiguity diagram takes on the form of the appropriate compressed pulse. Multiple repetitions of the responses can be noted in the figure along the velocity (Doppler) axis (i.e., along the cut at zero time shift). The repetition in velocity is symptomatic of the low value of PRF used. As was noted in Section 3.4, low PRF waveforms are likely to undersample the Doppler waveform, and this gives rise to ambiguous repeats of the responses in the velocity domain. The ambiguity diagram is a useful means of displaying and assessing the extent of the ambiguous repetition in both velocity and range domains of a pulsed waveform.

5.2 PEAK AND INTEGRATED SIDELOBE LEVELS

The time sidelobes in the ambiguity function help quantify the waveform in its ability to detect targets without interfering sidelobes. If the waveform is poorly designed, a large target could give rise to high sidelobes, which could mask a second nearby target or appear as an ambiguous repeat of the original target. The sidelobe performance can be quantified using two quantities: the *peak sidelobe level* (PSL) and the *integrated sidelobe level* (ISL). The PSL is defined as

$$PSL = 10\log_{10}\left[\frac{\text{Max sidelobe level}}{(\text{peak response})^2}\right] \tag{5-3}$$

The ISL is defined as

$$ISL = 10\log_{10}\left[\frac{\text{Total power in sidelobes}}{(\text{peak response})^2}\right] \tag{5-4}$$

The PSL is a useful measure when a single point target response is of concern, whereas the ISL is a more useful measure when distributed targets are of concern. Typical ISL values range from -10 to -20 dB.

CHAPTER 6

Clutter

Chapter Outline

6.1 CONCEPTS OF CLUTTER

Surface clutter is the unwanted return from the surface of the earth, be it land or sea. It includes static man-made objects and structures on the ground such as buildings, bridges, roads, railways, stationary vehicles, and electricity pylons. Thus, clutter is spread continuously over a wide area but may also have point features. It may also vary over neighboring regions with varying terrain type (e.g., agricultural land, forest, urban, desert, snowy tundra, mountainous or flat land) or sea surface (open ocean or coastal waters and their sea state, wind direction with respect to radar look angle). The use of the word *clutter* implies that it is an undesirable return. This contrasts with high-resolution ground-mapping radars such as airborne and spaceborne synthetic aperture radars (SARs) for which the ground and static man-made objects form the targets of interest. Often, the surface forming the clutter return is not completely stationary, for example, the swaying of trees in the wind or the wave action on the surface of the sea. In many cases, radars must discriminate against the clutter signal since it can limit the detection of targets of interest. Doppler processing techniques afford a high degree of rejection of surface clutter. Similarly, it is also often desirable to reject slow ground-moving targets (GMTs), such as slow-moving ground-based vehicles, since these are also not targets of interest. Hence, both surface clutter and GMTs may be smeared over a band of low Doppler shifts, and the clutter rejection filter must be suitably designed. Clutter rejection processing and radar waveform parameters are heavily influenced by the clutter environment and the radar platform parameters. These will be considered later, but first some basic concepts concerning surface clutter must be reviewed.

Volume clutter arises from unwanted returns occupying a volume in the detection space of the radar. Volume clutter usually results from a multitude of small scatterers filling a volume in the atmosphere such as rain, mist and fog, ice, dust, insects, and birds or even at interfaces between layers of the atmosphere differing in refractive index. It is commonplace for volume clutter to have an appreciable Doppler shift as the clutter often drifts at the wind speed. The atmospheric effects giving rise to clutter also have high transmission losses and so contribute to increased atmospheric losses; however, the clutter effect is of interest here.

6.2 | BACKSCATTER COEFFICIENT

The radar cross section (RCS) of a clutter surface is characterized by its backscatter coefficient, σ°. The backscatter coefficient, σ°, is the equivalent RCS per unit area illuminated and has units of m^2/m^2, which is often expressed on a decibel scale as dBsm or dBm^2 (decibels with respect to 1 square meter).

The RCS of volume clutter has a backscatter coefficient, η, which is the equivalent RCS per unit volume illuminated and has units of m^2/m^3 and may also be expressed on a decibel scale, as for σ°.

6.2.1 Terrain Backscatter

The backscatter coefficient for terrain depends on the following:

- Type of surface
- Roughness
- Frequency
- Grazing angle
- Polarization

6.2.2 Surface Roughness

Roughness is a loose term but can be quantified using the Rayleigh roughness criterion. With reference to Figure 6-1, consider two adjacent patches in the surface that differ by a height, h. If θ_g is the grazing angle, the path difference between reflections from the upper and lower surfaces differs by $2h \sin\theta_g$. If this difference is zero, the two contributions arrive in phase and the surface can be called smooth. If the phase difference is sufficiently small then the surface can still be called smooth. The upper limit on the acceptable phase difference is arbitrary but is normally taken as $\pi/2$, which gives rise to cancellation between the two contributions. A surface is called smooth if

$$2h \sin\theta_g \leq \lambda/4 \tag{6-1}$$

A phase difference of $2x\pi$ would also appear to be smooth, where $x = 1, 2, 3 \ldots$. In practice, a surface has irregularities that give a spread of values for h. Some points in the surface give zero phase difference, and others do not. The value of h should now be some average measure such as the root mean square (RMS) value. The smoothness of a surface is determined by the size of any irregularities relative to the wavelength. A particular surface may be smooth at one frequency but rough at another.

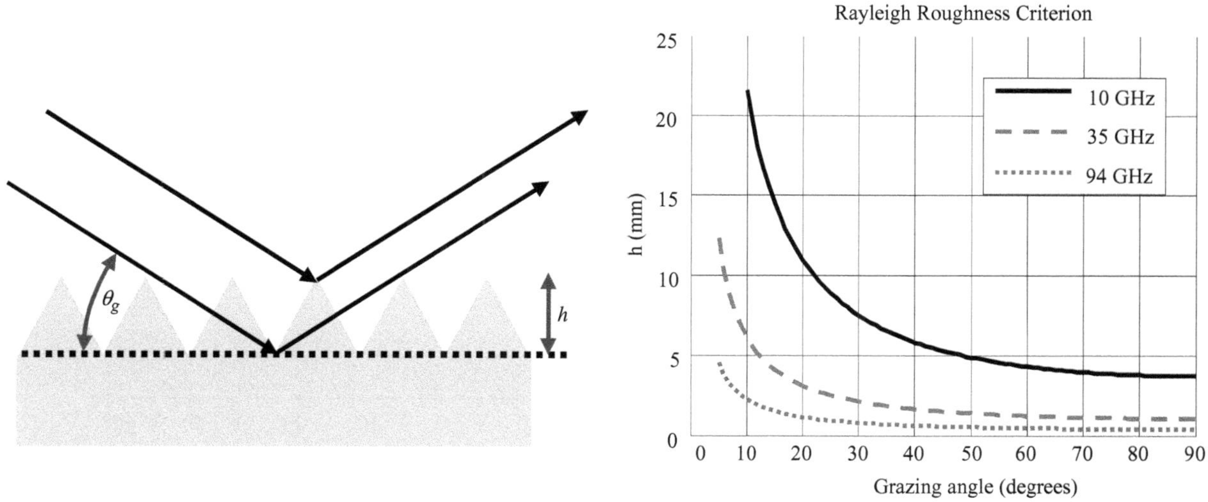

FIGURE 6-1 ■ Rayleigh Roughness Criterion: Reflections from Rough Surface (left); Roughness Threshold versus Grazing Angle (right)

The situation is depicted in Figure 6-1, where, on the left, the rays from the peaks and troughs of a rough surface are illustrated and, on the right, the smooth–rough threshold is graphed for 10 GHz, 35 GHz, and 94 GHz as a function of grazing angle. The region under each line is considered smooth and that above the line rough.

Smooth surfaces tend to give rise to *specular reflection*, that is, a reflected signal in which the angle of reflection equals the angle of incidence. The reflection from rough surfaces tends to be dominated by *diffuse reflection*, in which the reflected signal is scattered over a wide angular range. Surfaces dominated by specular reflection tend to have a large RCS over a narrow angular range and low RCS outside this angular range, whereas diffuse reflection gives rise to moderate RCS over a wide angular range. From Figure 6-1 it is obvious that a finer level of surface detail assumes a greater influence on its scattering properties at higher frequencies. Most naturally occurring surfaces would be considered rough at centimetric wavelengths, except perhaps a calm water surface. Many man-made materials have a smooth finish, even at millimetric wavelengths. The variation with grazing angle means that all surfaces tend to be smooth at very low grazing angles. Much the same can also be said of discrete targets as well as clutter surfaces.

6.2.3 Sea Surface Backscatter

For sea clutter, the backscatter coefficient is determined by the following:

- Frequency
- Grazing angle
- Polarization
- Sea state
- Wind speed
- Radar look angle with respect to the wind and waves

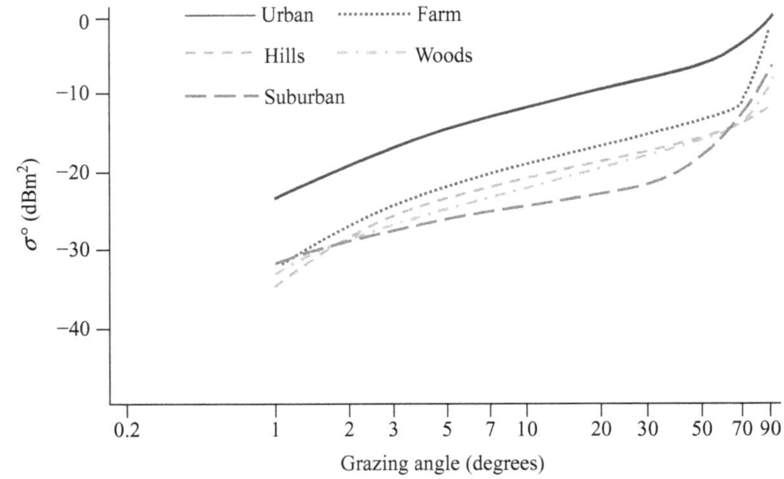

FIGURE 6-2 ■ Land Clutter Backscatter Coefficient versus Grazing Angle

The backscatter from a patch of the surface corresponding to one resolution cell arises from many independent scatterers within the patch. Micromotion of the clutter coupled with subtle changes in the radar look angle can give rise to considerable variation in the clutter return from a surface patch. Consequently, the backscatter coefficient exhibits statistical variation and may be described by a probability density function (PDF). Clutter has both a spatial variation from cell to cell and a temporal variation within each cell.

6.2.4 Terrain Backscatter as a Function of Grazing Angle

Examples of the backscatter coefficient variation with grazing angle are provided for various terrain types in Figure 6-2. All the types of surface are open to interpretation, and each of the lines should be replaced by a band at least 10 dB wide. Buildings in urban areas have a large total surface area and introduce many corner reflectors. At low grazing angles it becomes difficult to give a meaning to the angle. The grazing angle is defined in terms of the beam axis. If the elevation beamwidth is 6° and the grazing angle of the main beam boresight is 8°, then the far edge of the beam makes an angle of 5° with the surface and the forward edge of the beam makes an angle of 11°. There is also considerable variation in the backscatter coefficient over this range as well as the problem of variations in the orientation of the surfaces. One may also observe from Figure 6-2 that the backscatter coefficient increases dramatically as the grazing angle approaches 90°. The ground clutter return becomes particularly strong close to vertical incidence.

6.2.5 Sea Surface Backscatter as a Function of Grazing Angle

Sea clutter backscatter coefficients tend to follow the form of Figure 6-3 in which the break points depend on wavelength, polarization, and sea state [9]. In the first region, before the first break point, σ° increases as the frequency and roughness increase. The roughness can increase simply because the variations in surface height increase or because the grazing angle has increased. A smooth, or approximately smooth, surface predominantly reflects energy in the forward direction, away from the radar. In this region σ° varies in the region of a fourth power of the grazing angle. Beyond some rather ill-defined angle, σ° increases slowly with increasing frequency. It still increases with increasing roughness. Horizontal

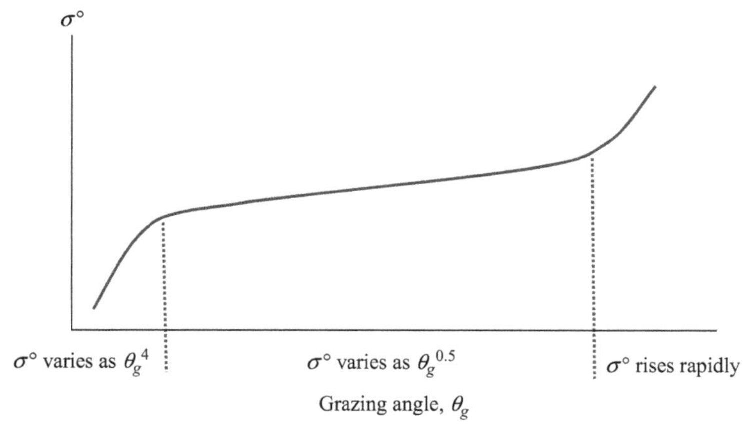

FIGURE 6-3 ■ Sea Clutter Backscatter Coefficient versus Grazing Angle

polarization has a stronger dependence on frequency than vertical polarization. In this region the variation with grazing angle is weak and is something like a square root of the grazing angle. At higher grazing angles there is another break point beyond which σ° has a very weak dependency on frequency and decreases with increasing roughness. As the beam is approaching the vertical there is strong reflection from facets facing the radar. The rougher the surface, the greater the degree of diffuse scattering away from the radar and the smaller the specular reflection returning to the radar.

It is always possible to find an expression that fits some experimental data. Such an expression fits only that particular data and may or may not be an adequate description of other data. Using it to predict what may be found should another set of data be obtained is little more than wishful thinking. The backscatter coefficient depends on so many things that an expression is useless if there is no indication of the conditions prevailing when the measurements were made. The break points between the regions are poorly defined, and it is pointless to attempt to define it using a parameter, the sea state, which is an estimated integer.

The only meaningful approach is to measure the surface, sea or land, using the frequency, polarization, and radar parameters of interest. The values obtained are valid only for those particular parameters. Attempts have been made to model the surface; the upper limit model by the Georgia Institute of Technology predicts the highest value of backscatter coefficient that should be found and is presented in Figure 6-4. One of the model parameters is the complex relative permittivity of sea water. Note the difference in the complex relative permittivity from the L-band to C-band. With low wind speed and wave heights the backscatter should be significantly less that the maximum. The measured and modeled sea clutter backscatter coefficient as a function of grazing angle is presented in Figure 6-4.

Summarizing a number of trials for different wind speeds, in this case at X-band using vertical polarization, shows that the backscatter coefficient is in the region of -35 to -50 dBm^2, as illustrated in Figure 6-5.

Horizontal polarization gives a similar picture to that of vertical polarization except for the deeper, broader region of very a low backscatter coefficient close to zero grazing angle. This arises because the boundary conditions state that the E field component parallel to the surface of a pure conductor must be zero at the conductor surface. Although water is not a pure conductor, its conductivity is high, which suppresses the backscatter of a horizontally polarized signal from a flat, calm sea surface. This case is illustrated in Figure 6-6.

FIGURE 6-4 Upper Limit Model and Measured Sea Surface Backscatter Coefficient versus Grazing Angle at X-Band and C-Band

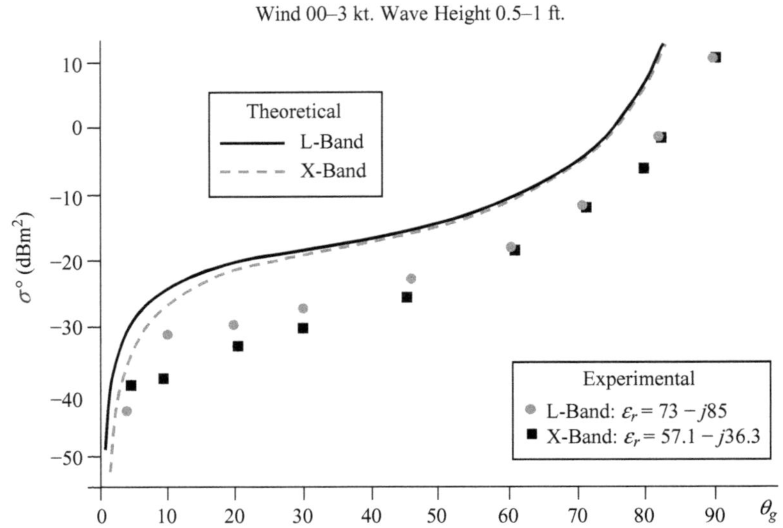

FIGURE 6-5 Upper Limit Model for Sea Surface Backscatter Coefficient versus Grazing Angle at X-Band and Vertical Polarization

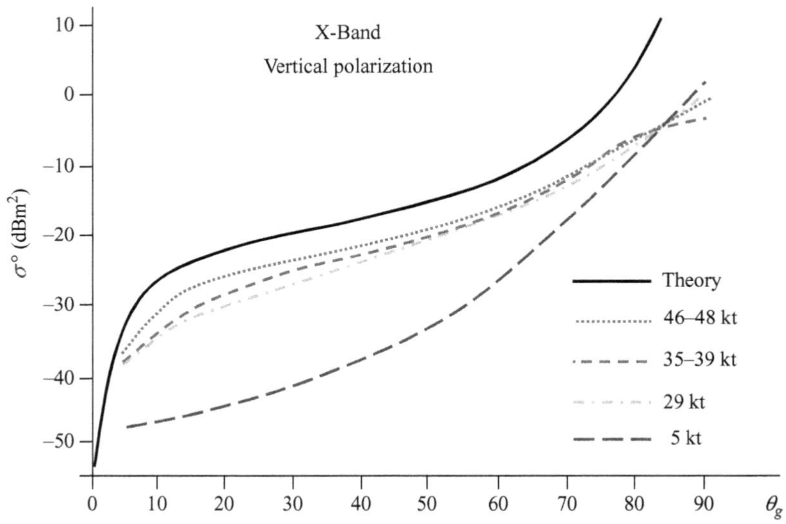

FIGURE 6-6 Upper Limit Model for Sea Surface Backscatter Coefficient versus Grazing Angle at X-Band and Horizontal Polarization

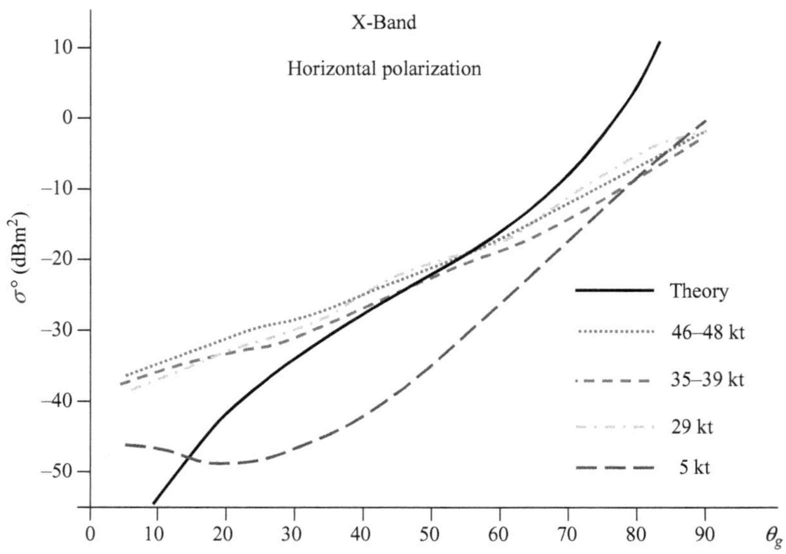

6.3 STATISTICAL MODELS OF CLUTTER

6.3.1 Gaussian (Rayleigh) Clutter

Many different PDFs and models of clutter decorrelation in time have been used to describe the clutter from different surface types. Indeed, clutter is often named after the distribution it follows; however, caution must be exercised when referring to a clutter PDF since one must be sure of exactly what quantity adheres to the named PDF. One common clutter PDF is termed *Gaussian* clutter and is a popular model as much for the mathematical convenience of using the Gaussian PDF as anything else. Gaussian clutter is so called because it results in a clutter voltage in either the real or imaginary part of the baseband voltage signal, which exhibits a Gaussian distribution. Note that the baseband of Doppler sensing radars is synthesized into two channels: one representing the real part of the signal and the other the imaginary part of the signal. The two channels therefore preserve the complex representation of the baseband signal, which is necessary for the subsequent processing stages. This will be described further in Chapter 8. Since the real and imaginary parts of the clutter voltage have a Gaussian PDF, their vector sum represents the magnitude of the complex signal and has a Rayleigh PDF. Some texts refer to this clutter as *Rayleigh* clutter for this reason. Furthermore, since the signal power is proportional to the voltage magnitude squared, the clutter power has an exponential PDF. Radar cross section is a ratio of powers (recall the definition of RCS, section 2.5.1, equation (2-28)), so the PDF appropriate to the backscatter coefficient of Gaussian clutter is, in fact, an exponential function.

Gaussian clutter is an accurate representation of clutter that arises from a large number of independent scatterers that are all equally weighted. It is appropriate to wooded land at grazing angles > 5°, sea clutter over large-resolution cells, sea clutter over small-resolution cells at a grazing angle > 5°, as well as rain clutter and chaff. The complex clutter voltage, V_C, is given by

$$V_C = x \cos \omega t + j\, y \sin \omega t$$

in which x and y are Gaussian variables with a standard deviation, s_c and $j = \sqrt{-1}$.

The PDF describing the magnitude of the clutter voltage envelope is

$$p\left(|V_C|\right) = p(V_C) = \frac{V_C}{s_c^2} \exp\left(-\frac{V_C^2}{2s_c^2}\right)$$

The PDF defining the backscatter coefficient of Gaussian clutter power is

$$p(\sigma^\circ) = \frac{1}{\sigma_0^\circ} \exp\left(-\frac{\sigma^\circ}{\sigma_0^\circ}\right) \tag{6-2}$$

where σ_0° is the standard deviation in the backscatter coefficient, σ°.

6.3.2 Rician Clutter

Another often used clutter PDF model is *Rician* clutter. This refers to the Rician PDF, which describes the backscatter coefficient (or clutter signal power). Rician clutter is applicable in circumstances where there are many independent scatterers plus a dominant steady component that may typically arise from tree trunks, pylons, buildings, or rocks. It is also a good representation of the temporal characteristics of land clutter. The clutter voltage (magnitude of complex voltage) is similar to a steady signal plus Rayleigh distributed noise, which is a Rician distribution. The baseband spectrum is Gaussian plus a steady direct current (DC) value. The distribution of power (clutter radar cross section) is an expression involving the power returned from the steady part of the scene and the power returned from the variable part of the scene. The balance between these two components varies from one situation to another. Let

P = total instantaneous power returned = steady power + instantaneous variable power

The total power is a measure of the clutter radar cross section.

$P_s + P_v$ = steady power + mean variable power

$m^2 = P_s/P_v$ = ratio of steady power to mean variable power.

The complex clutter voltage is given by

$$V_C = (A_r + x)\cos\omega t + j\ (A_i + y)\sin\omega t$$

in which $A = A_r + jA_i$ is a complex term providing an offset.

The PDF defining the backscatter coefficient of Rician clutter power is

$$p(\sigma^\circ) = p(P) = \frac{1+m^2}{P_s + P_V}\exp(-m^2)\exp\left(-\frac{P(1+m^2)}{P_s+P_V}\right) I_0\left[2m\sqrt{\frac{P(1+m^2)}{P_s+P_V}}\right] \tag{6-3}$$

If the steady component is zero then the voltage distribution would be expected to become Rayleigh, which leads to an exponential power distribution. Putting $P_s = 0$ and $m^2 = 0$ gives an exponential distribution for $p(\sigma^\circ)$.

The Rician PDF for a forested scene at X-band blown by a 10 mph wind is illustrated in Figure 6-7. When the wind speed falls, the number of fixed scatterers increases and

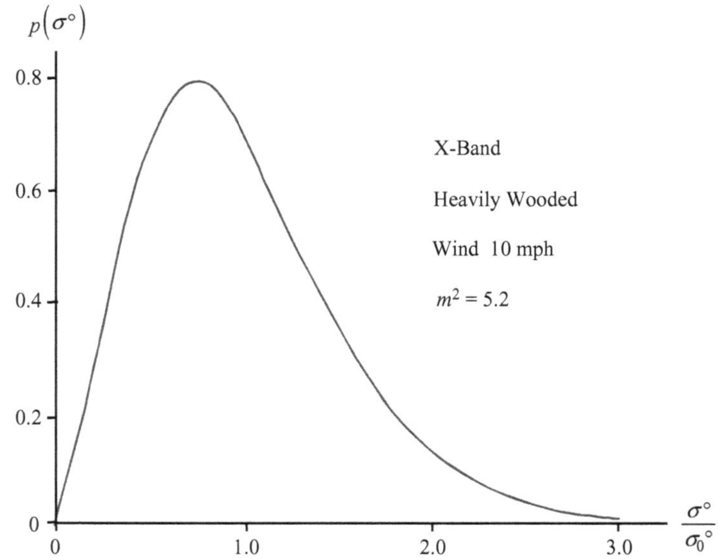

FIGURE 6-7 Clutter Backscatter Coefficient PDF for Forest, Moderate Wind at X-Band

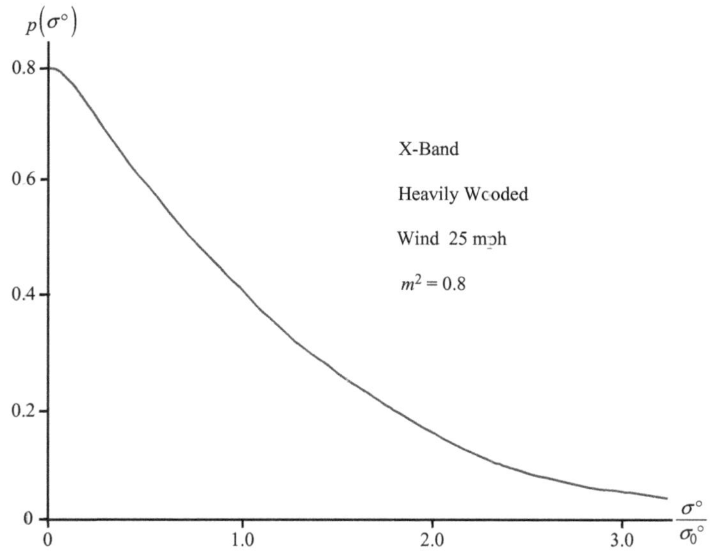

FIGURE 6-8 Clutter Backscatter Coefficient PDF for Forest, Strong Wind at X-Band

the number of fluctuating scatterers decreases. The steady power increases and the mean variable power decreases, leading to an increase in the parameter m. The effect is to shift the peak to the right. When compared with the case of an exponential power distribution, there is a greater chance of clutter values exceeding the mean, but the more extreme values (> 3) are less likely. This has implications for the setting of any threshold used to separate target and clutter.

In a heavily forested scene with a strong wind blowing, the moving components dominate. The Rician PDF for a forested scene at X-band blown by a 25 mph wind is illustrated in Figure 6-8. The steady component is small in comparison, so the ratio of steady power to mean variable power is low. If the steady component is zero, then the variable component is left, which has a Rayleigh voltage distribution and an exponential power distribution. Here the steady component is not zero but is small, so the distribution is only approximately exponential.

In a rocky terrain there is a high proportion of fixed scatterers, so the ratio of steady to variable powers is high. The power distribution is more Gaussian in shape and has a much smaller spread than the previous examples. The Rician PDF for a rocky scene at X-band blown by a 10 mph wind is illustrated in Figure 6-9. If a Rician distribution is appropriate, the situation could arise that the appropriate detection threshold (1) changes with time as the wind speed changes, (2) changes during a scan as the radar illuminates different types of terrain, and (3) changes at different ranges. These three examples illustrate the major problem associated with clutter. In the case of noise there is only one distribution, and its parameter is constant (noise power). In the case of Rician clutter, parameters can change and cause major changes in the shape of the distribution, even though all shapes belong to the same distribution. There is the further complication that different types of clutter may have different underlying distributions.

6.3.3 Log-Normal Clutter

A further distribution that has proved popular, particularly in the description of the clutter backscatter coefficient of the sea surface, is the *log normal* distribution, which is

FIGURE 6-9 Clutter Backscatter Coefficient PDF for Rocky Terrain, Moderate Wind at X-Band

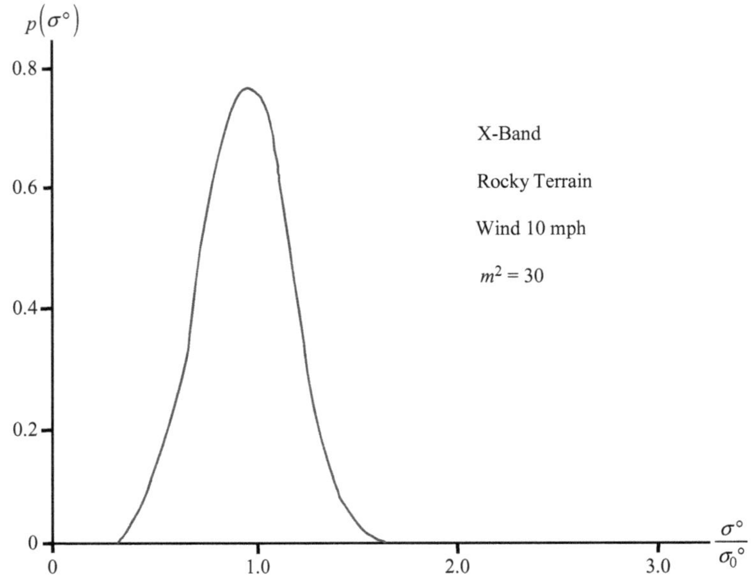

illustrated in Figure 6-10. In a Gaussian distribution the variable extends to infinity in both directions (Figure 6-10, top plot). The median value divides the area under the distribution curve into two equal areas. Because the curve is symmetrical, the mean and median are equal. In the case of a variable that cannot take negative values, such as with the backscatter coefficient, then the Gaussian distribution could never apply. However, if logarithms are taken, then values from 0 to ∞ yield logarithms covering the range $-\infty$ to $+\infty$, and a Gaussian distribution could apply (Figure 6-10, middle plot). With logarithms, equal steps in $\log(x)$ mean equal ratios in x. It is common to use a logarithmic scale to cover a wide range of values. If the distribution of $\log(x)$ is Gaussian (i.e., *normal*), then the variable is said to have a log normal distribution (Figure 6-10, bottom plot). If the logarithmic scale is converted to a linear scale, the effect is to progressively stretch the scale as it moves to the right. The result is a distorted picture that has a long tail on the right (positively skewed). The long tail moves the mean away from the median, and the ratio of mean to median, which is 1 for a Gaussian distribution, is now greater than 1. The mean-to-median ratio is a measure of the distortion suffered by the original Gaussian distribution and depends on the standard deviation. Note that in statistics the measure of asymmetry is termed the *skewness* and is an example of a higher-order moment.

There are cases in which the measured distribution of a target's radar cross section or the clutter backscatter coefficient has higher values more frequently than the Rayleigh or Rician distributions would indicate. In such cases the log-normal, with suitable choice of parameters, can be a better description. A sea surface often gives more large values than would be expected from a Rayleigh distribution. These larger values can be associated with the breaking of the tops of the waves. The greater the sea state, the greater the number of these larger values and the longer the tail of the distribution (greater mean-to-median ratio). The mean-to-median ratio is plotted for various sea states as a function of the grazing angle in Figure 6-11.

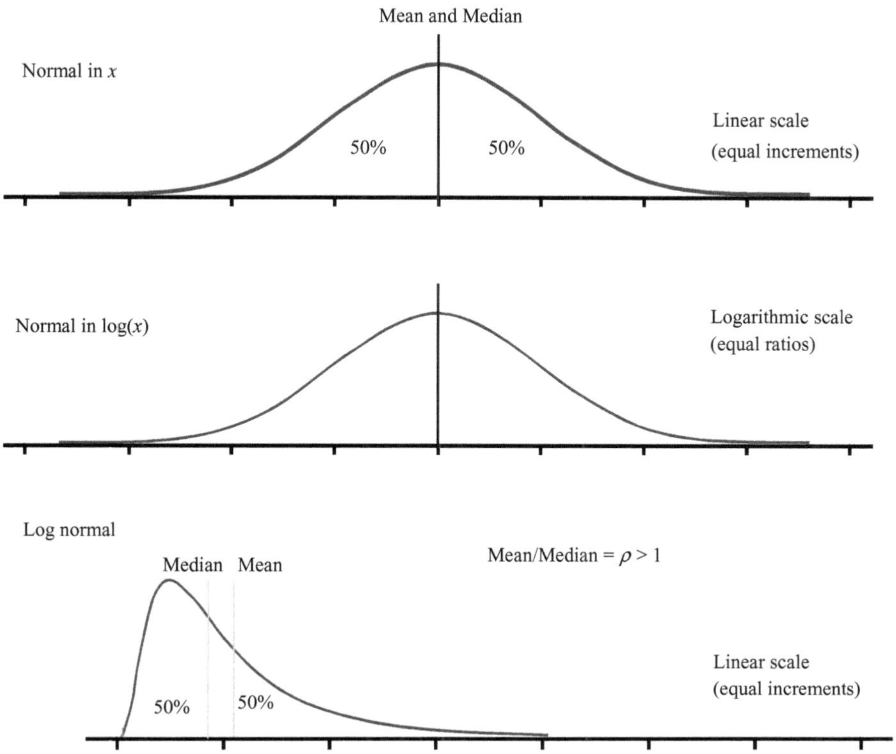

FIGURE 6-10 The Log-Normal Distribution: The Normal Distribution (top); Normal in log(x) (middle); Log-Normal Distribution (bottom)

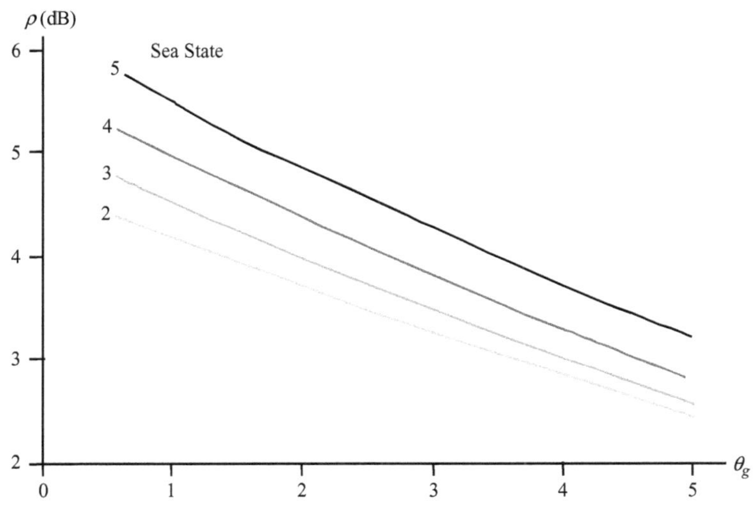

FIGURE 6-11 Mean-to-Median Ratio of Log-Normal Distribution for Sea Clutter versus Grazing Angle and Sea State

6.3.4 Weibull Clutter

The *Weibull* distribution, another PDF used in describing land and sea clutter, lies between log-normal and Rayleigh (i.e., exponential power PDF). The PDF is given by

$$p(\sigma^\circ) = \frac{b}{\alpha}\left(\frac{\sigma^\circ}{\alpha}\right)^{b-1} \exp\left(-\left(\frac{\sigma^\circ}{\alpha}\right)^{b}\right) \tag{6-4}$$

The mean value is given by

$$\sigma_0^\circ = \alpha\Gamma\left(\frac{b+1}{b}\right) \tag{6-5}$$

and the median value is given by

$$\sigma_m^\circ = \alpha\sqrt[b]{\ln(2)} \tag{6-6}$$

where b is the shape parameter, α is a scale parameter, and Γ represents the gamma function given by

$$\Gamma(n) = \int_0^\infty x^{n-1}\, e^{-x}\, dx \tag{6-7}$$

The mean-to-median ratio is given by

$$\text{mean:median ratio} = \frac{\sigma_0^\circ}{\sigma_m^\circ} = \frac{\Gamma(1+b^{-1})}{(\ln 2)^{1/b}} \tag{6-8}$$

For $b = 1$ the distribution becomes a Rayleigh, which may therefore be regarded as a particular case of the Weibull distribution. (It is also a particular case of the Rician.) Like the Rician, changing the parameter, in this case b, changes the shape of the curve, which has implications for setting any threshold. The Weibull distribution is illustrated in Figure 6-12 for various values of b. Values of $b = 1$ to 3 are typical.

Squaring signals halves the shape parameter b; therefore, a voltage distribution with a Rayleigh shape has $b = 2$ and therefore an exponential power distribution with a shape parameter of $b = 1$. Land clutter power distribution typically has a shape parameter around $b = 1$ but could go as high as $b = 1.3$ in rocky terrain (and is liable to have a DC offset too and thus look more Rician). Sea clutter ranges from states with $b = 1$ (which looks like thermal noise) to $b = 0.3$ (in very rough seas). As the shape parameter reduces, the

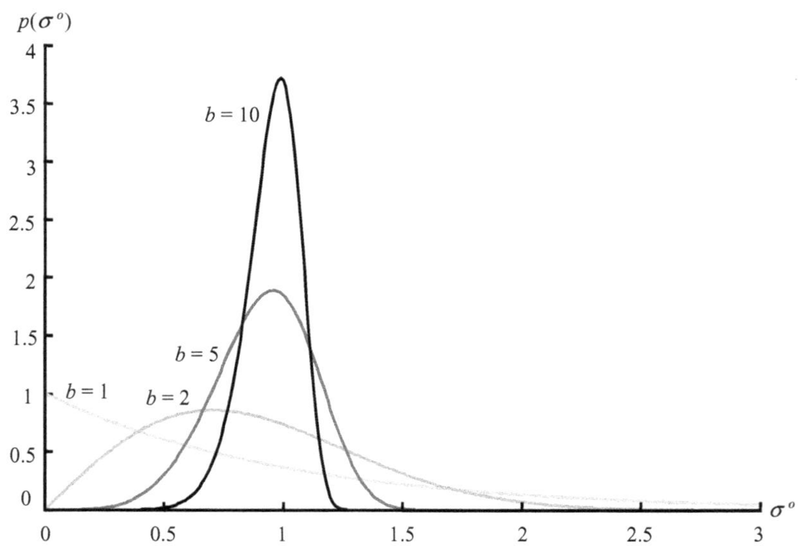

FIGURE 6-12 Weibull Distribution

"tail" of the distribution gets longer and therefore the threshold must be raised to maintain a low false alarm rate.

6.3.5 *K*-Distributed Clutter Model

The *K-distribution* is used in describing land and sea clutter. Its voltage PDF is given by

$$p\left(V_C\right)=\frac{2\alpha}{\Gamma(v)}\left(\frac{\alpha V_C}{2}\right)^{v} K_{v-1}\left(\alpha V_C\right) \tag{6-9}$$

where V_C represents the amplitude of the clutter voltage envelope, K_v is a second-order modified Bessel function, and v is a shape parameter. (Note that this is not the same shape parameter used in the Weibull distribution). For $v > 10$ the distribution approximates a Rayleigh and therefore can be used to describe the voltage envelope of a noise signal. Typically, the shape parameter lies in the range 0.1 to 10, but values less than 0.1 are not unheard of in very rough sea conditions. The reciprocal of the shape parameter may also be specified sometimes for very rough conditions.

6.3.6 Cumulative Probability Density Functions

When assessing the trade between a suitable threshold level, P_{FA}, and the clutter statistics, it is useful to consider the *cumulative (probability) density function* (CDF) of radar cross section. Mathematically, the CDF of a variable x is the integrated area under a PDF curve, $p(x)$, which lies below a given value of x as the value of x is swept from $-\infty$ to $+\infty$. In the context of radar detection, the value of x may be considered the threshold value for declaring a target detection, so the integrated area under the curve of $p(x)$ lying below the threshold value represents $1 - P_{FA}$. That is,

$$1-\mathrm{CDF}=P_{FA} \tag{6-10}$$

The CDF of the *K-distribution* PDF is given by

$$\mathrm{D}(\sigma^\circ)=1-\frac{2}{\Gamma(v)}\left(\frac{\alpha\sigma^\circ}{2}\right)^{v} K_{v}\left(\alpha\sigma^\circ\right) \tag{6-11}$$

The CDF of the Weibull PDF is given by

$$\mathrm{D}(\sigma^\circ)=1-\exp\left(-\left(\frac{\sigma^\circ}{\alpha}\right)^{b}\right) \tag{6-12}$$

This is graphed in Figure 6-13 for various values of the Weibull shape parameter, b. The graph is normalized to the mean of the clutter backscatter coefficient. The vertical axis indicates the probability that any random observation of the clutter would be less than the target (threshold) clutter value indicated along the horizontal axis. Thus, for $b = 1$, approximately 64% of all observations are less than the mean. For $P_{FA} = 10^{-6}$ and $b = 1$, a threshold of 13.8 times the mean (11.4 dB) would be required. For $b = 0.3$, a threshold of 683 times the mean (28.3 dB) would be required. This indicates how subtle differences in the clutter statistics demand considerably different detection thresholds if the P_{FA} is to remain constant.

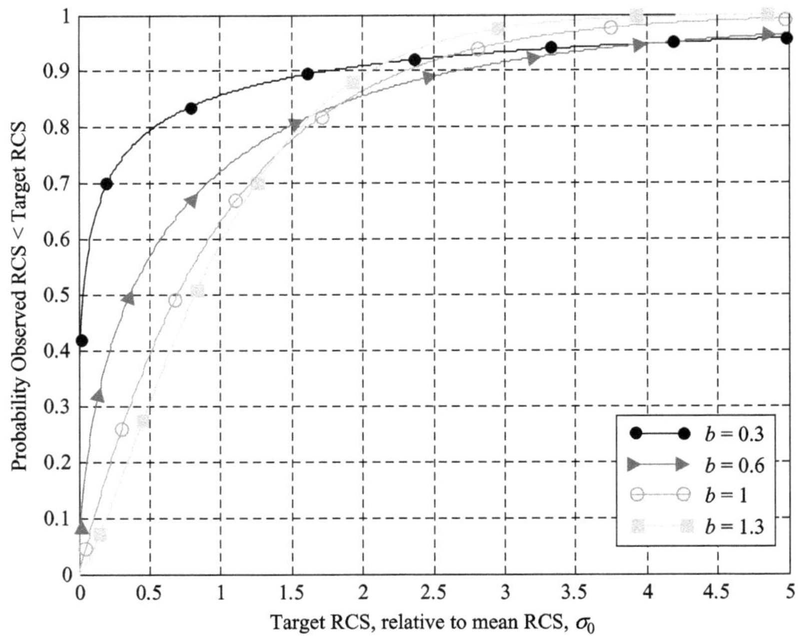

FIGURE 6-13 CDF of Weibull Clutter

6.4 CLUTTER FOR AN AIRBORNE RADAR

6.4.1 The Airborne Scenario

Surface clutter from an airborne radar arises due to the transmission and reception via the antenna sidelobes and, if the main beam boresight is depressed below the horizon, via its main beam. This gives rise to sidelobe clutter (SLC) and main beam clutter (MBC), both of which are spread in the time (range) and Doppler (velocity) dimensions. A typical airborne case is illustrated in Figure 6-14 in which the radar platform is moving along a straight line path parallel to the ground at a velocity, V, and the antenna is depressed below the horizontal.

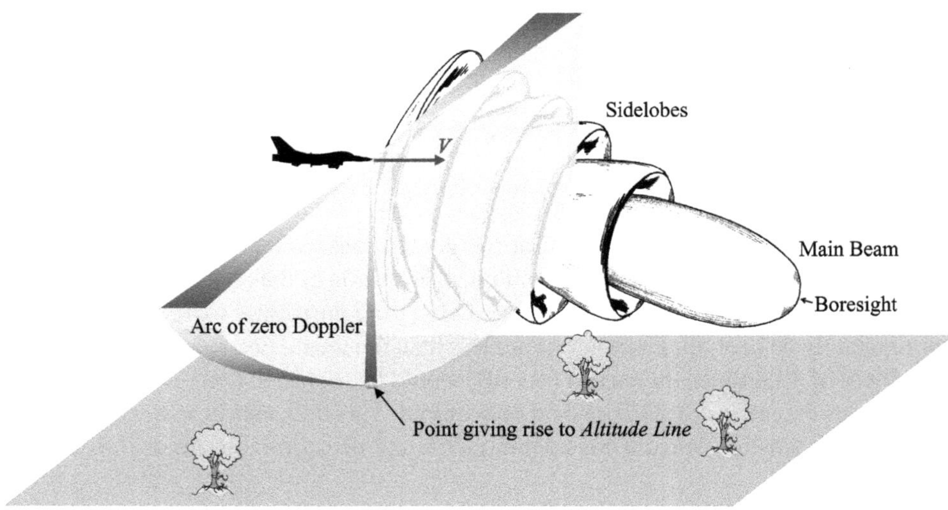

FIGURE 6-14 Airborne Radar in Look-Down Scenario

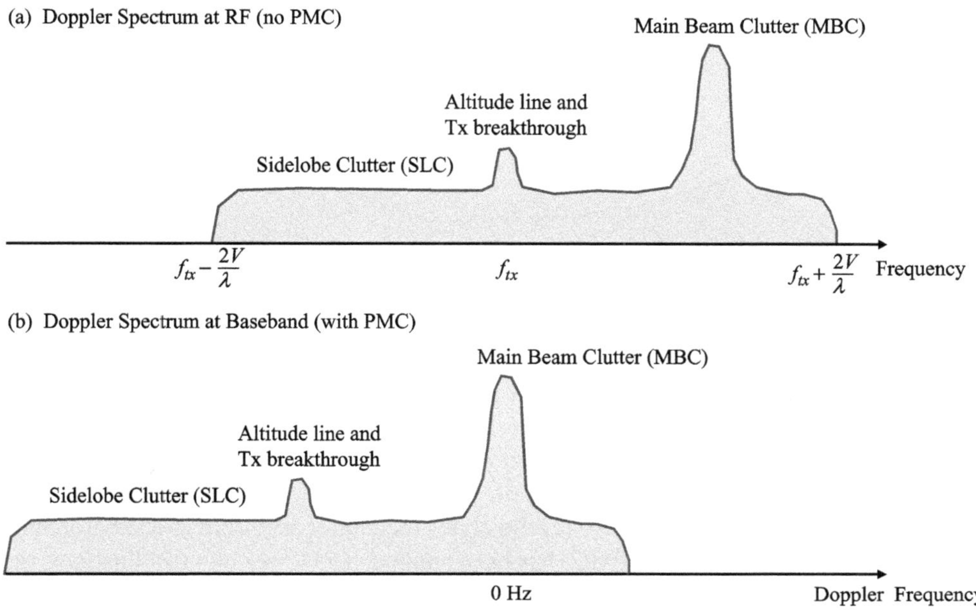

FIGURE 6-15 Surface Clutter from Airborne Radar in the Doppler (Velocity) Domain: (a) At RF, no PMC; (b) At Baseband with PMC

The main beam is seen to intersect the ground ahead of the radar and gives rise to strong MBC. The sidelobes intersect the ground all around the radar and give rise to SLC, which is greatly spread in both range and Doppler.

6.4.2 Clutter Spread in the Range and Velocity Dimensions

The spectrum of MBC and SLC is illustrated in Figure 6-15. Since the main beam boresight is close to the velocity vector of the platform, as in Figure 4-2, MBC is centered at a high positive Doppler frequency given by (4-11). Some small portion of the sidelobes are directed along the velocity vector of the platform; these give rise to a weak clutter return from some distant point on the horizon having a Doppler shift commensurate with the full forward velocity of the radar, V. The majority of the sidelobes do, however, illuminate the ground closer to the radar and therefore at angles further displaced from the velocity vector. This results in SLC spread over lower Doppler frequencies. Similarly, antenna sidelobes and backscatter lobes result in SLC spread over a range of negative Doppler shifts up to the Doppler shift associated with the full negative platform velocity, $-V$. SLC is generally much weaker than MBC because the two-way gain of the sidelobes is very much lower than the two-way gain of the main beam. SLC is spread over a continuum of Doppler shifts commensurate with a spread in velocities of $+V$ to $-V$. Therefore, the greater the platform velocity, the greater the spectral spread of SLC. Some sidelobes point over an arc that is perpendicular to the velocity vector, which contributes to SLC having zero Doppler shift. The arc of zero Doppler extends left and right of the platform track in a semicircle passing underneath the radar. This includes the point on the ground directly underneath the radar. This point is the closest ground point to the radar and so contributes SLC at the shortest possible range, which is equal to the radar altitude. This point also illuminates the ground at a grazing angle of 90°, and, as can be seen from Figure 6-2 to Figure 6-6, the backscatter coefficient reaches a maximum at 90°. SLC arising from this point on the surface is somewhat stronger than the remaining SLC on account of

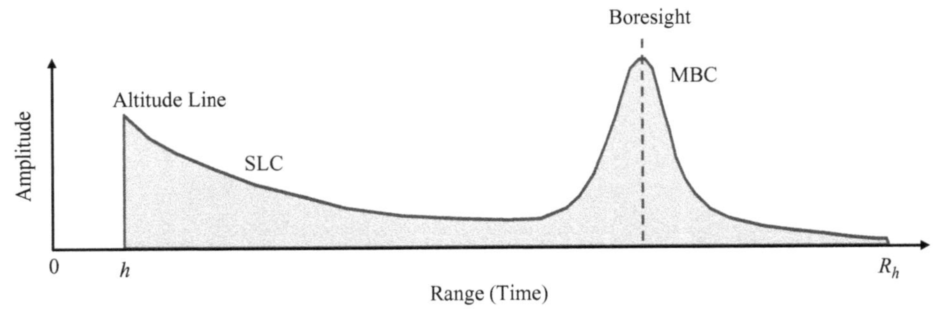

FIGURE 6-16 Surface Clutter from Airborne Radar in the Time (Range) Domain

its minimal range and high backscatter coefficient. A peak is therefore encountered in the SLC occurring at zero Doppler and a range equal to one's altitude, known as the *altitude line*. Furthermore, if there is any residual breakthrough of transmitter power into the receiver, this, too, falls at zero Doppler (but may be range independent) and would contribute to the altitude line. Figure 6-15 illustrates the spectrum of MBC and SLC from an airborne platform. The upper plot (a) illustrates the clutter spectrum at RF, before any platform motion compensation (PMC) has been applied. The lower plot (b) illustrates the (double-sided) baseband clutter spectrum once PMC has been applied.

Figure 6-16 illustrates the clutter in the range (or time) dimension. The earliest clutter return is the altitude line. Clutter corresponding to greater ranges couples to the radar from shallower grazing angles and grows weaker with increasing range on account of the lower values of the backscatter coefficient encountered at lower grazing angles and the losses over the greater range. Eventually, one arrives at the slant range of the main beam boresight to the surface. MBC is very powerful around this range. Beyond that, SLC rapidly diminishes since this corresponds to SLC arising from the surface ever closer to the horizon and therefore ever closer to a grazing angle of 0°. SLC ends at the range to the horizon. The range to the horizon, R_h, in kilometers, from a radar at an altitude of h meters is given approximately by

$$R_h \approx 4.1\sqrt{h} \tag{6-13}$$

6.4.3 The Ideal Clutter Map

Marrying together the clutter in the range and velocity domains yields a two-dimensional display of the clutter power distributed over range/velocity space. The depiction of clutter power as a function of range and velocity is known as an *ideal clutter map* [10]. An example of an ideal clutter map for a typical airborne radar flying at an altitude of 5000 m and at a velocity of 250 m/s in a 10.8° look-down and 0° azimuth look-ahead attitude is given in Figure 6-17. The range to the horizon calculated from (6-13) is 290 km; however, only the clutter over the first 100 km is plotted here. This gives the raw clutter power as a function of range and velocity before any PMC has been applied. MBC is seen as the intense clutter region (plotted as white, lower right) centered at a velocity of 246 m/s and a range of 27 km and corresponds to the main beam footprint on the ground. At shallow grazing angles the main beam footprint is highly extended in range; this is evident from the manner in which MBC is smeared over a large range bracket. MBC becomes more highly confined in range at higher grazing angles. The vertical striations and other patches in the clutter map are the SLC and are indicative of the sidelobe structure of the antenna pattern. SLC is seen to exist over a ±250 m/s velocity bracket in keeping with the speed of the radar platform. The SLC starts with the altitude line seen as a thin line at zero velocity

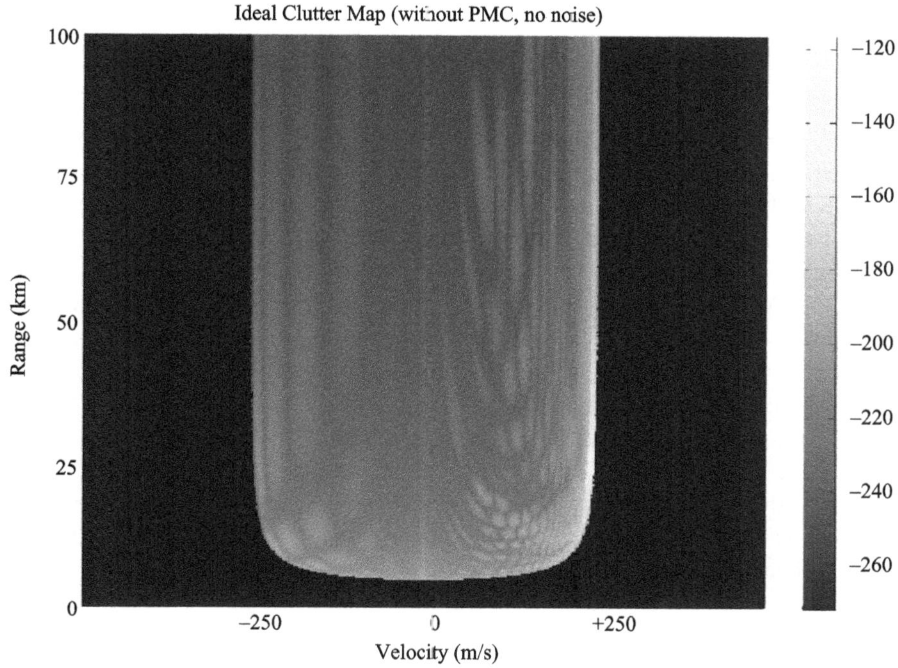

FIGURE 6-17 Ideal Clutter Map

starting at a range of 5 km, which corresponds to the platform altitude; SLC continues on out to the maximum range of 100 km and would indeed continue out to the horizon with ever decreasing power.

Surface clutter often sets the limit on target detection. In Sections 2.2 and 6.3, it was noted how the threshold level relative to the noise plus clutter sets P_{FA}. Since the clutter varies considerably over the range/velocity detection space of the radar, so too would the false alarm rate if a fixed threshold level were to be used. Furthermore, the clutter levels and statistics are highly dependent on parameters such as altitude, scan angle, and surface type. It would be desirable to maintain a constant false alarm rate by employing a variable threshold level appropriate to the local clutter levels and statistics. Such a detection mechanism is termed a *constant false alarm rate* (CFAR) detector and is based on an adaptive threshold. Standard CFAR detection theory is covered in greater detail in Chapter 7, and additional considerations specific to pulse Doppler radar are covered in Chapter 14.

6.4.4 Spectral Spread of MBC

One may observe that MBC is spread over a certain Doppler band rather than being confined to precisely zero Doppler. Three effects lead to the MBC bandwidth.

6.4.4.1 Effect of Main Beam Beamwidth

First, the main beam has a certain beamwidth, so clutter is coupled to the radar over the angular range of the main beam. PMC can offset only the main beam boresight to zero Doppler. Clutter received via the main beam but not on its boresight has a slightly differing look angle with respect to the platform velocity vector and therefore a slightly differing radial velocity and Doppler shift. The wider the beamwidth, the greater the bandwidth of MBC. The Doppler shift along the main beam boresight is given by equation (4-11). The limits on the Doppler bandwidth of MBC, f_{dMBC}, can be found by amending equation

(4-11) to take the beamwidth into account.

$$f_{dMBC} = \frac{2V \cos\left(\theta_{az} \pm \theta_{Baz}/2\right) \cos\left(\theta_{el} \pm \theta_{Bel}/2\right)}{\lambda} \tag{6-14}$$

in which θ_{az} and θ_{el} are the antenna boresight steering angles with respect to the velocity vector in azimuth and elevation, respectively, and θ_{Baz} and θ_{Bel} are the antenna beamwidths in azimuth and elevation, respectively. (Generally, $\theta_{Baz} = \theta_{Bel} = \theta_B$ for an antenna that is symmetrical in both planes.) From (6-14), one may notice that the spread of f_{dMBC} is also dependent on the azimuth and elevation scan angles; thus, as an antenna scans, the bandwidth of MBC grows and shrinks. The minimum bandwidth occurs when the scan angles are both zero ($\theta_{Baz} = \theta_{Bel} = 0°$), whereas the maximum occurs when the antenna boresight is scanned perpendicular to the platform velocity ($\theta_{Baz} = 90°$, $\theta_{Bel} = 90°$). Consider, for example, a radar operating on a wavelength of 3 cm from a platform of velocity 250 m/s in straight and level flight at an altitude of 10,000 m observing a region at a (slant) range of 100 km. This necessitates a steering angle θ_{el} of 5.7° below the horizontal. If the antenna beamwidth in both planes is $\theta_{Baz} = \theta_{Bel} = 4°$, then in the look-ahead position $\theta_{az} = 0°$ and $f_d = 16{,}584$ Hz on boresight but $f_d = 16{,}506$ Hz on the edge of the beam, giving $f_{dMBC} = 78$ Hz. If, on the other hand, the radar is scanned to the broadside, $\theta_{az} = 90°$ (all other factors being equal), then $f_d = 0$ Hz on boresight but $f_d = \pm 581$ Hz on the leading and trailing edges of the beam, giving $f_{dMBC} = 1162$ Hz. Therefore, the clutter rejection notch filter must be designed to operate over the appropriate range of antenna scan angles or must be adaptive to the radar scan angle. Furthermore, the PMC correction must track changes in scan angle in real time. It is worth noting that the MBC bandwidth is also dependent on platform velocity; fire control radars onboard supersonic strike aircraft and active radar seekers on missiles give rise to a considerable spread in MBC and SLC.

6.4.4.2 Internal Clutter Motion

The second effect that results in a broadening of MBC is internal clutter motion (ICM). Many clutter surfaces are not absolutely stationary and are in fact in a state of motion and so impart Doppler shifts on the clutter returns. ICM may be modeled as a spread of velocities that have a Gaussian PDF [11], which leads in practice to phase and amplitude modulation of the received pulse train. The probability density function of the velocity modulation, $p(v)$, can be described as

$$p\left(v\right) = \frac{1}{\sigma_{cv}\sqrt{2\pi}} \exp\left(\frac{-v^2}{2\sigma_{cv}^2}\right) \tag{6-15}$$

where v is clutter velocity in m/s, and σ_{cv} is the standard deviation of the clutter internal motion in m/s. The relative power in each velocity cell may be calculated by integrating (6-15) over each velocity cell interval, with the 0 m/s cell being defined as having limits of $\pm\Delta V_r/2$, where ΔV_r is the velocity resolution. Equation (6-16) details the integration process for each cell of index i, with $i = 0$ indicating the central cell, and $\pm i$ stepping through the neighboring cells (i is an integer), resulting in the power contribution in each velocity cell, P_i.

$$P_i = \int_{i-\Delta V_r/2}^{i+\Delta V_r/2} p\left(v\right) dv \tag{6-16}$$

A clutter standard deviation of 1.5 m/s appears to be appropriate for the deviation expected from rain and chaff [12]. With a 1.5 m/s standard deviation and Gaussian profile, the central velocity cell retains approximately 38% of the original power for a typical CPI of 7 ms (this yields $\Delta f_d = 143$ Hz, $\Delta V_r = 2.14$ m/s for a 10 GHz radar), with the remaining power spread over the close adjacent velocity cells. A 1 m/s standard deviation leads to the central velocity cell retaining approximately 56% of the power and provides a match to the upper end of anticipated heavy sea clutter. A 0.2 m/s standard deviation would be appropriate for agricultural land clutter. The spectrum cannot be truly Gaussian because that would imply the presence of infinite, or at least very high, frequency components. In practice, however, the Gaussian assumption is sufficiently valid over levels of interest.

6.4.4.3 Uncorrected Platform Instabilities

Instabilities in the radar platform that are not accurately corrected using PMC also cause MBC to spread. Short-term accelerations along all three axes or rotational movements can give rise to variations in the relative velocity resolved along the main beam boresight in accordance with equation (4-10) and hence variations in the Doppler shift given by (4-11). While PMC techniques aim to correct for these, short-term errors can be incurred leading to a Doppler spreading of MBC. This tends to be worst for high-velocity platforms in a turbulent flight path.

6.5 CLUTTER DECORRELATION

6.5.1 Clutter Decorrelation with Time

One means of combating noise is to use a long integration time and so exploit the correlation properties of target signals and noise (Section 2.4.3). Noise decorrelates on a timescale equal to the reciprocal of the noise bandwidth, which is much less than the time between pulses. The noise samples from one pulse to the next are uncorrelated, but the target signal samples are correlated, giving a signal-to-noise ratio (SNR) improvement. Similarly, the signal-to-clutter ratio (SCR) may be improved over time because of the correlation of target signals but the partial decorrelation of clutter due to its ICM. If it is assumed that the clutter spectral power density, $S(f)$, is Gaussian, that is,

$$S(f) = \exp\left[-\ln 2\left(\frac{2f}{\Delta f_C}\right)^2\right] \tag{6-17}$$

where Δf_C is the 3 dB bandwidth of the clutter spectrum, one may conveniently compute the autocorrelation of clutter as

$$r_{xx}(t') = \int_{-\infty}^{\infty} S(f)\, e^{j2\pi f t'} df = \exp\left[-\frac{(2\pi\, t'\, \Delta f_C)^2}{11 \cdot 2}\right] \tag{6-18}$$

The autocorrelation function and the power spectral density are a Fourier transform pair. The spectrum is Gaussian, so its inverse transform is also Gaussian. As time passes the autocorrelation becomes smaller but, in theory, never becomes zero. The wider the bandwidth of the clutter, the more rapid its decorrelation.

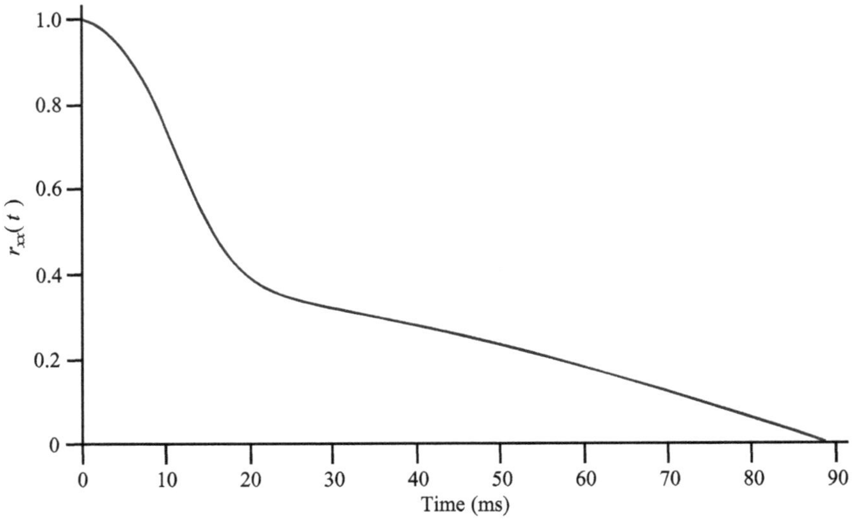

FIGURE 6-18 Decorrelation of Clutter from a Forested Scene

For a bandwidth of $\Delta f_C = 60$ Hz, the autocorrelation is halved from its peak of unity after 8 ms. This means that the clutter has a high correlation from pulse to pulse. Taking 0.1 as a suitable threshold indicating decorrelation, it takes 14 ms, much of the dwell time of a search radar, to decorrelate. For clutter, integration is far less effective as it is for noise. The decorrelation of clutter from a forested region is illustrated in Figure 6-18. Some features in the ground clutter decorrelate relatively quickly, whereas other features decorrelate more slowly. The correlation falls steadily at first and then more slowly.

The decorrelation of sea clutter is illustrated in Figure 6-19: Decorrelation of Sea Clutter. For sea clutter, there is a rapid decorrelation for the first 10 ms followed by a much slower decorrelation over a time scale of seconds. The initial drop is caused by the movement of surface ripples driven by the wind. These fine details move quickly, and the phase relationships change rapidly. The slow change is caused by the mass movement of the wave structure. At a particular location the water surface is elevated, sinks, and then

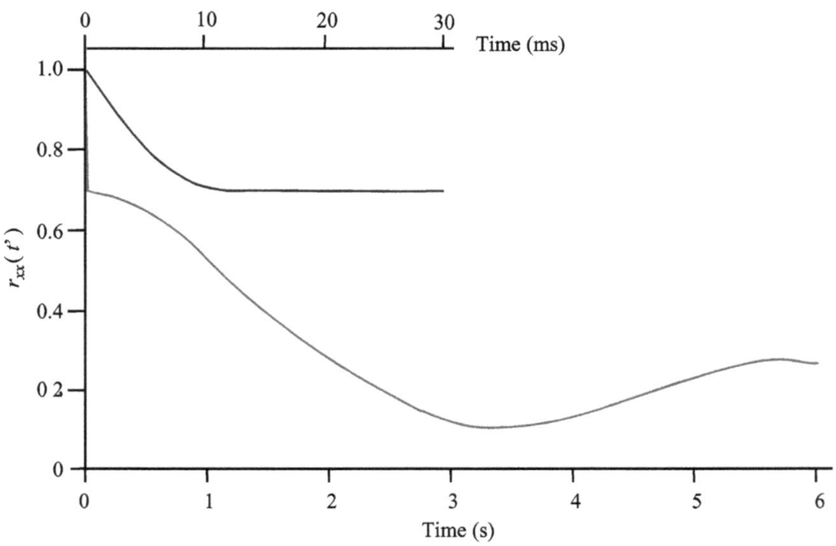

FIGURE 6-19 Decorrelation of Sea Clutter

rises again. The timescale for this sequence is much the same at all positions, so after a period of time the surface has a similar structure to that seen at an earlier time and the correlation starts to rise. The surface details repeat in a general way with a period equal to that of the swell of the sea surface, but there are inevitably small differences so the second correlation peak is significantly lower than the first.

6.5.2 Clutter Decorrelation Using Frequency Agility

One tactic that may be used to force a more rapid decorrelation of clutter is frequency agility. There is a high degree of decorrelation of clutter observed at differing frequencies; the degree of clutter correlation is inversely proportional to the bandwidth of the frequency agility. A radar may improve the SCR by operating over a succession of frequencies within the beam dwell time.

6.6 | LOW PULSE REPETITION FREQUENCY RADAR RESPONSE TO CLUTTER

6.6.1 Low Pulse Repetition Frequency Response to Clutter in the Range Domain

Low PRF radars are defined as having a sufficiently low PRF so as to avoid any range ambiguity. The range (time) characteristic of clutter given in Figure 6-16 is therefore repeated within each successive receiving period without any ambiguity, as in Figure 6-20. SLC is worst in the early range cells and can be eliminated simply by blanking returns in these range cells. It may be difficult to blank MBC, particularly as this may be smeared over many range cells at low grazing angles. Blanking MBC would render the radar blind over typical ranges of interest and so is often not a viable proposition.

6.6.2 Low PRF Response to Clutter in the Velocity Domain

Each and every frequency component in the spectrum of a pulsed radar signal results in its own clutter spectrum. In this way the clutter spectrum is repeated at intervals equal to the PRF, as shown in Figure 6-21. The point was made in Section 4.6 that a long maximum unambiguous range demands a low PRF and that Doppler ambiguities may arise as a result. For example, from (3-6) if $R_{mu} = 200$ km, a PRF no greater than 750 Hz is needed. It was also noted in Section 6.4.4 that the bandwidth of MBC may be in the order of several tens of Hertz to several hundreds of Hertz. The upshot is that MBC may occupy a bandwidth similar to the PRF and therefore to the interval over which the clutter spectrum is repeated. This is evident in Figure 6-21, in which ambiguous repeats of the clutter spectrum are spread across a wide band. Filtering of MBC using a clutter notch filter would now filter

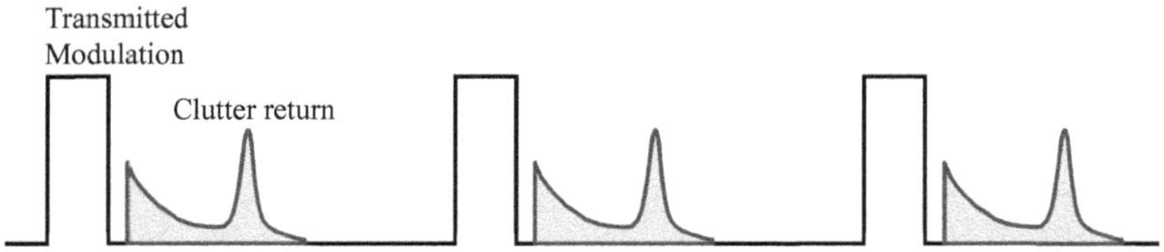

FIGURE 6-20 Low PRF Response to Clutter in Range Domain

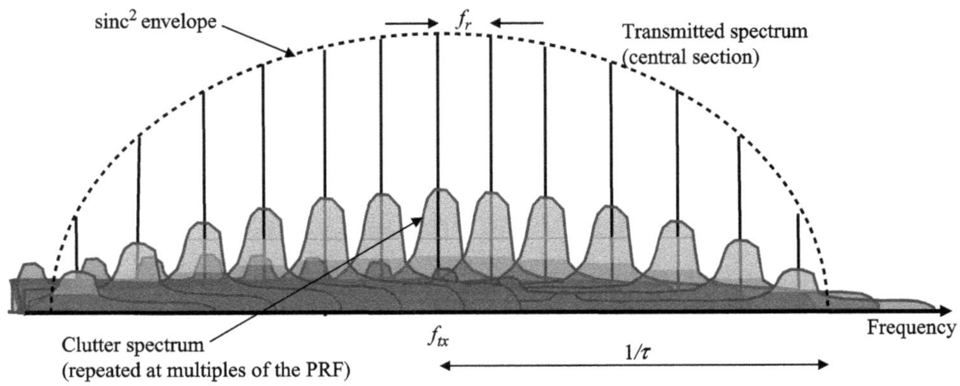

FIGURE 6-21 Low PRF Response to Clutter in the Doppler Domain

away most, if not all, of the Doppler band. Such filtering would result in broad bands of blind velocities centered on the corresponding Doppler shifts equal to multiples of the PRF; this, again, is not a viable proposition.

Low PRF waveforms suffer from high levels of MBC that cannot be rejected in the time or frequency domains. Many targets of interest have to compete with high levels of MBC to be detected, which leads to poor detection performance. Furthermore, Doppler ambiguities are likely. Hence, the use of low PRF waveforms is not viable for most pulse Doppler radar applications, particularly airborne applications. Low PRF waveforms are mainly used when target range information is required but velocity information is not required, particularly in the absence of clutter or scenarios in which the clutter return is very low. Airborne early warning radars use low PRF waveforms for the detection and ranging of distant targets in an above-horizon search mode.

6.6.3 Moving Target Indicator Radar for Clutter Rejection

In the early days of radar, low PRF—and its associated clutter problems—was prevalent. Attempts to combat the problem were based on the use of filters using delay line cancelers fitted in the baseband signal path that rejected zero Doppler shifted returns and a narrow band of low Doppler shifts. These filters are repetitive in the frequency domain such that MBC and its repetition at multiples of the PRF are also filtered away. Only signals exhibiting a significant Doppler shift were admitted to the detection process. This was known as moving target indicator (MTI) radar. Early MTI systems did not measure the Doppler shift; they merely blocked the clutter return. MTI radar is still used today, albeit with higher PRFs and Doppler filtering to subdivide the clutter in the velocity domain (see Section 6.7). An MTI filter is still a common feature of modern pulse Doppler systems since it eliminates the worst of the clutter at an early stage of the processing such that subsequent processing stages do not have to handle high levels of MBC.

6.7 CLUTTER-LIMITED DETECTION RANGE

The analysis presented in Sections 2.1 to 2.4 considered the detection performance of a radar limited by thermal noise alone. In many cases, clutter becomes the limiting factor. Discrete targets may be detected only when the SCR or, if both noise and clutter are present at similar levels, the signal-to-noise-plus-clutter ratio (SNCR), is sufficiently high. Clutter statistics differ from those of noise, and the specifications on detection performance dictate values of SCR, which are typically higher than the SNR requirements associated with noise-limited detections. Certain clutter surfaces, particularly a rough sea, are described by a very long-tailed probability density function, so there is a very real possibility of incurring a large clutter spike that could either trigger a false alarm or mask a small discrete target. Unlike the noise-limited case, the SCR is not improved by using a higher-power transmitter or higher gain/larger aperture antenna since these parameters affect the desired target signal return and the clutter return equally. In a clutter-limited situation, the SCR is determined by the ratio of the RCS of the desired target to the RCS of the clutter.

6.7.1 Range Resolution-Limited Case

For an airborne radar in a look-down mode, two simplistic cases can be considered. The first of these pertains to a range resolution-limited case and is applicable to low grazing angles. The scenario is presented in Figure 6-22. The shaded region on the surface is the area that couples clutter to one range cell. It is approximately rectangular, having a down-range dimension given by the radar range resolution resolved through the grazing angle and projected onto the surface. Its cross range dimension is given by the spread of the main beam over the slant range to the surface.

The area of the clutter patch falling in one range cell is therefore

$$R\theta_{Baz}\Delta R \sec\theta_g$$

The clutter RCS from this patch is therefore given by the product of its area and its backscatter coefficient

$$\sigma^\circ R\theta_{Baz}\Delta R \sec\theta_g$$

If a discrete target of RCS $= \sigma$ sits in this area, then the SCR is given by the ratio of the target RCS to the clutter RCS, that is,

$$SCR = \frac{\sigma}{\sigma^\circ R\theta_{Baz}\Delta R \sec\theta_g} \tag{6-19}$$

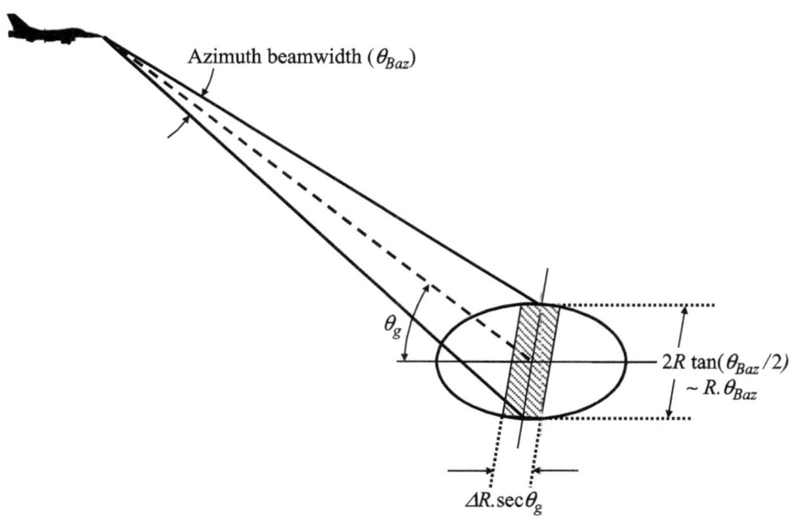

FIGURE 6-22 Clutter from Airborne Radar in Look-Down—Range Resolution-Limited Case

The maximum range corresponds to the minimum value of SCR necessary to meet a specified detection performance. Hence, rearranging (6-19) gives

$$R_{max} = \frac{\sigma}{\sigma^{\circ}(SCR)_{min}\theta_{Baz}\Delta R \sec\theta_g} \tag{6-20}$$

Equation (6-19) indicates that the SCR is inversely proportional to range. This contrasts with the noise-limited case in which SNR $\propto 1/R^4$. Equation (6-19) provides a simplistic estimate of the clutter-limited range as given by (6-20) but should be used with caution as many factors have been omitted in its derivation. The limitations of (6-20) are as follows:

- It omits SLC falling in the same range cell.
- It omits the decorrelation of clutter in time (and target decorrelation).
- It assumes low PRF operation, that is, that no ambiguous clutter returns fall in the same range cell.
- It assumes that the same backscatter coefficient may be applied across the whole area, which may encompass a range of grazing angles and surface types.
- It assumes there is no Doppler filtering (Doppler filtering subdivides the area into regions of differing Doppler shift).

This last point is particularly relevant as it indicates that clutter can be subdivided yet further by imposing another discriminant: Doppler. Indeed, pulse Doppler radar is valuable from the point of view that it can both determine target velocity and minimize the clutter problem by virtue of its subdivision in velocity, a point alluded to in Chapter 1. In general, the clutter problem can be minimized using fine resolution in angle, range, and velocity.

6.7.2 Beamwidth-Limited Case

The second look-down case is the beamwidth-limited case and is applicable to very high grazing angles. It is best illustrated by considering the vertical incidence case, as in Figure 6-23. In this situation, the whole surface within the beam footprint falls within one range cell.

If the main beam has a circular sectional shape then the main beam footprint is an elliptical area that becomes a circular patch at vertical incidence, $\theta_g = 90^{\circ}$. The area of the clutter patch is now

$$\approx \pi R^2 \left(\frac{\theta_{Baz}}{2}\right)\left(\frac{\theta_{Bel}}{2}\right)\operatorname{cosec}(\theta_g)$$

and the SCR is given by

$$SCR = \frac{\sigma}{\sigma^{\circ}\pi R^2 \left(\frac{\theta_{Baz}}{2}\right)\left(\frac{\theta_{Bel}}{2}\right)\operatorname{cosec}(\theta_g)} \tag{6-21}$$

As before, the maximum range corresponds to the minimum acceptable SCR necessary to meet a specified detection performance. Hence, rearranging (6-21) gives

$$R_{max}^2 = \frac{\sigma}{\sigma^{\circ}\pi\,(SCR)_{min}\left(\frac{\theta_{Baz}}{2}\right)\left(\frac{\theta_{Bel}}{2}\right)\operatorname{cosec}(\theta_g)} \tag{6-22}$$

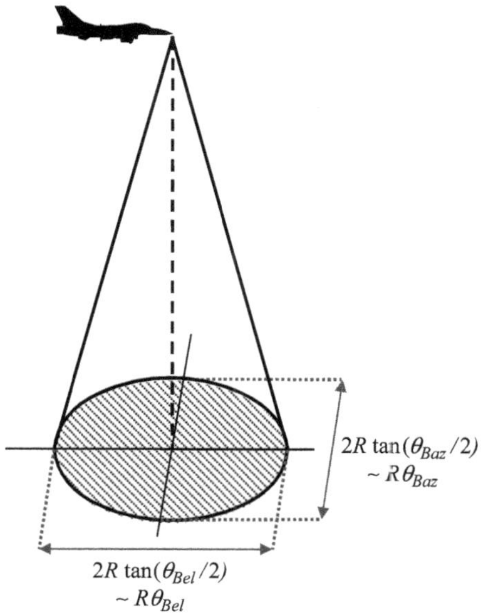

FIGURE 6-23 Clutter from Airborne Radar in Look-Down—Beamwidth-Limited Case

In the beamwidth-limited case, $SCR \propto 1/R^2$. The derivation of (6-22) is subject to the same limitations as previously listed for (6-20).

The transition between the range resolution-limited case and the beamwidth-limited case occurs for a grazing angle given by

$$\tan\theta_g = \frac{R\theta_{Bel}}{\Delta R} \tag{6-23}$$

6.8 | CLUTTER FOR A SURFACE-BASED RADAR

In many respects the clutter situation for a surface-based radar is a simplification of the airborne case. Some surface-based radars may be on moving platforms and may even be mounted at some considerable height above the clutter surface. Radars mounted onboard ships may be both moving and at the height of the mast above the sea surface and can therefore be treated in the same way as airborne radars. The exception, of course, is that the platform velocity and altitude are very much more modest. However, ground-based air surveillance radars are typically stationary and at ground level. In such cases, all the SLC and MBC fall at zero Doppler. There is no broadening of the clutter spectrum by virtue of the beamwidth since the platform velocity is zero; however, ICM may give rise to some clutter spectral broadening. Considerable clutter broadening can be experienced in high sea states due to the wave action. The rejection of MBC and SLC using a clutter notch filter now becomes a viable proposition.

If the main beam is pointing toward the surface, a strong clutter return can be expected. This may arise when detecting or tracking targets close to the horizon and can be quite troublesome in hilly terrain. The main beam may well illuminate a distant hillside at a high grazing angle, which results in a strong clutter return. Sidelobes are always directed at the ground, even if the main beam is aimed skyward; this is unavoidable. SLC is strong due to the proximity of the ground and is present from zero range outward. Overwhelming

SLC may well be present over the first few range cells, and some form of blanking may be necessary.

Low PRF waveforms may be viable for some ground-based pulse Doppler radar applications, particularly if the maximum target velocities are modest. In such cases, Doppler filtering does not help minimize the clutter problems, since the clutter is narrow band. However, low PRFs may be ambiguous in Doppler and exhibit a series of broad blind velocities corresponding to the rejection of MBC and its repetition in the Doppler domain. This may be overcome using multiple PRFs whose blind velocities are staggered from each other. This ensures that the whole Doppler band is visible in at least one PRF and is the traditional approach taken by MTI radar. Alternatively, higher PRFs are necessary, particularly for airborne radars, and high or medium PRF systems become the order of the day and are the subject of Part 2 of this book.

6.9 VOLUME CLUTTER

The problem is not volume clutter that is spatially separated from the target return but clutter that overlaps the target return and thus competes with the target for detection. The target return has a down-range length in space of $c\tau$, where τ is the processed pulse width of the target return. Any clutter return also has a range extent of $c\tau$. Clutter further away from, or nearer to, the target by $\Delta R = \frac{c\tau}{2}$, where ΔR is one range resolution cell, is not in the same resolution cell and so does not compete with the target. Clutter competing with the target lies in a range $c\tau$, centered on the target. Clutter very slightly closer to the target gives a return that overlaps the target return by a very small percentage of $c\tau$. Clutter separated from the target by $1/4\, c\tau$ overlaps the target return by $1/2\, c\tau$, or 50%. Clutter at the same range as the target overlaps by 100%. The range of competing clutter covers $c\tau$, but the overlap varies from 0% to 100% with an average of 50%. This is equivalent to having clutter over the range $1/2\, c\tau$ giving 100% competition, so the effective range of the clutter is $1/2\, c\tau$. The antenna beam diverges to a width of $\sim R\theta_{Baz}$ and a height $\sim R\theta_{Bel}$, where θ_{Baz} and θ_{Bel} are the azimuth and elevation beamwidths in radians, respectively. The volume of clutter that falls within one range resolution cell in any one beam position is illustrated in Figure 6-24. This has a volume given by

$$\text{Volume} = \frac{c\tau}{2} R^2 \pi \tan\frac{\theta_{Baz}}{2} \tan\frac{\theta_{Bel}}{2} \tag{6-24}$$

where $\tan\frac{\theta_B}{2} \approx \frac{\theta_B}{2}$ for small values of θ_B.

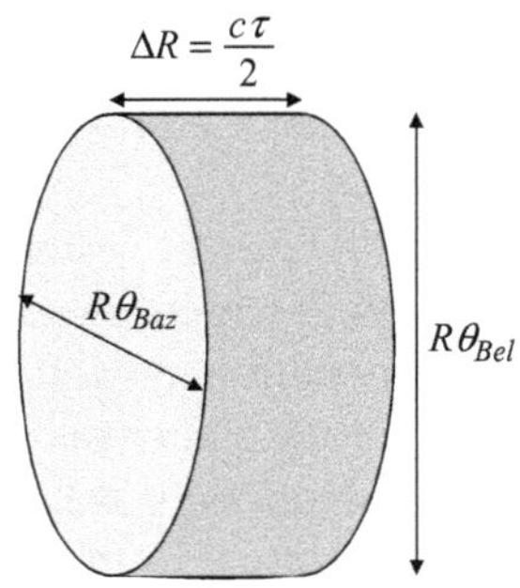

FIGURE 6-24 Volume Clutter in One Resolution Cell

Assuming that this volume is filled with independent scatterers that are uniformly distributed, the clutter RCS is given by

$$\text{Clutter RCS} = \eta \frac{c\tau}{2} R^2 \pi \tan \frac{\theta_{Baz}}{2} \tan \frac{\theta_{Bel}}{2} \qquad [\text{m}^2]$$

The signal-to-clutter ratio is now given by

$$SCR = \frac{2\sigma}{\eta c\tau \; R^2 \pi \tan \frac{\theta_{Baz}}{2} \tan \frac{\theta_{Bel}}{2}} \tag{6-25}$$

Rearranging this gives the clutter-limited range

$$R^2_{max} = \frac{2\sigma}{\eta c\tau \; (SCR)_{min} \, \pi \tan \frac{\theta_{Baz}}{2} \tan \frac{\theta_{Bel}}{2}} \tag{6-26}$$

This is an R^2 law, as is the case of beam-limited surface clutter. Assessing the amount of competing clutter is basically a problem of resolution in range, azimuth, and elevation. The square law variation with range means that as the range decreases the target does not rise above the clutter as fast as it rises above the noise. Rain may well fill the beam uniformly in azimuth but is unlikely to do so in elevation, unless the elevation beamwidth is small (e.g., tracking radar). If the beam is filled only fractionally, then the clutter return is reduced in proportion. Even if rain fills the beam vertically it often decreases in intensity with height. The target is somewhere in the main beam. Clutter is also in the main beam and in the sidelobes. For the clutter the effective beamwidth is greater than the 6 dB (two-way) beamwidth.

The expression for the clutter return should also include atmospheric loss and beam shape loss. In practice, the uncertainty of the backscatter coefficient and the contribution from sidelobes (ignored here) does not justify their inclusion but is best quantified via computer modeling. In the calculation of the signal-to-clutter ratio, the losses are the same for both signal and clutter. Note that radiated power and antenna gain are also the same for each and so have no influence on the ratio.

A significant source of volume clutter is rain, not least because of its prevalence in certain climates. An empirical expression for the backscatter coefficient for rain is

$$\eta(\text{dBm}^2) = -112 + 40 \log f_{\text{GHz}} + 16 \log r_{\text{mm/hr}} \tag{6-27}$$

where f is the frequency in GHz, and r is the rain fall rate in mm/hour.

Various graphs exist, as in Figure 6-25, but they vary somewhat. Rain consists of a large number of drops. The contribution of each drop depends on its diameter relative to the wavelength and the dielectric properties of water, which depend on temperature and frequency. Rain has a range of drop sizes, and the distribution of drop sizes depends on meteorological conditions. The same rain rate in different parts of the world can give different backscatter coefficients.

One means of combating rain clutter is the use of circular polarization [5]. Circularly symmetrical targets reverse the sense of circular polarization; hence, left-handed circular polarization is scattered predominantly as right handed on reflection from rain drops and vice versa. Circular polarization reflecting from a perfectly spherical raindrop should not be seen by the receiving antenna. In practice, the drops are not perfectly spherical, and

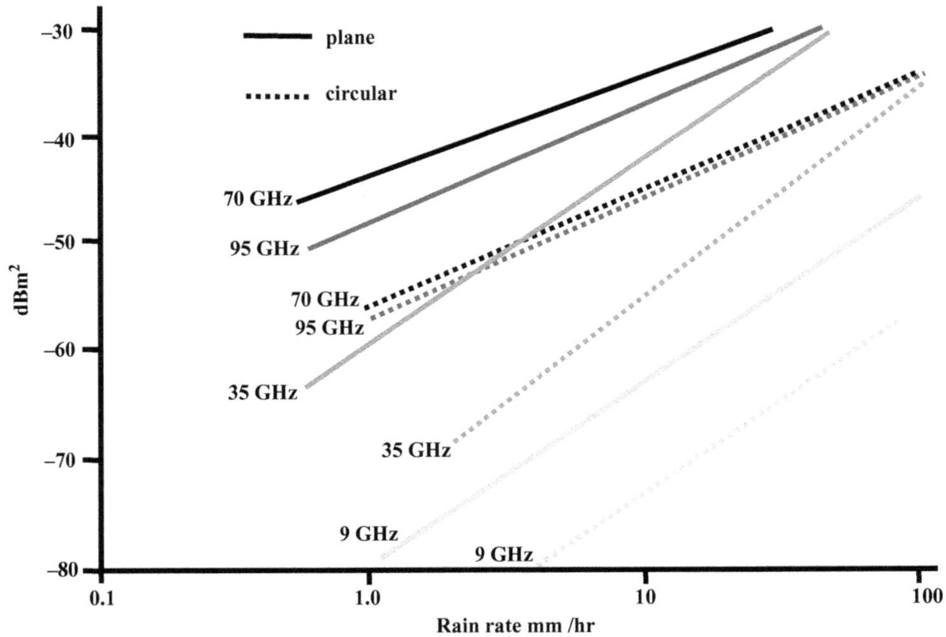

FIGURE 6-25 ■ Rain Backscatter Coefficient

multiple reflection mechanisms occur within a raindrop to interact with the polarization. The hand of polarization changes on reflection from a conducting surface. All odd numbers of reflections are seen; all even numbers of reflections are not seen. Both the rain and target reflections decrease, with the rain clutter returns decreasing by more than those of the target. The overall effect is that the signal-to-clutter ratio is improved, but with a weaker absolute target return. There is an increased chance that the signal does not cross the threshold, so the probability of detection is degraded.

CHAPTER 7

Pulse Doppler Processing

Chapter Outline

7.1 THE PROCESSING CHAIN

The processing of pulse Doppler radar signals is commonly depicted as a sequence of stages known as the *processing chain*. Signal processing is applied to enhance the signal-to-noise ratio (SNR) to enhance detection performance, to discriminate between true targets and interference, clutter, or jamming, and to extract the desired information about targets that, at the very least, includes target range and velocity. While the exact order and processing operations differ from radar to radar, they are all likely to include the stages depicted in Figure 7-1.

Moving target indicator (MTI) radar and its role in providing rejection of main beam clutter (MBC) were mentioned in Section 6.6.3. MTI clutter rejection will be covered in greater detail in Section 7.2. Range gating and pulse compression was covered in Sections 3.3.4 and 3.9, respectively. The mainstay of pulse Doppler processing is the fast Fourier transform (FFT); it is covered in detail in Section 7.3 and forms the bulk of this chapter. Some radars perform some non-coherent integration (in addition to the coherent integration implicit in FFT processing) and this will also be covered in Section 7.3. Constant false alarm rate (CFAR) detection was mentioned in Section 6.4.3. Standard CFAR techniques and theory are covered in Section 7.4, and some additional material arising from recent research on pulse Doppler systems is also given in Chapter 14. Finally, target extraction will be covered in Chapter 12, rather than here, as it is inextricably linked with the correlation of target range and velocity and the problems of *ghosting*. These issues are dealt with in future chapters on medium pulse repetition frequency (PRF) in Part 2 of this book, but mention is made of it here merely for the sake of completeness.

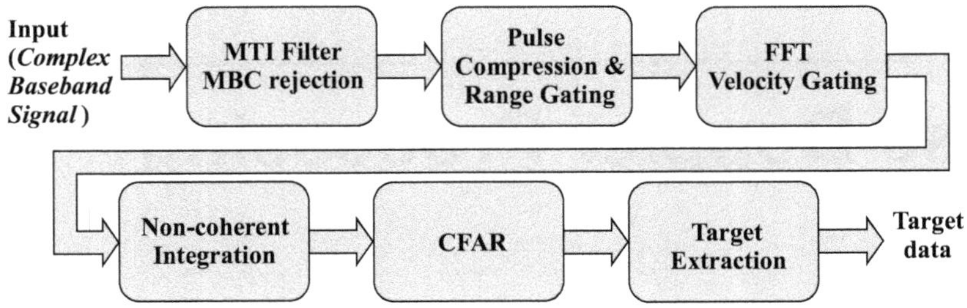

FIGURE 7-1 The Processing Sequence

7.2 MTI REJECTION OF MBC

MTI radar originally evolved from low PRF systems in which it was desirable to reject stationary, or low-velocity, clutter signals and admit only those returns from moving targets. The target discrimination was based on the presence or absence of a Doppler shift on returning echoes. The Doppler shift was demodulated by down-converting returns to a baseband output using a homodyne or, better still, a superheterodyne receiver (Chapter 8). In MTI systems, no attempt is made to measure the Doppler shift, and hence velocity, of target returns; the goal is simply to admit returns with a Doppler shift and reject those without one. This Doppler discrimination is accomplished using an MTI *clutter cancellation filter* in the baseband signal path [5].

MTI radar is still used today and is not necessarily confined to low PRF systems. MTI clutter cancellation filters are often used as a first stage of processing in modern high and medium PRF radars to strip out clutter at an early stage in the processing chain.

7.2.1 The Baseband Signal

In Section 4.6 it was noted that the baseband waveform of a pulsed radar consists of a sequence of pulses that sample the Doppler cycle. The phase difference between successive pulses (samples) is given by

$$\Delta\varphi = 2d\frac{2\pi}{\lambda} \qquad \text{repeat of (4-4)}$$

where d is the distance by which the range changes in the time between pulses. Since from (4-3) one may write $d = V_r/f_r$, substituting this into (4-4) gives

$$\Delta\varphi = \frac{4\pi V_r}{\lambda f_r} \tag{7-1}$$

Equation (7-1) indicates that the phase shift is a function of the relative velocity of the target (V_r), the PRF of the radar (f_r) and the wavelength of the transmitted radio frequency (RF) carrier signal (λ) and hence its frequency.

When a stationary target is detected, it results in zero Doppler shift, so the baseband waveform becomes a succession of samples of a direct current (DC) level. The pulse samples of a Doppler shifted moving target and of a stationary target are shown in Figure 7-2. The top graph shows the pulses arising from a moving target. These pulses are amplitude modulated by the Doppler cycle; hence, they sample the Doppler cycle at $\Delta\varphi$ intervals of phase. The lower graph shows the pulses arising from a stationary target such as a clutter return. These pulses sample a waveform of zero frequency (i.e., a DC level); hence, they have a constant amplitude.

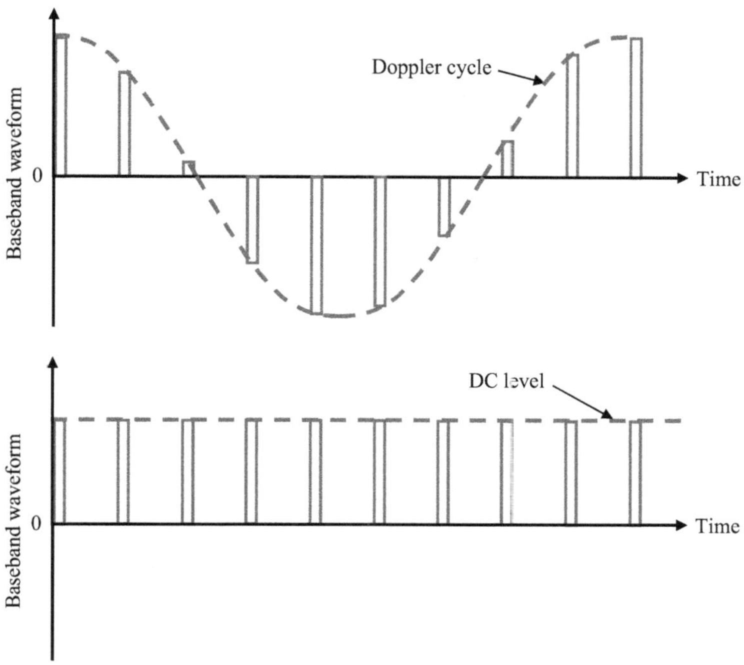

FIGURE 7-2 Pulse Samples of Moving Target (top) and Stationary Target (bottom)

The sampling frequency is given by the PRF, f_r, of the radar. For a moving target, it is possible to imagine an extreme case in which one sample is received for every complete cycle of the Doppler waveform. In this case, $f_d = f_r$ and the phase difference between samples, $\Delta\varphi = 360°$ or 2π radians. Similarly, if exactly two samples are received for every complete cycle of the Doppler waveform, $f_d = f_r/2$, and the phase difference between samples, $\Delta\varphi = 180°$ or π radians. In general, the relationship between the Doppler frequency and the phase difference between pulses can be expressed as

$$f_d = f_r \frac{\Delta\varphi}{2\pi} \qquad (7\text{-}2)$$

7.2.2 Single Delay Line Canceler

The concept on which clutter cancellation is implemented is that pulses from moving targets vary in amplitude whereas those from stationary targets are at a constant level. The simplest form of clutter cancellation is therefore to compare the amplitudes of successive pulses. This is done in a single delay line canceler, which is illustrated in Figure 7-3. This circuit divides the baseband (video) signal into two paths: one of which is delayed by one pulse repetition interval (PRI $= T_r$) and the other is undelayed. The delayed and undelayed paths are then applied to a differencing amplifier. The first of two pulses incident on the circuit is applied to the delay line input. The second of two pulses is incident on the circuit input a time T_r later by which time the first pulse emerges from the delay line. At this point the two pulses have now been brought into time coincidence and form the two inputs to the differencing amplifier. The differencing amplifier outputs a voltage corresponding to the difference between the two pulse voltages. In this way the circuit has formed the difference between successive pulses. This process continues as more pulses are incident on its input, and the canceler circuit continues to output a series of voltage pulses equal to the difference between successive pairs of input pulses. It is also worth noting that the

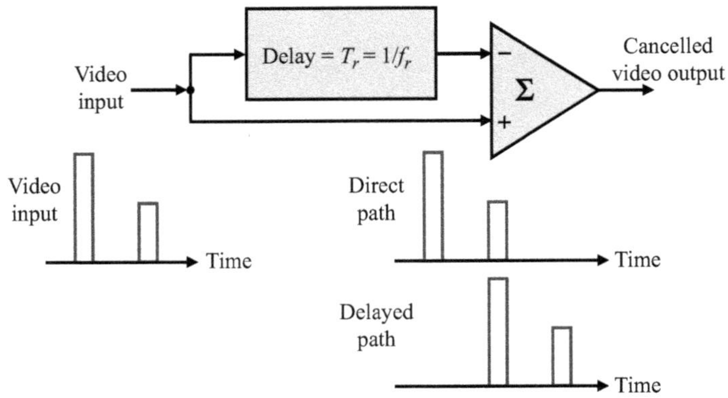

FIGURE 7-3 Single Delay Line Canceler

single delay line canceler circuit considered here does not produce a valid output at the time of the first pulse; that is, it takes two incident pulses before one obtains a valid output.

As seen in Section 4.5.1, the continuous wave (CW) transmitted signal may be expressed as

$$A_1 \sin(\omega_{tx} t) \tag{7-3}$$

The received signal is therefore

$$A_2 \sin\left[(\omega_{rx} t) - 2\omega_{tx}\frac{R}{c}\right] \tag{7-4a}$$

where $\omega_{rx} = \omega_{tx} \pm \omega_d$, and the $2\omega_{tx}\frac{R}{c}$ term represents the phase shift over the two-way journey to the target and back. Therefore, the received signal may alternatively be expressed as

$$A_2 \sin\left[(\omega_{tx} \pm \omega_d)\, t - 2\omega_{tx}\frac{R}{c}\right] \tag{7-4b}$$

The received signal is coherently mixed with the coupled portion of the transmitted signal in a multiplicative mixer. This results in a homodyne down-conversion of the received signal to a baseband output. The mixer output is therefore

$$= A_1 \sin(\omega_{tx} t)\, A_2 \sin\left[(\omega_{tx} \pm \omega_d)\, t - 2\omega_{tx}\frac{R}{c}\right]$$

$$= \frac{A_1 A_2}{2}\cos\left[(2\omega_{tx} \pm \omega_d)\, t - 2\omega_{tx}\frac{R}{c}\right] + \frac{A_1 A_2}{2}\cos\left[(\pm\omega_d t) - 2\omega_{tx}\frac{R}{c}\right]$$

The first term runs at twice the transmitted frequency and is easily filtered off, leaving just the component given by the second term, which runs at the Doppler frequency. Thus, the filtered baseband output is

$$= \frac{A_1 A_2}{2}\cos\left[(\pm\omega_d t) - 2\omega_{tx}\frac{R}{c}\right]$$

The peak amplitude is somewhat arbitrary, so it is more convenient to let $A_3 = \frac{A_1 A_2}{2}$. The baseband signal may also be termed the *video* signal and forms the video input signal to the canceler circuit. The envelope of the video input to the canceler, V_V, is therefore given by

$$V_V = A_3 \cos\left[(\pm\omega_d t) - 2\omega_{tx}\frac{R}{c}\right] \tag{7-5}$$

Since the radar is pulsed, the value of V_V at the times of the pulses is important. Consider the values of V_V at times t_1 and t_2, which coincide with two successive pulses.

$$V_{V1} = A_3 \cos\left[(\pm\omega_d t_1) - 2\omega_{tx}\frac{R}{c}\right] \tag{7-6a}$$

and

$$V_{V2} = A_3 \cos\left[(\pm\omega_d t_2) - 2\omega_{tx}\frac{R}{c}\right] \tag{7-6b}$$

The canceler brings these two pulses into time coincidence and forms the difference between them. Hence, the canceled video output, V_{VC}, is given by

$$V_{VC} = V_{V2} - V_{V1} = A_3 \cos\left[(\pm\omega_d t_2) - 2\omega_{tx}\frac{R}{c}\right] - A_3 \cos\left[(\pm\omega_d t_1) - 2\omega_{tx}\frac{R}{c}\right]$$

$$V_{VC} = -2A_3 \sin\left[\pm\tfrac{1}{2}(t_2 - t_1)\omega_d\right] \sin\left[\pm\tfrac{1}{2}(t_2 + t_1)\omega_d - 2\omega_{tx}\frac{R}{c}\right]$$

Since the time difference between pulses $(t_2 - t_1) = T_r = 1/f_r$, $\omega_d = 2\pi f_d$ and $\omega_{tx} = 2\pi f_{tx}$, these expressions may be substituted into the equation for V_{VC} to obtain

$$V_{VC} = -2A_3 \sin\left(\pm\pi\frac{f_d}{f_r}\right) \sin\left(\pm\pi f_d(t_2 + t_1) - 4\pi f_{tx}\frac{R}{c}\right) \tag{7-7}$$

The second sine term, $\sin\left(\pm\pi f_d\ (t_2 + t_1) - 4\pi f_{tx}\frac{R}{c}\right)$, indicates that the canceled video output has a sine wave envelope that runs at the Doppler frequency, f_d, together with a range-related phase shift $4\pi\ f_{tx}\frac{R}{c}$, which is incurred over the two-way path out to the target and back. The $-2A_3 \sin\left(\pm\pi\frac{f_d}{f_r}\right)$ term is the peak amplitude of the sine wave envelope and is dependent on the ratio of the Doppler frequency to the PRF. Again, it ought to be stressed that the actual canceled video output is in fact a sequence of pulses running at the radar PRF whose envelope adheres to the expression of (7-7). Therefore, the canceled video output pulses still provide samples of the Doppler waveform, but the amplitude of these pulses is scaled by the first sine term in (7-7). The response of the canceler is given by the modulus of the $-2A_3 \sin\left(\pm\pi\frac{f_d}{f_r}\right)$ term, that is, $|V_{VC}|$.
When $f_d = 0$,

$$\left|-2A_3 \sin\left(\pm\pi\frac{0}{f_r}\right)\right| = 0$$

so there is no output for zero Doppler shifted target returns and static clutter is rejected. The canceler response reaches a maximum value of $2A_3$ when

$$\sin(\pi f_d/f_r) = 1$$
$$\pi f_d/f_r = \pi/2$$

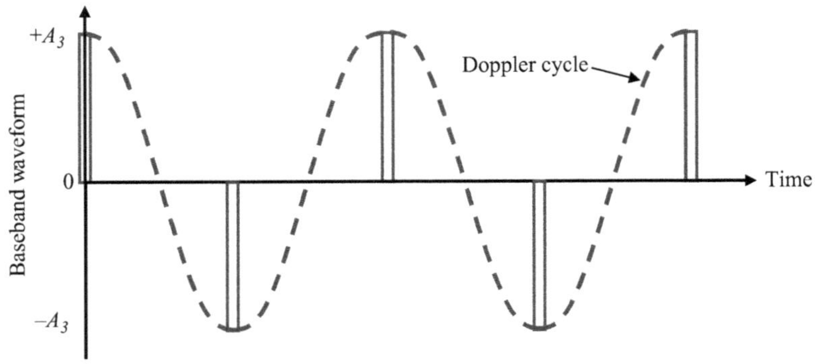

FIGURE 7-4 ■ Peak Response of Canceler

that is, when

$$f_d = f_r/2$$

and when the range, R, is such that the phase shift term $2\omega_{tx}\frac{R}{c}$ places successive returns at positive and negative peak values of the Doppler cycle. This case is illustrated in Figure 7-4. The voltage levels of pulses vary between $+A_3$ and $-A_3$ and hence give rise to the maximum difference of $2A_3$. These pulses sample the Doppler cycle at $\Delta\varphi = 180°$ phase intervals.

A plot of $\left|2A_3 \sin\left(\pm\pi\frac{f_d}{f_r}\right)\right|$ against f_d represents the frequency response of the canceler. The response is graphed in Figure 7-5 and illustrates the zero canceler output at zero Doppler shift and the peak output of $2A_3$ when $f_d = f_r/2$. It may also be noted that the filter response repeats along the frequency axis. Since the signal being sampled is the Doppler frequency, f_d, and the sampling frequency is given by the PRF, f_r it must be appreciated that inadequate sampling leads to ambiguity. Ambiguous repeats of the canceler response therefore occur when $f_d > f_r/2$ and is evident by the repetitive nature

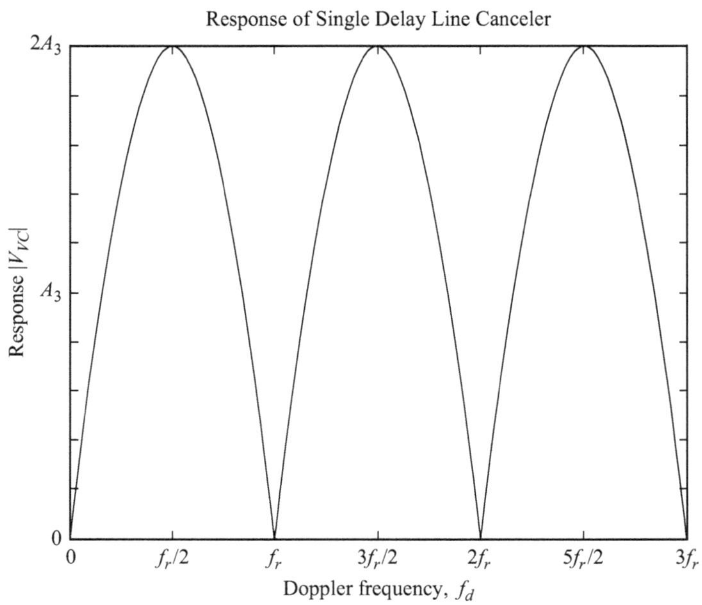

FIGURE 7-5 ■ Response of Single Delay Line Canceler

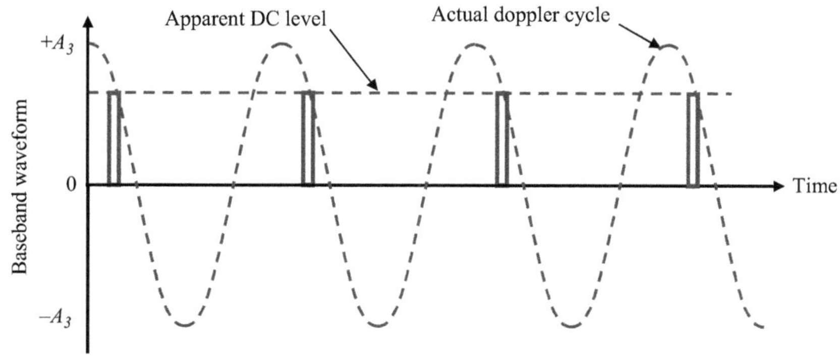

FIGURE 7-6 Ambiguous Baseband Samples at $f_d = f_r$

of Figure 7-5. The rejection of static clutter falling at $f_d = 0$ is repeated at $f_d = mf_r$, where $m = 1, 2, 3 \ldots$. In Section 6.6.2 it was noted that the clutter spectrum is repeated at Doppler shifts equal to multiples of the PRF; this is illustrated in Figure 6-21. Thus, the previously described single delay line canceler circuit not only rejects clutter at zero Doppler shift, but also its repeats at Doppler shifts equal to multiples of the PRF. The repeated rejection nulls in the canceler response are easily understood from the phase sampling of the Doppler waveform concept. If the Doppler shift is a multiple of the PRF, consecutive pulses fall at the same point of phase on the Doppler cycle and therefore have the same amplitude and are rejected by the canceler. The case of $f_d = f_r$ is plotted in Figure 7-6. Similarly, the peak in the canceler response repeats when pulses sample the Doppler cycle at $\Delta\varphi = (180 + 360m)^\circ$ phase intervals, that is, when both $f_d = (2m + 1) f_r/2$ and the appropriate range phase shift is incurred.

The canceler response gives rise to several problems.

(i) Inadequate Rejection Bandwidth

The filter characteristic plotted in Figure 7-5 has a narrow stopband. This would result in inadequate rejection of clutter possessing a low Doppler shift and of slow-moving targets, which are of no interest. For example, 30 dB of rejection is obtained only when $|f_d| \leq 0.01 f_r$. Furthermore, as the antenna beam scans across a static target, a scan modulation is imparted onto a succession of pulses that would be perceived as a small Doppler shift and so would not be completely rejected by the canceler. Scan modulation arises due to the variation in the antenna main beam gain as it scans across a fixed target. It may often be necessary to have a broader rejection band, for which a more elaborate canceler circuit is required. These are described in the ensuing sections.

(ii) Blind Velocities

In addition to rejecting the ambiguous repeats of clutter centered at Doppler shifts equal to multiples of the PRF, the canceler also rejects genuine targets whose actual Doppler shift happens to fall at a multiple of the PRF. Hence, there is a succession of blind velocities, V_b, given by

$$V_b = \frac{\lambda m f_r}{2} \tag{7-8}$$

For low PRF systems this can be quite problematic because these blind velocities are closely spaced across the range of likely target velocities. The problem also exists in medium PRF systems, although it is less severe than in low PRF radars as the PRFs are typically much higher so fewer blind velocities are incurred. In both low and medium PRF systems the problem of blind velocities is overcome by alternating the operation of the radar over several PRFs suitably chosen such that the blind velocities in one PRF do not coincide with the blind velocities of another PRF. By staggering the blind velocities away from each other, target visibility is maintained across the Doppler band without compromising the clutter rejection. This subject is dealt with further in Chapter 11 in the context of medium PRF pulse Doppler systems. For traditional MTI systems a system of changing the PRF on each alternate pulse was developed and is known as *staggered PRFs*. The use of staggered, multiple PRFs requires switching the appropriate delay in the canceler circuit to match the PRI currently used. Early MTI radars developed ingenious ganged switching mechanisms to do this; however, these need not be considered any further here, particularly as modern digital signal processing techniques have rendered the problem of imparting an appropriate delay onto data a trivial one. The blind velocity problem does not arise in high PRF systems since the first blind velocity, corresponding to $m = 1$, occurs beyond the maximum velocity of interest.

(iii) Blind Phases

It has been noted that the maximum response is obtained when $f_d = f_r/2$ and for range phase shifts resulting in pulses returning coincident with the positive and negative peaks of the Doppler cycle, as illustrated in Figure 7-4. However, a marginally different range phase shift would result in pulses returning coincident with the zero crossing points of the Doppler cycle. This gives a zero output and, therefore, target blindness and is known as the *blind phase* problem. Hence, the condition $f_d = f_r/2$ may result in the peak response from the canceler or a zero response from the canceler, or anything in between, depending on the range phase shift term given by $4\pi f_{tx}\frac{R}{c}$. The canceler response can vary between these limits as the target range changes; however, a target may sit in a blind phase spot for some time. This problem can also occur in low, medium, and high PRF radars, including pulse Doppler radars, which down-convert signals to a zero frequency baseband output. The blind phase problem is overcome using a *quadrature detector* circuit, which is used to down-convert signals into two baseband channels which are in phase quadrature with each other (see Section 8.4.1). The two channels are known as the *I-channel* (In-phase channel) and the *Q-channel* (Quadrature-phase channel) and can be thought of as representing the real and imaginary parts of baseband signals. Thus, the two channels yield the full complex representation of the baseband signal. The magnitude of the complex signal is proportional to the sum of the squares of the voltages in each channel, ($\sqrt{I^2 + Q^2}$), and is a constant, irrespective of the range phase shift term. Thus, by considering the full complex baseband signal, the blind phase problem is overcome. The quadrature detector and its properties are described more fully in Section 8.4.1.3.

7.2.3 Double Delay Line Canceler

The rejection bandwidth can be increased using a doubled delay line canceler. This is simply two single delay line cancelers in series and is depicted in Figure 7-7.

The double delay line canceler does not produce a valid output until the third pulse has been applied to its input, so three pulse inputs must be considered in its analysis. The first two of these occur at times t_1 and t_2 and have voltage levels V_{V1} and V_{V2} given by equations (7-6a) and (7-6b), respectively. The third video input pulse occurs at time t_3 and

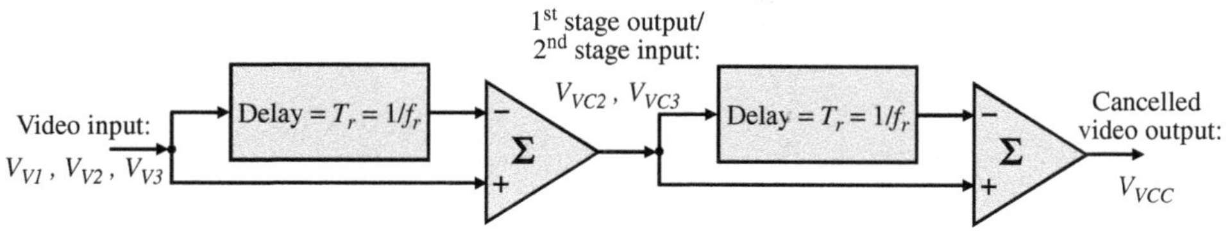

FIGURE 7-7 Double Delay Line Canceler

has a voltage level V_{V3} as given in (7-6c), which follows the pattern of (7-6a) and (7-6b):

$$V_{V3} = A_3 \cos\left[(\pm\omega_d t_3) - 2\omega_{tx}\frac{R}{c}\right] \tag{7-6c}$$

This results in two legitimate output pulses from the first stage occurring at times t_2 and t_3. These pulses have voltage levels V_{VC2} and V_{VC3} and may be found by substituting the appropriate times into (7-7). Thus,

$$V_{VC2} = V_{V2} - V_{V1} = -2A_3 \sin\left(\pm\pi\frac{f_d}{f_r}\right)\sin\left(\pm\pi f_d\,(t_2 + t_1) - 4\pi f_{tx}\frac{R}{c}\right) \tag{7-9a}$$

$$V_{VC3} = V_{V3} - V_{V2} = -2A_3 \sin\left(\pm\pi\frac{f_d}{f_r}\right)\sin\left(\pm\pi f_d(t_3 + t_2) - 4\pi f_{tx}\frac{R}{c}\right) \tag{7-9b}$$

These two pulses form the input to the second stage of the circuit, which results in an output pulse of voltage V_{VCC} occurring at time t_3 given by

$$\begin{aligned} V_{VCC} &= V_{VC3} - V_{VC2} \\ V_{VCC} &= (V_{V3} - V_{V2}) - (V_{V2} - V_{V1}) \\ V_{VCC} &= V_{V3} - 2V_{V2} + V_{V1} \end{aligned} \tag{7-10}$$

Substituting (7-9a) and (7-9b) into (7-10) results in

$$V_{VCC} = -4A_3 \sin^2\left(\pm\pi\frac{f_d}{f_r}\right)\cos\left(2\pi\, f_d\left(t_3 - \frac{1}{f_r}\right) - 4\pi\, f_{tx}\frac{R}{c}\right) \tag{7-11}$$

Equation (7-11) indicates that the canceled output from the double delay line canceler still has an envelope adhering to the Doppler cycle together with a range-related phase shift. The response of the circuit is given by the $4A_3 \sin^2\left(\pm\pi\frac{f_d}{f_r}\right)$ term. The response is graphed in Figure 7-8 and illustrates the zero canceler output at zero Doppler shift and the peak output of $4A_3$ when $f_d = f_r/2$. As with the single delay line canceler, the double delay line canceler is also prone to the blind phase problem and exhibits ambiguous repeats of its response for $f_d > f_r/2$. Since the response is given by a $\sin^2$ function, the canceler has a broader rejection notch around $f_d = 0$ and its ambiguous repetition at $f_d = mf_r$. This results in broader bands of clutter cancellation and blind velocities. In this case, 30 dB of rejection is obtained when $|f_d| \le 0.057 f_r$, which is considerably broader than the case of the single delay line canceler.

7.2.4 Multiple Delay Line Cancelers

More complex delay line cancelers can be produced by including further stages of the single delay line canceler in series. The inclusion of each additional stage does the following:

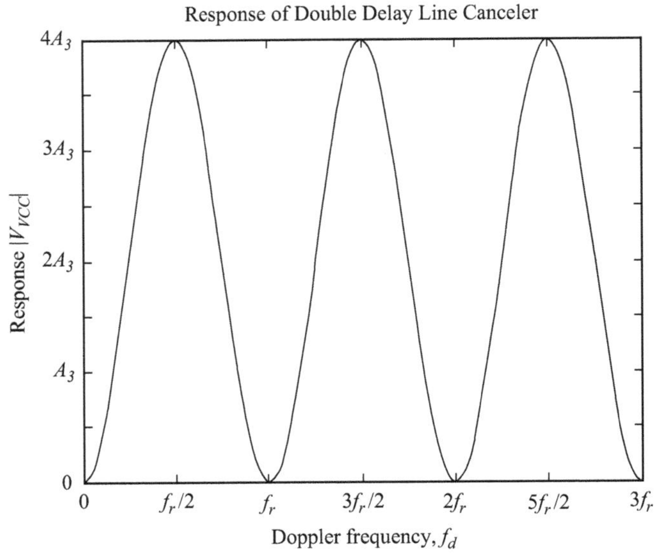

FIGURE 7-8 Response of Double Delay Line Canceler

- Increases the power to which the sine term response of the filter is raised, thus broadening the rejection bands at $f_d = 0, mf_r$
- Requires one more pulse on the input before a legitimate canceled output is obtained
- Increases the peak response at $f_d = f_r/2$ by a further factor of two.

For example, a triple delay line canceler therefore has a response given by $8A_3\ \sin^3\left(\pm\pi\frac{f_d}{f_r}\right)$ and requires four input pulses before a legitimate output is obtained.

7.2.5 Three-Pulse Canceler

An alternative form of canceler operating on three input pulses is the *three-pulse canceler*, shown in Figure 7-9. This circuit brings three successive pulses into time coincidence using two delay lines matched to the pulse repetition interval (PRI). The voltages of each pulse are weighted and added. The weighted sum forms the canceled output. The weighting factors are $\times 1$ for the first pulse (V_{V1}), $\times{-2}$ for the second pulse (V_{V2}), and $\times 1$ for the third pulse (V_{V3}). The canceled video output is therefore given by $V_{V1} - 2V_{V2} + V_{V3}$, which is the same as that for V_{VCC} given for the double delay line canceler in equations (7-10) and (7-11). The three-pulse canceler has the exact same characteristics as the double delay line canceler; it takes three input pulses before a legitimate output is obtained and has the same $\left|4A_3\ \sin^2\left(\pm\pi\frac{f_d}{f_r}\right)\right|$ response.

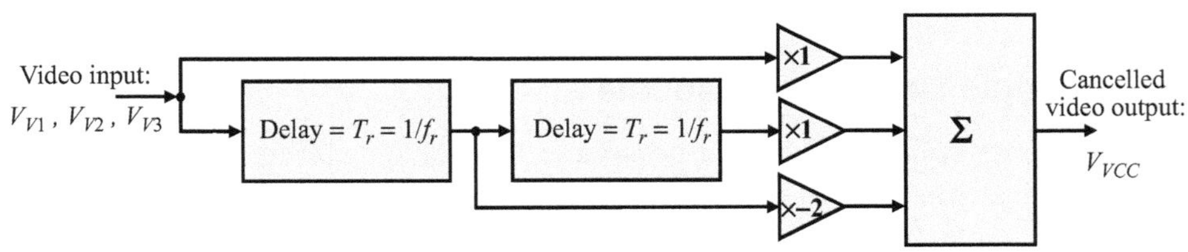

FIGURE 7-9 Three-Pulse Canceler

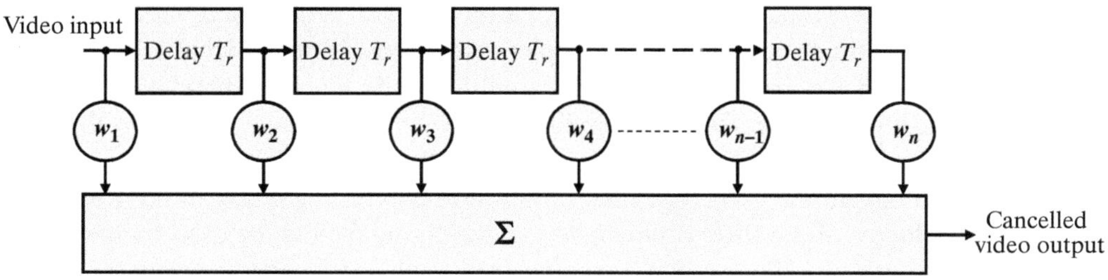

FIGURE 7-10 Transversal Filter

7.2.6 Transversal Filters

The three-pulse canceler is an example of a *transversal filter*. A more general form of the transversal filter is given in Figure 7-10. The filter can be made as large as required and therefore may operate on as many pulses and is desired. For example, a transversal filter with four stages constitutes a four-pulse canceler and has equivalent properties to a triple delay line canceler, that is, a response given by $8A_3 \sin^3\left(\pm\pi \frac{f_d}{f_r}\right)$. In general, for a $\sin^n\left(\pm\pi \frac{f_d}{f_r}\right)$ response one would use an n delay line canceler or an $(n+1)$ pulse canceler, both of which require n delay lines and n input pulses before a legitimate output is obtained.

One set of weighting factors that may be used in a transversal filter to reproduce the response of the delay line type filters considered thus far is given by the binomial coefficients of alternating sign [5]. For an $(n+1)$ pulse canceler the ith weighting factor is given by

$$w_i = (-1)^{i-1}\frac{n!}{(n-i+1)!\,(i-1)!} \qquad i = 1, 2, 3, \ldots n+1 \tag{7-12}$$

An easy way to derive the weighting factors is to construct Pascal's triangle with alternating signs:

n	**w_i sequence**							
2	1	−1						
3	1	−2	1					
4	1	−3	3	−1				
5	1	−4	6	−4	1			
6	1	−5	10	−10	5	1		
7	1	−6	15	−20	15	6	1	
8	1	−7	21	−35	35	21	−7	1
etc.								

A transversal filter with binomially weighted coefficients gives a broad passband frequency response, albeit with no sidelobes in the stopband. In many circumstances, a narrower filter response is required, but this incurs sidelobes. The weights may be adjusted for any desired filter response such as a Dolph–Chebyshev 40 dB response (i.e., stopband ripple/sidelobes of −40 dB), which is popular in several designs [13, 14].

The bandwidth of the filter response is given by the inverse of the time for which input pulses are required. This is simply restating the relationship that the Doppler resolution is the inverse of the integration time, that is, $\Delta f_d = \frac{1}{t_{int}}$, which was equation (4-15), first encountered in Section 4.5.2. For an n-element transversal filter, the integration time

$t_{int} = n \times T_r = n/f_r$ and the Doppler resolution, $\Delta f_d = f_r/n$. Larger filters require longer integration times but yield finer Doppler resolutions. If the weighting factors, w_i, adjust the *phase* of the signal as well as its amplitude, the filter passband can be tuned such that it may be centered at any desirable frequency in the range of zero to the PRF $[0, f_r]$ [5]. An example of this is the digital filter operation described in Section 7.3.2 culminating in equation (7-18). The filter would also exhibit repetition in the frequency domain at multiples of the PRF. If now n sets of weighting factors are used to operate on the n pulses, a bank of n filters could be synthesized, each one tuned to a center frequency displaced from its neighbors by f_r/n [5]. The bank of filters would thus cover the whole Doppler band with a resolution of f_r/n. This is known as a *discrete Fourier transform* (DFT) (Section 7.3.2), and it lends itself conveniently to the digital processing of signals. The DFT is a transform of n time samples into n (or more generally, $m \neq n$) frequency samples. There is a mathematical equivalence between the DFT of time data and the output of a bank of band-pass filters, but several conditions are required for this to be accurate: the filter frequency response has to be that of the window used; the center frequencies have to correspond to the DFT frequencies; and the outputs of the filter bank have to be sampled at the correct time. An example of the filtering characteristics for an eight-point DFT (i.e., $n = 8$) is shown in Figure 7-11.

The filtering provided by the DFT introduces velocity gating that may be used to discriminate between clutter of differing velocities and to measure target velocity. Such filtering techniques are employed on modern airborne MTI radars as they afford superior detection performance to single delay line clutter filters alone. The responses from targets of interest, moving at a fixed velocity, possess a fixed Doppler frequency. Such targets are largely confined to a single filter response, although there may be some spread into the neighboring filters if the exact Doppler shift falls partway between the center frequencies of two neighboring filters. However, on the whole, most of the target signal is detected within one filter. Clutter, however, may be spread over a wide range of Doppler frequencies. It was noted in Section 6.4.2 how sidelobe clutter (SLC) received by a fast-moving airborne radar can be spread over a wide Doppler band. It is quite possible for the target response to fall within the band of SLC. MBC may be removed using a single or double delay line canceler, as appropriate, but the SLC remains. In the absence of any further filtering of the Doppler band, the target is now competing with *all* the SLC for detection. If, however, the Doppler

FIGURE 7-11 8 Point Discrete Fourier Transform Filter Characteristics

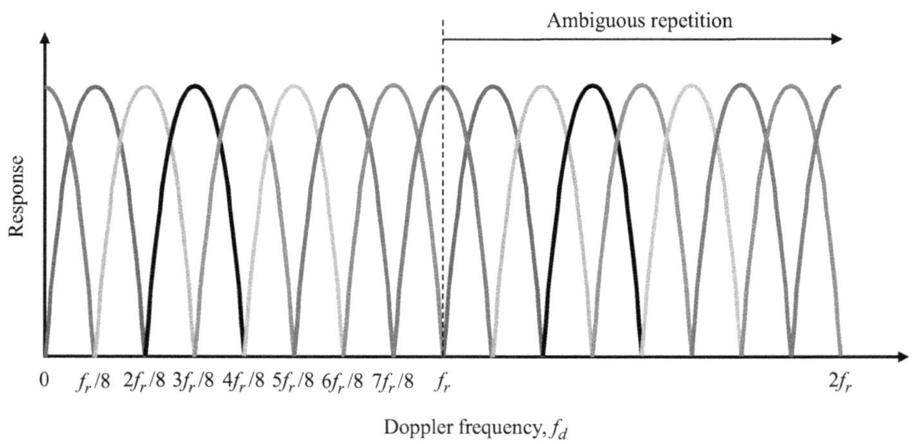

band is subdivided into fine Doppler cells, then only the portion of SLC falling within the filter occupied by the target competes with the target for detection (plus any SLC falling within the ambiguous repetition of the filter response at higher Doppler frequencies). The SLC is therefore divided out across all filters, whereas the target response is confined to just one filter. This results in a much improved signal-to-clutter ratio (SCR) and hence a better detection performance. The improvement is enhanced for finer Doppler cells (larger n, longer integration times). Fine velocity and range resolution improve detection performance in conditions of high clutter but at the expense of time and bandwidth, respectively.

7.2.7 Digital MTI Cancellation

The advent of high digitization rates and large shift registers has enabled MTI cancellation to be carried out digitally. The process is illustrated in Figure 7-12. The radar receiver output is typically digitized into 12-bit (or more) resolution at a rate consistent with the desired range resolution and loaded into the canceler. The canceler is a large digital store, such as a shift register, capable of holding all the data samples received over one pulse repetition interval. Each location in the shift register holds a sample received within a single range cell. The flow of data in the store is controlled by a clock that also governs the analog-to-digital converter (ADC). On each clock cycle, the data in the register shuffle down one position and a new sample is loaded in at the input. Over the course of one PRI, the register is filled with samples, each corresponding to the data in the range cells spanning the PRI. The first and last locations in the register correspond to the same range cells in successive receiving intervals; these samples are therefore one PRI apart from each other. Data held at these locations are tapped out to a subtractor, the output of which is the canceled signal. As fresh data continue to load into the input of the register, old data are lost from the register end. With each clock cycle, fresh data corresponding to the current range cell are loaded into the first position in the store, and data corresponding to the same range cell on the previous receiving period are held in the last position in the

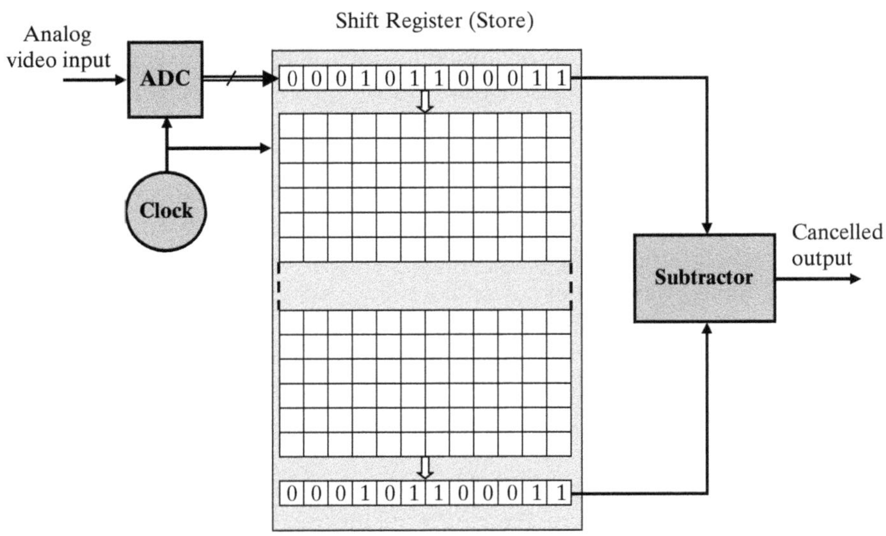

FIGURE 7-12 Digital MTI Cancellation

store. The subtractor continues to provide the canceled output for each range cell in its turn. The flexibility of digital designs like this permits variable PRIs to be accommodated.

Naturally, there are variations on this theme. Larger stores could accommodate two, three, or more PRIs worth of data. Tapped outputs corresponding to the same range gate in successive receiving periods may be weighted and added to yield the response of, for example, double or triple delay line cancelers or transversal filters of any desired length, the length limited only by the capacity of the data store. It ought to be stressed that the digital MTI cancellation techniques described here are simply digital implementations of the analog cancelers considered earlier. The analysis of the analog cancelers may equally be applied to their digital counterparts. The digitization of signals can introduce a degree of quantization error and may also introduce errors associated with the discrete time sampling of the signal [5]. Notwithstanding, in general the digital processing of the signals offers greater flexibility and performance in the processing than can be attained using conventional analog cancelers.

Before the subject of clutter cancellation is laid to rest, the possibilities offered by *space time adaptive processing* (STAP) and the techniques of the *displaced phase center antenna* (DPCA), which may be thought of as a special form of STAP [15], must be acknowledged. These techniques may be implemented on modern active electronically steered phased array antennas (section 8.5.4) and operate on several channels of data arising from the antenna subarrays over a processing interval comprising several receiving periods [16]. Clutter and interference are suppressed by forming the weighted sum of data samples corresponding to a consistent range (and velocity) cell over several receiving periods. The weights may be calculated on the basis of data in nearby range bins and subsequently adaptively refined to minimize clutter and interference and maximize desired target returns. STAP offers excellent clutter rejection for airborne radars operating a *ground-moving target indication* (GMTI) mode that seeks to map the position of moving ground targets.

7.2.8 Quantifying MTI Performance

In quantifying the performance of clutter cancellation, the frequency response of the clutter canceler and the spectrum of the clutter should be considered. These vary from one radar to the next and are highly dependent on the clutter environment. An example is illustrated in Figure 7-13, where $S_C(f)$ represents the clutter spectral power density, and $H(f)$ represents the canceler spectral (voltage) response. The example in Figure 7-13 depicts a single delay line canceler response.

The radar's ability to suppress clutter is known as the *clutter attenuation* (CA). This is defined as ratio of clutter power into the canceler to the clutter power leaving the canceler and can be calculated by integrating the clutter power spectrum, $S_C(f)$, over the entire spectrum and dividing by the clutter power modified by the canceler response, $H(f)$, also over the whole spectrum. This is summarized by

$$CA = \frac{\textit{Clutter In}}{\textit{Clutter Out}} = \frac{\int_0^\infty S_C(f)df}{\int_0^\infty S_C(f)|H(f)|^2 df} \tag{7-13a}$$

A glance at Figure 7-13 shows that both the canceler response and the clutter spectrum are repetitive at the PRF. Therefore, the clutter attenuation over the whole spectrum can be found by examining the band from zero to f_r, and the upper limit of the integrals may

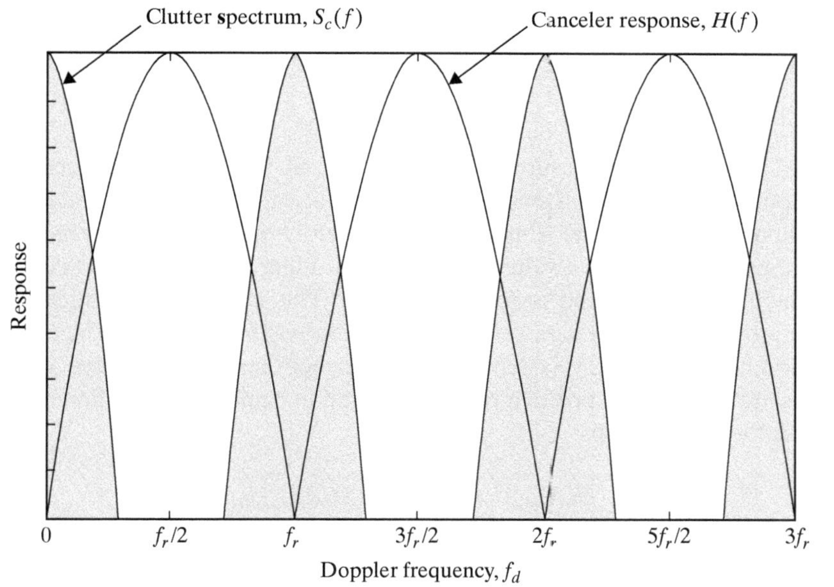

FIGURE 7-13 Canceler Response and Clutter Spectrum

be adjusted accordingly.

$$CA = \frac{Clutter\ In}{Clutter\ Out} = \frac{\int_0^{f_r} S_C(f)df}{\int_0^{f_r} S_C(f)|H(f)|^2 df} \tag{7-13b}$$

This integral is frequently impossible to solve analytically because $S_C(f)$ is generally unknown (and even if it is known, no analytical solution to the integrals in (7-13b) can be derived) and recourse is usually made to numerical approximation. Further analysis on the computation of the clutter attenuation is given by Nitzberg [5].

A useful measure of the effect of MTI on the radar's performance in clutter is given by the *improvement factor* (IF) of the MTI system. It is defined as the ratio of SCR at the output of the canceler to that at its input [5]. In practice, this ratio is a function of the target Doppler, so the improvement factor is the mean value averaged over all Doppler frequencies of interest. Thus,

$$IF = \frac{(SCR)_{out}}{(SCR)_{in}} = \frac{\left(\frac{S}{C}\right)_{out}}{\left(\frac{S}{C}\right)_{in}} \text{ averaged over all } f_d \tag{7-14}$$

where S and C represent the target signal power and clutter power, respectively.

The clutter attenuation and improvement factor are related by

$$IF = CA \times \overline{\left(\frac{S_{out}}{S_{in}}\right)} \tag{7-15}$$

where S_{out} is the desired signal on the canceler output, and S_{in} is the desired signal on the canceler input. The term (S_{out}/S_{in}) represents the target signal gain and has a value of 2 for single delay line cancelers and 6 for double delay line cancelers. Practical values for IF can be as high as 50–60 dB.

A certain SCR is required from the canceler output to allow target detection with the required probabilities of detection and false alarm. The precise value depends on the clutter statistics. However, assuming that the clutter has Gaussian statistics, the required SCR in a clutter-limited radar is the same as the required SNR in a noise-limited radar. The

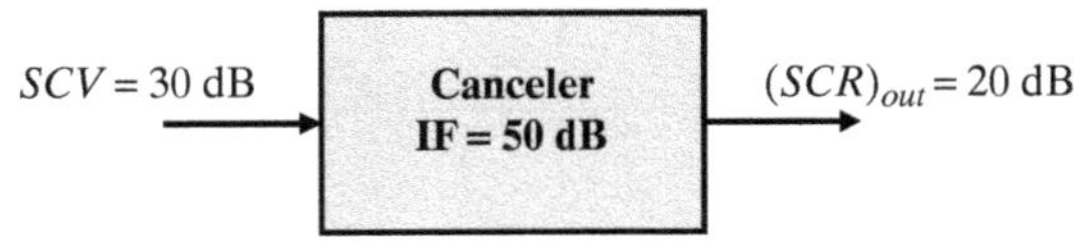

FIGURE 7-14 ■ Quantifying MTI Performance

IF of the MTI system is a measure of how much the SCR decreases between the output and the input of the canceler. Hence, it is possible to calculate the SCR required at the canceler input. In practice, the IF is usually sufficiently high to make the required input SCR less than unity (negative value in decibels), leading to the ability to detect targets that are much weaker than the surrounding clutter. This ability to dig targets out of the clutter is a property of the system known as *subclutter visibility* (SCV). The only point to bear in mind here is that SCV is defined as C/S rather than more intuitive S/C. This is done to ensure that SCV is positive when expressed in decibels. The following equation summarizes the relationships:

$$SCV_{(\mathrm{dB})} = IF_{(\mathrm{dB})} - (SCR)_{out\,(\mathrm{dB})} \tag{7-16}$$

Typical systems allow SCV in the region of 30 dB or more. Thus, targets can be 1/1000 the power of the clutter echo yet still be detected. An example of the relationships is illustrated in Figure 7-14.

Transversal filters using the binomial weighting factors are close to the optimum designs giving the maximum IF and maximum CA.

7.3 FFT PROCESSING

Previous chapters and sections have established the following principles:

- The baseband output of a pulse Doppler radar consists of a sequence of pulses that sample the Doppler cycle, as shown in Figure 4-7 or Figure 7-2.
- The sampling frequency is the radar PRF, f_r.
- The baseband pulses sample the Doppler cycle at regular intervals of phase, $\Delta\varphi$, given by $\Delta\varphi = \frac{4\pi V_r}{\lambda f_r}$ which is a repeat of equation (7-1).
- Target returns are range gated prior to Doppler filtering.
- A clutter notch filter such as an MTI canceler is typically used to reject stationary and slow-moving target returns prior to Doppler filtering. This processing stage is not strictly necessary and may be omitted if there are only modest clutter levels.
- Doppler discrimination using a Doppler filter bank (Figure 4-5) subdivides the Doppler band into a set of contiguous Doppler cells and, hence, a set of velocity cells (Figure 4-6). This may be implemented using transversal filters (Figure 7-10).
- Section 7.2.6 showed how a transversal filter produced a weighted sum of a set of baseband samples. The response of the filter resulted in a passband of bandwidth f_r/n, where n is the number of samples. Multiple filters differing in the phase weighting applied to each sample could be produced resulting in a set of Doppler filters (Figure 7-11).

The basic signal processing operation to yield a bank of Doppler filters is the DFT. Since the DFT operation must be repeated for every range cell on digitized data of 12-bit resolution or better and must keep up with the real-time throughput of data, the desire is to minimize the computational load. FFT algorithms originally evolved to reduce the number of processing operations, particularly multiplication operations, and thus enable the real-time Doppler processing necessary for modern systems. However, more recently the increases in computing power enable large FFTs to be computed in real time with little problem, although the advantage of the reduced number of computations is still an attractive feature.

7.3.1 Intuitive Approach to the DFT

The DFT is best explained by example. Consider a radar operating at a frequency of 10 GHz and a PRF, f_r, of 32 kHz. From (1-1), its wavelength, λ, is 0.03 m or 3 cm. If this radar detects an approaching target of radial velocity, V_r, of 40 m/s, then from (7-1) the phase difference between successive pulses, $\Delta\varphi$, is $\pi/6$ or 30°. From (4-2) or (7-2) the Doppler frequency is deduced to be 2667 Hz. Let us suppose that 32 pulses are received ($n = 32$); each pulse constitutes a sample of the baseband Doppler waveform. The acquisition of 32 samples takes 32 times the PRI, that is, $32 \times 1/32000 = 1$ ms. In the time between each pulse, then from (4-3) the range to the target reduces by $V_r/f_r = 40/32000 = 1.25$ mm, which is $\lambda/24$ and is consistent with a phase change over the two-way path of 30°. The phase of the pulse samples can be depicted on a phasor diagram as in Figure 7-15, which shows the first six samples with regular 30° phase increments in a clockwise direction. For convenience, the numbering starts at zero. The rotation of the phasor samples in a clockwise direction takes place at the Doppler frequency; the *clockwise* direction denotes a *positive* Doppler frequency. Sample 0 sits in the 12 o'clock position, that is, at a phase angle of 0°. Sample 1 sits at 30°, sample 2 at 60°, and so on. The phase of each sample is therefore given by $k \times 30°$, where $k = 0, 1, 2, \ldots (n-1)$ and is an index referring to the sample number within the sequence.

The processing backs off the phase of each sample by $k \times \Delta\varphi'$ and forms the vector sum, where $\Delta\varphi'$ is a trial value of a phase increment. Let us assume that the trial value of the phase increment equals the actual phase difference between samples, i.e., 30°. Sample 0 receives no back-off in phase and so still resides in the 12 o'clock position. Sample 1 is backed off by 30° and so now also coincides with the 12 o'clock position. Sample 2 is backed off by $2 \times 30° = 60°$ and so now also resides at the 12 o'clock position. Since the trial value of the phase increment equals the actual value all 32 samples are backed off to the 12 o'clock position. The vector sum is now equal to 32 times the amplitude of each individual sample. This process of imparting a phase back-off at multiples of a trial phase value followed by forming the vector sum is a test for a particular frequency. This operation therefore yields a filter, which in this case is tuned to 2667 Hz. Filters

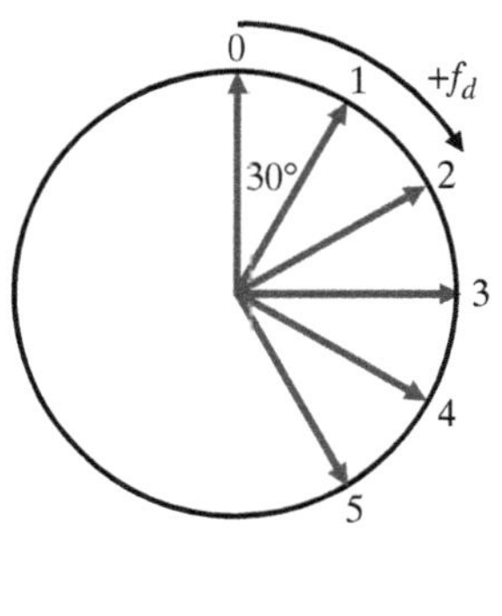

FIGURE 7-15 Phasor Representation of Samples

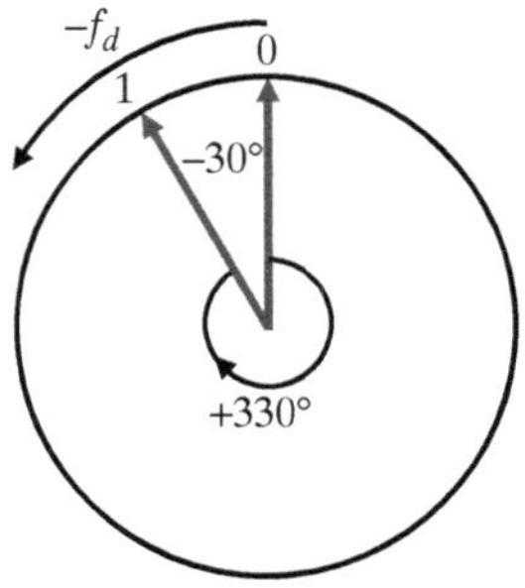

FIGURE 7-16 Negative Doppler Phasors

tuned to different frequencies can be produced by trying different values of the trial phase increment, $\Delta\varphi'$. Each different trial value of $\Delta\varphi'$ constitutes a test to see if the samples fit a particular frequency. Had the trial value of phase increment not been equal to the actual value (30°), then the samples would have been backed off to random positions around the phasor diagram, and their vector sum would have been a low value. Thus, there is a large output when samples fit the trial frequency and a small output when they do not. The combination of several trial values of phase increments yields a bank of Doppler filters.

Had a receding target of radial velocity $V_r = -40$ m/s been detected, the phasors would have exhibited a –30° increment, and the rotation of the phasors would have been in an anticlockwise direction, denoting a negative Doppler frequency. The test for a phase increment $\Delta\varphi = -30°$ is implemented by a trial value $\Delta\varphi' = +330°$, that is, $360° - 30°$, as illustrated in Figure 7-16. Thus, filters covering positive Doppler shifts can be implemented by trial values of phase between 0° and 180°, and filters covering negative Doppler shifts can be implemented by trial values of phase between 180° and 360°. The maximum unambiguous magnitude of the phase shift between samples, $|\Delta\varphi|_{mu} = 180°$ or π radians and occurs when the Doppler frequency is a half the PRF, that is, $f_d = f_r/2$. Thus, trial values of phase between 0° and 180° give rise to filters covering the Doppler band of 0 Hz to $+f_r/2$, whereas trial values of phase between 180° and 360° give rise to filters covering the Doppler band $-f_r/2$ to 0 Hz. The total width of the Doppler band is thus equal to the PRF since this is the sampling frequency. It is normal for the processing to apply trial phase values that are multiples of $360°/n$. In the general case this gives rise to n Doppler filters distributed over the Doppler band, with each filter having a bandwidth of f_r/n.

Since in this example $n = 32$, the trial phase values are multiples of 11.25°. This sets a phase resolution of 11.25°, which in turn leads to a Doppler resolution, from (7-2), of 1000 Hz. This is to be expected since the width of each Doppler filter is $f_r/n = 32,000/32 = 1000$ Hz. This is also consistent with the Doppler resolution being given by the inverse of the integration time, recalling (4-15) $\Delta f_d = \frac{1}{t_{int}}$. Using the numbers of this example $\Delta f_d = 1/1$ ms $= 1$ kHz. In this example, the first 16 filters cover the positive Doppler domain, and the last 16 cover the negative Doppler domain. The zero-th Doppler filter is centered at zero Doppler and may be blanked to reject MBC. The 16th filter is actually tuned to phase increments of $16 \times 11.25° = 180°$ or a Doppler frequency of +16 kHz. (By the same token, this filter also covers a Doppler frequency of −16 kHz.) The 17th filter is tuned to a phase increment of 191.25° (or −168.75°), which corresponds to a Doppler frequency of −15 kHz. In the example of the closing target at 40 m/s, the target is seen largely in the third Doppler filter, which is actually tuned to phase increments of $3 \times 11.25° = 33.75°$ or to a Doppler frequency of $3 \times 1000 = 3000$ Hz. This filter

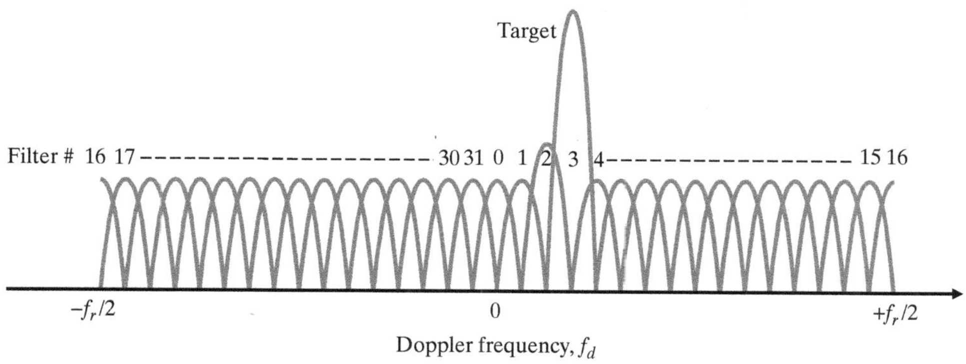

FIGURE 7-17 DFT Doppler Filtering

is closest to the actual values of $\Delta\varphi = 30^\circ$, $f_d = 2667$ Hz. However, since the actual target Doppler is below the center frequency of the third Doppler filter, it also resides on the upper edge of the second Doppler filter, so a small response in this filter is also seen. In the example of the receding target at -40 m/s the target would largely be seen in the 29th Doppler filter, which is actually tuned to phase increments of $29 \times 11.25^\circ = 326.25^\circ$ (equivalent to -33.75°) or to a Doppler frequency of -3000 Hz. Again, this filter falls closest to the actual values, although some low level response in the 30th filter would also be expected. The filter arrangement across the Doppler band is illustrated in Figure 7-17, together with the filter numbering and the response to the example target closing at 40 m/s.

One may notice that the target produces a large response in the third Doppler filter and a smaller response in the second Doppler filter because its actual Doppler shift is partway between the two. This is quite typical of most targets since it is unlikely that the actual target Doppler shift falls precisely at the center frequency of one filter. The system can refine its Doppler accuracy on the balance of power across neighboring filter responses. This yields a Doppler accuracy, and hence velocity *accuracy*, of about 1/10 its Doppler, or velocity, *resolution*. The straddling of a target across two or even three consecutive Doppler filters is analogous to the range cell straddling phenomena discussed in Section 3.8.1.

7.3.2 Analytical Approach to the DFT

The following expression for the baseband video output was derived in Section 7.2.2, equation (7-5):

$$V_V = A_3 \cos\left[(\pm\omega_d t) - 2\omega_{tx}\frac{R}{c}\right] \qquad \text{repeat of} \quad (7\text{-}5)$$

The amplitude term, A_3, and the phase shift term, $2\omega_{tx}R/c$, are somewhat arbitrary, so for convenience let $A_3 = 1$ and $2\omega_{tx}R/c = 0$. It was further noted in Section 7.2.2 that it is necessary to form quadrature channels, named the I and Q channels, which represent the real and imaginary parts of the complex baseband signal. Therefore, the complex baseband signal may be expressed as

$$= \cos(2\pi f_d t) + j\sin(2\pi f_d t)$$
$$= \exp(j2\pi f_d t)$$

This signal is sampled at times equal to multiples of the PRI. Hence, the samples of the complex baseband signal are given by

$$X(k) = \exp(j2\pi f_d kT_r) = \exp(j2\pi f_d k/f_r) \tag{7-17}$$

where $k = 0, 1, 2 \ldots (n-1)$ and is an index referring to the number of each sample in a sequence of n samples. Each sample is drawn from a consistent range cell over n consecutive receiving periods; in this way returns at a constant range are analyzed.

A filter is formed by taking the phase weighted sum of these samples. The phase weightings are given by

$$W^{f\cdot k} = \exp(-jfk2\pi/n) \tag{7-18}$$

where $f = 0, 1, 2 \ldots (n-1)$ and is an index referring to the number of each filter and n represents the total number of filters or the *point* size of the DFT. This weighting factor imparts a phase shift of $fk2\pi/n$ radians on the appropriate sample defined by k. The point size is often set equal to the number of samples, although strictly this need not necessarily be the case. The weighting factors, $W^{f\cdot k}$, are often referred to as the *twiddle factors*.

A bank of filters is formed by taking the weighted sum

$$S(f) = \sum_{k=0}^{n-1} X(k)\, W^{f\cdot k} \tag{7-19a}$$

For example for an eight-point DFT, $n = 8$, $k = 0, 1, 2, 3, 4, 5, 6, 7$, and $f = 0, 1, 2, 3, 4, 5, 6, 7$. The samples are represented by

$$\begin{aligned} X(k) = {} & \exp(0),\ \exp(j2\pi f_d/f_r),\ \exp(j4\pi f_d/f_r),\ \exp(j6\pi f_d/f_r), \\ & \exp(j8\pi f_d/f_r),\ \exp(j10\pi f_d/f_r),\ \exp(j12\pi f_d/f_r),\ \exp(j14\pi f_d/f_r) \end{aligned}$$

The case of the zero-th filter is not particularly interesting since when $f = 0$ all the twiddle factors reduce to unity, $W^{0\cdot k} = \exp(0) = 1$, implying a zero phase shift.
Consider the phase weights for filter 1 ($f = 1$):

$$\begin{aligned} W^{1\cdot k} = {} & \exp(0),\ \exp\left(-j\frac{2\pi}{8}\right),\ \exp\left(-j\frac{4\pi}{8}\right),\ \exp\left(-j\frac{6\pi}{8}\right),\ \exp\left(-j\frac{8\pi}{8}\right), \\ & \exp\left(-j\frac{10\pi}{8}\right),\ \exp\left(-j\frac{12\pi}{8}\right),\ \exp\left(-j\frac{14\pi}{8}\right) \\ W^{1\cdot k} = {} & 1,\ \exp\left(-j\frac{\pi}{4}\right),\ \exp\left(-j\frac{\pi}{2}\right),\ \exp\left(-j\frac{3\pi}{4}\right),\ \exp(-j\pi), \\ & \exp\left(-j\frac{5\pi}{4}\right),\ \exp\left(-j\frac{3\pi}{2}\right),\ \exp\left(-j\frac{7\pi}{4}\right) \end{aligned}$$

or alternatively

$$\begin{aligned} W^{1\cdot k} = {} & 1,\ \exp\left(-j\frac{\pi}{4}\right),\ \exp\left(-j\frac{\pi}{2}\right),\ \exp\left(-j\frac{3\pi}{4}\right),\ \exp(-j\pi), \\ & \exp\left(j\frac{3\pi}{4}\right),\ \exp\left(j\frac{\pi}{2}\right),\ \exp\left(j\frac{\pi}{4}\right) \end{aligned}$$

The weighted sum for filter 1 ($f = 1$) is given by

$$S(1) = \sum_{k=0}^{n-1} X(k) W^{1\cdot k}$$

A convenient way this operation can be depicted is to use matrix multiplication. A single filter can be formed by the multiplication of a row matrix whose elements are the time samples of the complex baseband signal with a column matrix whose elements are the phase weighting factors. The dimensions of the matrices must match so the number of time samples must equal the number of phase weights ($= n$). Thus, for filter 1 ($f = 1$), the operation becomes

$$S(1) = (X(0)\ X(1)\ X(2)\ X(3)\ X(4)\ X(5)\ X(6)\ X(7)) \begin{pmatrix} W^{1\cdot 0} \\ W^{1\cdot 1} \\ W^{1\cdot 2} \\ W^{1\cdot 3} \\ W^{1\cdot 4} \\ W^{1\cdot 5} \\ W^{1\cdot 6} \\ W^{1\cdot 7} \end{pmatrix}$$

Similarly, for filter 2 ($f = 2$), the phase weights are

$$\begin{aligned} W^{2\cdot k} &= \exp(0),\ \exp\left(-j\frac{4\pi}{8}\right),\ \exp\left(-j\frac{8\pi}{8}\right),\ \exp\left(-j\frac{12\pi}{8}\right),\ \exp\left(-j\frac{16\pi}{8}\right), \\ &\quad \exp\left(-j\frac{20\pi}{8}\right),\ \exp\left(-j\frac{24\pi}{8}\right),\ \exp\left(-j\frac{28\pi}{8}\right) \\ W^{2\cdot k} &= 1,\ \exp\left(-j\frac{\pi}{2}\right),\ \exp(-j\pi),\ \exp\left(-j\frac{3\pi}{2}\right),\ \exp(-j2\pi), \\ &\quad \exp\left(-j\frac{5\pi}{2}\right),\ \exp(-j3\pi),\ \exp\left(-j\frac{7\pi}{2}\right) \end{aligned}$$

and the operation to yield filter 2 becomes

$$S(2) = (X(0)\ X(1)\ X(2)\ X(3)\ X(4)\ X(5)\ X(6)\ X(7)) \begin{pmatrix} W^{2\cdot 0} \\ W^{2\cdot 1} \\ W^{2\cdot 2} \\ W^{2\cdot 3} \\ W^{2\cdot 4} \\ W^{2\cdot 5} \\ W^{2\cdot 6} \\ W^{2\cdot 7} \end{pmatrix}$$

Similarly, the phase weights and matrix operations to derive further filters may be derived.

To produce a bank of such filters, a row matrix whose elements are the time samples of the complex baseband signal must be multiplied with a rectangular matrix in which the elements of each column are the appropriate phase weightings for each filter. Each column of the rectangular matrix is populated with elements giving the appropriate phase weightings for a different filter. Therefore, the number of columns sets the number of filters and is known as the *DFT point size,* or simply the *DFT size*. Although it is not strictly necessary for the number of filters to match the number of samples, it is often desirable to set these equal to each other, in which case the matrix of phase weightings becomes square of dimensions $n \times n$. In such cases $f = 0, 1, 2, \ldots (n-1)$. The matrix operation now looks like

$$S(f) = (X(0), X(1), X(2), \ldots X(n-1)) \begin{pmatrix} W^{0\cdot 0} & W^{1\cdot 0} & W^{2\cdot 0} & \ldots & W^{(n-1)\cdot 0} \\ W^{0\cdot 1} & W^{1\cdot 1} & W^{2\cdot 1} & \ldots & W^{(n-1)\cdot 1} \\ W^{0\cdot 2} & W^{1\cdot 2} & W^{2\cdot 2} & \ldots & W^{(n-1)\cdot 2} \\ \vdots & \ldots & \ldots & \ldots & \vdots \\ W^{0\cdot(n-1)} & \ldots & \ldots & \ldots & W^{(n-1)\cdot(n-1)} \end{pmatrix} \tag{7-19b}$$

This yields a row matrix of n elements representing the response in the n Doppler filters, that is,

$$S(f) = (S(0),\ S(1),\ S(2), \ldots S(n-1))$$

This indicates how the n time samples are Fourier transformed into n Doppler frequency cells via the multiplication of the phase weighting matrix. This operation constitutes an n-point DFT and requires n^2 multiplications and n additions.

Matrix mathematics is a useful and powerful method of representing the transformation of data from one domain to another, in this case the Fourier transform of samples from the time domain to samples in the frequency domain.

7.3.3 The Fast Fourier Transform

The DFT requires n^2 multiplications and n additions and so represents a heavy computational burden. Methods are sought to reduce the number of computational steps. This is feasible since the high level of repetition in the phase weighting factors can be exploited. If one takes $W = \exp(-j2\pi/n)$, then from (7-18) the power to which W is raised is the product fk. The matrix transformation representing the phase weightings for the eight-point DFT can therefore be expressed as

$$[\mathbf{W}] = \begin{pmatrix} W^0 & W^0 & W^0 & W^0 & W^0 & W^0 & W^0 & W^0 \\ W^0 & W^1 & W^2 & W^3 & W^4 & W^5 & W^6 & W^7 \\ W^0 & W^2 & W^4 & W^6 & W^8 & W^{10} & W^{12} & W^{14} \\ W^0 & W^3 & W^6 & W^9 & W^{12} & W^{15} & W^{18} & W^{21} \\ W^0 & W^4 & W^8 & W^{12} & W^{16} & W^{20} & W^{24} & W^{28} \\ W^0 & W^5 & W^{10} & W^{15} & W^{20} & W^{25} & W^{30} & W^{35} \\ W^0 & W^6 & W^{12} & W^{18} & W^{24} & W^{30} & W^{36} & W^{42} \\ W^0 & W^7 & W^{14} & W^{21} & W^{28} & W^{35} & W^{42} & W^{49} \end{pmatrix}$$

Simplifications can be made since $W^0 = 1$. Furthermore, the index can be expressed as *modulo* (n) since when $fk = n$, the phase weighting of W^n imparts a 2π radian phase shift equivalent to zero phase shift. Hence, for the example of the eight-point DFT, $W^8 = W^0 (= 1)$, $W^9 = W^1$, $W^{10} = W^2$, etc. The matrix of phase weightings can now be simplified to

$$
[\mathbf{W}] = \begin{pmatrix}
1 & 1 & 1 & 1 & 1 & 1 & 1 & 1 \\
1 & W^1 & W^2 & W^3 & W^4 & W^5 & W^6 & W^7 \\
1 & W^2 & W^4 & W^6 & 1 & W^2 & W^4 & W^6 \\
1 & W^3 & W^6 & W^1 & W^4 & W^7 & W^2 & W^5 \\
1 & W^4 & 1 & W^4 & 1 & W^4 & 1 & W^4 \\
1 & W^5 & W^2 & W^7 & W^4 & W^1 & W^6 & W^3 \\
1 & W^6 & W^4 & W^2 & 1 & W^6 & W^4 & W^2 \\
1 & W^7 & W^6 & W^5 & W^4 & W^3 & W^2 & W^1
\end{pmatrix} \quad (7\text{-}20)
$$

The phase weightings imparted by the elements of the **[W]** matrix are, from (7-18):

W^0 : 0 radians (0°)

W^1 : $\pi/4$ radians (45°)

W^2 : $\pi/2$ radians (90°)

W^3 : $3\pi/4$ radians (135°)

W^4 : π radians (180°)

W^5 : $5\pi/4$ radians (225°)

W^6 : $3\pi/2$ radians (270°)

W^7 : $7\pi/4$ radians (315°)

However, it may be seen that the π radian (180°) phase shift from W^4 gives rise to an inversion of the signal; therefore, $W^4 = -W^0 = -1$. Furthermore, the $5\pi/4$ radian (225°) phase shift from W^5 can be replaced by a $-3\pi/4$ radian (−135°) phase shift, so $W^5 = W^{(4+1)} = W^4 \times W^1 = -W^1$. Similarly, the $3\pi/2$ radian (270°) phase shift from W^6 can be replaced by a $-\pi/2$ radian (−90°) phase shift, so $W^6 = W^4 \times W^2 = -W^2$. Last, the $7\pi/4$ radian (315°) phase shift from W^7 can be replaced by a $-\pi/4$ radian (−45°) phase shift, so $W^7 = W^4 \times W^3 = -W^3$.

Making these substitutions, (7-20) may therefore be rewritten as

$$
[\mathbf{W}] = \begin{pmatrix}
1 & 1 & 1 & 1 & 1 & 1 & 1 & 1 \\
1 & W^1 & W^2 & W^3 & -1 & -W^1 & -W^2 & -W^3 \\
1 & W^2 & -1 & -W^2 & 1 & W^2 & -1 & -W^2 \\
1 & W^3 & -W^2 & W^1 & -1 & -W^3 & W^2 & -W^1 \\
1 & -1 & 1 & -1 & 1 & -1 & 1 & -1 \\
1 & -W^1 & W^2 & -W^3 & -1 & W^1 & -W^2 & W^3 \\
1 & -W^2 & -1 & W^2 & 1 & -W^2 & -1 & W^2 \\
1 & -W^3 & -W^2 & -W^1 & -1 & W^3 & W^2 & W^1
\end{pmatrix} \quad (7\text{-}21)
$$

Thus, the whole operation is represented by (7-22):

$$S(f) = X(k)[\mathbf{W}] \quad (7\text{-}22a)$$

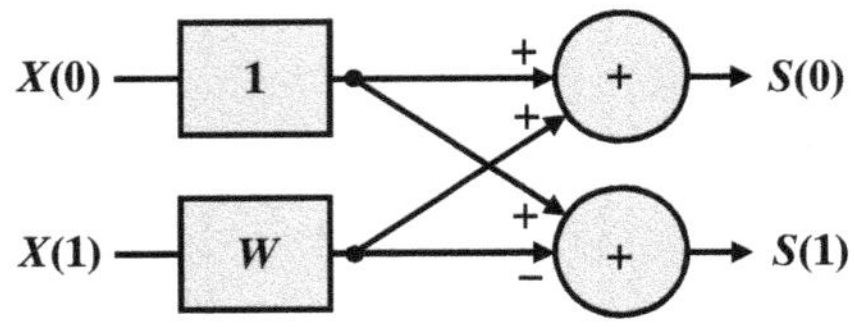

FIGURE 7-18 Basic Butterfly Operation

Since the matrix **[*W*]** is *symmetric*, its rows and columns may be interchanged with no effect on the outcome. Therefore, it is somewhat more convenient to expand $S(f)$ and $X(k)$ as column matrices, so writing this out in full results in

$$\begin{pmatrix} S(0) \\ S(1) \\ S(2) \\ S(3) \\ S(4) \\ S(5) \\ S(6) \\ S(7) \end{pmatrix} = \begin{pmatrix} 1 & 1 & 1 & 1 & 1 & 1 & 1 & 1 \\ 1 & W^1 & W^2 & W^3 & -1 & -W^1 & -W^2 & -W^3 \\ 1 & W^2 & -1 & -W^2 & 1 & W^2 & -1 & -W^2 \\ 1 & W^3 & -W^2 & W^1 & -1 & -W^3 & W^2 & -W^1 \\ 1 & -1 & 1 & -1 & 1 & -1 & 1 & -1 \\ 1 & -W^1 & W^2 & -W^3 & -1 & W^1 & -W^2 & W^3 \\ 1 & -W^2 & -1 & W^2 & 1 & -W^2 & -1 & W^2 \\ 1 & -W^3 & -W^2 & -W^1 & -1 & W^3 & W^2 & W^1 \end{pmatrix} \begin{pmatrix} X(0) \\ X(1) \\ X(2) \\ X(3) \\ X(4) \\ X(5) \\ X(6) \\ X(7) \end{pmatrix} \tag{7-22b}$$

A useful method to illustrate the matrix multiplication of (7-22) is the *Butterfly diagram*. The basic operation of a butterfly diagram is illustrated in Figure 7-18. Two time samples, $X(0)$ and $X(1)$, are inputted on the left. The two samples are weighted: the first by unity (this imparts no change in its amplitude or phase); and the second by a phase shift set by the twiddle factor, W. The sum and difference of the weighted samples is then derived to form the two outputs: $S(0)$ and $S(1)$. This is an example of a two-point FFT operation.

The basic butterfly operation can be extended and repeated to handle 4, 8, 16... inputs and outputs. In fact, an FFT of point size equaling any power of two can be realized using repeated stages of the butterfly operation. This process known as *decimation in time*. The example illustrated in the matrix operations of (7-22) is that of an eight-point FFT. Since $8 = 2^3$, the butterfly diagram must be repeated over three stages. The butterfly representation of an eight-point FFT is illustrated in Figure 7-19.

The inputs to stage one are the time samples. The stage one computations comprise four two-point FFT operations and must be completed first; its outputs form the inputs to the second stage. The second stage computations comprise two four-point FFT operations and must next be completed and its outputs passed to the third stage. The outputs of the third stage are the frequency samples. Larger FFTs require additional stages; each stage must be completed in its turn with the outputs of a preceding stage forming the inputs to the subsequent stage. The computation can be implemented on FFT point sizes only equal to a power of two and is known as *Radix*-2, that is, $n = 2^x$, where $x = 1, 2, 3, \ldots$ (a positive integer), and the processing requires x stages of computation. In this way an FFT is implemented using $n \log_2(n)$ multiplication and addition operations. In the case of the eight-point FFT illustrated in Figure 7-19, one requires 24 multiplications and 24 additions. This is far fewer than the n^2 multiplications and n additions required for the DFT (= 64 multiplications and 8 additions, in the case of the eight-point DFT) and represents a considerable reduction of computational effort and therefore enables a comparable increase in computational speed, particularly for large FFT point sizes.

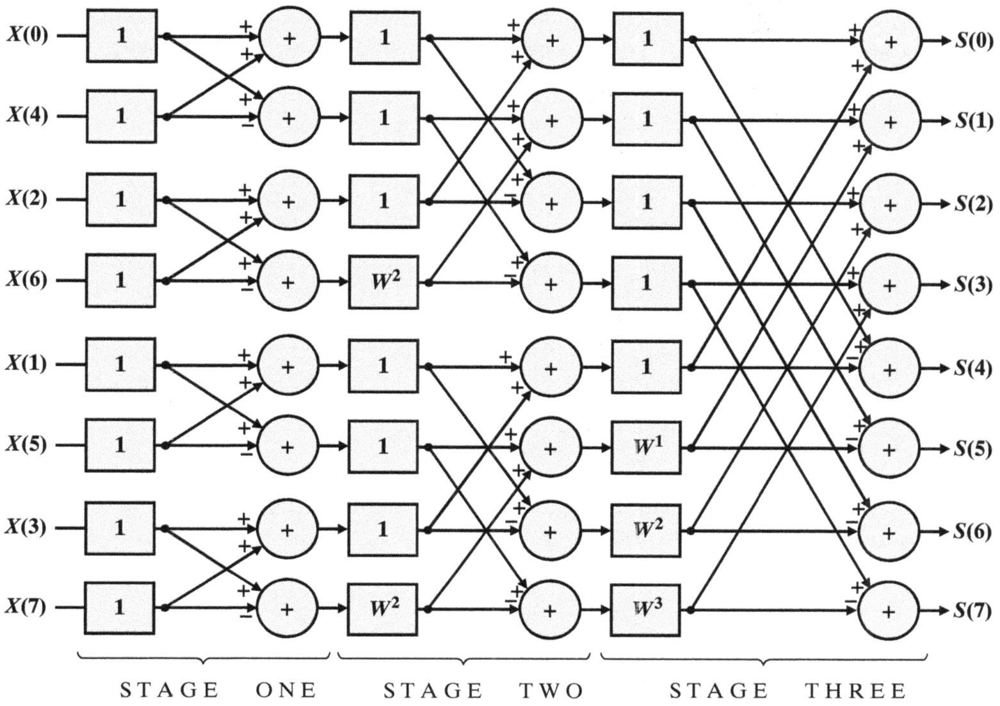

FIGURE 7-19 Decimation in Time (Butterfly) Operation for an Eight-Point FFT

In spite of apparent simplicity of the transform given in (7-22b), the transform matrix **[*W*]** may be simplified yet further using the *Cooley–Tukey* algorithm [17]. In general, the transform matrix **[*W*]** has dimensions of $n \times n$ and is populated by non-zero elements. This algorithm simplifies **[*W*]** by breaking it down into a number of simpler matrices that are diagonal, or very nearly so. Other simplification algorithms are available, and there are variants of the original 1965 Cooley–Tukey algorithm; however, this algorithm paved the way for rapid digital signal processing such as the FFT processing of returns in pulse Doppler radar.

The Cooley–Tukey algorithm is best illustrated by presenting the process for the eight-point FFT operation of (7-22b). First, the matrix transformation is divided into the odd- and even-numbered samples.

$$S(f) = \begin{bmatrix} S(0) \\ S(1) \\ S(2) \\ S(3) \\ S(4) \\ S(5) \\ S(6) \\ S(7) \end{bmatrix} = \begin{bmatrix} 1 & 1 & 1 & 1 \\ 1 & W^2 & -1 & -W^2 \\ 1 & -1 & 1 & -1 \\ 1 & -W^2 & -1 & W^2 \\ 1 & 1 & 1 & 1 \\ 1 & W^2 & -1 & -W^2 \\ 1 & -1 & 1 & -1 \\ 1 & -W^2 & -1 & W^2 \end{bmatrix} \begin{bmatrix} X(0) \\ X(2) \\ X(4) \\ X(6) \end{bmatrix} + \begin{bmatrix} 1 & 1 & 1 & 1 \\ W^1 & W^3 & -W^1 & -W^3 \\ W^2 & -W^2 & W^2 & -W^2 \\ W^3 & W^1 & -W^3 & -W^1 \\ -1 & -1 & -1 & -1 \\ -W^1 & -W^3 & W^1 & W^3 \\ -W^2 & W^2 & -W^2 & W^2 \\ -W^3 & -W^1 & W^3 & W^1 \end{bmatrix} \begin{bmatrix} X(1) \\ X(3) \\ X(5) \\ X(7) \end{bmatrix}$$

This results in the sum of two shorter FFTs: one processing even-numbered samples; and the other processing odd-numbered samples.

Next, the order of some of the frequency cells and the appropriate rows of the transform matrices are swapped around. This does not affect the outcome; it merely rearranges the order of the frequency cells.

$$\begin{bmatrix} S(0) \\ S(2) \\ S(1) \\ S(3) \\ S(4) \\ S(6) \\ S(5) \\ S(7) \end{bmatrix} = \begin{bmatrix} 1 & 1 & 1 & 1 \\ 1 & -1 & 1 & -1 \\ 1 & W^2 & -1 & -W^2 \\ 1 & -W^2 & -1 & W^2 \\ 1 & 1 & 1 & 1 \\ 1 & -1 & 1 & -1 \\ 1 & W^2 & -1 & -W^2 \\ 1 & -W^2 & -1 & W^2 \end{bmatrix} \begin{bmatrix} X(0) \\ X(2) \\ X(4) \\ X(6) \end{bmatrix} + \begin{bmatrix} 1 & 1 & 1 & 1 \\ W^2 & -W^2 & W^2 & -W^2 \\ W^1 & W^3 & -W^1 & -W^3 \\ W^3 & W^1 & -W^3 & -W^1 \\ -1 & -1 & -1 & -1 \\ -W^2 & W^2 & -W^2 & W^2 \\ -W^1 & -W^3 & W^1 & W^3 \\ -W^3 & -W^1 & W^3 & W^1 \end{bmatrix} \begin{bmatrix} X(1) \\ X(3) \\ X(5) \\ X(7) \end{bmatrix}$$

Consider the transform matrix for the odd-numbered samples. This matrix can be factorized as follows:

$$\begin{bmatrix} 1 & 1 & 1 & 1 \\ W^2 & -W^2 & W^2 & -W^2 \\ W^1 & W^3 & -W^1 & -W^3 \\ W^3 & W^1 & -W^3 & -W^1 \\ -1 & -1 & -1 & -1 \\ -W^2 & W^2 & -W^2 & W^2 \\ -W^1 & -W^3 & W^1 & W^3 \\ -W^3 & -W^1 & W^3 & W^1 \end{bmatrix} = \begin{bmatrix} 1 & 0 & 0 & 0 & 0 & 0 & 0 & 0 \\ 0 & W^2 & 0 & 0 & 0 & 0 & 0 & 0 \\ 0 & 0 & W^1 & 0 & 0 & 0 & 0 & 0 \\ 0 & 0 & 0 & W^3 & 0 & 0 & 0 & 0 \\ 0 & 0 & 0 & 0 & 1 & 0 & 0 & 0 \\ 0 & 0 & 0 & 0 & 0 & W^2 & 0 & 0 \\ 0 & 0 & 0 & 0 & 0 & 0 & W^1 & 0 \\ 0 & 0 & 0 & 0 & 0 & 0 & 0 & W^3 \end{bmatrix} \begin{bmatrix} 1 & 1 & 1 & 1 \\ 1 & -1 & 1 & -1 \\ 1 & W^2 & -1 & -W^2 \\ 1 & -W^2 & -1 & W^2 \\ -1 & -1 & -1 & -1 \\ -1 & 1 & -1 & 1 \\ -1 & -W^2 & 1 & W^2 \\ -1 & W^2 & 1 & -W^2 \end{bmatrix}$$

For convenience let

$$\boldsymbol{A} = \begin{bmatrix} 1 & 1 & 1 & 1 \\ 1 & -1 & 1 & -1 \\ 1 & W^2 & -1 & -W^2 \\ 1 & -W^2 & -1 & W^2 \end{bmatrix}, \boldsymbol{B} = \begin{bmatrix} 1 & 1 & 1 & 1 \\ W^2 & -W^2 & W^2 & -W^2 \\ W^1 & W^3 & -W^1 & -W^3 \\ W^3 & W^1 & -W^3 & -W^1 \end{bmatrix} \text{ and } \boldsymbol{D} = \begin{bmatrix} 1 & 0 & 0 & 0 \\ 0 & W^2 & 0 & 0 \\ 0 & 0 & W^1 & 0 \\ 0 & 0 & 0 & W^3 \end{bmatrix}$$

The whole operation can now be represented as

$$S(f) = \begin{bmatrix} \boldsymbol{A} \\ \boldsymbol{A} \end{bmatrix} X(even) + \begin{bmatrix} \boldsymbol{B} \\ -\boldsymbol{B} \end{bmatrix} X(odd)$$

Furthermore, due to the previous factorization, this can be written as

$$S(f) = \begin{bmatrix} \boldsymbol{A} \\ \boldsymbol{A} \end{bmatrix} X(even) \quad + \quad \begin{bmatrix} \boldsymbol{D} & \boldsymbol{0} \\ \boldsymbol{0} & \boldsymbol{D} \end{bmatrix} \begin{bmatrix} \boldsymbol{A} \\ -\boldsymbol{A} \end{bmatrix} X(odd)$$

$$S(f) = \begin{bmatrix} \boldsymbol{A} \\ \boldsymbol{A} \end{bmatrix} X(even) \quad + \quad \begin{bmatrix} \boldsymbol{DA} \\ -\boldsymbol{DA} \end{bmatrix} X(odd)$$

where

$$0 = \begin{bmatrix} 0 & 0 & 0 & 0 \\ 0 & 0 & 0 & 0 \\ 0 & 0 & 0 & 0 \\ 0 & 0 & 0 & 0 \end{bmatrix}$$

is the 4 × 4 zero matrix.

This indicates that the $\boldsymbol{A}$ matrix is common to the processing of both odd- and even-numbered samples. The matrix $\boldsymbol{D}$ is diagonal and so cannot be simplified any further. However, the following simplification can be made to $\boldsymbol{A}$.

The $\boldsymbol{A}$ matrix can be subdivided into four 2 × 2 matrices.

$$\boldsymbol{A} = \left[\begin{array}{cc:cc} 1 & 1 & 1 & 1 \\ 1 & -1 & 1 & -1 \\ \hdashline 1 & W^2 & -1 & -W^2 \\ 1 & -W^2 & -1 & W^2 \end{array}\right]$$

Alternatively,

$$\boldsymbol{A} = \begin{bmatrix} \boldsymbol{C} & \boldsymbol{C} \\ \boldsymbol{E} & -\boldsymbol{E} \end{bmatrix}, \text{ where } \boldsymbol{C} = \begin{bmatrix} 1 & 1 \\ 1 & -1 \end{bmatrix} \text{ and } \boldsymbol{E} = \begin{bmatrix} 1 & W^2 \\ 1 & -W^2 \end{bmatrix}$$

However, matrix $\boldsymbol{E}$ factorizes as

$$\boldsymbol{E} = \begin{bmatrix} 1 & W^2 \\ 1 & -W^2 \end{bmatrix} = \begin{bmatrix} 1 & 1 \\ 1 & -1 \end{bmatrix} \begin{bmatrix} 1 & 0 \\ 0 & W^2 \end{bmatrix}$$

$$\boldsymbol{E} = \boldsymbol{CF}$$

where

$$\boldsymbol{F} = \begin{bmatrix} 1 & 0 \\ 0 & W^2 \end{bmatrix}$$

Thus, the $\boldsymbol{A}$ matrix can be written as

$$\boldsymbol{A} = \begin{bmatrix} \boldsymbol{C} & \boldsymbol{C} \\ \boldsymbol{E} & -\boldsymbol{E} \end{bmatrix} = \begin{bmatrix} \boldsymbol{C} & \boldsymbol{C} \\ \boldsymbol{CF} & -\boldsymbol{CF} \end{bmatrix}$$

which further simplifies to

$$\boldsymbol{A} = \begin{bmatrix} C & 0 \\ 0 & C \end{bmatrix} \begin{bmatrix} I & 0 \\ 0 & F \end{bmatrix} \begin{bmatrix} I & I \\ I & -I \end{bmatrix}$$

where $0 = \begin{bmatrix} 0 & 0 \\ 0 & 0 \end{bmatrix}$, the 2 × 2 zero matrix, and $\boldsymbol{I} = \begin{bmatrix} 1 & 0 \\ 0 & 1 \end{bmatrix}$, the 2 × 2 identity matrix.

The $\boldsymbol{C}$ matrix is of dimension 2 × 2 and cannot be simplified any further. Similarly, the F matrix is diagonal and cannot be simplified any further.

While the matrix operations representing the transform may now appear to be highly complex, all the matrices are diagonal or very nearly diagonal. The simplification of the transform leads to an increase in computational speed.

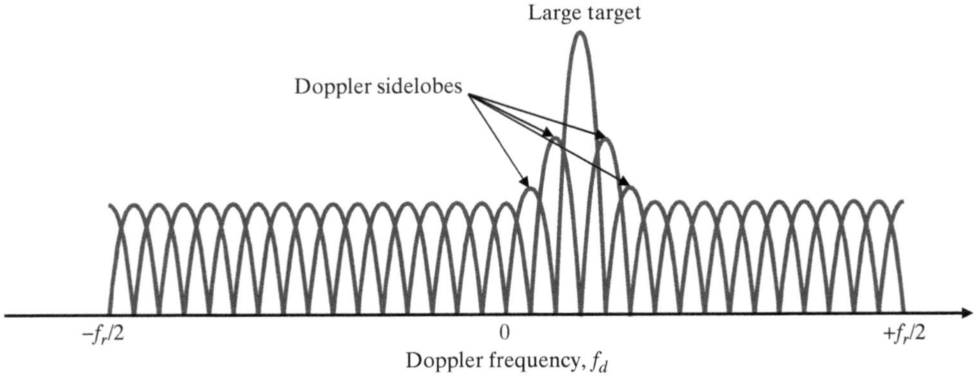

FIGURE 7-20 Doppler Sidelobes in a Uniformly Weighted FFT

7.3.4 The Windowed Discrete Fourier Transform

DFT processing in which every sample is equally amplitude weighted results in a sinc function response in the frequency domain having maximum Doppler sidelobes 13.2 dB below the main peak. Thus, the sidelobes associated with a large target response fall in neighboring Doppler cells and may mask smaller targets in these cells or even trigger the indication of a false target. The problem is illustrated in Figure 7-20. It is therefore desirable to reduce the sidelobes by applying a tapered weighting to the time samples in what is known as a *windowed DFT*. The weighting not only reduces sidelobes, however, but also reduces the peak response, broadens the Doppler filter slightly (i.e., a slight loss of Doppler/velocity resolution), and causes a small increase in the noise bandwidth. Thus, the matrix transformation used in DFT processing must impart amplitude as well as phase weightings to the samples. These effects are summarized for several of the more popular window functions used in DFT Doppler processing in Table 7-1. The amplitude weighting function, $w(k)$, is a function of the sample number defined by k.

When DFT processing is used, the noise bandwidth in which detections are made is the width of each Doppler cell. For a uniform DFT window function this results in $B_N = f_r/n$. However, the DFT process entails the coherent integration of n samples. Noise in these samples is uncorrelated and is subject to an integration (power) gain of n. Therefore, the total noise power in each DFT cell is equivalent to that obtained from a noise bandwidth of f_r. Other DFT window functions increase the noise bandwidth by the factors given in the right-hand column of Table 7-1.

Weighted or *windowed* DFTs have several applications in radar:

1. The far field radiation pattern of an antenna is given by the two-dimensional Fourier transform of its aperture illumination function. The far field pattern is characterized by a main beam and sidelobes. The sidelobes are an undesirable feature since they are a source of sidelobe clutter, among several other problems. Sidelobes may be minimized by using a tapered aperture illumination function, however, this is accompanied by a slight loss in main beam gain and an increase in the beamwidth of the main beam, which influence the magnitude and Doppler spread of main beam clutter, respectively. This will be considered further in Chapter 14.
2. The spectrum of a pulsed radar is given by the Fourier series expansion of the pulsed waveform. The suppression of sidebands over a wide bandwidth is often necessary for spectral containment and mutual interference concerns. In Section 3.5.2, the point

TABLE 7-1 ■ Summary of Common DFT Window Functions [18]

Name (Alterative Names)	Function	Peak Response	Maximum Sideband Level wrt to peak (dB)	Increase in Noise Bandwidth, B_N
Uniform	$w(k) = 1$ for $0 \le k \le n$	1.00	−13.2	1.00
Hamming (raised cosine)	$w(k) = 0.54 - 0.46\cos\left(\frac{2\pi k}{n-1}\right)$ for $0 \le k \le n$	0.53	−41.8	1.37
Hann (Hanning, raised cosine)	$w(k) = 0.5 - 0.5\cos\left(\frac{2\pi k}{n-1}\right)$ for $0 \le k \le n$	0.48	−31.5	1.50
Bartlett (triangular)	$w(k) = \frac{2k}{n-1}$, for $0 \le k \le (n/2) - 1$ $w(k) = 2 - \frac{2k}{n-1}$, for $n/2 \le k \le n$	0.48	−26.5	1.33
Blackman	$w(k) = 0.42 - 0.5\cos\left(\frac{2\pi k}{n-1}\right) + 0.08\cos\left(\frac{4\pi k}{n-1}\right)$ for $0 \le k \le n$	0.41	−58.1	1.73
Blackman-Harris	$w(k) = 0.35875 - 0.48829\cos\left(\frac{2\pi k}{n-1}\right) + 0.14128\cos\left(\frac{4\pi k}{n-1}\right) - 0.01168\cos\left(\frac{6\pi k}{n-1}\right)$ for $0 \le k \le n$	0.35	−92.0	2.01
Cosine (sine)	$w(k) = \cos\left(\frac{k\pi}{n-1} - \frac{\pi}{2}\right)$ for $0 \le k \le n$	0.62	−22.9	1.24
Gauss	$w(k) = \exp\left[-\frac{1}{2}\left(\frac{k-(n-1)/2}{\sigma(n-1)/2}\right)^2\right]$ $\sigma \le 0.5$ (here $\sigma = 0.4$), $0 \le k \le n$	0.48	−44.9	1.45 ($\sigma = 0.4$)
Lanczos (sinc)	$w(k) = \text{sinc}\left(\frac{2k}{n-1} - 1\right)$ for $0 \le k \le n$	0.57	−26.3	1.31

was made that this can be accomplished if the pulse rise and falls times are extended by shaping the profile of the pulses using a tapered function. This results in a slight loss of peak transmitted power and an increase in the bandwidth of the central spectral lobe (Figure 3-7). Matched filter reception now requires a somewhat different filter response; otherwise, a degree of mismatch is incurred that further degrades the detection performance.

3. The response of a matched filter in performing pulse compression is characterized by the inclusion of unwanted sidelobes in the time, or range, domain. These can be suppressed by introducing a degree of mismatch in the filter response, as described in

Section 3.9.6. The mismatch can be implemented by applying a weighting function to the frequency response of the filter. This has the side effect of reducing the peak response of the compressor, which leads to a reduction in detection performance and extending the duration of the compressed output and thus degrading the range resolution.

Points 1 and 3 are very similar. They both indicate that, in general, the reception of target returns incurs sidelobes in the range domain as well as the velocity domain. The sidelobe response of a radar waveform in both range and velocity domains is described by its ambiguity function (Chapter 5).

Each application tends to favor certain windowing functions; the functions given in Table 7-1 are popular in DFT pulse Doppler processing but are not commonly used in the other applications previously described.

7.3.5 FFT Processing Losses

Under ideal circumstances, FFT processing entails the coherent processing of n returns. The best signal processing gain expected is therefore a factor of n. However, in practice, FFT processing yields a somewhat lower processing gain. The factor by which the actual processing gain falls below the maximum is termed the *processing loss*. FFT processing losses arise for the following reasons:

- Doppler cell straddling
- Filter shape losses
- Windowed DFT losses
- Broadened equivalent noise bandwidth

Doppler cell straddling occurs when the Doppler shift of a target falls partway between the center frequency of two Doppler cells (discussed in Section 7.3.1 and illustrated in Figure 7-17). This leads to the target echo falling toward the edge of the passband of two consecutive Doppler cells. The target echo power is now smeared over the two Doppler cells; no individual cell receives the full target power. The reduced power in any individual cell gives rise to a straddling loss.

Filter shape loss is akin to the Doppler cell straddling in that a target response that does not coincide with the peak of a filter results in a reduced output from that filter. The lower response of a filter away from its peak value represents the filter shape loss.

The widow functions described in Section 7.3.4 result in a reduced peak response, as detailed in Table 7-1. This constitutes the windowed DFT loss.

Since the main lobe response is wider when a windowing function is applied, slightly more noise is admitted into each Doppler cell. The increase in the equivalent noise bandwidth, B_N, is given in Table 7-1. SNR takes a double hit when a windowing function is applied; the reduced main response leads to a reduction in target signal strength, S, and the increase in B_N admits more noise, N.

Total FFT processing losses of between 2 dB and 4 dB are usually incurred. A loss of around 3 dB is typical, that is, $E_i(n) = 0.5$ and $L_i(n) = 2$, which leads to $I_i(n) = n/2$.

7.3.6 Zero Padding and DFT Size

Previously, the point was made that the number of time samples is often set equal to the DFT size, which in turn is set to a power of two. Sometimes, a given PRF and coherent processing interval (CPI) duration do not yield exactly 2^x time samples. In these circumstances, the set of time samples can be made up to the next power of two through the inclusion of additional samples that are all set to zero. This is known as *zero padding*. The speed and convenience of FFT processing using 2^x samples can now be enjoyed. When zero padding is used, the integration gain is set by the number of non-zero samples, but the DFT size is determined by the total number of samples, which would be somewhat larger than the number of non-zero time samples.

While many radars match the number of time samples to the DFT size, this need not be the case, and, indeed, modern radar systems often set these parameters to be different. It is often desirable to adjust the number of time samples to fix a common CPI duration for all CPIs since this results in a common Doppler resolution. Advances in modern computing power no longer demand radix-2 operations, and data may be processed in real-time using any arbitrary DFT size (below a certain maximum).

7.4 | CONSTANT FALSE ALARM RATE DETECTION

7.4.1 The Limitations of Fixed Threshold Detection

Section 2.2 described a simple thresholding system to declare target detections in the presence of thermal noise. This assumed the noise-limited detection case in which only white Gaussian noise is present at the receiver input. Narrow-band filtering within the receiver distorts the noise voltage envelope to a Rayleigh probability density function (PDF). This noise voltage envelope, plus any target signals, is then compared with a fixed threshold voltage to decide whether a target is present. A noise spike could also exceed the threshold and trigger a false alarm. The "target present/absent" decision is made in each range cell and in each velocity cell and at each beam position (angular cell). The false alarm rate (FAR) is a function of the relative levels of the threshold voltage and noise voltage envelope in each range/velocity/angle cell. In the circumstances previously described this relationship is

$$T_{FA} = \frac{1}{FAR} = \frac{1}{B_N} \exp\left(\frac{v_T^2}{2v_N^2}\right) \tag{7-23}$$

in which T_{FA} is the mean time between false alarms, v_T is the threshold voltage, v_N is the root mean square (RMS) noise voltage, and B_N is the noise bandwidth. The FAR relates to P_{FA} by equation (2-8). Furthermore, for a given SNR, P_D may be deduced, as described in Section 2.2.1 or by equation (2-31) in Section 2.5.5. So for known, fixed values of v_T and v_N, the radar detection performance can be completely described.

The $\left(\frac{v_T^2}{2v_N^2}\right)$ term in (7-23) is known as the *threshold-to-noise ratio (TNR)* and is a power ratio. The false alarm rate is very sensitive to TNR. The threshold level is scaled from the noise level

$$v_T^2 = k_1\left(2v_N^2\right) \tag{7-24}$$

where k_1 is a multiplier that sets the TNR. The choice of the k_1 multiplier therefore sets P_{FA} and P_D.

The use of a fixed threshold voltage is all very well and good when the noise statistics (e.g., PDF, mean levels) remain fixed over all range, velocity, and angular cells. This is not likely to be the case for noise jamming, other noise-like forms of interference and clutter. In particular, noise jamming tends to vary over different angles of reception, and, as can be seen from Figure 6-17, clutter varies considerably over both range and velocity axes; it is also angle dependent. Furthermore, clutter statistics may exhibit spatial variations such as a ramp transition from a region of calm sea through regions of progressively rougher seas or the abrupt transition from land to sea clutter that would be encountered along a coastline. In such circumstances a fixed threshold would lead to considerable variations in P_{FA} (and P_D) with range, velocity, and angle. Should a threshold voltage set on the basis of the worst-case conditions be used, it would undoubtedly be too high for the majority of the time, and this would lead to degraded P_D over much of the detection space of the radar.

7.4.2 The CFAR Concept

If the TNR, $k_1 = \left(\frac{v_T^2}{2v_N^2}\right)$, can be held constant, then the FAR is also constant. Thus, if v_N were to change, the threshold, v_T, must vary in sympathy with it. This yields a constant false alarm rate (CFAR) function but only in a very limited sense because here it is assumed that the input is thermal noise in which only its RMS level varies; all other statistical quantities, notably its PDF, remain the same. This is seldom the case with clutter whose statistics can vary considerably from one terrain type to another (see Section 6.3). Nevertheless, this does establish the important principle of requiring an adaptive threshold.

To obtain a true CFAR function, it is necessary to adapt the threshold to the local noise-plus-interference-plus-clutter statistics. For each cell in which a "target present" decision is made, a CFAR circuit must decide whether the voltage in that test cell exceeds the voltages in neighboring cells by some suitable margin. Here it is assumed that the noise/interference/clutter statistics in the neighboring cells is homogenous and is representative of the test cell. The problems of this approach are (1) what is the validity of the assumption of homogeneity in the statistics between the test cell and neighboring cells, (2) how many neighboring cells should be used to judge what those statistics might be, and (3) given the limits on how well the statistics may be estimated from the cells within the CFAR circuit what is a suitable margin to apply in the "target present" decision-making process?

A CFAR circuit may be implemented across range cells, velocity cells, angular cells, or combinations of these. The size of the circuit (i.e., how many cells it includes and the time it takes to update the data in these cells) determines the circuit's transient and steady-state responses to clutter/interference and therefore the radar detection performance. A large variety of CFAR circuits are in common usage: some set the threshold on the basis of the statistics of data in its cells (e.g., the product of a constant and the mean), some set the threshold on the basis of a logical function, and some take an ordered statistics approach. Hybrid methods combine two such approaches. Each method has its relative merits, and limitations and may be well-suited to certain types of clutter/interference variations and less well-suited to others.

7.4.3 CFAR Loss

Practical CFAR circuits provide an estimate of the noise/interference/clutter statistics on the basis of a limited sample set whose homogeneity is unknown [5]. The ideal CFAR circuit would provide an accurate estimate of these statistics, but this can be attained only using an infinite number of samples of homogenous data. This would ultimately lead to the situation of the fixed threshold detector working with a fixed level of thermal noise, as described earlier. This "ideal" situation enables a threshold to be set on the basis of the desired P_{FA}, which, in turn, dictates the P_D for a given SNR. However, in practice, the "ideal" CFAR detector cannot be built. The noise/interference/clutter statistics are estimated from a number of samples within reference cells of the CFAR circuit, and a threshold is derived on the basis of this estimate. However, since there is an error on the estimate, the threshold has to be set somewhat higher to allow a safety margin in recognition of the error on the estimate. As the number of cells included within the CFAR circuit increases it becomes less likely that the data within it are homogenous. If (and it may be a big *if*) the data are homogenous, the estimate is more accurate. Fewer cells are more likely to capture homogenous data, but too few cells leads to an inaccurate estimate of the local conditions and an inappropriate threshold to be set. Either P_{FA} or P_D suffers as a result of an inappropriate threshold. This leads to the concept of *CFAR loss* [5].

Recall (2-31), which gives the SNR requirement for a desired combination of P_D and P_{FA} for a Swerling case 1 and 2 type target detected in thermal noise alone:

$$SNR = \frac{\log P_{FA}}{\log P_D} - 1 \qquad \text{repeat of (2-31)}$$

This corresponds to the ideal case of CFAR detection described already. Since a practical CFAR circuit inevitably leads to degradations in P_D or P_{FA}, these degradations may be recovered with a superior signal-to-noise ratio. A better signal-to-noise ratio, denoted SNR_M, where M is the number of sample cells, is now required in a practical CFAR detector to yield the same P_D and P_{FA} performance as that of the ideal case (e.g., thermal noise only). The ratio of SNR_M now required and the SNR given by (2-31) is the CFAR loss.

7.4.4 Cell Averaging CFAR (CACFAR)

Cell averaging CFAR (CACFAR) is most commonly implemented in the range dimension in which a threshold appropriate for each range cell is derived as the scaled mean of the powers in range cells immediately before and after the test cell. It is normal to blank the range cell immediately before and after the test cell since a target in the test cell could straddle neighboring cells and corrupt the estimate of the mean. A schematic of the CACFAR circuit is illustrated in Figure 7-21. The circuit acts on the detected signals. Here

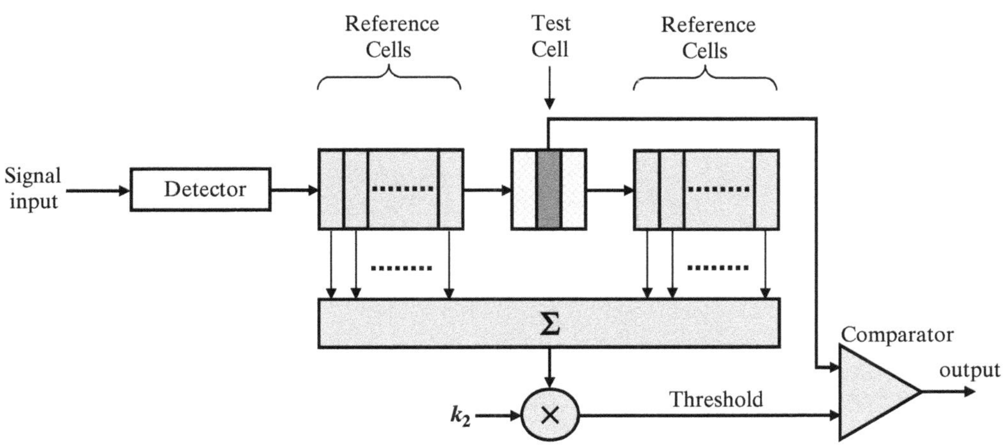

FIGURE 7-21 CACFAR Circuit

it is normal to use a square law detector that outputs a signal proportional to the square of the input magnitude, that is, proportional to the power of the input signal. (Variations exist in which the detector output depends linearly on the magnitude of the input, or its voltage.) The CACFAR circuit is a finite impulse response (FIR) structure in which serial range data are applied to the input of a tapped delay line. Each tap corresponds to one range cell; the central tap is the test cell, and the taps either side of it are blanked. The serial input data are converted to a parallel form. The mean of the reference cells is formed and multiplied by a constant to derive the threshold for that test cell. Data progress one tap in every range cell with the oldest range cell data dropping out the end and a new range cell entering at the input.

The estimated mean power arising from noise, targets, interference, and clutter across the reference cells is given by [5]:

$$\overline{P} = \frac{1}{M}\sum_{m=1}^{M} P_m \tag{7-25}$$

where P_m is the power in the mth cell, and a total of M reference cells are used. The threshold is therefore

$$T_{CACFAR} = k_2\overline{P} = \frac{k_2}{M}\sum_{m=1}^{M} P_m \tag{7-26}$$

T_{CACFAR} is an adaptive threshold, and k_2 is a multiplier ($\neq k_1$) chosen to set a desired false alarm rate. A target detection is declared when the power in the test cell exceeds the threshold. Let P_{TC} denote the power in the test cell. From (7-26), the target detection criterion can be expressed as

$$P_{TC} \rangle\ T_{CACFAR} = \frac{k_2}{M}\sum_{m=1}^{M} P_m \tag{7-27}$$

This can be rearranged as

$$\frac{P_{TC}}{\frac{1}{M}\sum_{m=1}^{M} P_m} \rangle\ k_2 \tag{7-28}$$

This indicates that the test cell is normalized by the mean value of the reference cells. Since the reference cells capture clutter returns from a range bracket, they may be regarded as providing a reference window against which the central test cell is normalized. A detection is declared when the normalized test cell exceeds the multiplier, k_2.

For homogenous samples and large M, the CFAR loss in dB (to a first-order approximation) is given by

$$L(\text{dB}) = 10\log_{10}\left(\frac{SNR_M}{SNR}\right) \tag{7-29}$$

For statistically homogenous powers, a true CFAR function is obtained. The CFAR loss in dB reduces linearly with the logarithm of the number of cells. The loss becomes zero as M approaches infinity since $SNR_\infty \rightarrow SNR$. CFAR loss is also a function of the required P_{FA} but is only weakly dependent on the required P_D.

However, for inhomogenous powers, the CFAR function breaks down. This can happen when the reference cells are dominated by clutter returns since clutter powers can undergo a ramp variation with range or a step transition at a particular range. If clutter power changes linearly with range, then a ramp profile of clutter power would sit within the reference window and CACFAR would provide a high estimate of the mean power and set a correspondingly high threshold. The false alarm rate would therefore decrease in proportion to the rate of change of clutter power with range. The higher threshold in response to a clutter ramp also reduces the probability of detection, P_D. Ordinarily, in a homogenous clutter environment, a larger number of reference cells minimizes the CFAR loss and hence improves P_D. However, in a inhomogenous clutter environment such as the clutter ramps considered here, a larger number of reference cells entails a wider reference window, so ramps in the clutter power reside within the CFAR circuit for longer. In this case, the depression in P_D now lasts longer. Thus, increasing the number of reference cells pulls P_D in both directions. This results in there being an optimum number of reference cells to yield the lowest SCR necessary to achieve a specified detection performance that is a function of the gradient of the clutter ramp. The optimum number varies from 8 cells (for a clutter gradient of 2 dB/cell) to approximately 32 cells (for a clutter gradient of 0.25 dB/cell) [5]. Once a steady state has been restored and any such changes have cleared the reference window, the CFAR function would also be restored. The design of a CACFAR circuit for one clutter environment (usually the homogenous case) is suboptimal when exposed to alternative clutter environments.

Similar irregularities characterize the transient behavior of a CACFAR circuit. Step changes in clutter can be incurred at abrupt transitions between clutter types such as a coastline or a boundary between agricultural land and forest. A step change is also incurred at the transition from a noise-limited environment to one dominated by clutter; such a transition would occur around the altitude line of airborne radars. Variations in false alarm rate and CFAR loss would be seen over the duration for which the step resides within the reference cells. When the lower clutter level resides within the test cell but the step to the higher level is accommodated within the reference window, some of the reference cells see the higher level, and the estimate of the mean would be high. This results in a high threshold and hence a low P_D and P_{FA}. On the other hand, when the high clutter level resides within the test cell but the lower level is accommodated within the reference window, the estimate of the mean would be low. This is particularly troublesome as the low threshold results in a high P_{FA}, and a flood of false alarms can be expected. These two situations are illustrated in Figure 7-22.

A further potential problem for CACFAR is the effect of having additional targets within the reference cells. The additional target signals in the reference cells corrupt the estimate of the mean noise/interference/clutter levels and result in an increase in the mean of the reference window and thus an increase in the threshold. This, in turn, causes a decrease in the probability of detection of targets in the test cell. Normally, one sets a threshold (or threshold multiplier, k_2) to obtain a given P_{FA} and P_D performance on the basis of homogenous signals. The P_D asymptotically approaches its design value as the number of reference cells, M, approaches infinity. Should further targets be present in the reference window, P_D asymptotically approaches a lower value as M approaches infinity than was obtained previously and constitutes a CFAR loss. This contribution to the CFAR loss is a function of P_D, number of targets within the reference window, the mean target radar cross section (RCS), and the statistics of the target fluctuations. One counter to this problem is to apply *censoring*. This technique deletes the largest reference cell voltage, or the two largest or more, depending on the number of additional targets there are assumed to be within the reference window. Censoring incurs an increase in CFAR

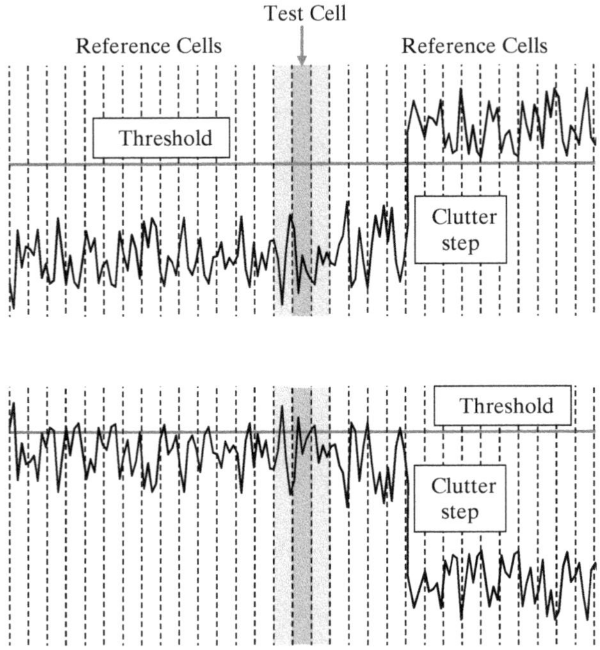

FIGURE 7-22 CACFAR Response to Clutter Step: Low Clutter Level in Test Cell (top); High Clutter Level in Test Cell (bottom)

loss in homogenous interference/clutter and breaks down if there are more targets in the reference window than assumed.

CACFAR is the optimum detection mechanism in homogenous environments such as noise and continuous noise jamming. CACFAR clearly has problems in the presence of certain clutter scenarios or multiple targets, and there is no way of knowing whether CACFAR, or any other form of CFAR, is optimal in non-homogenous situations. Fortunately, a wide range of alternative CFAR strategies and hybrid techniques offer some advantages over CACFAR.

The typical implementation of CACFAR is that the voltages from each range cell tap of Figure 7-21 are proportional to the power of signals in each range cell. This follows as a consequence of using a square law detector on the input. Alternative detector types give rise to variations on this theme. For example, a linear detector results in the voltages on each tap being proportional to the voltage in each range cell. A square law detector plus logarithmic amplifier results in voltages on each tap that are proportional to the power in decibels in each range cell. This is of particular interest because it results in an estimate of the *geometric mean*. Recently, there has even been interest in CACFAR detectors in which the voltage on each tap is proportional to the square root of the voltage in each range cell. Each variation has advantages in differing regions of the ROC curve.

The earlier analysis is based on an estimate of the *arithmetic mean* of the reference cell powers. However, one may alternatively form the *geometric mean* of the reference cell powers, in which case a target detection criterion becomes

$$\frac{P_{TC}}{\left(\prod_{m=1}^{M} P_m\right)^{1/M}} \rangle k_3 \tag{7-30}$$

where k_3 is the threshold multiplier, and the power in the test cell is normalized with respect to the geometric mean of powers in the reference cells. Since the product of a set of variables can be derived on the basis of the sum of their logarithms, (7-30) can be rearranged as

$$\log(P_{TC}) - \frac{1}{M}\sum_{m=1}^{M}\log(P_m) > \log(k_3) \tag{7-31}$$

This is easily implemented using a logarithmic amplifier followed by a square law detector on the CACFAR input. The geometric mean CACFAR detector provides a better estimate of the mean clutter power in conditions of a clutter ramp. The minimum SCR necessary to achieve a specified detection performance is now independent of the gradient of the clutter ramp and continues to reduce as the total number reference cells, M, increases. However, the minimum SCR requirement is greater than that of the arithmetic mean estimator in homogenous environments and leads to a greater CFAR loss. This is an illustration of the general trend that target detection and CFAR losses worsen for CFAR detectors designed to operate in a wider range of clutter conditions [5].

7.4.5 Greatest-of CFAR (GOCFAR)

This is very similar to CACFAR except that the mean of reference cells before the test cell and after the test cell is computed and the greater of these is selected and scaled to set the threshold. A schematic of the GOCFAR circuit is illustrated in Figure 7-23, in which k_4 represents the appropriate threshold multiplier.

The GOCFAR detector avoids the problem of high P_{FA} that can occur at step transitions in clutter when the higher level resides within the test cell (lower plot of Figure 7-22), since the threshold would always be set on the basis of the half of the reference window accommodating the higher clutter level. However, it also sets the higher threshold when the lower clutter level resides within the test cell, so it tends to exaggerate the depression in P_D and P_{FA} experienced by a CACFAR detector. The GOCFAR circuit also results in a small increase in CFAR loss in the presence of homogenous interference or clutter. Normally, the avoidance of the flood of false alarms resulting from the former case is preferable to the loss in detection performance of the latter. Censoring may also be applied in each half of the GOCFAR circuit to counter the problems of multiple targets. The logic of selecting

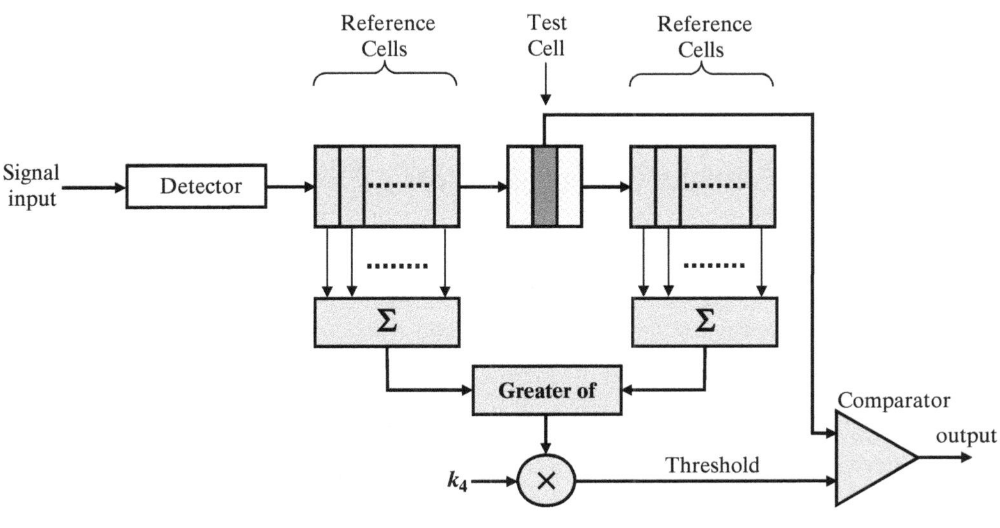

FIGURE 7-23 GOCFAR Circuit

the greater of the two halves, coupled with the chance of targets falling in the half of lower clutter, tends to corrupt the estimate of the mean a little less than for CACFAR. Therefore, GOCFAR is inherently a little more resilient to multiple targets than CACFAR.

It may be possible to enjoy the best of both CACFAR and GOCFAR by adopting a hybrid approach. If the two means in the GOCFAR circuit are considerably different, then it may reasonably be assumed that a step transition is present in the reference window and GOCFAR processing is conducted. If the two means are similar, then the mean of both halves of the reference window may be computed and CACFAR detection is obtained.

7.4.6 Two-Parameter CFAR

The CFAR detectors considered thus far derive a threshold on the basis of an estimate of the mean level within the reference window. An alternative strategy is to derive the threshold as a function of the *median* level within the reference window. The estimate of the median level is less sensitive to the presence of multiple targets and has a low CFAR loss in the presence of homogenous interference or clutter. It transpires that both the mean and median values have some merit when computing a suitable threshold within a CFAR circuit.

In Section 6.3, a variety of statistical models were presented that have been used to describe the magnitude of clutter returns. Some clutter sources are described by a highly skewed PDF. Here, it is useful to characterize these PDFs using the mean: median ratio or by using higher-order statistical moments such as *skew*. Therefore, CFAR circuits seeking to detect targets in clutter-limited scenarios in which the clutter PDF is highly skewed should set a threshold as a function of two (or more) statistical parameters. Where two statistical quantities are used, the detector is known as a *two-parameter CFAR* circuit. Generally, the two parameters of primary interest are the mean and median. However, ever more complex CFAR detectors are being designed that compute a threshold as a function of an ever increasing number of statistical quantities.

A two-parameter CFAR detection strategy that has been used is to base threshold detections on the following condition:

$$\frac{\left(P_{TC} - \overline{P_M}\right)}{\sigma_M} \rangle\ T \tag{7-32}$$

where P_{TC} is the signal in the test cell, $\overline{P_M}$ is the mean of the signals in the CFAR window, σ_M is the standard deviation of the signals in the CFAR window, and T is the threshold value.

7.4.7 Ordered Statistic CFAR (OSCFAR)

The reference cells capture M samples of the noise/interference/clutter in the vicinity of the test cell. The ordered statistic CFAR approach rearranges these samples in ascending order and sets a threshold on the basis of a certain rank order. The threshold may be derived on the basis of the product of the required percentile with a multiplier, which is a function of the required probability of false alarms. Typically, a threshold based on a percentile within the range of 50% to 80% is used. The 50% rank corresponds to the median and can be estimated very accurately from the samples held in the CFAR circuit. A threshold may then be derived by multiplying the 50% rank sample by a large scaling factor; however, this also multiplies the error in the estimate of the median. On the other hand, if a higher percentile is chosen, such as 75% to 80%, the error in estimating this percentile is much higher, but a correspondingly lower multiplier is required. This results in a trade between the error in the estimate of the percentile and how that error is magnified by the multiplier. The optimum turns out to be in the range 75% to 80%, which corresponds to the minimum CFAR loss. For example, for the 70th percentile (x_{70}) of Rayleigh distributed clutter (complex) voltage, the threshold voltage is given by

$$v_T = x_{70}\sqrt{\frac{\ln(P_{FA})}{\ln(1-0.7)}} \tag{7-33}$$

The square root term in (7-33) is the threshold multiplier and is a function of the required value of P_{FA}.

7.4.8 Clutter Maps

Airborne radars are likely to operate over a wide variety of clutter surfaces. Due to the rapid platform velocity and changes in elevation angle, the clutter can vary rapidly with range and may be inconsistent from scan to scan. In particular, the variation of sea clutter can vary wildly both in the short-term and spatially. However, ground-based surveillance systems, especially those operating from fixed locations over land, often observe clutter that changes very gradually with range and that is relatively consistent from one scan to the next. In such cases the estimate of the background clutter level can be built up over several successive scans of the antenna. The CFAR reference window is now two-dimensional: range and scan number. These dimensions are actually both measures of time: the range coordinate is a short-term time window capturing data from a range bracket around the test cell, and the scan number is a much longer-term time window capturing repeats of the range data at intervals defined by the scan speed. This entails the long-term storage of data over several scans; however, the data storage requirement may be eased by feeding back estimates from the previous scan. The threshold is computed as the weighted average of the mean in the range dimension seen on the current scan and the mean level in the range dimension seen on the previous scan. In this way the FIR circuit of Figure 7-21 is incorporated into a larger infinite impulse response (IIR) circuit that feeds back the estimate from the previous scan to compute the threshold for the current scan. Since the estimate for the current scan is weighted by that of the previous scan that, in its turn, was weighted by the scan before that and so on, it may be evident that the current estimate is colored by all previous estimates with diminishing influence the further back in history one cares to go. The timescale over which past estimates continue to influence the current estimate depends on the weighting factors. If the data from the previous scan are heavily weighted, then the circuit makes an estimate of the background clutter on the basis of several previous scans. This estimate grows in accuracy as the weighting from the previous scan increases, and, as a consequence, an increasingly accurate threshold is set and the CFAR loss reduces. However, the validity of this operation is founded on the rather tenuous assumption that the clutter conditions do not change over the timescale of the scans that have a meaningful influence on the current estimate. Such a circuit is slow to respond to any changes in the clutter conditions. In fact, old data circulate forever with diminishing influence but are never truly flushed from the detector; however, they decay over repeated scans with a time constant determined by the weighting factor. Hence, there is a compromise between low CFAR losses and the circuit's response to changes in the clutter scene determined by the weighting factors.

Returning to the notion that a ground-based surveillance radar records a range profile at each antenna azimuth location, it is evident that data are conveniently displayed on a plan position indicator (PPI) display. This display format modulates the intensity of the display in proportion to the strength of received signals. Data are displayed in polar coordinates in which the radar location is taken to be at the center of the display; the distance from the display center is calibrated in terms of range, and the angle around the display corresponds to the azimuth angle. For ground-based radars, all clutter, both MBC and SLC, falls at or close to zero Doppler and is easily rejected using an MTI filter. However, in some applications it is useful for the operator to observe the clutter on the display. Land forces operating a battlefield surveillance radar often find it useful to switch out the MTI filter in undulating terrain since the display is now dominated by the clutter returns. This constitutes a *clutter map*, that is, a display of clutter using the polar coordinates of range and azimuth. Clutter maps in undulating terrain may have regions in which no clutter is displayed (only background noise and interference). The lack of clutter is indicative of dead ground such as an area in the shadow of a hill that could conceal enemy forces. If the radar cannot see into these areas, neither can a human observer, and he approaches them with caution! In the absence of any MTI filtering, a CFAR circuit may now apply a two-dimensional (range/azimuth) CFAR reference window in which the clutter may be sampled and an appropriate threshold derived. In this way, clutter is estimated in the vicinity of the test cell on the basis of many more samples than is possible using the range coordinate alone. This carries the advantage of more accurate estimates and lower CFAR losses but is based on the assumption of homogenous clutter within the reference window along both the range and azimuth axes.

It is now desirable to partition a clutter map into regions of homogenous clutter, with the clutter statistics of each region differing from those of neighboring regions. An example of this is described by Hughes and Lewis [19] in which a self-adaptive CFAR system is described to form regions in a range/azimuth clutter map of like clutter statistics. This is applied in a coastal scene containing many different clutter types and discrete targets. The CFAR function is closely coupled to a multiple-hypothesis tracker to control the number of false alarms, to reduce CFAR loss, and hence to enable the tracker to track targets as they move through a difficult scene with good integrity and few distractions from false tracks.

Some ground-based air surveillance radars receive data from several receiving antenna beams stacked vertically to derive target elevation data. Each beam in the elevation plane feeds a separate receiver and processing chain. Beams at the low elevations receive strong ground clutter, and the MTI filters and CFAR detectors used here must cater for these conditions. Beams at higher elevations may receive clutter due to atmospheric and weather effects and chaff but are not likely to be exposed to slow speed targets; their MTI filter and CFAR detector circuits must be designed accordingly. In this way it is possible to use multiple CFAR detectors commensurate with the conditions prevailing at different elevation cells.

Naturally, these ideas may be extended to radars on moving platforms, except that now the SLC is spread over velocities equal to $+/-$ the platform velocity. It now becomes pertinent to consider the distribution of clutter in range and velocity as well as azimuth and elevation. The distribution of clutter from an airborne radar in range and velocity forms a clutter map, albeit in range/velocity rather than range/azimuth, an example of which is given in Figure 6-17. In many airborne pulse Doppler applications it is commonplace to use a two-dimensional CFAR reference window in range/velocity space. The consequences of this coupled with medium PRF processing are considered further in Chapter 14.

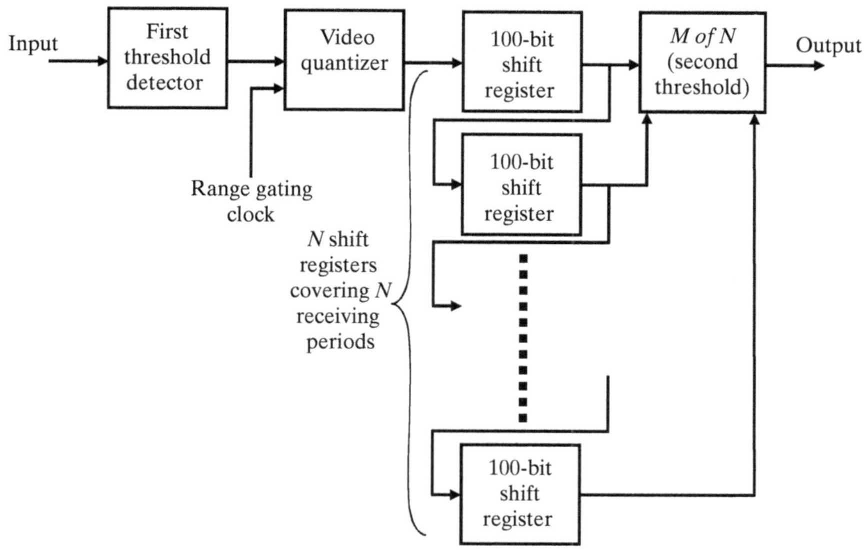

FIGURE 7-24 Binary Integrator

7.4.9 Binary Integrator

False alarms may be controlled by the process of binary integration (also known as a *pattern correlator*). Although this is not specifically a CFAR function, the binary integrator contributes to the detection performance of a radar and thus is dealt with here for convenience. It is easiest to consider a binary integrator operating on range gated data, although it can function on Doppler gated data or both range and Doppler gated data. The binary integrator is an example of a non-coherent integrator. It involves a two-threshold system and is depicted in Figure 7-24 in which 100 range cells across each receiving period are assumed. The first stage consists of a simple detection system that incurs many false detections. However, these are likely to occur in random range cells, whereas genuine target returns are detected in a consistent range cell. The output of the first threshold detector is standardized, often digitized, such that all threshold crossings result in a signal of constant amplitude (logic 1 or 0) and duration equal to one range cell width. This standard video signal is then clocked into shift registers a range cell at a time over N receiving periods. Should there be a threshold crossing in the same range cell in M or more receiving periods, then a target detection is declared. By careful selection of M and N it is possible to conclude that this degree of consistency should arise only for genuine targets. The M from N criterion is the second threshold and should rarely be triggered by false alarms.

The video quantizer outputs a 1 or a 0 due to the presence of a target with respective probabilities of P_D and $(1 - P_D)$ in each of the 100 range cells across N receiving periods. The probability of obtaining exactly m target detections out of the total of N receiving periods is given by

$$\frac{N!}{m!(N-m)!} P_D^m (1 - P_D)^{(N-m)} \tag{7-34}$$

The probability of target detection is thus the sum of the expression given at (7-34) for all values of m that exceed the minimum requirement of M and is given by

$$\sum_{m=M}^{N} \frac{N!}{m!\,(N-m)!} P_D^m\,(1 - P_D)^{(N-m)} \tag{7-35}$$

Similarly, the video quantizer outputs a 1 or a 0 due to the presence of noise/interference/clutter with respective probabilities of P_{FA} and $(1 - P_{FA})$ in each range cell across N receiving periods. The overall probability of false alarms is thus given by (7-35) with P_{FA} replacing P_D and is therefore given by

$$\sum_{m=M}^{N} \frac{N!}{m!\,(N-m)!} P_{FA}^{m}\,(1 - P_{FA})^{(N-m)} \tag{7-36}$$

For example, consider the case of a Swerling case 5 (nonfluctuating) target in the presence of Gaussian noise. If the first threshold sets $P_D = 0.8$ and $P_{FA} = 10^{-2}$ and the second threshold sets a criterion of *6 from* 10 to declare a target detection, then the overall $P_D = 0.996$ and the $P_{FA} = 2.1 \times 10^{-10}$. This would give a false alarm rate, with a 1 MHz noise bandwidth, of 2.1×10^{-4} or one false alarm every 79 minutes on average and is a considerable improvement on the original value of $P_{FA} = 10^{-2}$.

The binary integrator is particularly pertinent to medium PRF pulse Doppler systems that operate on an *M of N* basis.

7.4.10 Hybrid Approaches

It was noted in Section 7.4.5 that a hybrid CACFAR/GOCFAR approach may be taken to obtain the benefits of both methods and to minimize CFAR loss. Hybrid approaches are powerful when dealing with two quite different situations. The hybrid CACFAR/GOCFAR approach handles clutter steps and multiple targets well. An alternative hybrid approach may be the use of a fixed threshold detector coupled with conventional CACFAR and is useful in handling the transition from thermal noise to clutter. Censoring can also be applied in any of the CFAR detectors considered. Various statistical approaches have also been described and are often implemented using two-parameter CFAR detectors operating with two-dimensional reference windows. Modern processing power also enables the exact approach to be tailored to specific interference, clutter, and target situations. The difficulty is often how a change in conditions is detected to initiate a switch in CFAR detection strategy. No approach fits all circumstances; everything is a compromise. Hybrid and adaptive approaches can, however, help control CFAR loss.

In some radar applications such as automatic target recognition (ATR) in synthetic aperture radar (SAR) imaging systems, a dual threshold approach is often taken. The first threshold is implemented in a simple circuit and is designed to capture image chips of interest while eliminating the majority of false alarms. This screens the data for possible targets but inevitably admits some false alarms. A second threshold is then applied from a more complex detector to discriminate between targets of interest and false alarms.

The subjects of clutter (Chapter 6), MTI rejection (Section 7.2), and CFAR detection are all closely bound up with each other. Military radars are facing ever more complex jammers and lower RCS targets, yet the expectation for improved detection performance, better target discrimination, and covert operation in all environments is increasing. The ability to deliver this will be determined by the ingenuity in the design of waveforms and their associated processing.

CHAPTER 8

Radar Hardware

Chapter Outline

8.1 INTRODUCTION

The chief technological hurdles associated with the development of early radars were the radio frequency (RF) and microwave components. The invention of the magnetron independently in the United Kingdom and Japan in 1940 and its subsequent productionization in the United States proved pivotal because this brought about the generation of high-power centimetric wave signals and hence made microwave radar feasible. Radars were undoubtedly microwave systems, and the key to their success were the developments in the components that could generate and handle high-power microwave signals, the sensitive microwave components used within radar receivers, and the antenna systems they used. The developments in military radar systems drove the developments in microwave engineering, and it is fair to say that radar capability was dictated by the performance of its microwave front end. The few commercial applications of microwave technology that existed at that time were largely spin-offs from the military developments. This picture remained the case until the advent of low-cost, powerful computing became available in the mid-1980s.

The power and speed of modern digital signal processing (DSP) now enables designers and researchers to realize ideas that could never have been supported by analog circuits. Modern radar developments now tend to dwell on the extraction of more target data from higher numbers of targets and at faster rates, the increased automation of radar systems, radars that are adaptive to their operating conditions, and the control of multiple modes and waveforms to support a wide variety of tasks. In more recent times, radar capability has become limited by processing power, and radars are often perceived as digital devices. Allied to this, the microwave aspects of a radar system have tended to be overlooked. The digitizer has crept closer to the antenna, and for many in the field the microwave elements have become relegated to those of a transducer operating at the front end. Furthermore, computer clock speeds are now running at microwave frequencies; this has opened a new application of microwave engineering relevant to modern radars. Naturally, microwave engineering has not stood still over the last 25 years, and in the meantime the world has witnessed the proliferation of high-volume, low-cost, ultra-reliable commercial communications. The commercial sector now drives developments in microwave engineering, and military systems are the beneficiaries of this work.

Part 2 of this book is largely concerned with waveform design and the associated processing techniques for modern pulse Doppler radar systems. However, this chapter redresses the balance by considering the microwave engineering aspects that make it all happen. This chapter comprises sections that give an overview of transmitter and frequency synthesis technology, receivers, and antennas. These sections illustrate that radar capability is still fundamentally limited by the performance of these microwave elements. However, modern advances in microwave engineering, coupled with the advances in computer processing power, have brought about some rather spectacular innovations. This is especially true of active electronically scanned array (AESA) antennas and modern frequency synthesizers. It is not the intention of this chapter to provide exhaustive coverage of these technologies, which are amply covered in other texts, but to provide some pertinent theory and a summary of the state-of-the-art and thus to illustrate the important role these technologies play in realizing but ultimately limiting the capabilities of modern pulse Doppler radars.

8.2 RADAR TRANSMITTERS

This section describes two of the most important specifications concerning the microwave signals pertinent to pulse Doppler radars: (1) coherency; and (2) close-to-carrier noise. Also, an introduction will be provided to the devices that can be used as sources of microwave signals. Radars use microwave oscillators and synthesizers as a source of transmitted power or for providing low-power drive signals for amplification in a master oscillator power amplifier (MOPA) type transmitter. RF and microwave signals also act as local oscillators (LOs) within a superheterodyne receiver and provide digital clocking and timing signals. Most radar systems have one fundamental oscillator from which all other signals are derived. This ensures that any drift in the oscillator frequency, such as due to temperature changes, cause a compensating drift in all other signals such that all signals remain coherent with each other.

Most radar transmitters may be classified into two broad categories: (1) power oscillators; or MOPA types. Power oscillators generate the fundamental source of the microwave signal and do so at the full transmitted power. These may be high-power tube devices such as the magnetron (seldom, if ever, used in modern pulse Doppler radars) or low-power solid-state devices. The MOPA type transmitter is far more prevalent these days. In this scheme, a low-power microwave signal is derived from a solid-state source and subsequently is amplified by a power amplifier. The power amplifier may be a high-power tube device such as the traveling wave tube (TWT) or klystron or may be a solid-state amplifier, as is common place in transmit/receive (T/R) modules of a modern AESA system.

Radar systems transmit RF power, which is typically the following:

- High peak power
- Microwave

- Wide band (pulsed, phase, or frequency modulated)
- Frequency agile
- Coherent, both from pulse to pulse and phase locked to local oscillators running within the receiver
- Low noise (spectral purity)

Pulse Doppler radars place tight specifications on the coherency and spectral purity of the signals they use and these are addressed in the next two sections.

The RF power and the frequency at which it is produced are closely coupled parameters of oscillators and amplifiers. A common figure of merit used to describe this relationship is the Pf^2 product. This is formed by multiplying the power in watts with the frequency, in GHz, squared and is approximately constant for a given technology. Thus, the RF power available reduces very rapidly as frequency increases. It is also worth noting that cost also scales as Pf^2.

8.2.1 The Necessity for Coherency

Coherency is an important attribute to maintain for the measurement of the Doppler shift. Section 4.1 noted that a succession of pulse returns from a moving target exhibit a regular phase difference. This phase difference is a function of the wavelength, pulse repetition frequency (PRF), and relative velocity of the target. Sections 7.2 and 7.3 described how this effect can be exploited to provide rejection of stationary clutter and for the measurement of the target velocity, respectively. However, if successive pulses are transmitted having random phase differences with respect to each other, then the phase differences due to the target motion are disguised and no measurement of velocity is possible. Similarly, if successive pulse echoes are subject to random phase shifts as they propagate through the receiver, the phase differences due to the target motion are disguised and no measurement of velocity is possible. Therefore, the local oscillator signals used by the receiver must also be coherent and phase locked to the transmitted signal. This places a constraint on the phase stability requirements of a transmitted pulse train and local oscillators. Fast Fourier transform (FFT) processing measures the phase increment between successive pulse returns with a resolution of $360°/n$ over a time duration equal to the coherent processing interval (CPI), where n is the point size of the FFT, and t_{int} is the CPI time. Hence, a phase instability of $360°/n$ per t_{int} causes the target return to be smeared across two Doppler cells. Thus, a frequency error of $\frac{1}{nt_{int}}$ causes target smear into an adjacent Doppler cell. In practice, a frequency stability of less than this quantity is usually specified. Achieving this degree of frequency stability over the full operating temperature range of airborne military radars requires temperature stabilization of the oscillators and is far from trivial to achieve for radars using long integration times. Long integration times are required for fine Doppler resolution such as those operating on a low carrier frequency or in high-resolution synthetic aperture radar (SAR) systems.

Coherency between the transmitted signal and the local oscillators can be achieved using phase locking synthesis techniques (Section 8.3.3). A coherent pulse train can be derived from a stable continuous wave (CW) source and gating its output using a microwave switch, such as a PIN diode switch, to define the PRF and pulse width. Care must be taken in the amplification of the transmitted signal to minimize the variation in the phase response of the power amplifier. This can be particularly troublesome in high-power tube devices since thermal expansion and contraction of the tube can result in variations in the electrical path length through the device. The electrical path length may also vary in response to variations in the electron beam velocity, which in turn is dependent on the stability of the accelerating potential. This requires the high-tension electrical supplies to such tube devices to be very highly stabilized (typically, to within a few volts on a supply of several kV) and represents a significant design difficulty.

8.2.2 Close-to-Carrier Noise

Detailed spectrum analysis shows that no signal is a pure tone. Small and unintentional modulations manifest themselves as sidebands on either side of a carrier frequency. These modulations occur as random electrical noise and at discrete frequencies and are known as *close-to-carrier noise*. This type of noise results from both amplitude and phase (or frequency) modulations of the carrier, and it is often important to differentiate between the two. The basis of close-to-carrier noise measurement is spectrum analysis. This section describes some of the characteristics of close-to-carrier noise, a few considerations concerning the (phase) noise characteristics of a pulsed signal, and some notes regarding its measurement.

8.2.2.1 Spectrum Analysis

If a signal on a spectrum analyzer is viewed with sufficient resolution, it is evident that its response is not a single line at the signal frequency (within the constraints set by the resolution bandwidth). The signal is likely to be spread over a wider band, having both random noise and discrete frequency modulations associated with it, such as the spectrum in Figure 8-1.

These modulations arise due to the operation of the oscillators used or method of frequency synthesis employed in the generation of the signal together with other effects such as electromagnetic compatibility (EMC) problems (e.g., pick-up and cross-talk of fast switching logic), microphony, and variations in the insertion loss or electrical length of circuitry supporting the signal. It is often important to determine the spectral purity of a signal through a measurement of its close-to-carrier noise. It is also typical to determine the contributions due to amplitude and phase (or frequency) modulations (AM and ΦM, respectively) separately. A spectrum analyzer does not differentiate between AM and

FIGURE 8-1 Double Sideband Close-to-Carrier Noise Profile of a Signal

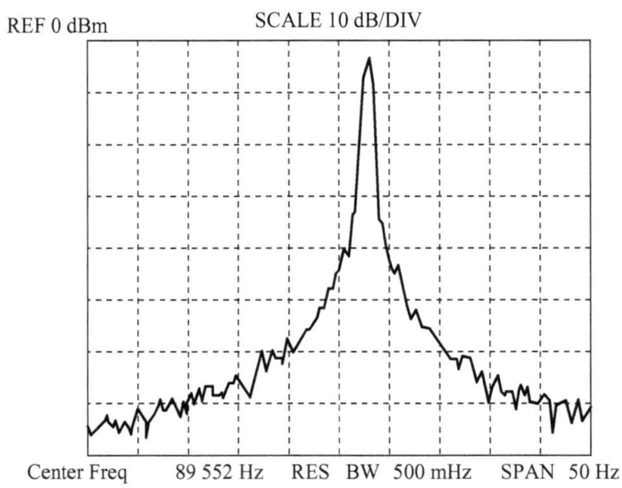

ΦM, so some form of demodulation of each is required prior to the spectral analysis of the baseband signal.

In specifying and measuring close-to-carrier noise, it is appropriate to consider the spectral noise power density in decibels with respect to the carrier level (dBc) as a function of the offset frequency away from the carrier frequency. By convention, the spectral noise power density is referenced to a 1 Hertz measurement bandwidth; hence, the units of close-to-carrier noise are dBc/Hz. However, it is often not practical to actually perform the measurement of close-to-carrier noise in a 1 Hertz bandwidth since the integration time of such a narrow-band filter would make the measurement exceedingly time-consuming. The measurement of close-to-carrier noise should ideally be made using a bandwidth equal to that of the radar detection system. In the context of pulse Doppler radar the appropriate measurement bandwidth is the Doppler filter bandwidth. For a measurement bandwidth of B_N, a correction of $10 \log_{10}(B_N/1)$ has to be made to the measured result to normalize this to a 1 Hz bandwidth. For example, a measured level of −100 dBc/100 Hz equates to −120 dBc/1 Hz. Allied to the noise spectrum, discrete modulations may be found at specific frequencies on the signal, resulting in low-level spurious lines in the close-to-carrier noise spectrum. The level of these spurious lines is not dependent on the measurement bandwidth. To resolve any spurious lines from the background noise, a certain signal-to-noise ratio (SNR) may be required. If the previous example is extended to assume that a signal-to-noise ratio of 5 dB is required, then spurious lines may be detected down to a level of −95 dBc; smaller lines would go undetected even though the noise is plotted some 25 dB lower at −120 dBc/Hz. Thus, the measurement bandwidth does determine the minimum level of discrete spurious modulations that can be detected. For pulse Doppler radar, no spurious lines should be detected above the measured noise level over a band of offset frequencies covering the total Doppler band of the radar. Automated close-to-carrier noise measurement test equipment runs a threshold a few decibels above the measured noise level to detect the presence of spurious lines, which must be displayed at their measured levels and not at a level corrected for the measurement bandwidth.

8.2.2.2 The Importance of Close-to-Carrier Noise

The specification, minimization, and measurement of close-to-carrier noise are important in radar and communication systems. Problems occur when the system tries to process a weak signal in the presence of a strong one, since the close-to-carrier noise of the strong one can mask the detection of the weaker one. Excessive close-to-carrier noise has the following effects:

- Causes co-channel interference in communication and radar systems
- Forms a clutter spectrum in pulse Doppler radar
- Contributes to the noise figure of a receiver
- Has spurious lines that can appear as false targets
- Produces errors in communication systems

8.2.2.3 Origin of Close-to-Carrier Noise

The close-to-carrier noise characteristics of any signal depend on the fundamental properties of the oscillator as well as the noise within the circuits that generate the signals, noise on the oscillator bias controls, and on the action of the synthesizer used to derive RF signals. Thermal noise is present in all conductors and oscillators and arises due to the thermal agitation of charge carriers [20] (see Section 8.4.2 for more details). Shot noise is also present in electron vacuum tubes and semiconductor PN junctions due to fluctuations in bias and hence the current flowing through such devices. Both thermal and shot noises have a constant spectral noise power density and are therefore known as *white* noise; both are also directly proportional to temperature. Thermal noise tends to be the dominant noise source and so is considered the most important. Flicker noise also occurs in solid-state components and vacuum tubes, although it is not well understood. Flicker noise has a spectral noise power density that varies as $1/f$, where f is the frequency and so is also known as $1/f$ noise or *pink* noise. It can be a strong source of noise at low frequencies but rapidly diminishes with increasing frequency since its spectral power density decays by 3 dB per octave. Flicker noise modulates RF oscillators, causing appreciable close-to-carrier noise at low offset frequencies. Flicker noise is apparent in the close-to-carrier noise plot of Figure 8-3 at offset frequencies up to about 100 Hz; it is characterized by the linear noise power profile of a constant negative gradient when plotted on logarithmic scales. Cross-talk, pick-up of interference, and microphony due to equipment vibration can also contribute close-to-carrier noise, particularly at discrete frequencies. These sources of noise may be minimized by good design practice such as circuit layout, by screening, and, in the case of microphony, by using antivibration mounts. Physical shock and vibration can be severe in many military applications, particularly aboard off-road vehicles and all airborne platforms.

8.2.2.4 AM Close-to-Carrier Noise Measurement

AM noise arises due to instabilities in the amplitude of a signal. The demodulation of AM signals is a trivial matter of using a diode envelope detector. For small signals the current passed by a Schottky diode is proportional to the applied voltage squared. Thus, the diode current is the demodulated AM, and this baseband signal can be passed directly to a spectrum analyzer. The analyzer displays the double sideband modulation envelope and has to be halved to yield a single sideband measurement. The carrier signal appears as a direct current (DC) component and the frequency of modulations displayed by the analyzer is the frequency offset of the modulation from the carrier. It is normally required to calibrate the amplitude of modulations in decibels with respect to the carrier level (dBc). The calibration requires an amplitude modulation of a known modulation depth, producing a discrete sideband at a known level, to be applied to a test signal or the carrier. This process is automated in modern test equipment.

8.2.2.5 Close-to-Carrier Phase Noise Measurement

Phase and frequency modulation are tantamount to each other, and no distinction is made between them with respect to noise measurement. Phase noise arises due to a phase jitter in the waveform. There are several methods by which close to carrier phase noise may be measured; they differ in the way the signal is demodulated. The most common technique is to phase lock the signal under test to a low noise reference. A functional block diagram illustrating the technique is shown in Figure 8-2.

This method depends on having a reference source, the voltage controlled oscillator (VCO) in Figure 8-2, and a spectrum analyzer that both have lower noise floors than the test signal. The phase-locked loop (PLL) operates by tuning its VCO to the same frequency as the RF input and with a phase such that the two inputs to the phase sensitive detector (PSD) are in quadrature. The PSD outputs a voltage proportional to the cosine of the phase difference between its two inputs and is thus zero volts for a phase difference of 90°. Any

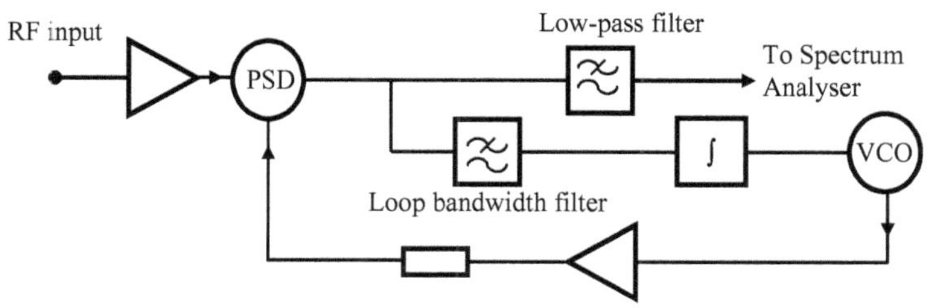

FIGURE 8-2 Phase Locked Loop for Phase Noise Measurement

voltage is indicative of the error in phase quadrature between the two inputs. The loop is locked when the error voltage is zero, that is, when phase quadrature is established. A PLL therefore has *no* frequency error, although there may be a phase error. Should a phase error occur, a PSD output voltage is generated, which after filtering and integration is applied to the tuning pin of the VCO to re-establish phase quadrature. The filtering, integration, and VCO tuning speed determine the reaction characteristics of the PLL. The two amplifiers isolate the VCO and test signal applied to the RF input from each other and thus prevent injection locking of one source to the other. The operation of the loop may break down (i.e., lock would be lost) if the quadrature error between the two PSD inputs is too great or if the rate of change of the phase error (a frequency) is too great. Rates of changes of phase errors (i.e., frequency difference between the two inputs) may be tracked by the VCO if they fall within the passband of the loop bandwidth filter.

Assume that phase quadrature has been established and that an additional phase error, $\Delta\phi$, is present on the test signal. The PSD output is now

$$\begin{aligned}&\propto \cos(90^\circ + \Delta\phi)\\&= \sin(\Delta\phi)\\&\approx \Delta\phi\end{aligned}$$

if $\Delta\phi$ is small. Thus, the PSD output is proportional to the phase noise of the RF test signal. In fact, the PSD output is proportional to the sum of the phase noises of the test signal and VCO, but the latter can be ignored if its phase noise is very much less than the former. The modulation frequency of any frequency modulation (FM) on the test signal gives rise to a rate of change of $\Delta\phi$ and can be measured as an absolute frequency on the spectrum analyzer. Automated test equipment scales the amplitude trace as a single sideband plot of spectral noise power density in decibels with respect to the carrier level. The vertical units are expressed as

$$\mathcal{L}(f) = \frac{P_{ssb}\,(per\ 1\ \text{Hz})}{P_S} \qquad [\text{dBc/Hz}] \tag{8-1}$$

where P_{ssb} is the single sideband power, and P_S *is* the carrier power.

$$\mathcal{L}(f) = 1/2\Delta\phi_{rms}^2(f) \tag{8-2}$$

The horizontal axis is the frequency offset from the carrier. A typical phase noise plot is shown in Figure 8-3. This illustrates the spectral phase noise power density of a test signal that tends to decrease with increasing offset frequency away from the central carrier frequency. This figure plots the single sideband close to carrier phase noise over a range of offset frequencies ranging from 1 Hz on the left-hand edge to 40 MHz on the right. Also evident in the figure is the presence of discrete spurious lines at offset frequencies equal

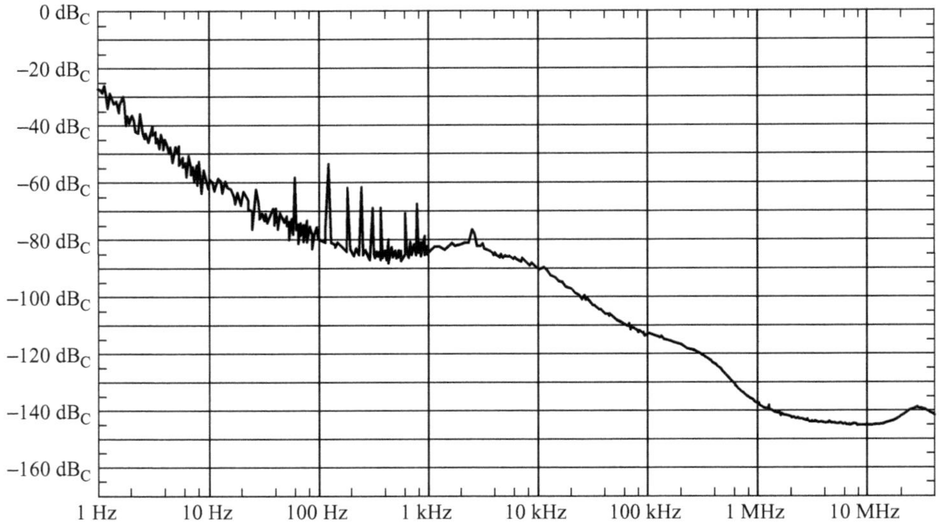

FIGURE 8-3 Typical Close-to-Carrier Phase Noise Plot

to the mains supply frequency and its multiples. Clearly, some low-level breakthrough of the mains supply is modulating the signal under test.

8.2.2.6 Limitations of the Phase-Locking Technique

1. The measurement needs to be calibrated against a test signal of known modulation index.
2. Frequencies within the loop bandwidth are tracked by the VCO. Measurements of phase noise at frequency offsets less than the loop bandwidth are not readily available. Although the loop bandwidth may be made very narrow (100 s Hz or a few kHz), phase noise specifications may extend to within a few Hertz of the carrier. The suppression of modulations by the PLL within its loop bandwidth can, however, be calibrated to provide measurements at offset frequencies within its loop bandwidth, although this is an awkward procedure. Fortunately, this is automated on modern test equipment.
3. Large phase or frequency modulations cause the PLL to lose lock on the test signal, so they cannot be tolerated. The technique is best suited to low noise signals.
4. This technique requires a reference source of considerably lower noise than the test signal.
5. Measurements may be made on a pulsed carrier provided that the PRF falls outside the loop bandwidth (otherwise lock may be transferred to the PRF line within the signal spectrum) and provided also that the pulsed signal exceeds a minimum duty ratio (otherwise the signal is not present for a sufficiently long time to allow the loop to lock to it). Therefore, there are limits on the lowest PRF and duty ratio acceptable.

8.2.2.7 Delay Line Frequency Discriminator Technique

A delay line and mixer operating as a phase sensitive detector have the combined effect of a frequency discriminator. The delay line transforms any frequency modulation into a phase fluctuation, and the PSD, with its inputs in phase quadrature, outputs voltage fluctuations which are directly proportional to the phase fluctuations. The functional block diagram is given in Figure 8-4.

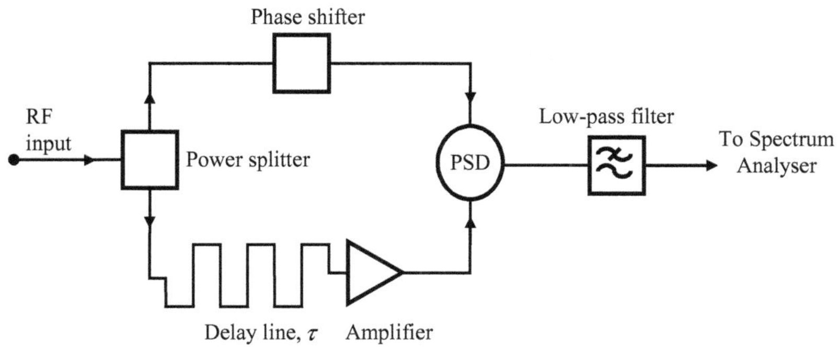

FIGURE 8-4 The Delay Line Frequency Discriminator Technique

The frequency-to-phase conversion has a $(\sin x)/x$ characteristic with a null at a modulation frequency, $f_m = 1/\tau$. Measurements are therefore limited to a modulation frequency $< 1/2\tau$. Greater sensitivity is achieved using a longer delay line, but this restricts the maximum modulation frequency that can be measured. In general, this technique is less sensitive than the phase-locking approach, but it does not require a low noise reference. The system is calibrated by the application of a test signal with a known frequency modulation.

8.2.2.8 Phase Noise on Pulsed RF Signals

The phase noise of a pulsed signal is that of the CW signal gated by the pulse envelope. In the frequency domain the resulting phase noise spectrum is the convolution of the CW noise spectrum and the pulsed spectrum. The spectrum therefore includes components at multiples of the PRF, with the CW phase noise spectrum folded about the PRF lines and is illustrated in Figure 8-5. The pulse width and duty cycle must be sufficient to allow the PLL to maintain its lock on the signal. This becomes increasingly awkward for duty cycles below 5%. The noise power spectrum of the pulsed signal is reduced by the square of the duty cycle compared with the CW spectrum. System sensitivity decreases as the duty cycle decreases.

The presence of the PRF lines in the spectrum can induce the PLL to transfer its lock to a PRF line, setting up a large phase error and incurring a loop settling time. To avoid this a low-pass antialiasing filter is fitted within the loop just before the loop bandwidth filter to filter out the PRF lines. The loop bandwidth cannot now be greater than the PRF.

The folded spectrum about the PRF lines results in an increase in the close-to-carrier noise since the single sideband (SSB) noise of the carrier is added to the SSB noise reproduced about the PRF line. This results in a 3 dB increase where the two are equal, that is, close to 1/2PRF offset from the carrier. This is illustrated in Figure 8-5 (Fig C on right).

PRF jitter is a source of close-to-carrier phase noise since for a coherent train of pulses the phase difference between successive pulses is a function of the time between the pulses and the carrier frequency only. Any jitter in the timing to control the gating of RF pulses results in jitter on the phase difference between pulses that manifests itself as close-to-carrier noise. Highly stable digital clocks are therefore required to derive pulsed waveforms of low close-to-carrier noise.

8.2.2.9 Setting the Specification

In some situations, a particularly large clutter return may be incident on the receiver. This clutter signal is a replica of the transmitted waveform together with its

FIGURE 8-5 Pulsed Carrier Phase Noise

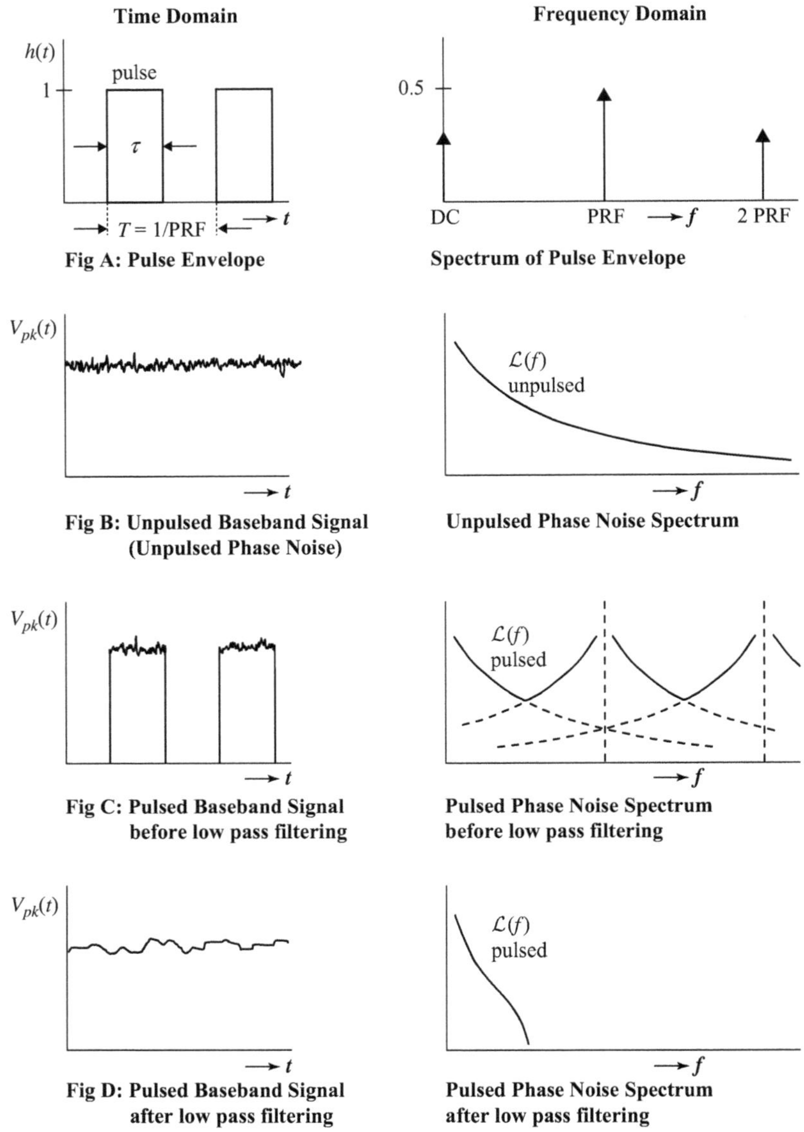

close-to-carrier noise and spurious lines. Presuming the clutter surface to be stationary, then the clutter signal returns have zero Doppler shift. (Any breakthrough of the transmitted signal into the receiver occurring during the transmitted pulse on account of a lack of receiver isolation would also appear as a clutter return at zero range and Doppler.) A moving target indicator (MTI) filter provides rejection of this signal over the bandwidth of its main beam clutter (MBC)/ground-moving target (GMT) rejection notch. However, the close-to-carrier noise extends across the Doppler band. Close-to-carrier noise at offset frequencies greater than the rejection notch bandwidth is not attenuated by the MTI filter. If this close-to-carrier noise exceeds the ambient thermal noise level, then it is the close-to-carrier noise that determines the system sensitivity and degradations in detection performance result. Furthermore, any spurious lines seen above the system noise floor could trigger false detections.

If the clutter attenuation (CA) of the MTI filter is set equal to the close-to-carrier noise level with respect to the carrier peak as measured in a bandwidth matching the Doppler resolution bandwidth, then the peak level of the clutter return falling at zero Doppler is attenuated down to the level of the surrounding close-to-carrier noise, which spreads into the Doppler band. Therefore

$$CA = \left(\frac{P_{ssb}(per\ 1\ \text{Hz})}{P_S}\right) \times \Delta f_d$$

It ought to be pointed out that the Doppler resolution bandwidth, $\Delta f_d = 1/t_{int} \times DCBF$, where *DCBF* is the Doppler cell broadening factor due to the FFT window function (see Section 7.3.4, Table 7-1). Medium PRF pulse Doppler radars exhibit Doppler ambiguity that results in multiple repeats of the close-to-carrier noise in the Doppler band. If the number of ambiguous repeats is given by the aliasing factor (AF), then the equation relating the clutter attenuation and close-to-carrier spectral noise power density becomes

$$CA = \left(\frac{P_{ssb}(per\ 1\ \text{Hz})}{P_S}\right) \times \frac{DCBF}{t_{int}} \times AF \tag{8-3}$$

Equation (8-3) can be used to specify the MTI clutter attenuation and close-to-carrier noise specifications at offset frequencies beyond the MTI filter rejection band but within the radar's Doppler band. The clutter attenuation specification must also be set to provide adequate rejection of the most powerful clutter return expected during operational use. Similarly, the level of close-to-carrier noise that spills over into the Doppler band should not exceed the thermal noise floor of the receiver ($= kT_0 B_N F$) when this maximum clutter return is incident on the receiver. Hence, the actual specification limits are dictated by the assumed maximum clutter return with which the radar must cope.

Close-to-carrier phase noise at offset frequencies corresponding to the transition band of the MTI filter receive some degree of rejection but not the full rejection obtained at zero Doppler. The permissible close-to-carrier noise levels here can increase with decreasing offset frequency to mirror the increasing MTI filter rejection in its transition band. At such low offset frequencies most oscillators and synthesizer generate higher levels of close-to-carrier noise. The exact levels and profile of close-to-carrier noise depends on the performance of the oscillators and the design of the frequency synthesizer used to derive the signals.

If the clutter surface is in motion (e.g., the sea surface), then Doppler broadening of the clutter occurs, as described in Section 6.4.4.2. This spreads the clutter return into the first few Doppler cells and tends to dominate over the close-to-carrier noise. In such cases it becomes rather pointless to specify very low levels of close-to-carrier noise at offset frequencies dominated by the internal clutter motion.

8.2.2.10 Example

It is perhaps best to illustrate the effects described in the preceding sections by way of example. Assume a Doppler radar to transmit a carrier frequency of 10 GHz and to have a close-to-carrier noise level of −120 dBc/Hz at an offset frequency of 2 kHz. Assume also that the Doppler resolution of the radar is 100 Hz($B_N = 100$); thus, the close-to-carrier noise level is more usefully expressed as –100 dBc/100 Hz at an offset frequency of 2 kHz. A Doppler shift of 2 kHz corresponds to a target velocity of 30 m/s and may be assumed to fall outside the rejection bandwidth of the MTI filter. Consider a look-down scenario in which a large clutter return and a target moving at 30 m/s are present, giving rise to a signal-to-clutter ratio (SCR) of −55 dB.[1] The large clutter return is centered at zero Doppler, so it is rejected by the MTI filter; however, the close-to-carrier noise associated with the clutter spreads into the Doppler band such that the 30 m/s target competes with the close-to-carrier noise at a 2 kHz offset frequency arising from the clutter return for detection. In this Doppler cell the SNR is −55 dB + 100 dB = +45 dB, in which the noise is clutter-related close-to-carrier noise.

Insufficient receiver isolation may also contribute some leakage of the close-to-carrier noise into the receiver and ultimately the Doppler processing. The noise leakage is given by

$$N_{CCN\_leakage} = P_{pk} \cdot CCN \cdot isolation$$

where P_{pk} is the peak transmitted power, CCN is the close-to-carrier noise level, and $isolation$ is the receiver isolation.

The returning signal power from a target of interest is given by the radar range equation (Section 2.1.1) as

$$S = \frac{P_{pk} G A_E \sigma}{(4\pi)^2 R^4} \quad \text{(single pulse detection)}$$

and is also directly proportional to the transmitted power. Therefore, the SNR in the Doppler cell occupied by the 30 m/s target is given by

$$SNR = \frac{P_{pk} G A_E \sigma}{(4\pi)^2 R^4} \Big/ P_{pk} \cdot CCN \cdot isolation = \frac{G A_E \sigma}{(4\pi)^2 R^4 \cdot CCN \cdot isolation}$$

For $G = 2500$, $A_E = 0.18m^2$, $\sigma = 5m^2$, and $R = 60$ km one obtains

$$SNR \approx \frac{10^{-18}}{10^{-10} \cdot isolation} = \frac{10^{-8}}{isolation}$$

Thus, to obtain an SNR of 100 (= 20 dB), an isolation of 10^{-10} or (−)100 dB is required (it may be assumed that the isolation, in decibels, is negative, so the negative sign is often omitted).

If the close-to-carrier noise leakage power, $N_{CCN\_leakage}$, is not to be the factor limiting the detection performance of the radar, then it must be considerably less than the thermal noise power in the Doppler cell. This places a limit on the maximum power that may be transmitted. The thermal noise power (equivalent input noise power) is given by kT_0B_NF (see Section 8.4.2). Therefore, the condition becomes

$$P_{pk} \cdot CCN \cdot isolation \leq kT_0B_NF$$

This constrains the transmitted power, $P_{pk} \leq kT_0B_NF/(CCN \cdot isolation)$

Assuming that $F = 4$ (6 dB) and recalling that $k = 1.38 \times 10^{-23}$ J/K and $T_0 = 290$ K, then the thermal noise power is 1.6×10^{-18} watts/100 Hz (= −148 dBm/100 Hz), and the transmitted power is limited to $P_{pk} \leq (1.6 \times 10^{-18})/(10^{-10} \times 10^{-10}) \leq 160$ watts, or, in decibels with respect to a milliwatt, $P_{pk} \leq -148$ dBm + 100 dB + 100 dB giving $P_{pk} \leq +52$ dBm, which is the same result.

---

[1]This is a purely arbitrary assumption for illustrative purposes only.

When a transmitted power of precisely 160 W is used, the close-to-carrier noise leakage and thermal noise powers are equal, giving rise to a 3 dB increase in the receiver noise floor. If a transmitted power in excess of 160 watts is used, then the leakage of close-to-carrier noise would limit the detection performance (at least this would be the case in the 2 kHz Doppler cell). To avoid this one must either improve the close-to-carrier noise performance of the transmitted signal or improve the receiver isolation or the Doppler resolution.

8.2.2.11 Phase Noise Cancellation

Suppose the (phase noise contaminated) transmitted signal is injected directly into the receiver subsystem. Since the receiver LO would be contaminated by the same phase distortion, assuming they have been generated in the same way, the down-conversion process results in cancellation of the phase noise and the resulting signal would be pure. In reality, target echoes are range delayed and therefore exhibit a phase change on reception and only partial cancellation of the phase noise occurs. In general, the greater the range of the target, the less the cancellation of phase noise.

Consider two phasors of equal amplitude but of phase difference θ. The vector difference between the two can be expressed as (1,0) – ($\cos\theta$, $\sin\theta$). The relative power of this difference is thus

$$\begin{aligned}&(1-\cos\theta)^2+\sin^2\theta,\\&=2-2\cos\theta\end{aligned}$$

For a target at range R, the time delay between transmission and reception is $2R/c$. For a phase noise component at a frequency offset f_o this means there is a phase change of $2Rf_o2\pi/c$ radians. Inserting this as the value of θ and expressing the cancellation in decibels, the cancellation factor is

$$cancellation(\text{dB}) = 10\log_{10}\left(2-2\cos\left(\frac{4\pi Rf_o}{c}\right)\right) \tag{8-4}$$

This cancellation is illustrated in Figure 8-6 [21].

At offsets beyond 100 Hz the use of common oscillators can actually enhance the phase noise problem, and one must rely on the absolute purity of the signals. Fortunately, it is easier to design good spectral purity at larger offsets. At very small offsets, where phase noise tends to be worst, the coherent cancellation can be a considerable help. Note that for the phase noise cancellation to hold in its entirety, the receiver local oscillator should be coherent with the transmitter signal.

8.2.3 Magnetron

The magnetron is an incoherent source of high-power microwave signals at the full transmitted power and at the PRF and pulse width required by the radar. It is controlled by a modulator that provides a series of negative extremely high tension (EHT) voltage pulses defining the PRF and pulse width. The microwave frequency of oscillation is determined by the internal dimensions of the magnetron.

8.2.3.1 Structure

The internal structure of a magnetron is shown in Figure 8-7. The outer anode is electrically grounded and is typically manufactured from a copper block. It has a set of (often eight)

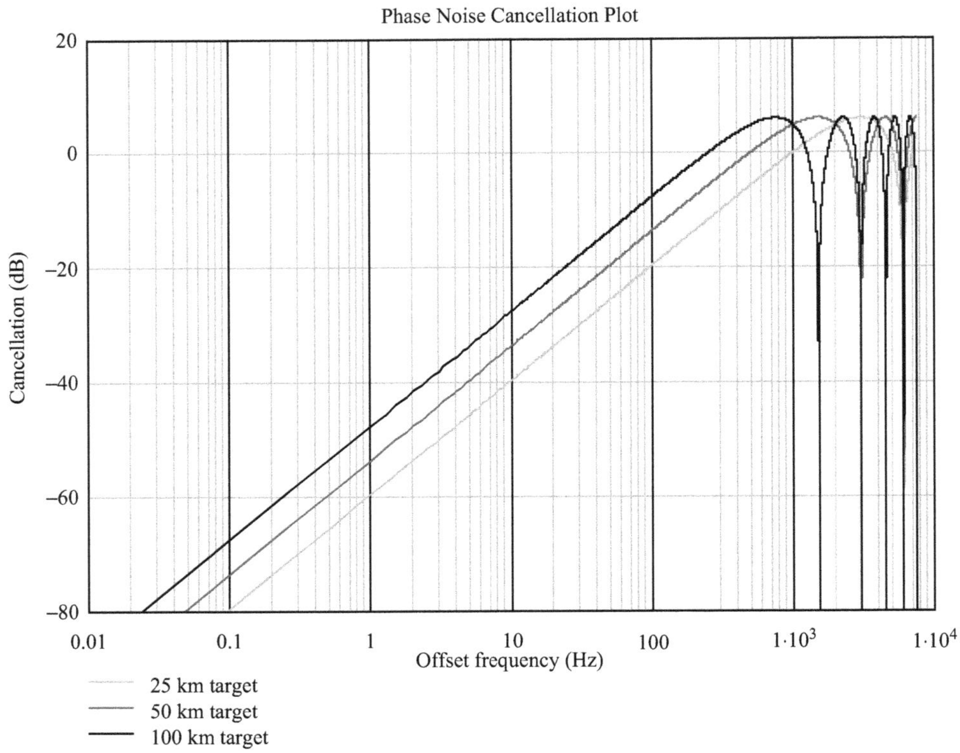

FIGURE 8-6 Phase Noise Cancellation in a Coherent System

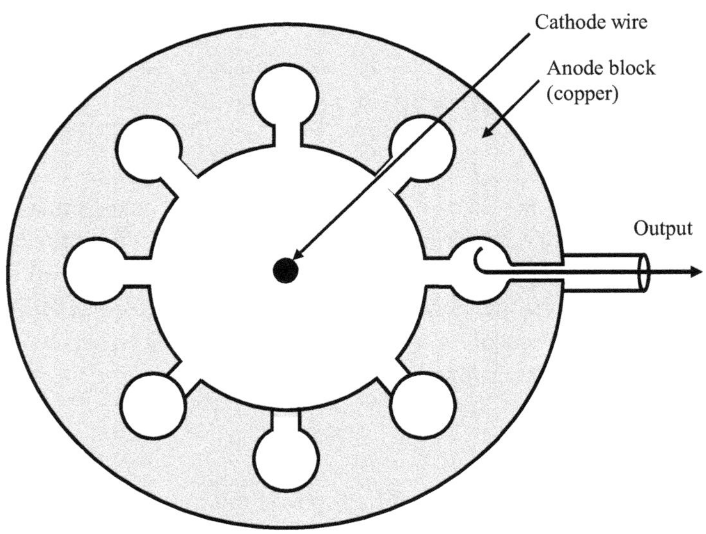

FIGURE 8-7 Basic Structure of a Magnetron

resonant cavities machined into it; these may be of various shapes, but the circular ones shown in Figure 8-7 are most common. A "rising sun" configuration of cavities is illustrated in Figure 8-9 and is more prevalent for magnetrons operating at 10 GHz and higher. The size of this structure determines the resonant frequency of the magnetron. Microwave power is taken via a probe in one of the anode cavities out into a waveguide run up toward the antenna. During operation the anode block runs hot and thus is fitted with cooling fins

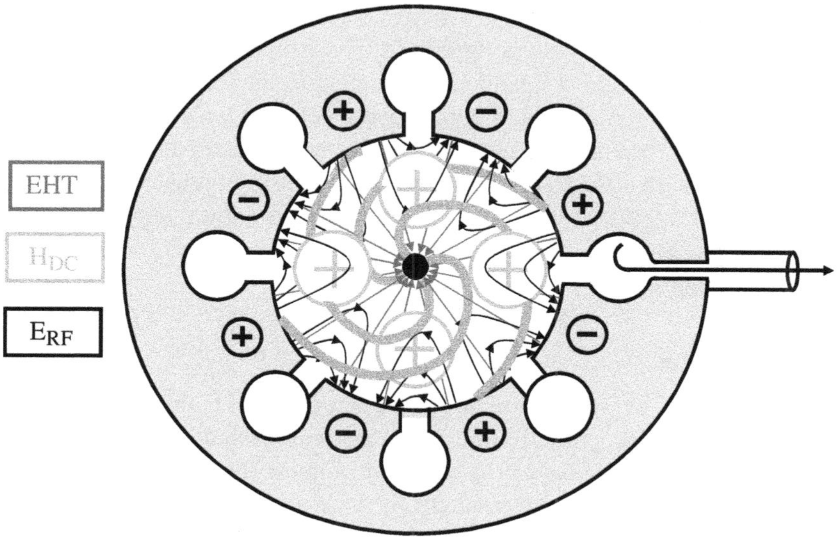

FIGURE 8-8 Magnetron Field Pattern

on the outside. Running up the middle is the cathode wire, the EHT pulses are applied to this wire. The cathode is coated in materials that emit electrons when heated. During startup the cathode is heated for this purpose, but after a short while the heating current may be switched off because it too runs hot. A large permanent magnet (not shown) provides a powerful DC magnetic field in the direction into the page. This contributes significant weight and bulk.

8.2.3.2 Operation

When the negative EHT is applied to the cathode, the electrons it emits are accelerated toward the anode block. Most of the acceleration takes place close to the cathode since the static E field generated by the EHT is strongest in this region, as shown in Figure 8-8. Since these electrons are moving perpendicular to the DC magnetic field they experience a force perpendicular to their motion such that they follow curved paths that terminate on the anode block. RF fields are excited in the resonant structures that extend into the space between the cathode and anode. These fields couple all the cavities together with the result that energy extracted from one cavity drains the whole structure. Alternate anode blocks are therefore π radians out of phase with their neighbors, as indicated by the + and − signs in the diagram. This mode of oscillation is therefore known as the π mode. The RF fields are initially weak, being excited by random noise at first. The RF E field (Figure 8-8) velocity modulates the majority of the electrons such that they form up into a series of spiral arms known as *spokes* that terminate on a positive block of the anode. Thus, in an eight-cavity magnetron, four spokes exist, and since each block reverses its phase each half-cycle, the spokes rotate around the device. As they do so they whip across the cavity apertures and kick new life into the RF oscillations via shock excitement. As the electrons in the spiral arms traverse from the cathode to the anode, they run against the RF E field and thus surrender energy to it. This mechanism produces considerable RF power. Electrons straying out of the spoke see a marginally different RF E field phase that either accelerates them or decelerates them such that they are forced back into the spoke. This effect is known as *velocity modulation,* and it facilitates a relatively efficient operation. Energy is initially supplied by the EHT, which is then converted into the kinetic energy of the electrons that then surrender it to the RF fields. A minority of electrons (not shown) are released from the cathode into a region of accelerating E field. These electrons extract energy from the RF E field, and hence they are accelerated. This increases the force from the DC magnetic field, and they curve more tightly and impact back on the cathode. This represents a loss mechanism; however, fortunately they do not form spokes and are quickly returned to the cathode. On impact with the cathode they generate heat ensuring that the device continues to run hot, and after a few minutes of operation the heater current applied to the cathode can be switched off. The loss mechanism due to these electrons is far outweighed by the gain of those in the desired spokes.

8.2.3.3 Limitations

A magnetron is pulsed by application and removal of the EHT; this effectively switches the device on and off. The phase of one pulse has no correlation with the phase of the previous pulse. A pulsed magnetron is therefore a non-coherent source of high-power microwave RF signals. Being non-coherent, it does not produce a neatly defined sinc function spectrum. It is therefore unsuitable for use in the transmitter of a coherent pulse Doppler radar. Although it is possible to externally lock a magnetron to a stable reference it is far better to use a device that is inherently coherent should coherency be required.

Since the frequency is determined by resonant structures, a magnetron is narrow band and limited to operation below about 40 GHz (above which the device becomes too small to manufacture). A magnetron may, however, be tuned by 5–10% using mechanical means to alter the resonant frequency of the cavity structure. The "crown of thorns" technique involves the insertion of rods into the cavity. Alternatively, a disk may be rotated over the cavities in a technique known as *spin tuning*.

A magnetron is susceptible to frequency pulling if it is connected to a mismatched load. Mismatch reflections re-enter the cavity and distort the field pattern leading to a detuning and loss of power. Load variations may be encountered in the rotating joints of a radar scanner. Thus, it is important to provide a good output match; otherwise, the magnetron may be detuned and suffer a loss of efficiency.

8.2.3.4 The Coaxial Magnetron

A coaxial magnetron (Figure 8-9) introduces a second resonant cavity coaxially surrounding the inner cavity structure and is thus much less susceptible to pulling. Further frequency changes are likely due to frequency pushing (change of RF with anode current) and frequency drift with temperature especially as the magnetron expands during warmup. The modulator must be very stable and consistent on a pulse-to-pulse basis to minimize frequency pushing. Typical EHT voltages are 10–20 kV. Long-term frequency changes also occur as the magnetron ages since the permanent magnets lose magnetism. The device is vulnerable to mishandling.

8.2.3.5 Modes of Oscillation

A magnetron could oscillate in other modes, as illustrated in the upper diagrams of Figure 8-10. This would lead to oscillation at a different frequency and a large reduction in efficiency. Echelon and ring strapping techniques are shown in the lower diagrams of Figure 8-10, in which alternate anode blocks are connected by a wire. Strapping ensures the π mode of oscillation by forcing each alternate anode block to the same potential.

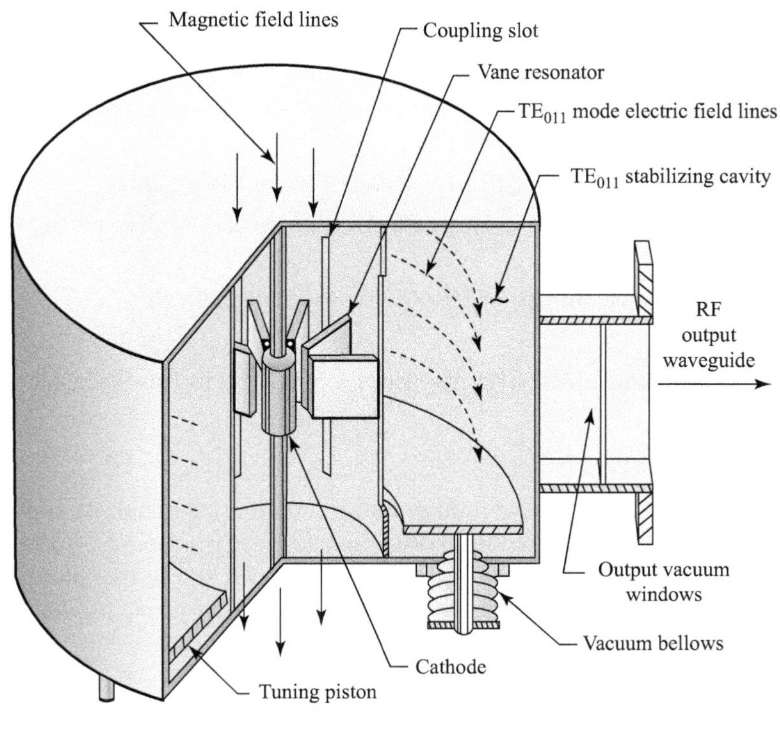

Coaxial magnetron

FIGURE 8-9 Coaxial Magnetron Illustrating "Rising Sun" Cavities

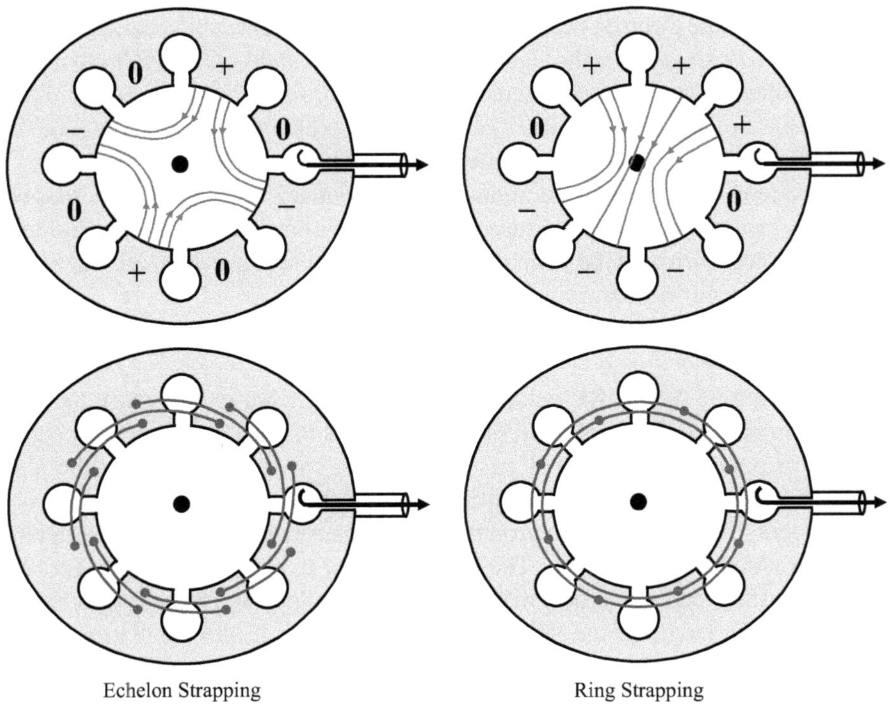

FIGURE 8-10 Other Magnetron Modes and Strapping

8.2.3.6 Performance

The following data typify the performance of a magnetron:

- Frequency: Usually the centimetric wave band but can produce reasonable peak powers into the millimetric wave band.
- Power: Centimeter wave band: 10 MW (peak), several kW (mean)
- Millimeter wave band: 100 kW peak at 35 GHz, 6 kW peak at 94 GHz, that is, $Pf^2 \approx 10^8$ to 5×10^7
- Efficiency: 80% (conventional) or 30–60% (coaxial magnetron)
- Tuning: 5–10%
- Tuning speed (spin tuned, 10 GHz magnetron): 500 MHz in 1 millisecond

8.2.4 Klystron

The klystron is mainly used as a very high-power amplifier, although its applications in radar are now rather limited. A low-level drive signal (usually pulsed) has to be applied to the klystron, which amplifies it to the full transmitter power and further pulses the signal to define the PRF and pulse width. The reflex klystron is a high-power oscillator but is old and seldom used and so is not considered any further.

8.2.4.1 Structure

The basic structure of a simple two-cavity klystron is shown in Figure 8-11. The device consists of an electron gun that fires a stream of electrons down its length at a very high energy. This requires a high EHT to provide the acceleration potential. The electron gun includes a grid plate that can turn the electron beam off by the application of a bias of a few hundred volts. The electron beam travels through a resonant cavity, called the *buncher* cavity, along a *drift space* then through a second cavity, called the *catcher* cavity, before striking the collector and being recirculated by the external biasing circuitry. The low-level RF drive is applied to the buncher cavity, after a preamplification stage (not shown) to a power level of a few watts. High-power RF is taken from the catcher cavity into a waveguide system. Both cavities are tuned to the frequency of transmission, that is, that of the drive signal. The whole structure is enclosed within a vacuum tube. Solenoid or permanent magnets surround the structure (not shown) to produce an axial magnetic field that confines the electron stream to the center of the tube.

8.2.4.2 Operation

As shown in Figure 8-12 the RF fields applied to the buncher cavity cause the electron stream to become velocity modulated; that is, electrons in the stream are subjected to an accelerating or retarding field as they pass through the cavity depending on the phase of the RF cycle when they made the transition through the cavity. Note that this does not change the average energy in the electron beam. This causes the electrons to form bunches as they move along the drift space. The catcher cavity is placed at the point where the bunches are most concentrated (they disperse after this point). The bunches of charge pass through the catcher cavity at the RF and thus shock excite it into oscillation. Since the electron bunches are very high energy, considerable RF power is imparted to the catcher cavity and is available on the output. The electrons have yielded much, but not all, of their energy to the catcher cavity and continue on to the collector. Since they still possess

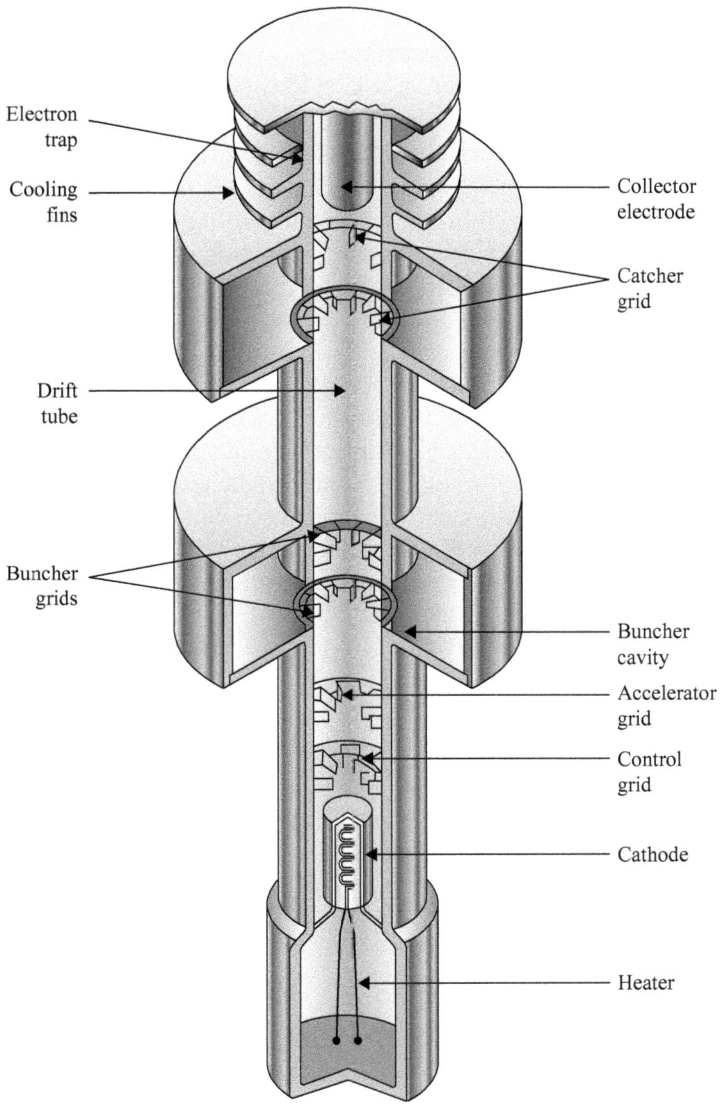

FIGURE 8-11 Structure of a Klystron

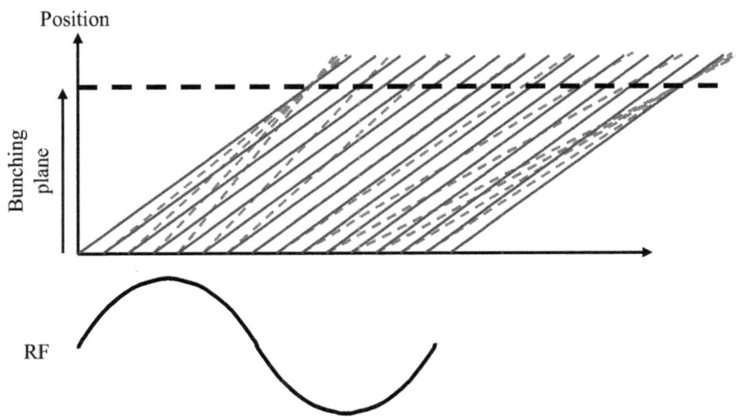

FIGURE 8-12 An Applegate Diagram Illustrating Velocity Modulation

considerable energy, soft X-rays are emitted at the collector as they are decelerated, and the collector must be cooled using external fins.

Figure 8-12 is an *Applegate diagram,* which illustrates the operation. The solid lines denote the path of electrons in the absence of the RF and the dashed lines indicate the path of the electrons once they have been accelerated and decelerated by the RF applied to the buncher. Bunches of electrons pass through the catcher cavity at a frequency defined by the RF. Pulsed operation is achieved by a combination of pulsing the RF drive and modulating the grid control. The RF drive pulse must be applied early since it takes a cavity several microseconds to achieve stable oscillation at the correct RF; this is determined by the cavity Q factor. The exact transmitted pulse width is determined by the grid bias waveform. Phase coherency between pulses is maintained. A klystron is also capable of CW operation.

The efficiency of a klystron depends on the concentration of the electron bunches. Most klystrons exhibit features to maximize the bunching effect. One commonly used technique is to insert one or two free cavities within the drift space. These oscillate in sympathy with the RF and further velocity modulate the electron beam and so improve the bunching. The axial magnetic field imparts a force on any electrons that attempt to move across its field. This overcomes the defocusing effect of the beam due to mutual repulsion and also ensures that the beam strikes the collector since otherwise the high-energy electrons would damage the klystron.

8.2.4.3 Limitations

Since the frequency of operation is determined by resonant cavities, the klystron is a narrow-band device. The bandwidth may be improved, from about 5%, by stagger tuning the free cavities, to around 10–20%. Note that frequency agility is provided by the drive signal. Hazards exist because of the X-rays from the collector and the high accelerating voltages used. The efficiency is rather poor since not all the energy should be extracted from the beam in the catcher cavity.

8.2.4.4 Performance

The following data typifies the performance of a klystron:

- Frequency of operation: Through the microwave and millimetric wave band, but most applications are low microwave frequencies (< 4 GHz) and ultra high frequency (UHF) (commercial broadcast). Performance degrades at higher frequencies.
- Gain: 60 dB or more
- Efficiency: 35–40%
- Power: 20 MW (peak), 350 kW (mean)
- Bandwidth: 5% (single tuned), 10–20% (stagger tuned)

8.2.4.5 The Extended Interaction Klystron, EIK

There is nothing preventing the design of klystrons being scaled to millimetric wave frequencies; however, the performance degrades considerably. At higher frequencies the following degradations can be expected:

- Reduced gain
- The requirement for stronger magnetic fields
- High cathode current density in miniature electron guns

High-frequency klystrons are characterized by a higher beam voltage and lower beam current compared with their lower frequency counterparts. Some of these problems of high-frequency operation can be overcome using the extended interaction klystron (EIK) or amplifiers (EIA) in which the narrow gaps through the cavities of an ordinary design are replaced by longer slow wave structures, commonly called ladder circuits. These devices are capable of operating frequencies of 25 to 220 GHz and at peak power levels of 2 kW to 40 W, respectively.

8.2.5 Traveling Wave Tube

The TWT, a high-power amplifying tube device, is popular in many radar and electronic warfare (EW) applications.

8.2.5.1 Structure

The structure of a TWT is shown in Figure 8-13. It consists of an electron gun firing a stream of electrons along the device toward a collector at the far end. The electron gun has a grid plate biased by a modulator so that the electron beam can be pulsed on and off. External circuitry recirculates the electrons from the collector back to the electron gun. The RF is applied to a slow wave structure shown in Figure 8-13 as the helix. Amplification of the RF takes place as it progresses along the helix via an interaction between the RF fields and the electron beam. The structure is accommodated within a vacuum tube. Solenoids or permanent magnets (not shown) surround the device to keep the electron beam focused.

8.2.5.2 Operation

The axial velocity of the RF along the slow wave structure is considerably slower than the speed of light. The velocity of the electrons in the beam is set to be slightly higher than the axial velocity of the RF. The RF E field velocity modulates the electrons and holds them in bunches that travel slightly faster than the RF on the slow wave structure. The electron bunches tend to be held in regions of the RF E field at which a null exists; however, because they are moving slightly faster than the RF the bunches pull forward into a region of E field that retards the electrons. The electron bunches surrender energy to the RF E field as they attempt to overtake it, so energy is transferred from the electron beam to the RF. As the RF is amplified, the bunching effect is reinforced, and further amplification takes place. Amplification takes place continuously along the length of the TWT but cannot go on indefinitely since eventually so much energy is extracted from the electron beam that it slows down to the same velocity as the axial velocity of the RF and

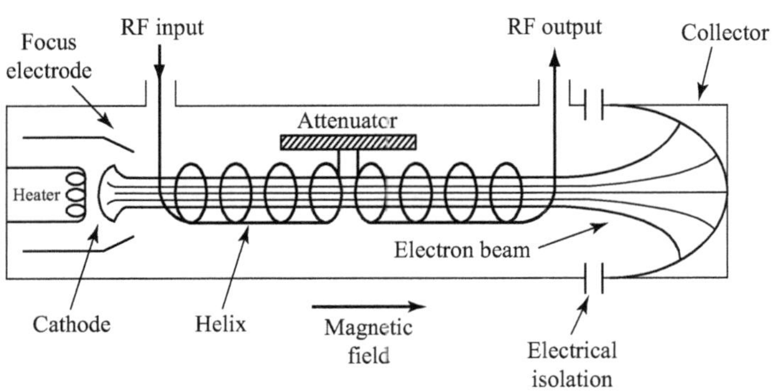

FIGURE 8-13 ■ Traveling Wave Tube

no further amplification takes place. Should the beam slow any more, it would extract energy from the RF.

A potential problem of the TWT is that reflections of the RF from a mismatched load can propagate back down the helix in the wrong direction. Although this reflection would not be amplified, if it is further reflected from the input end back in the forward direction there is a feedback situation that could cause undesired oscillations. This is countered by the use of an attenuating material on the inside of the TWT structure. This attenuates the reflected wave, thereby removing the feedback mechanism, and the forward traveling wave. The attenuation of the forward traveling wave is more than compensated for by the large forward gain.

There are alternative slow wave structures besides the helix. The state-of-the-art in TWTs is to use a cavity coupled slow wave structure, ccTWT, which achieves a $Pf^2 \approx 5 \times 10^7$.

As with the klystron, the RF usually requires some preamplification to a few watts. Pulsed operation is achieved by pulsing the RF drive and modulating the grid bias, and the phase coherency of the drive signal pulses is maintained. CW operation is also possible.

8.2.5.3 Performance

The following data typify the performance of a TWT:

- Frequency: 100 GHz plus
- Bandwidth: 10:1
- Power: 2–5 MW (peak), 1 kW (CW)
- Gain: 40–50 dB
- Efficiency: 20–40%
- Noise figure (low power): 2 dB

8.2.6 Comparison of High-Power Tubes

Many other high-power microwave tubes and variants of the magnetron, klystron and TWT are available but not considered here. The previous sections describe the most common types only.

There are few limitations of the TWT; it outperforms the magnetron and klystron in many respects. Its big advantage over other tubes is that since it does not use resonant structures it is very wideband and is the device most often used in EW applications. It also has a very low noise figure, which makes it the preferred choice in pulse Doppler radar transmitters. TWTs can operate well into the millimetric band (100 GHz). The TWT is not quite capable of the same output powers as the klystron; however, miniature TWTs are being developed for medium power amplification (a few 10s of watts) and may rival solid-state amplifiers if they can be made small and light enough.

Magnetrons have the advantage of producing high-power RF at an affordable price and are efficient. The modulators have to be able to switch the full EHT (kV) as opposed to the few 100 V bias voltages used to modulate the beam currents in TWTs and klystrons. Magnetrons also provide a phase incoherent RF source and have short- and long-term frequency stability problems that limit their use in modern radar systems.

Klystrons produce the highest power levels of any of the tubes considered but at the expense of size, weight, and radiation and high voltage hazards. They are used in high-power broadcast transmitters and in some high-power radar applications that can afford their size and weight, such as some naval radars.

8.2.7 Solid-State Oscillators

Solid-state oscillators are capable of producing a few tens of watts of RF power. They are used in radar transmitters when the power of several such devices (possibly hundreds or even thousands) is combined as occurs in AESA antennas or when a very small, short-range system is required, such as a missile seeker. Solid-state oscillators may also be used to provide the drive signals for high-power tubes and as local oscillators within superheterodyne receivers. At the heart of the power source is the transistor (or diode), which oscillates at the desired RF [20].

8.2.8 The Silicon Bipolar Transistor

Conventional silicon (Si)-only bipolar transistors continue to be used in cheap, high-volume production oscillator circuits below about 2 GHz. In general they have:

- Good power handling (a few 10s of watts peak)
- Rapid tuning over a wide bandwidth
- Low $1/f$ noise
- Can be locked to external sources to ensure coherency
- Robust and very reliable

The maximum frequency of operation is determined by the electron mobility in silicon, the base width, and parasitic reactances, often associated with its packaging. A base width of 0.1 μm yields a transit time of 20 ps, which in turn gives a usable bandwidth of up to around 10 GHz. The low $1/f$ noise makes Si bipolars a popular choice in low microwave frequency ($<$2 GHz) oscillators and VCO circuits used in indirect frequency synthesizers. However, at frequencies above about 2 GHz, their performance in oscillator and amplifier circuits is surpassed by other transistor types and semiconductor materials.

8.2.9 The Gallium Arsenide (GaAs) Field Effect Transistor (FET)

The GaAs metal-semiconductor field effect transistor (MESFET) has established itself as a favorite device in many microwave oscillators and amplifiers, especially where the combination of high operating frequency and low noise are required. The frequency limiting factors are gate length and carrier mobility, which is higher in GaAs than in Si and of course parasitic reactances. Compared with Si bipolars, MESFETs exhibit the following:

- Higher gain
- Higher frequency of operation
- Lower noise figures
- Higher output powers

GaAs MESFETs have also been used as RF sources in centimetric wave active antenna arrays. They can provide a few 10s watts of peak RF power at centimetric wave frequencies. Power and efficiency reduces as frequency increases.

8.2.10 Heterojunction Bipolar Transistors (HBTs)

Heterojunction devices are those formed by semiconductors of differing compositions. Si doped AlGaAs/undoped AlGaAs and InGaAs/InP are favorite materials. Pseudomorphic heterojunctions are those whose semiconductor materials are grown to provide matched crystal lattice constants. HBTs are used in amplifiers, oscillators (e.g., VCOs), and high-speed, very large-scale integration (VLSI) integrated circuits (ICs), since they are very high speed and have higher operating frequencies than both Si bipolars and GaAs FETs. They also boast higher powers and lower noise than GaAs FETs. An HBT can achieve 5 watts CW at 10 GHz and frequencies of operation well into the millimetric wave band, albeit with decreasing efficiency as frequency increases.

8.2.11 High Electron Mobility Transistor (HEMT)

The HEMT is another heterojunction device that may also be formed from pseudomorphic junctions (p-HEMT). They can be fabricated with very high carrier mobility and small gates, but their narrow channel widths lead to low power handling. They are very low noise and have the highest operating frequency of all the transistors (200 GHz). They are used as (local) oscillators, low noise amplifiers (LNAs), mixers, attenuators, and phase shifters.

8.2.12 Impact Ionization Avalanche Transit Time (IMPATT) Diode

This diode is an oscillating device that, when heavily reversed biased, uses impact ionization to create an avalanche of charge that drifts along a transit time region to produce negative resistance oscillations. (*Negative resistance* is the rather strange property by which the current through a device increases as the voltage across it decreases and is exhibited by certain diodes and transistors suitably biased and embedded into tuned circuits. It supports oscillation at the resonant frequency of the circuit.) The frequency of operation is determined by the length of the transit time region. As this can be made very small, an IMPATT diode can oscillate at up to 300 GHz; however, its DC to RF efficiency reduces with increasing frequency. The efficiency is around 8 to 15% in the millimeter wave band. They can be manufactured from Si and GaAs and operate in CW or pulsed modes. Individual diodes can output nearly 1 watt CW at 94 GHz. The Pf^2 of a single Si IMPATT diode is approximately 10^4.

The outputs of several diodes may be combined to form higher output transmitter powers. However, the efficiency of the combining networks reduces and their complexity increases, that is, as the number of diodes increases. There comes a point of diminishing returns that limits the total transmitter power of such a system to around a few tens of watts at 94 GHz. These devices are often used in millimetric wave active missile seekers and terminally guided submunitions. IMPATT diodes are broadband but noisy devices. They can be used in local oscillator and amplifier applications but need to be (injection-) locked to lower noise references, such as Gunn diodes.

8.2.13 Gunn Diode

The Gunn diode is another negative resistance device, in which at a critical bias condition a domain of low mobility electrons drift along the bulk of the diode. When they reach the anode, a current pulse is generated and another domain of low mobility electrons forms at the cathode and starts to drift across. The upper frequency limit is determined by the electron mobility and diode size. GaAs Gunn diodes can operate to 110 GHz, whereas InP Gunn diodes can operate up to around 200 GHz and achieve a Pf^2 in the order of 10^5.

Gunn diodes are lower noise than IMPATT diodes and are suitable for use as local oscillators and amplifiers. Gunn diode oscillators may be tuned using varactor diodes or mechanically by mounting them in a resonant cavity of variable dimensions and hence of variable resonant frequency.

8.2.14 In Summary

Several other transistors and diodes may be used to provide RF power, but the main ones have been covered here. Power at high microwave frequencies is always problematic. Power at millimetric wave frequencies may also be derived through frequency multiplication from lower frequencies. While this process is inefficient and results in high close-to-carrier noise and low powers, it may still be the best option for deriving RF power above 200 GHz.

8.3 | FREQUENCY SYNTHESIZERS

A frequency synthesizer is an RF circuit that generates an output signal derived from stable, low noise reference sources. Typically, there may be one or more microwave output signals each of which are capable of being tuned over several channel frequencies; these channel frequencies often have a regular spacing frequency. The output signals inherit the stability and low noise characteristics of the low frequency reference sources that are typically far superior to that obtainable from a free-running microwave oscillator. Frequency synthesizers also enable a large number of channel frequencies to be derived from a small number of oscillators. Two or more output signals may be locked together to ensure their phase coherency with each other.

Frequency synthesizers are used in radar to generate a low power sample of the transmitted signal, termed the *transmitter drive* or simply the *Tx drive*, with all the modulations required by the radar. This is then applied to a high-power amplifier such as a TWT in a MOPA-type transmitter. The synthesizer would also generate the first local oscillator (LO1) phase locked to the transmitter drive and applied to the front end down-converters used in the receiver. The LO1 signal frequency falls in the same band as the Tx drive but is offset from the Tx drive by a frequency equal to the system's first intermediate frequency (first IF). Both the Tx drive and LO1 signals are required to change channel together to maintain a constant first IF independent of the channel frequency. The synthesizer may also generate the second (and third) local oscillator signals used by the receiver in further down-conversion stages.

To achieve the most stable frequencies of the lowest close-to-carrier noise possible, all signals are referenced to a high-quality RF oscillator. A quartz crystal oscillator operating in the HF band of 3 to 30 MHz is often the preferred reference oscillator. Most current radars would use the very best reference oscillator as a fundamental clock to coordinate all system timings and as a reference for the synthesis of all RF and microwave signals. The use of a single fundamental reference oscillator ensures that any drift in its frequency affects *all* other signals such that they drift together. The drift in one signal is counteracted by the drift in the others, and full system coherency is maintained. While this suggests a single point of failure, such oscillators are very reliable and run problem-free for many years. The reference oscillator would be mounted in a thermostatically controlled cell to maintain a constant operating temperature and so minimize the effects of temperature related drift. For radars fitted in many military ground and airborne vehicles the reference oscillator would also be accommodated on antivibration (AV) mounts to minimize microphony effects that would modulate the oscillator output due to the pick-up of vibration and physical shock.

Parameters of importance in the design of frequency synthesizers for use in pulse Doppler radar systems together with typical performance requirements are as follows:

- Number of channel frequencies: 10 s to 100 s
- Channel spacing: 10 MHz
- Total band width: 5% of center frequency
- Channel switching speed: 1 μs
- Output power: 100 mW (20 dBm $\pm$ 0.5 dB)
- Long-term frequency stability: $\pm$100 kHz
- Harmonic suppression: -30 dBc
- Close-to-carrier phase noise: -100 dBc/Hz at 1 kHz offset
- Spurious lines: -70 dBc

Three types of synthesizers are commonly used (in addition to hybrids of these): (1) direct digital synthesizers (DDSs); (2) direct analog synthesizers (DASs); and (3) indirect synthesizers based on PLLs. These are considered in the following sections.

8.3.1 Direct Digital Synthesizers

Direct digital synthesizers (DDSs) do not operate in the microwave band, but they may be used in the synthesis of high-quality reference signals and modulation waveforms used in other synthesizers. A DDS has been considered for use as the frequency synthesizer for a HF radar since the frequencies (3 to 30 MHz) may be digitally synthesized directly [21].

The structure of a DDS is shown in Figure 8-14. It uses a clock to increment a counter, which addresses a read-only memory (ROM) that has been preloaded with samples of a sine wave or whatever waveform may be required. On each successive clock period the counter increments to the next address of the ROM and a sequence of samples representing the waveform are available on the ROM output. A digital-to-analog converter (DAC) converts the digital representation of the waveform from the ROM into an analog voltage level in a sequence of steps governed by the clock speed. Finally, smoothing of the output using a filter is applied to suppress any breakthrough of the clock frequency on the analog output. In its simplest form, the ROM may hold the samples corresponding to one cycle of a sine wave. As the process is repeated, repetitions of the basic sine wave are produced at some

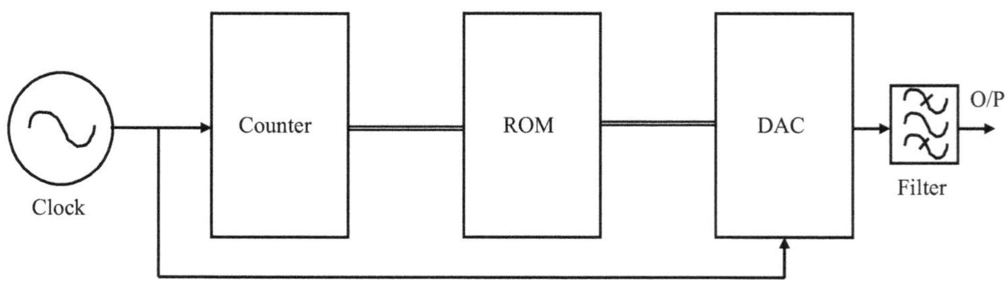

FIGURE 8-14 Direct Digital Synthesis

subharmonic of the clock frequency. Alternatively, with the availability of large ROMs, the ROM may hold samples of several sine wave cycles with an appropriate amplitude modulation applied, for example, for pulse shaping purposes. A further possibility is that a ramp or any other arbitrary waveform may be produced by preloading the appropriate samples into the ROM. The frequency of the output can be varied by changing the clock frequency. A change in clock frequency can be accomplished using a programmable divider to frequency divide the clock signal.

A DDS is extremely fast as it is mainly limited by the digital control. Very high switching speeds, comparable with direct analog schemes, are available. DDSs provide reasonably low phase noise (limited by its residual noise floor), even showing an improvement over the phase noise of the clock source itself. Exceptionally fine frequency resolution is available and is determined by the length of the counter; sub-Hz levels are easily achieved. On the negative side, a DDS has a limited usable bandwidth and poor spurious performance due to quantization and DAC conversion errors, which generally prohibits direct frequency multiplication of the DDS output up to microwave frequencies. Its highest frequency is limited by the Nyquist criterion to within one-half of the clock frequency. The practical design requires a low- or band-pass filter on its output to reconstruct the signal waveform, which further decreases the highest operation frequency to about 40% of the clock signal.

8.3.2 Direct Analog Synthesizers

Direct analog synthesisers (DASs) use several stable low-frequency oscillators, which are frequency multiplied, and mixed together to provide all the required channel frequencies. The signals are fed into a switching and filtering matrix to ensure that the correct harmonics of each signal are mixed together to produce the required channel. There are numerous possible designs for DASs, but an example is illustrated in Figure 8-15, in which

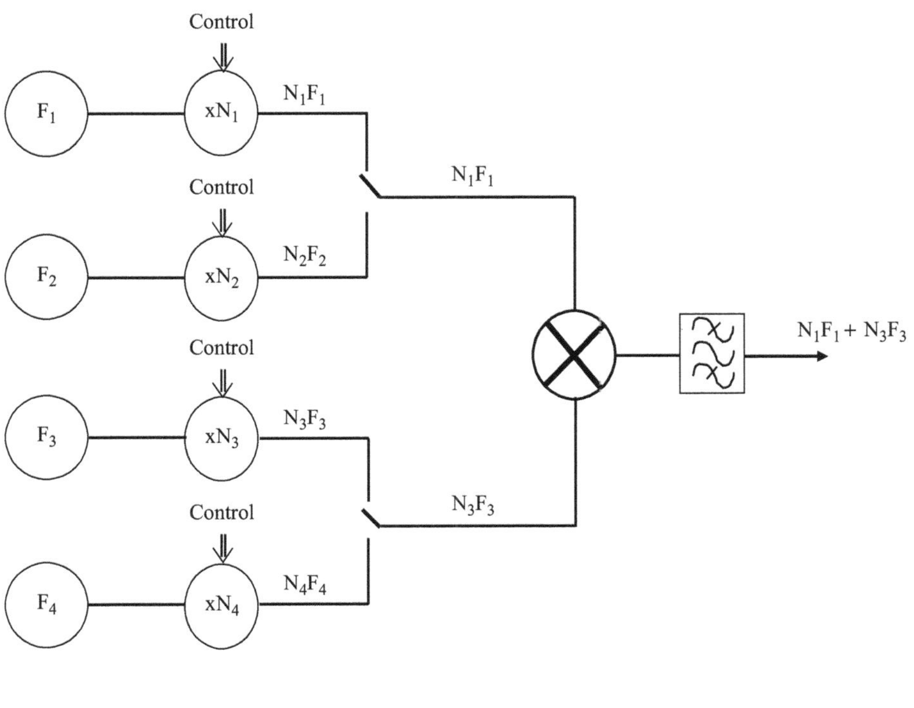

FIGURE 8-15 Direct Analog Synthesizer

four oscillators run at frequencies F_1, F_2, F_3, and F_4. These are frequency multiplied by factors N_1, N_2, N_3, and N_4, respectively. A switching matrix selects the desired products ($N_1 \cdot F_1$ and $N_3 \cdot F_3$), which are then combined in a mixer and a band-pass filter selects the sum of the intermodulation product ($N_1 \cdot F_1 + N_3 \cdot F_3$).

The difficulty with this approach is the filtering requirements to suppress spurious intermodulations. This problem is exacerbated by a requirement for many channels since this increases the complexity of the design. Channel change times can be very fast as this is governed by the speed of the switches. PIN diode switches can switch in a few nanoseconds. The suppression of unwanted signals is alleviated somewhat if DC power is removed from unused oscillators; however, this results in longer channel change times since the appropriate oscillators have to be powered up again when required. Phase coherency is ensured through the mixing process. Noise can be a problem since noise is increased in frequency multipliers by $10 \log_{10} N^2$, where N is the multiplication factor. Low-frequency oscillators such as quartz crystal oscillators are low noise and very stable (especially when temperature stabilized) but require high multiplication factors to take them to the microwave band. Direct synthesizers are not without their problems, but they are viable if only a few channel frequencies are required. Under these circumstances they are low noise and capable of fast channel change times.

8.3.3 Indirect Synthesizers, Phase-Locked Loops

A phased locked loop (PLL) circuit is a closed-loop circuit for the control of the phase of a VCO. The VCO is phase locked to low noise reference signals that define the VCO output frequency exactly. PLLs are used in RF and microwave frequency synthesis, the measurement of close-to-carrier noise (see section 8.2.2), and reception of weak signals of low SNR. A basic PLL used for frequency synthesis is illustrated in Figure 8-16.

The phase sensitive detector (PSD) derives the error in the phase between the VCO output (strictly, the divided loop IF signal) and Reference Oscillator 1 (input). The PSD is a balanced mixer supporting a DC output proportional to the cosine of the phase difference between its two inputs. When the PLL is locked, the two inputs to the PSD run at the same frequency and in-phase quadrature, that is, 90° to each other ($\cos 90° = 0$). Under locked conditions the intermediate frequency of the loop (i.e., the loop IF $= F_{out} - F_{ref2}$ and therefore $F_{ref1} = (F_{out} - F_{ref2})/N$. Rearranging this gives

$$F_{out} = N F_{ref1} + F_{ref2} \tag{8-5}$$

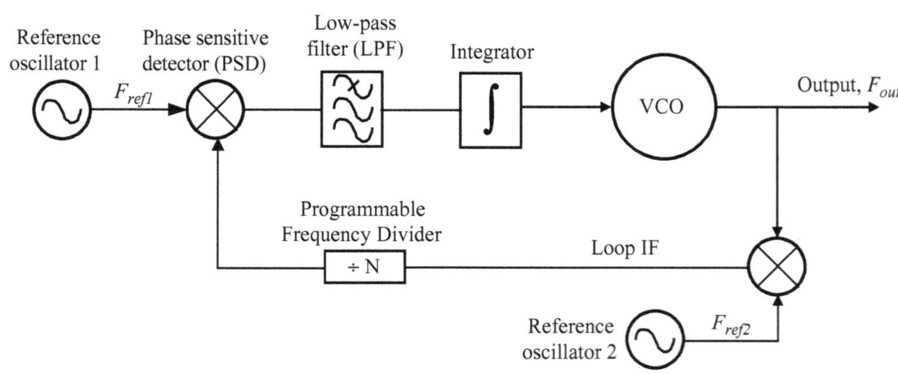

FIGURE 8-16 Basic PLL Synthesizer

A change of channel frequency can be brought about by changing either of the two reference frequencies or by changing the division ratio in the loop IF path; the latter is the usual technique. Should the division ratio change from N to $(N+1)$, the frequency, F_{out}, must change to restore the balance of equation (8-5). In going $N \rightarrow (N+1)$, there is initially a large phase error measured by the PSD that, after filtering and integration, retunes the VCO to $F'_{out} = (N+1)F_{ref1} + F_{ref2} = F_{out} + F_{ref1}$. Hence, the new output frequency differs from the old by F_{ref1}. Changing the division ratio by 1 causes a change in the output frequency by F_{ref1}; similarly, a change in the division ratio by 2 results in a change in output frequency by $2F_{ref1}$ and so on. Reference Oscillator 1 therefore defines the channel spacing frequency. These frequency relationships are maintained precisely when the loop is locked, although a small departure from the quadrature phase relationship between the two PSD inputs may exist. When locked, the integrator maintains the tuning voltage on the VCO (the integral of zero being a constant defined by the previous state of the loop). The frequency divider in the feedback path of the loop increases the noise within the loop (as does a frequency multiplier). In some cases a divider is omitted and a channel change is affected by a change in F_{ref1} or F_{ref2}.

When a reference frequency or the division ratio changes, the PSD outputs an error signal, which, after integration/filtering, retunes the VCO frequency to re-establish lock, provided that the error signal doesn't fall outside certain limits (the loop pull-in range). This loss of lock occurs when a large frequency step such as a change of channel is invoked. External circuitry needs to be included that continually monitors the PSD to ensure it remains locked. Should it detect a loss of lock, the external circuitry initiates its reacquisition by forcing the VCO to sweep through its band of operation. When the VCO is swept past its correct frequency, the error signal falls within the passband of the low-pass filter and the loop can be closed and would then reacquire lock of its own accord. This reacquisition mechanism would be invoked by any loss of lock. The VCO of an uncontrolled, out-of-lock PLL would be wandering aimlessly in frequency, whereas one trying to relock in this fashion would be typified by a frequency sweep. It is normal practice to safeguard the transmitter tube by gating out its RF drive during periods of lost lock and reacquisition. The rate at which the frequency sweep may be applied and therefore the speed with which a change of channel can be made are determined by the time constant for the whole loop. This is dominated by the low pass filter/integrator in the feedback path to tune the VCO. A wider loop bandwidth requires a shorter integration time and permits a faster change of channel. The loop's dynamic response is determined by its transfer function. Most PLL synthesizers are designed for a second- or third-order response with a wide loop bandwidth to minimize channel change times.

Either or both Reference Oscillator 1 and 2 may be derived from a DDS or a second PLL. Should any FM be required on the transmitted signal, then this can be imparted by modulating the frequency of Reference Oscillator 2. The loop's ability to track the FM faithfully and with minimal time slip is dependent on the transfer function of the loop and on its loop bandwidth in particular.

A free-running VCO has poor close-to-carrier noise performance at low offset frequencies, although the close-to-carrier noise improves at larger offset frequencies. Once embedded into a phase-locked loop, however, low frequency noise modulations arising from the VCO are suppressed by the action of the loop. Any VCO noise at an offset frequency below the loop bandwidth results in a corresponding modulation in the PSD output, which tunes the VCO to counteract its own noise. Hence, the loop suppresses VCO noise at offset frequencies within the loop bandwidth. This improves the spectral purity of the output at low offset frequencies. Any noise or discrete spurious phase and frequency modulations on the two reference oscillator signals would also modulate the PSD output signal. Should such modulations be within the loop bandwidth (this implies that they would be within the passband of the low-pass filter) then they would cause a corresponding modulation of the VCO output. However, higher-frequency phase noise and discrete spurious phase and frequency modulations are suppressed by the loop and do not modulate the VCO output. In this way, the close-to-carrier phase noise of the output is dominated by that of the VCO at offset frequencies greater than the loop bandwidth but by the reference oscillators at offset frequencies less than the loop bandwidth. Since the reference oscillators are low frequency, highly stable, and spectrally pure signals, it is desirable to allow these to determine the close-to-carrier noise characteristics of the synthesizer output at low offset frequencies. The loop bandwidth should therefore be set at an optimum point to take advantage of the low noise characteristics of the reference oscillators at low offset frequencies but then give way to the improving low noise characteristics of the VCO at the higher offset frequencies. Clearly, there exists a compromise in setting the optimum loop bandwidth to strike the best balance between the channel change time, tracking performance, and spectral purity.

PLL synthesizers that operate over wide bandwidths can sometimes output an RF power level that changes across the band because the response of the VCO is not sufficiently flat over its full bandwidth. An improved response can be obtained using a further feedback mechanism on the PLL output that senses the output power and provides a control signal to a variable attenuator, as illustrated in Figure 8-17.

In many radar applications it is necessary to derive LO signals that are coherent with the transmitter drive signal. This can be achieved using two PLLs, one of which is locked to the other, as illustrated in Figure 8-18. The lower loop synthesizes the LO1 signal, as previously described. A microwave switch on its output produces a pulsed LO1, if required. The LO1 signal provides the reference for the upper loop, which synthesizes the Tx drive signal. This produces a signal of frequency $F_{Tx} = F_{\mathrm{LO1}} - F_{IF}$ exactly, in which F_{IF} represents the system first IF. Slaving the one loop to the other ensures that the Tx drive is phase locked to the LO1 signal. This signal is also pulse modulated using a microwave switch on its output with the inverted logic control of the LO1 such that during the transmitted pulse the LO1 is off and hence the receiver is isolated. Both Tx drive and LO1 signals are coherent pulse trains. Any modulations and channel changes applied onto the lower loop are tracked by the upper loop such that both change channels together. Radar returns are down-converted using the F_{LO1} signal and result in an IF of $F_{IF} \pm f_d$. These are subsequently down-converted a second time using the LO2 signal, which runs

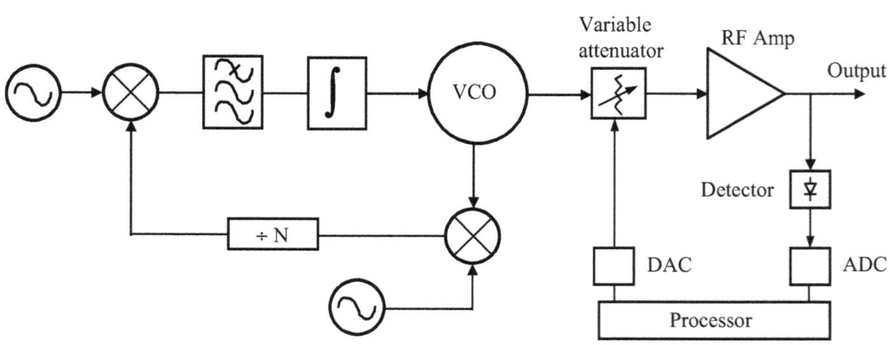

FIGURE 8-17 ■ PLL with Auto-Leveling

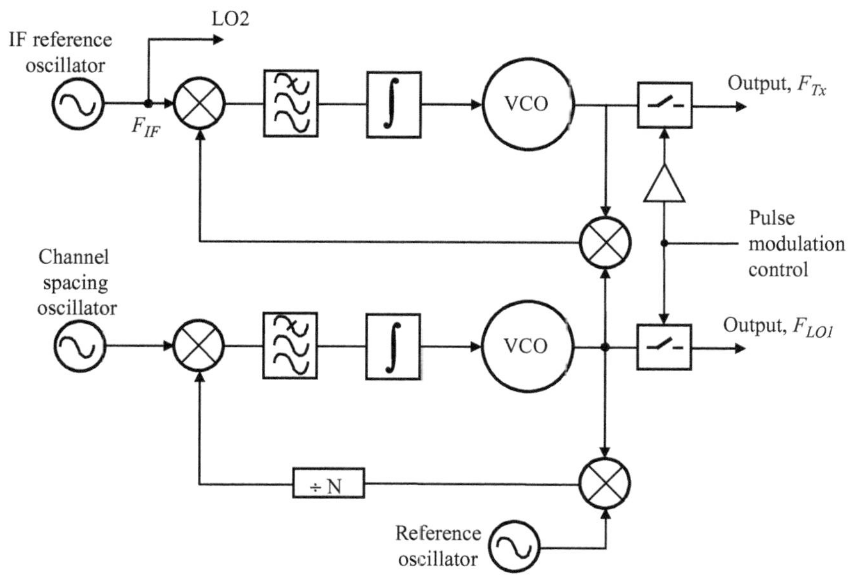

FIGURE 8-18 ■ Full Radar PLL Synthesizer

at a frequency of F_{IF}. This results in a video (or baseband) output at 0 Hz $\pm f_d$. This is a fully coherent system since all signals are phase locked to each other. This method of synthesis also ensures that drift in any one signal frequency results in a compensating drift in the other signals in order to maintain a baseband output at 0 Hz $\pm f_d$.

Indirect synthesizers based on PLLs offer excellent frequency stability and very low levels of close-to-carrier noise and spurious signals owing to the low-pass filter action of the loop. They can synthesize very many channel frequencies with fine channel spacing without incurring significant increases in complexity and cost. However, they exhibit a relatively long channel change time in the order of microseconds as opposed to the nanosecond switching time of DDS and DAS synthesizers.

8.4 RADAR RECEIVERS

The vast majority of radar receivers are the *superheterodyne* (also known simply as the *superhet*) design employing two or three stages of down-conversion (i.e., a *double* or *triple superhet*) since these are the most sensitive types. This section describes the superhet receiver and also presents some analysis of receiver noise figure.

8.4.1 The Superheterodyne Receiver

A radar receiver is conveniently represented as a series of components, as in Figure 8-19, in which the signal input is at the top left and the signal output is at the bottom right. Such a receiver is termed a *superheterodyne* receiver because the incoming signals are translated down in frequency in a process known as *down-conversion* or *heterodyning* to a lower frequency in a *mixer* using a signal known as the *local oscillator*. In Figure 8-19, there are in fact two stages of down-conversion, and hence this is specifically known as a *double superheterodyne* receiver. The down-conversion process is beneficial because it is easier and cheaper to provide the necessary amplification and filtering of radar returns at a frequency band well below the microwave band of operation but well above 0 Hertz

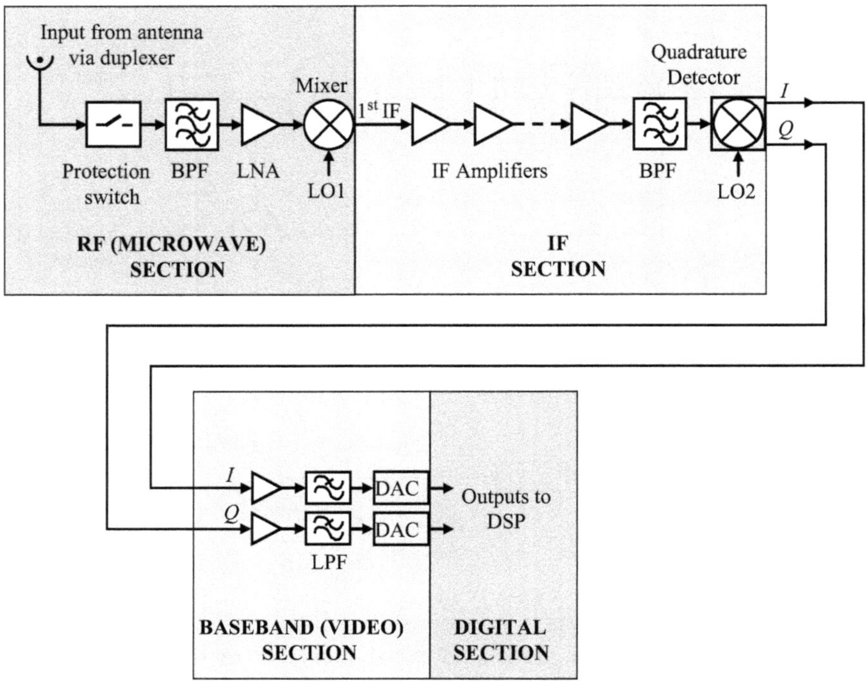

FIGURE 8-19 Double Superheterodyne Receiver

to avoid strong flicker noise close to DC. Incoming signals at the transmitted RF (± any Doppler shift) are applied to a mixer that is a highly non-linear circuit of diodes and results in the incoming RF being translated down in frequency to the IF. The LO signal runs in the same frequency band as the incoming RF but is offset from the RF by the system IF. The LO provides sufficient power to bias the diodes of the mixer; it also provides a frequency reference that defines the system's IF. As discussed in Section 8.3.3, if there is a change in the channel frequency of the RF, there would be a corresponding change in the LO frequency such that echoes are down-converted to a fixed IF, common to all channels of operation. In this way, the amplification and filtering of returns can be applied at the IF, which is independent of the channel frequency.

A more detailed description of a superheterodyne receiver now follows in the order of the signal flow indicated in Figure 8-19, in which the division between the frequency bands of the various parts of the receiver is grayscale coded. It is also common to refer to the receiver *front end* and its *back end*. These are rather vague terms, but *front end* nominally refers to the microwave and first IF sections, whereas the *back end* typically refers to the baseband and digital sections.

Received signals from the antenna are directed toward the receiver via the radar duplexer, as shown in Figure 8-20. The duplexer is a three-port microwave component such as a circulator, which accepts the transmitted signal from the transmitter and directs it toward the antenna and also accepts received signals from the antenna and directs them toward the receiver. The duplexer allows the same antenna to be used for both transmission and reception and provides typically 20 to 25 dB of isolation between the transmitter and receiver. During the transmitted pulse, about 1% (−20 dB) of the transmitted power is typically reflected by the rotary joints in the antenna scanner and the antenna itself due to small impedance mismatches. This reflected energy is also directed into the receiver by the duplexer. A damaging degree of transmitted power due to duplexer leakage and

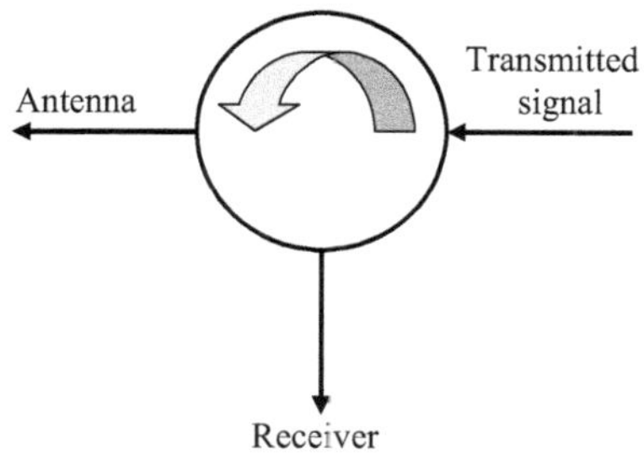

FIGURE 8-20 Radar Duplexer Based on a Three-Port Circulator

mismatch reflections can find its way into the receiver during the transmitted pulse, and some form of receiver protection is required.

This protection is afforded by a switch that isolates the receiver during the transmitted pulse. Some older radars used a waveguide device known as a Transmit Receive cell (*TR cell*) or a pre-Transmit Receive cell (*pre-TR cell*) in conjunction with a PIN diode limiter. Most modern systems use a solid-state switch based on PIN diodes. (A PIN diode is so called because it is formed from a layer of P-type, intrinsic, and N-type semiconductors. It does not switch at microwave frequencies but exhibits a diode impedance $\propto 1$/forward bias. A PIN diode can therefore attenuate a microwave signal as a function of its bias voltage. This enables the PIN diode to be used as a variable microwave attenuator, a limiter, or, by switching the bias between a low impedance and a high impedance state, a switch.) The RF pulse of the transmitted signal does not end instantaneously, indeed pulses may be shaped, as described in Section 3.5.2, but even if the possibility of pulse shaping is ignored it can take several tens of nanoseconds for the trailing edge of a high-power RF pulse to decay away to a safe level. Therefore, the gating control to the protection switch must allow for the decay of the RF to a safe level before closing the path to the receiver. Furthermore, the switch itself has a finite switching time. It would typically take approximately 20 ns for a high-power PIN diode switch to go from its high isolation state to its low insertion loss state. This contributes a further delay before the RF path to the receiver is fully closed and echo power can be received. The total delay can add up to about 100 ns after the trailing edge of the transmitted pulse. Similarly, the protection switch must be in its high isolation state before the buildup of the rising edge of the transmitted pulse. Finally, the timing of the protection switch timings must also account for the propagation delays from the transmitter to the receiver input. The transmitted pulse must be nested within the gating control of the protection switch, as illustrated in Figure 8-21; the degree of

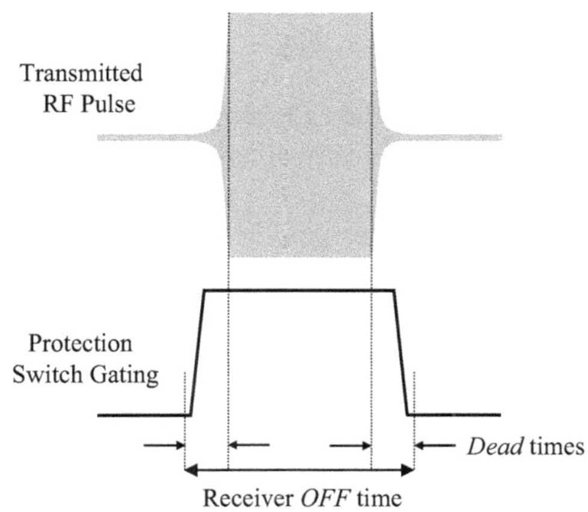

FIGURE 8-21 Protection Switch Timings

nesting implements a safety margin but at the expense of dead time; that is, the radar is neither transmitting nor receiving. This dead time contributes to the minimum range and the eclipsing losses of the radar, as discussed in Sections 3.3.2 and 3.3.3.

A single switch close to the receiver input protects the receiver front end from damage or from being driven into saturation but does not attenuate transmitter breakthrough adequately. Any remaining transmitter breakthrough would be processed as a close-range target of zero Doppler. Further isolation may be provided by additional switches further down the receiver chain. The mixer can act as a switch if the local oscillator is pulsed since down-conversion takes place only when the LO1 signal is applied. However, care must be taken when using the mixer/LO1 in this fashion since any reactance within the mixer may cause it to ring and generate spurious intermodulation products. Additional switches in the IF path (not shown in Figure 8-19) complete the job.

The next element in the receiver chain is a band-pass filter designed to pass the frequencies of all channels of operation and attenuate all out of band interference. In particular, this filter should reject the image band of the receiver and harmonics of the transmitted signal. Transmitters based on high-power tubes are often driven into saturation, so the transmitted output exhibits harmonic distortion. These harmonics are well outside the RF band of interest and may also be outside the band of the protection switch, so they must be rejected by the filter in the interests of suppressing out of band interference and, possibly, receiver protection. This filter is the last element providing protection against damaging signals for the sensitive components that are to follow.

It is beneficial to provide a stage of high gain and low noise figure as early in the receiver as possible. Ideally, this should come before significant RF losses are incurred but only once damaging high-power signals have been removed. This is provided by the LNA that operates across the microwave band of operation. When it comes to the design of microwave amplifiers, the provision of high gain and low noise figure are at odds with each other; maintaining good performance in both areas can be achieved only over a narrow bandwidth in the order of 5% of the center frequency of operation. Microwave amplifiers are specialist and costly components, so the bulk of the amplification takes place in the IF section.

Down-conversion in the front-end mixer then takes place in which the LO1 signal is applied with sufficient power (7 to 18 dBm may be required) to bias the diodes in the mixer. This process translates the carrier frequency of the microwave input (termed the *RF* in the context of mixers) to a lower frequency (i.e., the IF), in which the IF is defined by

$$F_{IF} = F_{LO} - F_{RF} \tag{8-6}$$

where F_{IF} is the IF frequency, F_{LO} is the LO frequency, and F_{RF} is the (microwave) *RF* frequency. Mixers can be used to up-convert signals to a higher carrier frequency, but this is seldom required in a radar receiver.

There are many different mixer designs. Most mixers used in radar receivers are multiplicative mixers since they form the product of the incoming RF and LO signals. If the RF signal is given by $\sin(2\pi F_{RF}t)$ and the LO is given by $\sin(2\pi F_{LO}t)$, then to a first degree of approximation the output is given by

$$\begin{aligned} &= \sin(2\pi F_{LO}t)\sin(2\pi F_{RF}t) \\ &= \frac{1}{2}\cos(2\pi(F_{LO} - F_{RF})t) - \frac{1}{2}\cos(2\pi(F_{LO} + F_{RF})t) \end{aligned}$$

The second term in this expression is out of band and easily filtered off. The first term is the required product and is the desired IF output

$$= \frac{1}{2}\cos(2\pi F_{IF} t) \tag{8-7}$$

In truth, higher-order products are also generated. These fall outside the IF band and are easily rejected using a filter. The previous simple analysis illustrates how the down-conversion process results in a halving of the voltage on the IF output, which corresponds to a 6 dB power loss. In practice, a conversion loss of 8 dB is more typical.

From the previous analysis it is evident that F_{LO} may be above or below F_{RF}, and indeed the choice of whether $F_{LO} = F_{RF} + F_{IF}$ or $F_{LO} = F_{RF} - F_{IF}$ is purely arbitrary. The frequency planning of most radars is that $F_{LO} = F_{RF} + F_{IF}$ is chosen. Rearranging this gives $F_{RF} = F_{LO} - F_{IF}$; however, any noise and signals at a frequency of $F_{LO} + F_{IF}$ are also down-converted to the IF. This latter band is known as the *image frequency* since it appears as a mirror image of the desired RF about the LO signal. Indeed, any noise or interference within the channel bandwidth centered on the image frequency could degrade the performance of the radar. The image band and the typical frequency planning of a radar are illustrated in Figure 8-22; the center frequency of F_{RF} and F_{LO} is indicated by the heavy line, and the spread of channel frequencies is indicated by the lighter lines. Only the central carrier component of each signal is shown in Figure 8-22 for clarity; in reality each signal would exhibit a range of spectral components on its upper and lower sidebands, as illustrated in Figure 3-7 for a simple pulsed waveform.

The RF can operate on a range of channel frequencies distributed over the RF band. Similarly, the LO1 also operates on a band of channel frequencies occupying the same bandwidth but displaced from the RF band by F_{IF}. This means that the image band varies with channel; however, the image band is always offset from the desired F_{RF} by $2 \times F_{IF}$. If the spread in the image band does not overlap the RF band, then the front-end microwave band-pass filter can pass the RF band and provide some rejection of the image band. This is only possible when

$$RF\ band < 2F_{IF} \tag{8-8}$$

The greater the margin of this inequality, the greater the separation of the RF band and its corresponding image band and thus the more feasible it becomes to reject the image band in the front-end filter, whose profile is shown by the dashed line enveloping the RF band in Figure 8-22. This is made feasible through the choice of a high first IF.

The incoming RF is down-converted to the IF in a manner that preserves all the modulations on the RF signal. Modulations on the LO may also be superimposed onto the IF output, including, most crucially, the close-to-carrier noise on the LO. The most

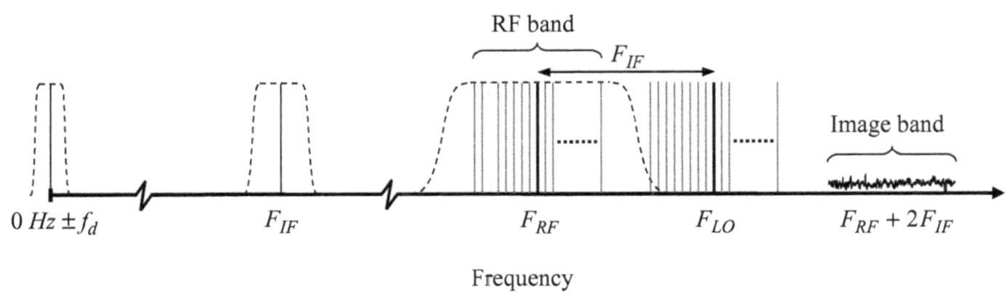

FIGURE 8-22 Frequency Planning

basic mixer is a single-ended diode [20]. This mixer has a high conversion loss and no suppression of LO noise and so is seldom used in a radar receiver. A balanced mixer comprises two diodes and boasts a superior input match and improved RF to LO isolation than the single-ended diode. It also provides a first-order cancellation of AM close-to-carrier noise from the LO. A double balanced mixer is a more elaborate circuit of four diodes that improves the RF to LO isolation yet further and has lower conversion loss than that of the balanced mixer but at the expense of input matching. Most modern radar receivers now use a variant of this known as an *image rejection double balanced mixer*, which, as the name suggests, down-converts RF on the desired side of the LO and provides about 20 to 30 dB of rejection of RF in the image band. So even if the inequality represented by equation (8-8) is not met and the RF band-pass filter is unable to provide any image band rejection, some at least may be afforded via an image rejection mixer.

On down-conversion to the first IF, the receiver has to impart plenty of gain. A huge amount of gain is required in a radar receiver. This is far more than that required in a typical communications application because the received signals are so weak from having undergone the $range^4$ losses of the two-way path; a power gain of 100 dB or more is typically required. It is worth noting that radar sensitivity is not dependent on the amount of gain; receiver gain does not feature in the analysis of radar detection performance or the radar range equation (Chapter 2). However, the performance of the amplifiers imparting this gain is key to preserving the wide dynamic range required of a radar receiver. Using a logarithmic amplifier and front-end amplifiers capable of handling high input powers before saturation occurs is key to the maintenance of a wide dynamic range. The front end of a radar receiver must typically exhibit a dynamic range of 90 to 100 dB. The highest input power seen in normal use is typically the transmitter breakthrough or extreme cases of high-power clutter returns (or jamming/interference).

The *dynamic range* of a receiver (or an individual amplifier used within the receiver) is the difference between the maximum and minimum input powers it can handle [20]. Both the maximum and minimum powers can be defined in several ways, depending upon the priority attached to various effects that can occur toward the extremes. For radars, the minimum input power is normally set by the minimum detectable signal power required to achieve a specified level of radar detection performance (see Section 2.3). For high input powers, receivers and amplifiers are driven into saturation, which in extreme cases can lead to component destruction. Saturation gives rise to gain compression and the generation of spurious intermodulation products. If gain compression is to be avoided then it is reasonable set the upper limit to the 1 dB compression point, $P_{1\,\mathrm{dB}}$. This is defined as an input power sufficient to cause a loss in gain of 1 dB with respect to the linear gain of the receiver/amplifier. If spurious intermodulations are to be avoided one may set the upper limit to correspond to two-thirds the third-order intercept point, $3OIP$ or $IP3$ or even IP^3. The third-order intercept point represents the input power level of two CW signals required to generate a third-order intermodulation product of output power equal to that of the two CW signals. Usually, it is not practical to drive a receiver/amplifier this hard because physical damage is likely long before this power level is attained. However, two-thirds of $IP3$ represents the point when the third-order intermodulation products first become visible above the ambient noise floor. Since receivers are a series of components, the situation in which just one of these stages is responsible for limiting the entire radar should be avoided. Ideally, it is better to ensure that several stages start to reach their limits together. The upper limit on the whole receiver must then be derived from the performance of its constituent stages cascaded together.

The IF section of the receiver also includes a band-pass filter that is designed to pass the signal bandwidth and reject interference on neighboring channels of operation. In the majority of cases it is not feasible to build a filter encompassing just one channel width at the full microwave frequency of operation since this would demand an unrealistically high filter *quality factor (Q-factor)*[2]. However, the demands on filter Q-factor and transfer function are far more realistically achieved if designed for a center frequency of F_{IF} (rather than F_{RF}). The traditional approach to radar would be to design this filter to be matched to the transmitted waveform (see Section 3.6) and may be a dispersive analog compression filter in the case of pulse compression systems. More recently, the matched filter is typically implemented in the digital signal processing. Nevertheless, it is wise to provide some filtering at this point if only to reject interference on neighboring radar channels.

This raises an important point regarding the selection of a suitable value of F_{IF}. On the one hand, a high value of IF is desirable because it affords a degree of image band rejection. In fact, it generally becomes more feasible to reject out-of-band interference for higher values of the IF center frequency. Ideally, one would want an IF center frequency that was several times the signal bandwidth. Failure to ensure this results in spectral components that would extend down to DC and beyond. Spectral components at an offset of $2F_{IF}$ from the central response appear to be "reflected" about DC and fold back into the IF passband and are a potential source of interference. This can be avoided if F_{IF} is set sufficiently high such that the offending spectral components are well below the ambient noise level in the IF passband. However, the amplifier noise figure and gain tend to improve at lower frequencies. Filters of lower Q-factors commensurate with lower center frequencies are easier to realize. Furthermore, it is also fair to say that costs scale with the Pf^2 product (Section 8.2), which favors the use of a low frequency IF. However, throughout the development of communications and radar equipment, certain frequencies have been habitually used as the system IF (e.g., 10.7 MHz is much used as an IF center frequency by many communications systems), and as a consequence component availability and cost are good at these frequencies, which further promotes the use of these frequencies. There is clearly a case for both a high-frequency IF and a low-frequency IF. The solution to this is to employ both, or to use several stages of down-conversion at successively lower IF center frequencies. A typical modern radar receiver would use a high first IF for the purposes of image rejection and the rejection of out-of-band interference and a much lower second IF, using a second mixer and LO2, for implementing the amplification and channel filtering at an affordable price. For example, an airborne fire control radar may operate on an RF band of 9.0 to 9.5 GHz, its first IF may be centered at 500 MHz (a range of 250 to 800 MHz would be typical), and its second IF may be centered at 10 MHz (a range of 1 MHz to 20 MHz would be typical). In this way a radar may use a double (or even a triple [20]) superheterodyne receiver using two (or three) IF sections, each of a decreasing center frequency.

The radar receiver provides a final stage of down-conversion to a video or baseband output (in effect an IF centered at zero frequency), which is subsequently digitized and directed to the processor. Most pulse Doppler radars follow this pattern; however, a few,

---

[2]The Q-factor is a measure of the resistive losses in a resonant circuit and gives rise to the identity that the Q-factor is the ratio of the center frequency to the bandwidth of a filter, i.e. its fractional bandwidth.

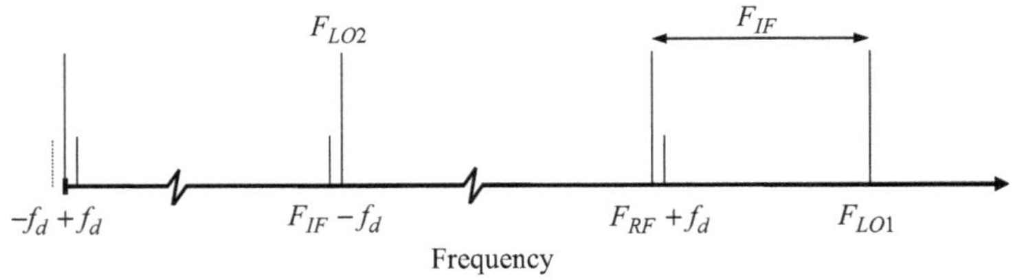

FIGURE 8-23 Approach/Recede Ambiguity

mostly older ones, do not. These provide a final stage of down-conversion to a non-zero IF prior to digitization and processing. Figure 8-19 and Figure 8-22 illustrate a system whose second stage of down-conversion results in a baseband output centered at 0 Hertz. The baseband signal is filtered to pass the Doppler band over $f_d = \pm f_r/2$ and reject any higher frequency components that could result in ambiguous responses within the Doppler band of the baseband signal. Consequently, this filter is commonly called the antialiasing filter. Once digitized, the signals enter the processing chain described in Chapter 7. To preserve the dynamic range and to minimize quantization error, digitization is conducted using 16-bit (or more) resolution, which also results in a dynamic range of $20 \log_{10}(2^{16}) = 96$ dB.

There are two problems associated with the down-conversion to a baseband output centered at 0 Hertz: an *approach/recede ambiguity;* and *blind phases*.

8.4.1.1 Approach/Recede Ambiguity

Consider the case illustrated in Figure 8-23 in which the central carrier component for a closing target (positive Doppler shift) is shown by the shorter line at a frequency of $F_{RF}+f_d$. Note the inversion of the apparent Doppler of this target response when down-converted to the first IF. This arises because F_{LO1} is *above* the RF band, so the frequency difference between the target response ($= F_{RF} + f_d$) and F_{LO1} is $F_{IF} - f_d$, that is, smaller than F_{IF}. When down-converted using a second local oscillator of frequency $= F_{LO2} = F_{IF}$, the target response at baseband falls at $-f_d$. However, the negative frequency domain has no physical meaning, and the target response appears folded back into the positive frequency domain at $+f_d$. Thus, it is indistinguishable from an opening target response of frequency $= F_{RF} - f_d$ in the RF band as this also results in a response at $+f_d$ in the baseband. An ambiguity exists as to whether the target is approaching or receding.

8.4.1.2 Blind Phases

This is identical to the blind phase problem that can occur in MTI radars and was described in Section 7.2.2 (iii). Blind phases can occur if the pulse returns fall at the zero crossing points of the baseband Doppler cycle and result in zero response in the baseband. The problem is illustrated in Figure 8-24. It arises when the Doppler shift equals half the PRF,

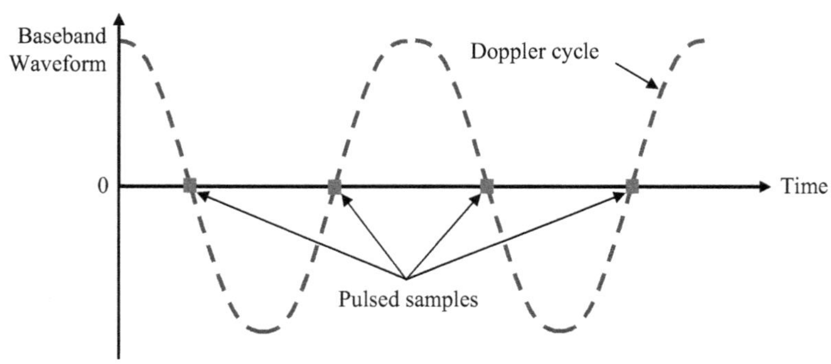

FIGURE 8-24 Blind Phases

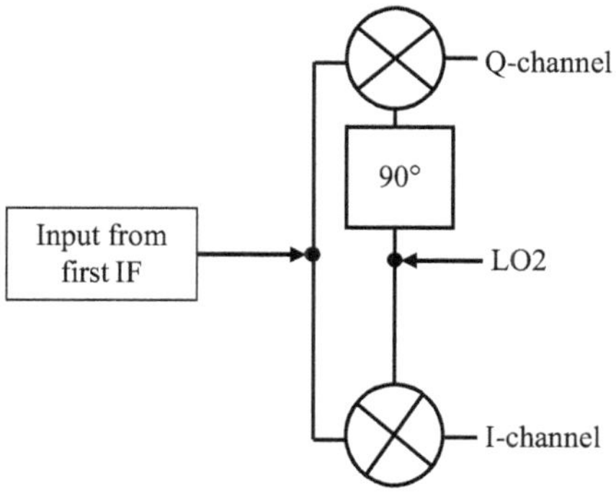

FIGURE 8-25 Quadrature Detector

$f_d = f_r/2$, or, more generally, when $f_d = (2n+1)/f_r$, where $n = 0, 1, 2, \ldots$ and when the two-way phase shift over the range, R, given by $2\omega_{tx}\frac{R}{c}$ equals a multiple of π. The combination of Doppler shifts and ranges that give rise to this problem does not persist for a long time but, nevertheless, may persist sufficiently long for the radar to lose the target detection.

8.4.1.3 The Quadrature Detector

Both of these problems can be overcome by performing the down-conversion to a 0 Hertz baseband signal using a *quadrature detector*, illustrated in Figure 8-25. The quadrature detector splits the incoming signal into two paths and applies them in equal amplitude and phase to two phase sensitive detectors (PSDs). The local oscillator is split and applied to the PSDs in equal magnitudes but in phase quadrature (90°). This generates two baseband outputs: one to which the local oscillator was applied *in phase*, termed the *I-channel;* and one to which the local oscillator was applied in *quadrature phase,* known as the *Q-channel*. The phase of the local oscillator dictates the phase of the baseband output such that the I-channel output is proportional to $\sin(2\pi f_d t)$ and the Q-channel output is proportional to $\cos(2\pi f_d t)$. The I- and Q-channels convey orthogonal components of the baseband signal and can conveniently be thought of as the real and imaginary parts of a complex signal. Processing of the baseband signal using an FFT operation requires the full complex representation of the signal since the magnitude and phase of the baseband signal is preserved only through the combination of its real and imaginary parts. The FFT of any single channel does not result in a legitimate transformation.

This circuit overcomes the approach/recede ambiguity by comparing the phase of the I- and Q- channels. Individual samples from the I- and Q-channels are illustrated in Figure 8-26. The magnitude of the response is given by the vector addition of the real and imaginary components, that is, $\sqrt{I^2 + Q^2}$, and exhibits a phase angle of $\tan^{-1}\left(\frac{Q}{I}\right)$.

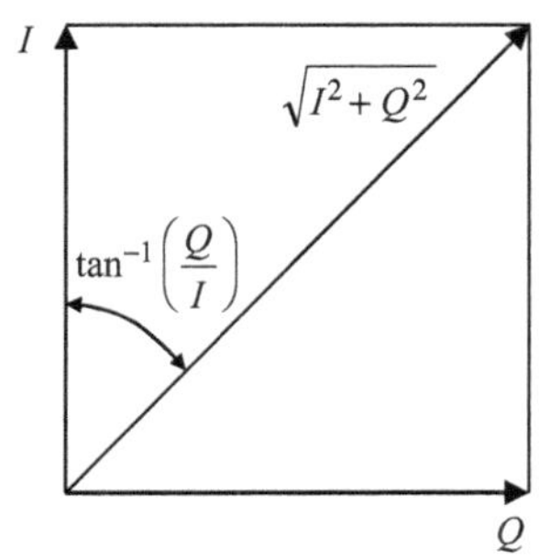

FIGURE 8-26 I- and Q-Channel Samples

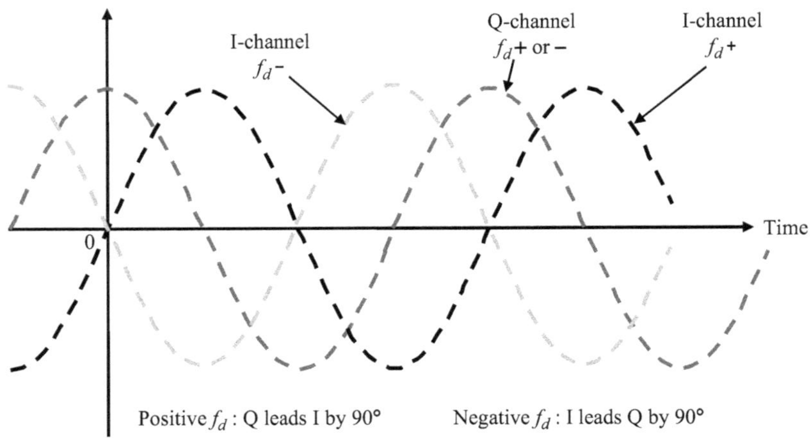

FIGURE 8-27 I- and Q-Channel Responses for Positive and Negative Doppler

Successive pulse samples give rise to phasors that progressively rotate around the phasor diagram at the Doppler frequency, as described previously in the context of discrete Fourier transform (DFT) and FFT processing (Section 7.3.1).

It was previously noted that the I-channel output is proportional to $\sin(2\pi f_d t)$ and the Q-channel output is proportional to $\cos(2\pi f_d t)$. If the Doppler shift, f_d, is positive, then the I-channel output is proportional to $\sin(2\pi f_d t)$ and the Q-channel output is proportional to $\cos(2\pi f_d t)$. If, on the other hand, the Doppler shift, f_d, is negative then the I-channel output is proportional to $\sin(-2\pi f_d t) = -\sin(2\pi f_d t)$ and the Q-channel output is proportional to $\cos(2\pi f_d t)$. The inverted sine wave response from the I-channel and the non-inverted cosine response from the Q-channel results from the odd and even function nature of sines and cosines, respectively. These waveforms are plotted in Figure 8-27.

For a positive Doppler, the Q-channel leads the I-channel by 90° and successive phasors (Figure 8-26) rotate clockwise, whereas for negative Doppler the I-channel leads the Q-channel by 90° and the successive phasors rotate counterclockwise. These two cases can easily be resolved by examining the sign of the phase angle: for positive Doppler $\tan^{-1}\left(\frac{Q}{I}\right)$ is positive, whereas for negative Doppler $\tan^{-1}\left(\frac{Q}{I}\right)$ is negative.

The quadrature detector is actually a hard-wired implementation of a two-point DFT; one frequency bin covers negative Doppler, and the other bin covers the positive Doppler. The weightings (Figure 7-18) are equal amplitudes and 0° and 90° of phase.

The quadrature detector overcomes the blind phase problem since a blind phase in one channel corresponds to the peak output in the other. This is illustrated for the case of positive Doppler in Figure 8-28.

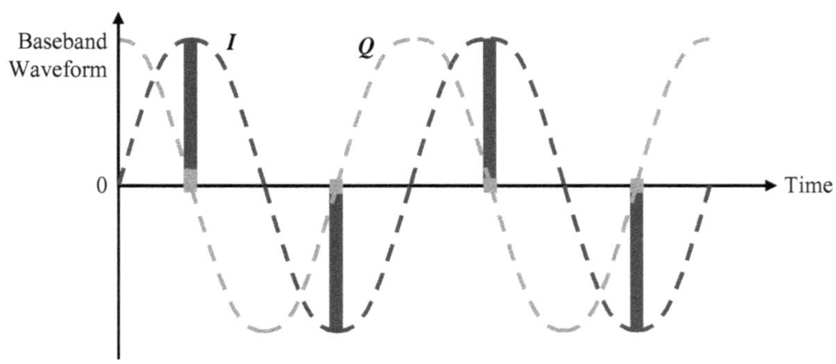

FIGURE 8-28 Overcoming Blind Phases

The magnitude of the response is given by

$$\sqrt{I^2 + Q^2} = \sqrt{\sin^2(2\pi f_d t) + \cos^2(2\pi f_d t)}$$

which is a constant since $\sin^2(2\pi f_d t) + \cos^2(2\pi f_d t) = 1$. Therefore, the magnitude of the response does not vary even though individual channels (I- or Q-) may drift toward a blind phase. Any amplitude imbalance or error in phase quadrature between the two channels results in a lack of channel isolation because a portion of the I-channel is coupled into the Q-channel and vice versa.

8.4.1.4 Signed IF

For systems whose output is centered at a non-zero frequency in which the center frequency offsets the whole Doppler band into the positive frequency domain, the approach/recede ambiguity and blind phase issues do not arise. There is no need to use quadrature detection as the magnitude and phase of the output signal is preserved. For example, if it is anticipated that Doppler shifts between ± 100 kHz may be encountered, the output may be centered at a frequency of 325 kHz. In these circumstances the Doppler band resides between 225 kHz and 425 kHz. The digitization of signals must be fast enough to capture frequencies as high as 425 kHz, and the subsequent FFT processing must be applied across the band 225 kHz to 425 kHz. Although seldom used these days, this technique, known as *signed IF*, has been used in the past. It must be accompanied by some very careful frequency planning since spectral components outside the Doppler band of interest may fold back about DC and fall into the Doppler band. If, in the previous example, a PRF of 200 kHz is used, the first PRF harmonic in the lower sideband falls at 325 kHz $-$ 200 kHz $=$ 125 kHz, which is fine since this falls outside the Doppler band. However, the third harmonic PRF line in the lower sideband falls at 325 kHz $-$ (3 $\times$ 200 kHz) $= -275$ kHz, which folds back into the positive frequency domain at $+275$ kHz and presents itself as a false target having an apparent Doppler shift of -50 kHz. The problem is illustrated in Figure 8-29 in which the indices $+1$, -1, -2, -3 and -4 refer to the order of PRF harmonic. The complexity of the problem increases if multiple PRFs are used, and it rapidly becomes very awkward to find frequency and PRF combinations that avoid false targets.

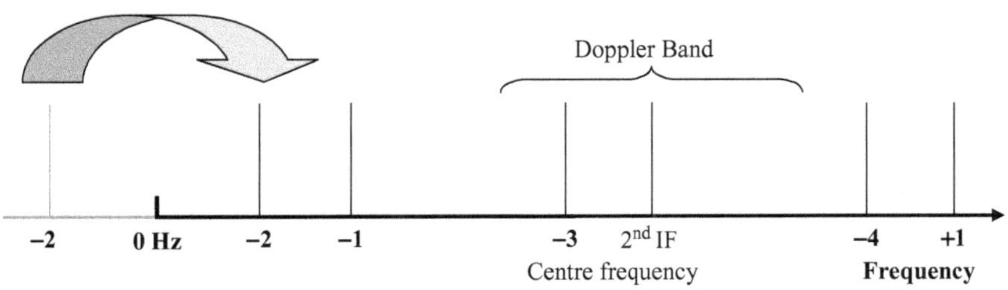

FIGURE 8-29 False Targets Due to PRF Harmonics in Signed IF

8.4.2 Receiver Noise Figure

8.4.2.1 Introduction

The term *noise* refers to unwanted electrical signals that are always present in electrical systems. Noise superimposed on a signal tends to obscure or mask the signal; it limits the ability of a radar to detect targets and the accuracy with which target parameters (e.g., range, angle, velocity) can be measured. Noise arises from a variety of sources, both man-made and natural. Man-made noise includes sources such as switching transients and other radiating electromagnetic signals. Natural noise includes electrical circuit and component noise, atmospheric disturbances, and galactic noise. Some noise signals are received from the outside world via the antenna, and some are internally generated within the receiver [20].

Good engineering design can eliminate much of the noise or its undesirable effects through filtering, shielding, modulating, and selecting an optimum receiver layout and antenna site. However, one natural source of noise—*thermal* or *Johnson* noise—cannot be eliminated. Thermal noise is caused by the thermal motion of electrons in all conductors and semiconductors and so is present in all resistive components and transmission lines. Thermal noise and noise figure together with their limiting effects on radar detection process were considered in Section 2.3.

8.4.2.2 Thermal Noise

If all sources of clutter, external noise, and interference could be excluded from the radar, thermal noise would still be present on the input to a radar receiver. Thermal noise is generated naturally in all conductors and semiconductors and ultimately sets the limit on the receiver sensitivity. It arises due to the random thermal agitation of electrons in conductors and gives rise to the flow of random currents and therefore random noise voltages across resistances. The thermal noise voltage has a Gaussian probability density function (PDF). The primary spectral characteristic of thermal noise is that its power spectral density is the same for all frequencies of interest in most RF systems. For this reason, this noise is referred to as *white* noise.

The perfect receiver is one that produces no noise and is depicted in Figure 8-30.

The thermal noise voltage generated across a resistance, R_S, is [22]

$$e_n = \sqrt{4kTB_N R_S}$$

where k is Boltzmann's constant $= 1.38 \times 10^{-23}$ (J/K), T is temperature in degrees Kelvin (K), and B_N is the noise bandwidth (Hz).

The noise power into the receiver, P, is given by

$$P = \frac{V^2}{R} = \left(\frac{e_n R_{in}}{R_S + R_{in}}\right)^2 \Big/ R_{in}$$

when the input is matched $R_S = R_{in} = R$ and the maximum power is transferred to the input. The noise power in $= \left(\frac{e_n R}{2R}\right)^2 \Big/ R$

$$= \frac{e_n^2}{4R}$$

$$= \frac{4kTB_N R}{4R}$$

$$= kTB_N$$

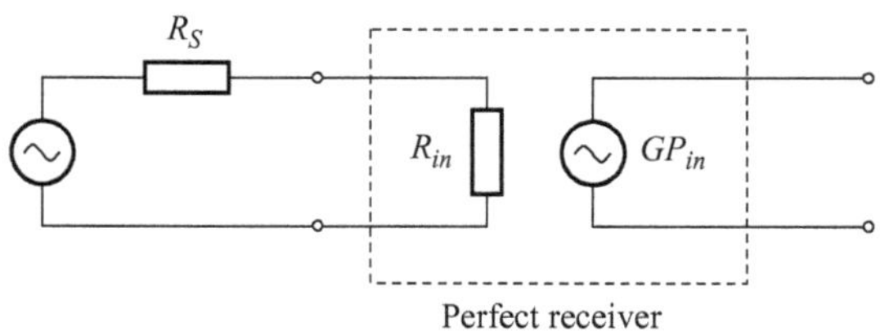

FIGURE 8-30 The Perfect Receiver

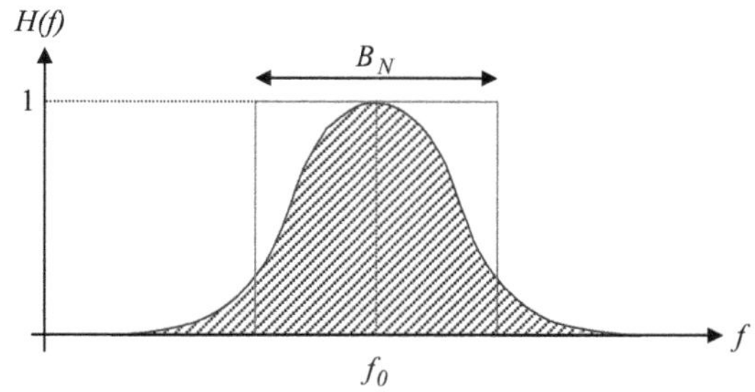

FIGURE 8-31 Noise Bandwidth

Therefore, under matched input conditions the thermal noise power, N_i, is given by

$$N_i = kTB_N \tag{8-9}$$

8.4.2.3 Noise Bandwidth

Noise bandwidth is formally defined as

> *The bandwidth of a rectangular filter whose noise power output is the same as that obtained from the actual frequency response of the system.*

This can be expressed mathematically as

$$B_N = \frac{\int_{-\infty}^{+\infty} |H(f)|^2}{|H(f_0)|} \tag{8-10}$$

where $H(f)$ is the frequency response of the system (e.g., receiver, filter), f_0 is the frequency of maximum response (usually midband), and $H(f_0)$ is the maximum response of the system and is usually normalized to unity (Figure 8-31).

8.4.2.4 Friis Formula

To make equitable comparisons between noise figures of different devices, a reference must be chosen. Friis suggested that noise figure be defined for a noise source at a standard temperature; $T_0 = 290$ K is taken for reference purposes [23]. This gives rise to a spectral noise power density of

$$N_i = kT_0$$
$$N_i = 4 \times 10^{-21} \text{ W/Hz or } 4 \times 10^{-18} \text{ mW/Hz or } -174 \text{ dBm/Hz}$$

This forms a fundamental lower limit on the thermal noise spectral power density.

8.4.2.5 Noise Figure

The receiver components contribute some additional noise onto the signal path and may impart losses that reduce the power of the target signal (losses do not affect the noise level since a level of at least kT_0B_N is always present). Both these effects degrade the signal-to-noise ratio as a target return progresses along the receiver chain. The added noise and loss effects are characterized by the *noise figure* or *noise factor*,[3] F, of the receiver.

[3]Both terms are commonly used; however, the term *noise figure* is used here.

Noise figure is defined as

The factor by which the noise on the output of an ideal (i.e., noiseless) device must be increased to give the same noise level out of the actual device.

Let S_i = signal power on input, N_i = noise power on input, S_o = signal power on output, and N_o = noise power on output.

Under ideal conditions, N_i is matched thermal noise only, that is,

$$N_i = kT_0 B_N \tag{8-11}$$

The gain of the receiver,

$$G_{RX} = S_o / S_i \tag{8-12}$$

Therefore, the signal strength on the output, $S_o = G_{RX} S_i$.

In the case of an ideal, noiseless receiver the noise on the output would be $G_{RX} N_i$. However, from the definition of noise figure, the actual noise on the output,

$$N_o = F G_{RX} N_i \tag{8-13}$$

Substituting (8-11) and (8-12) into (8-13) gives

$$N_o = kT_0 B_N F (S_o / S_i) \tag{8-14}$$

Rearranging (8-14) for F gives

$$F = \frac{(SNR_i)}{(SNR_o)} \tag{8-15}$$

where $(SNR_i) = S_i / N_i$ denotes the input SNR, and $(SNR_o) = S_o / N_o$ denotes the output SNR.

Since it is common to use the decibel scale when referring to noise figure, equation (8-15) can be expressed in dB as

$$F(\text{dB}) = SNR_i(\text{dB}) - SNR_o(\text{dB}) \tag{8-16}$$

This indicates that the noise figure is the ratio of the input SNR to the output SNR (or their differences in dB), and this relationship is often taken as the definition of noise figure. This relationship stems from the earlier definition based on the assumption that the input noise is matched thermal noise only. Equation (8-15) does not hold when external noise or noise-like interference is received by the radar, so caution must be exercised if using (8-15) as a definition of noise figure. However, equation (8-15) is a useful illustration of the fact that the noise figure of a receiver degrades the signal-to-noise ratio. The perfect receiver would have a noise figure, $F = 1$ (0 dB), in which case the input and output signal-to-noise ratios would be equal. A practical receiver has $F > 1 (> 0 \text{ dB})$ and so has a lower output SNR than its input SNR.

This definition and the associated signal flow are depicted in Figure 8-32.

In the example of Figure 8-32, $F = 40 \text{ dB} - 30 \text{ dB} = 10 \text{ dB}$ and $G = -40 \text{ dBm} - (-60 \text{ dBm}) = 20 \text{ dB}$.

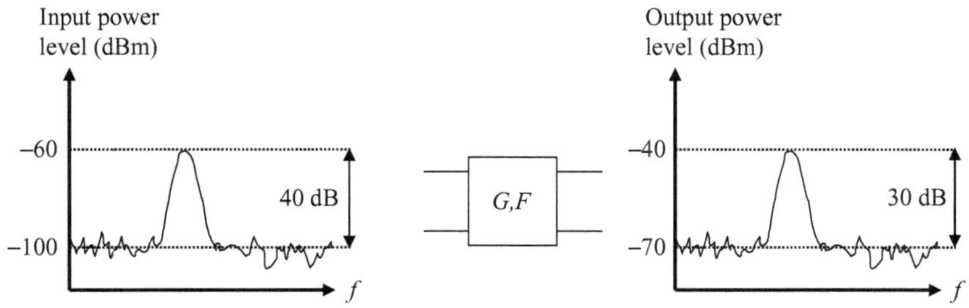

FIGURE 8-32 Noise Figure

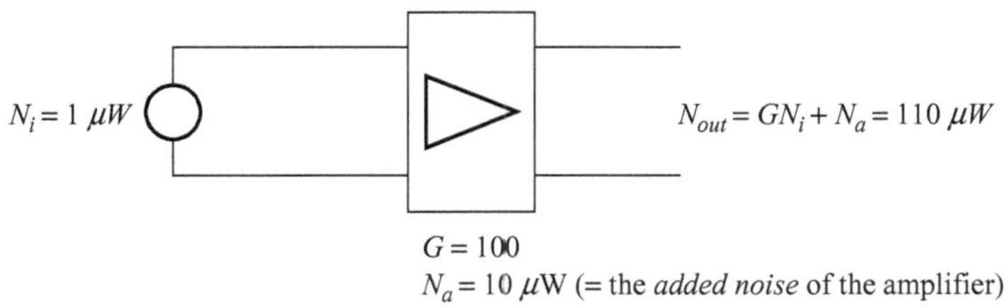

FIGURE 8-33 Added Noise of an Amplifier

8.4.2.6 Added Noise

A useful concept in quantifying the noise performance of a component such as an amplifier or an entire receiver is that of its *added noise* (N_a), which can be thought of as the additional noise imparted by the component onto the signal path. It is normal to consider that this is an equivalent amount of noise added to the output of the device. Figure 8-33 depicts an amplifier of gain $G = 100$ and an added noise, $N_a = 10\ \mu$W, with matched thermal noise of $N_i = 1\ \mu$W present on its input.

Alternatively, a noisy amplifier can be represented as a noiseless amplifier in series with a noise source $N_{ai} = \frac{N_a}{G} = \frac{10\ \mu\text{W}}{100} = 0.1\ \mu$W, as depicted in Figure 8-34.

The noise figure of the amplifier is given by

$$F = \frac{(SNR)i}{(SNR)o} = \frac{(SNR)i}{GS_i/[G(N_i + N_{ai})]} = \frac{S_i/N_i}{GS_i/G(N_i + N_{ai})} = \frac{N_i + N_{ai}}{N_i} = 1+\frac{N_{ai}}{N_i} = 1+\frac{N_{ai}}{kTB_N}$$

where

S_i = signal power at the amplifier input
N_i = noise power at the amplifier input
N_{ai} = amplifier added noise referred to the input port
G = amplifier gain

Since $N_{ai} = N_a/G$, the relationship between noise figure and added noise may be expressed as

$$F = 1 + \frac{N_a}{GkTB_N} \tag{8-17a}$$

or conversely as

$$N_a = (F - 1)GkTB_N \tag{8-17b}$$

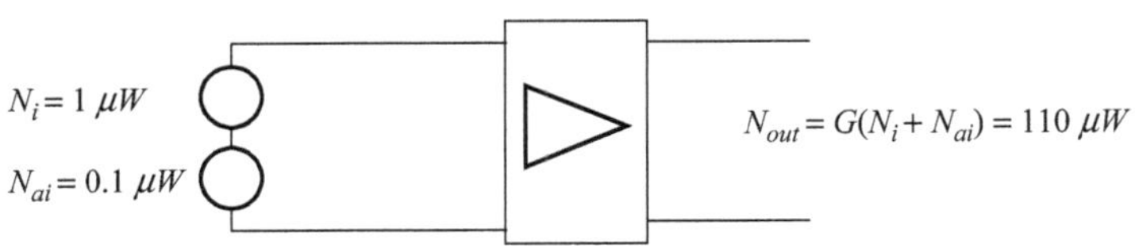

FIGURE 8-34 Alternative Depiction of Added Noise

8.4.2.7 Noise Temperature

The *equivalent noise temperature*, T_e, of a receiver is the fictitious temperature by which a matched source resistor would have to be raised (over and above the reference temperature, $T_0 = 290$ K) to give the same output noise power from a perfect receiver as is obtained from the actual receiver.

Let the noise out of perfect receiver $= GkT_0B_N$ (assuming a source resistance temperature $= T_0$), the noise out of an actual receiver $= FGkT_0B_N$, and the noise out of perfect receiver $= Gk(T_0+T_e)B_N$ (assuming a source resistance temperature $= T_0+T_e$). Equating the second and third of these quantities, as suggested by the previously definition, gives

$$Gk(T_0 + T_e)B_N = FGkT_0B_N$$
$$T_0 + T_e = FT_0$$

Therefore,

$$F = \frac{T_0 + T_e}{T_0} = 1 + \frac{T_e}{T_0} \tag{8-18a}$$

or

$$T_e = T_0(F - 1) \tag{8-18b}$$

The temperature increase, T_e, contributes extra thermal noise equivalent to the added noise

$$N_a = GkT_eB_N \tag{8-19}$$

It is also common to refer to the *total system noise temperature*, T_S, in which

$$T_S = T_0 + T_e \tag{8-20}$$

The total system noise temperature is related to the noise figure by

$$T_s = FT_0 \tag{8-21}$$

8.4.2.8 Cascaded Noise Figure

In many cases, it is desired to calculate the noise figure of several stages in series with each other since RF components are often cascaded in series with others along a receiver [23]. Consider the two-stage system illustrated in Figure 8-35.

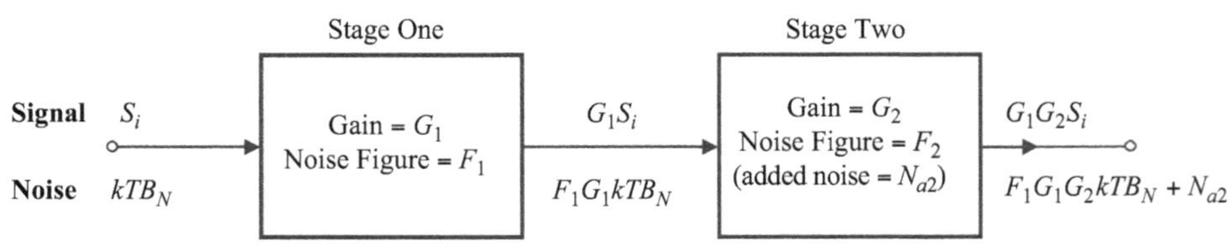

FIGURE 8-35 Cascaded Noise Figure

Substituting the quantities in Figure 8-35 into (8-15) gives

$$F = \frac{S_i/kTB_N}{G_1G_2S_i/(F_1G_1G_2kTB_N + N_{a2})}$$

$$F = \frac{F_1G_1G_2kTB_N + N_{a2}}{G_1G_2kTB_N}$$

$$F = F_1 + \frac{1}{G_1}\left(\frac{N_{a2}}{G_2kTB_N}\right)$$

Applying (8-17a) to the second stage gives

$$F_2 = 1 + \frac{N_{a2}}{G_2kTB_N}$$

Rearranging

$$F_2 - 1 = \frac{N_{a2}}{G_2kTB_N}$$

Substituting this into the expression for F gives

$$F = F_1 + \frac{F_2 - 1}{G_1} \tag{8-22}$$

This can be extended for any number of stages to give

$$F = F_1 + \frac{F_2 - 1}{G_1} + \frac{F_3 - 1}{G_1G_2} + \frac{F_4 - 1}{G_1G_2G_3} + \cdots \tag{8-23}$$

In truth, it is unnecessary to require three or more stages of calculation since the first two stages can be computed using (8-22) to calculate their combined noise figure, which can then be regarded as lumped together as a new first stage. The new first stage and old third stage can then be solved as a two-stage system using (8-22) again. In this way, multiple stages can be analyzed by solving (8-22) repeatedly.

Equations (8-22) and (8-23) illustrate the fact that the first-stage noise figure adds directly to the total noise figure. However, the noise figures of subsequent stages are reduced by the gain of all preceding stages. It is therefore important to ensure that the first stage of any receiver is ideally one having a low noise figure and a high gain. In microwave amplifier design, the requirements of low noise figure and high gain tend to clash with each other. The noise figures of subsequent stages become less significant the further down a receiver chain one cares to go.

8.4.2.9 Noise Figure of a Passive Device

A passive device exhibits loss, L, which is the inverse of gain ($L = 1/G$). For an input signal of S_i the output signal would be S_i/L. For matched thermal noise on the input of kTB_N, a noise power would still be observed on the output of kTB_N since this is always present in all passive components. From (8-15), the noise figure of a passive device is therefore

$$F = \frac{(SNR_i)}{(SNR_o)} = \frac{S_i/kTB_N}{S_i/LkTB_N} = L$$

and

$$F = L \tag{8-24}$$

In other words, the noise figure of a passive component is equal to its loss.

Front-end losses have a particularly damaging effect on the noise figure of a receiver since from (8-22) it was noted that the first-stage noise figure (and hence *loss*) adds directly to the total. Furthermore, if the first stage exhibits loss rather than gain, the contributions from subsequent stages are enhanced rather than reduced. Hence, it is of paramount importance to minimize front-end loss in any receiver. Crucially, any losses preceding the first gain stage must be minimized. In receiver design, it is very important to minimize the plumbing losses and insertion loss of the duplexer, front-end filter, and protection switch, which have to precede the low noise amplifier. It is highly desirable to place a low noise amplifier before the front-end mixer because an RF/IF conversion loss of around 8 dB is typically incurred.

As has been shown, the performance of the receiver front end and of the low noise amplifier in particular is fundamental to system sensitivity and accuracy.

8.5 RADAR ANTENNAS AND ARRAYS

This section gives a brief overview of aperture type antennas and array antennas used for radar applications. For more detail, the reader is referred to the excellent text of Balanis [24]. It concludes by summarizing some of the capabilities brought about by active electronically steered array antennas.

8.5.1 Basic Concepts

An antenna is a transducer for launching or receiving electromagnetic (EM) waves. One key point regarding antennas is their property of reciprocity; that is, they behave as a receiver in exactly the same way that they do as a transmitter. Since the antenna is the last component in the transmission of the signal from the radar and the first component in the reception of echoes, it is doubly important; in fact, antenna properties feature twice in radar range equation equation (2.6) (in Section 2.1.1). The antenna provides two basic functions:

1. It provides an impedance match between the transmission line of its feed and free space for the efficient transmission and reception of RF.
2. It provides directionality. On transmission most of the power is concentrated into a beam directed along its boresight. Similarly, as a receiver, it receives energy principally from the direction of this beam. The directionality of an antenna is illustrated by it polar diagram, an example of which is given in Figure 8-36.

FIGURE 8-36 Antenna Polar Diagram

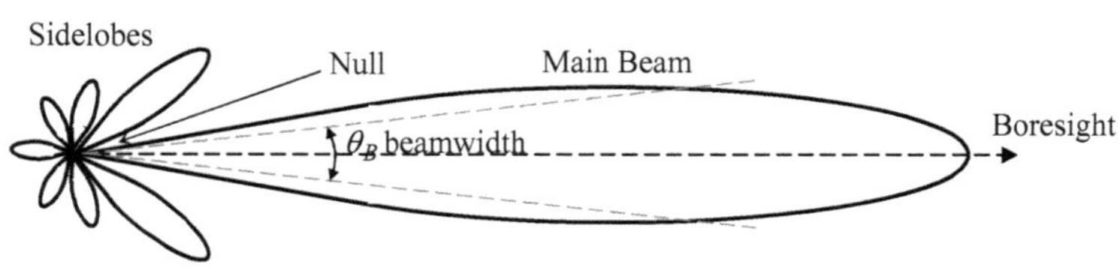

The polar diagram can be expressed as the magnitude of the electric field (E field or simply **E**) or as power density in polar coordinates. Since power density is proportional to the square of the E field, plots of **E** and power density are similar in form but do exhibit subtle differences. It is important to know which quantity is plotted. The polar diagram of an antenna is characterized by a main beam and sidelobes with nulls at angles between these features. Furthermore, the polar diagram of an antenna need not have circular symmetry, as implied by the two-dimensional view presented in Figure 8-36. An example of a three-dimensional depiction of an antenna polar diagram is given in Figure 6-14.

The key properties of the antenna are as follows:

- Gain
- Beamwidth
- Sidelobe level (with respect to the main beam)
- Radiation pattern
- Capture area
- Radiation resistance and matching
- Polarization

Each is described in more detail in the ensuing sections.

8.5.1.1 Antenna Gain

Antenna gain is always expressed with reference to an isotropic radiator, which radiates equally in all directions. Such an antenna is of little practical use and cannot be made; however, its polar diagram may be imagined to be a sphere, or a circle if represented on a two-dimensional page. Figure 8-37 compares the polar diagrams of an isotropic radiator and a directional antenna fed with the same power.

Antenna gain is defined as

$$\text{Gain} = \frac{\text{Power density along maximum of directive antenna (its boresight)}}{\text{Power density from isotropic antenna fed with same power}} \tag{8-25}$$

Note:

- Power density $\propto$ (Electric field strength)^2
- The ratio is usually expressed in dB
- Gain allows for antenna efficiency

Typical values range from 50 dB (very high gain) to negative (ratio is less than 1).

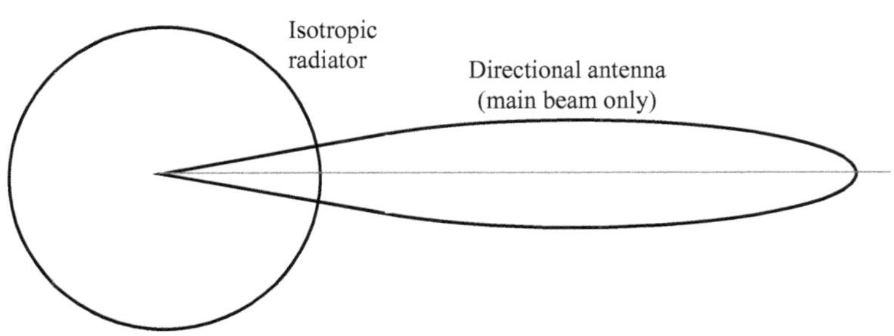

FIGURE 8-37 Antenna Gain

8.5.1.2 Beamwidth

The beamwidth, θ_B, is the angle subtended by the main beam. The "ideal" beamwidth assumes the power density inside the beam to be equal to the maximum and zero outside and is surprisingly accurate. Clearly, the main beam is not shaped like this, so by convention the beamwidth is taken to be equal to the angle between the two points on the contour of the main beam where the power density is half the maximum boresight value (or 0.707 the E field strength). This gives the 3 dB beamwidth.

The beamwidth is related to the wavelength, λ, and antenna dimension, D, by

$$\theta_B = \frac{\lambda}{D} \text{ radians} \quad \text{or} \quad \theta_B = \frac{57\lambda}{D} \text{ degrees} \tag{8-26}$$

For a rectangular antenna, the beamwidth is given by $\theta_B \approx 60\frac{\lambda}{D}$ degrees, whereas for a circular antenna of diameter, D, the beamwidth is given by $\theta_B \approx 51\frac{\lambda}{D}$ degrees. The actual beamwidth of an antenna is a function of its geometry and the power distribution across its aperture. Thus the combination of short wavelength and large antenna yields a narrow beamwidth. The angular resolution of a radar is determined by the beamwidth of the ***receiving*** antenna.[4] The beamwidths in the azimuth and elevation planes may differ. This occurs when the width and height of the antenna differ. The beamwidth and gain are also related by the following approximation:

$$G \approx \frac{42000}{\theta_{AZ}\theta_{EL}} \text{ (rectangular antenna)}$$

$$G \approx \frac{27000}{\theta_{AZ}\theta_{EL}} \text{ (circular antenna)}$$

where θ_{AZ} and θ_{EL} are the azimuth and elevation beamwidths in degrees, respectively.

It is also worth noting that the angle between the main beam boresight and the first null is approximately equal to the beamwidth.

8.5.1.3 Sidelobe Level

Sidelobes have various undesirable effects. Jamming power may be directed at a radar via its antenna sidelobes, and large echoes from the ground (clutter returns) may be received even when the main beam is directed away from the ground. A uniformly illuminated rectangular antenna has sidelobes whose gain is only 13.2 dB less than that of the main beam. Values of at least −30 dB are typically required, and by careful design values as low as −40 dB (and exceptionally −60 dB) can be achieved.

Antennas also have a back lobe, or radiation in the direction opposite to the main beam, as illustrated in Figure 8-36. Antenna sidelobes and back lobes are greatly influenced by the presence of a radome. For forward facing radars accommodated in airborne vehicles the radome shape is dictated by aerodynamic considerations and is seldom sympathetic to the polar diagram of the antenna within.

8.5.1.4 Radiation Pattern

The radiation pattern of an antenna is a plot of **E** versus angle (or $\mathbf{E}^2$ versus angle) and may be expressed in polar or Cartesian coordinates. Strictly, this plot refers to the *far*

[4]The exception to this is Synthetic Aperture Radar (SAR), for which the angular resolution is determined by the two-way beam pattern, i.e. the combination of the transmit and receive beam patterns and space-time coding techniques.

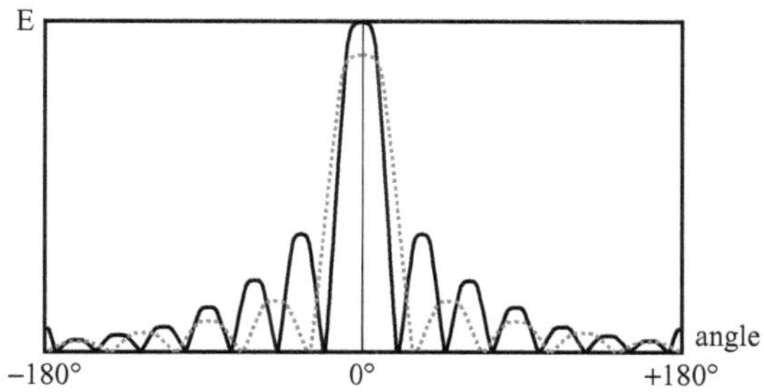

FIGURE 8-38 Radiation Pattern (Cartesian Coordinates)

field of the antenna; that is, it is at a sufficiently large distance away from the antenna such that wave fronts emerging from the antenna may be considered as plane waves, or parallel straight contours of equal phase. By convention, the far field threshold is taken as the range $\geq 2D^2/\lambda$ from the antenna. At closer ranges the antenna operates in its *near* or *induction field,* and the E field varies considerably with distance and the means by which energy couples to an obstacle differs from the far-field case. A typical radiation pattern (polar coordinates) is shown in Figure 8-36; the equivalent in Cartesian coordinates is shown in the solid line in Figure 8-38. Sidelobe levels may be reduced by tapering the E field (or power) profile across the antenna aperture. The profile of the E field defines the *aperture illumination function.* The far field radiation pattern is the Fourier transform of the aperture illumination function. For a uniformly illuminated rectangular aperture, the first sidelobe is 13.2 dB lower than the main beam and is shown in the solid line plot of Figure 8-38. Beamwidth and gain depend on the shape of the aperture and its illumination function. Lower sidelobes can be obtained by using a tapered illumination function. This can be brought about by transmitting gradually less power from the edges of the antenna or array. Tapered illuminations tend also to increase the beamwidth, to reduce main beam gain, and to reduce the sidelobe level in much the same way as windowed FFT functions affect the Doppler filter shape (Section 7.3.4). These effects are illustrated in the dotted line plot of Figure 8-38. The broadened main beam leads to a degradation in angular resolution and angular tracking accuracy and a wider beam footprint that captures more main beam clutter.

8.5.1.5 Tapered Illumination Functions

Tapered illumination functions reduce sidelobe levels in exactly the same way as windowed FFT functions reduce Doppler sidelobes. Popular antenna illumination functions include the Hamming, Hann, cosine, and Gaussian functions whose characteristics are summarized in Table 7-1. In addition to these functions, various Taylor functions are also popular because they set the peak sidelobe level and the number of sidelobes that attain the peak level, defined by the $\bar{n}$ (*n-bar*) number, before subsequent sidelobes at greater angular offsets decay away. Table 8-1 summarizes the features of the Taylor 35 dB and Taylor

TABLE 8-1 Summary of Taylor Illumination Functions [25]

Weighting Function	Peak Gain (dBi)	Gain Loss wrt Uniform (dB)	3 dB beamwidth (deg)	Peak SLL (dB)	rms SSL (dB)
Uniform	33.53	0	3.08	−17.50	−32.12
Taylor 35 dB ($\bar{n} = 2$)	32.41	1.111	3.74	−34.82	−47.20
Taylor 45 dB ($\bar{n} = 2$)	31.70	1.822	4.04	−43.63	−54.82

45 dB illumination functions (plus the uniform illumination function for reference) for a circular array antenna of diameter, $D = 19\lambda$.

8.5.1.6 Effective Aperture

An antenna captures a signal that it passes to the receiver (with a certain efficiency), whose power depends of the area of the antenna. The effective aperture, A_E, is defined as

$$A_E = \frac{\text{Power received by matched receiver}}{\text{Power density flowing past antenna}} \tag{8-27}$$

Due to antenna inefficiencies the effective aperture area is typically about 2/3 its physical area. The effective aperture area is related to its gain by

$$G = \frac{4\pi A_E}{\lambda^2} \tag{8-28}$$

This provides further illustration that a high gain antenna requires a small wavelength and large antenna (A_E).

8.5.1.7 Radiation Resistance and Matching

The power radiated by an antenna, P_r, is given by $P_r = I^2 R_r$, where I is the root mean square (RMS) current in the antenna, and R_r is the radiation resistance. This radiation resistance is a fictitious resistance that accounts for the radiated power in a transmitting antenna. It is normally very much greater than the DC resistance, R_l, that would be measured with an ohmmeter. The antenna impedance, Z_A, is given by

$$Z_A = R_l + R_r + X_A$$

where X_A is the reactive part of its impedance. Ideally, Z_A should be matched to the characteristic impedance of the feed for the maximum power transfer to the antenna.

The DC resistance gives rise to a loss of power, so an efficiency is defined given by the ratio $R_r/(R_r + R_l)$. Since R_r is normally much greater than R_l, the efficiency is high and is equal to 100% for a lossless antenna ($R_l = 0$).

8.5.1.8 Polarization

Polarization refers to the orientation of the electric field. This is determined by the antenna structure, feed, and orientation. For the best reception, it is important that a receiving antenna accepts the correct polarization as the incoming signal. Typically, signals may be plane polarized in either a vertical or horizontal orientation, as shown in Figure 8-39.

Signals may also possess circular polarization, whereby the electric field orientation undergoes a 360° variation in one wavelength. This may be left-handed circular polarization, for a counterclockwise rotation of **E** or right-handed for a clockwise rotation of **E**.

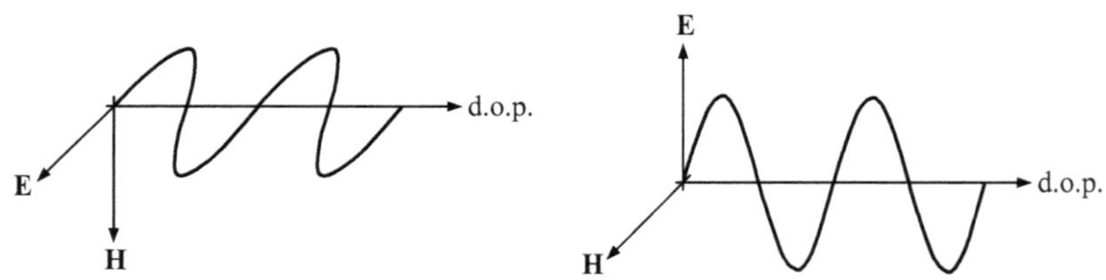

FIGURE 8-39 Plane Polarization: Horizontal (left); Vertical (right)

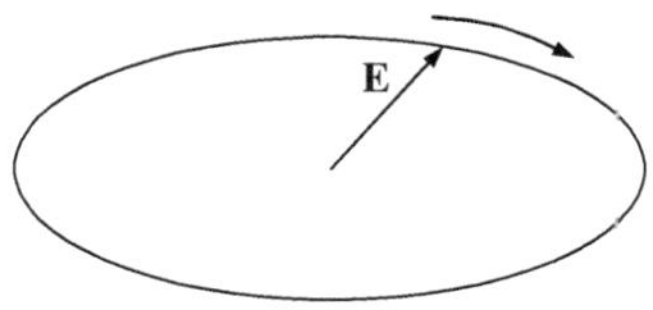

FIGURE 8-40 Elliptical Polarization

In general, signals can be regarded as being elliptically polarized (left- or right-handed), as shown in Figure 8-40. In elliptical polarization, the E field strength varies as it rotates. Elliptical polarization becomes plane polarization when the one axis of the ellipse equals zero; it becomes circular when both axes are equal.

8.5.2 Radar Aperture Antennas

This section considers a range of traditional aperture antennas which have radar applications.

8.5.2.1 Waveguide Horns

An open-ended section of waveguide would radiate into the space around it rather inefficiently and so acts as a crude antenna; however, it would be poorly matched and the beamwidth would be quite wide and the gain low. An improvement on this is the waveguide horn. The horn is formed by flaring the end of a waveguide feed. The flare provides a better match and a slightly increased gain and reduced beamwidth. The horn may be sectoral, pyramidal, or conical. It is commonly used as a primary feed for reflectors and lenses. Some waveguide horns are shown in Figure 8-41. A better match may be obtained by using curved surfaces to flare the horn.

The gain of a horn is $= \frac{4\pi A}{\lambda^2}$, where A = 0.6 a.b (sectoral and pyramidal horns) or A = 0.4 D^2 (conical horn). A range of standard gain horns are manufactured having gains of around 20 dBi.

The beamwidth is $= \dfrac{80\lambda}{a}$ (E field tapered across "a" dimension)

$= \dfrac{50\lambda}{b}$ (E field uniform across "b" dimension)

or, for a conical horn, beamwidth $= \frac{70\lambda}{D}$ (vertical) or $= \frac{60\lambda}{D}$ (horizontal) for vertical polarization. Typical beamwidths achieved are around 20° in each plane.

The optimum length for a horn, L, is governed by the fact that the variation of phase from center to edge of the horn aperture should not exceed 90°; otherwise sidelobes become large. This yields $L \approx \frac{D^2}{2\lambda}$. Wider flare angles than this condition allows are possible if a lens is inserted in the aperture of the horn to correct the phase.

A waveguide horn boasts very modest performance, inadequate for most radar applications; however, it may be a useful feed for a parabolic reflector antenna.

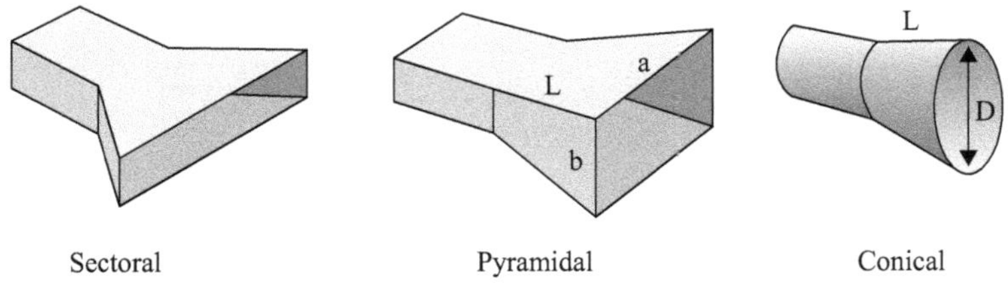

FIGURE 8-41 Waveguide Horns

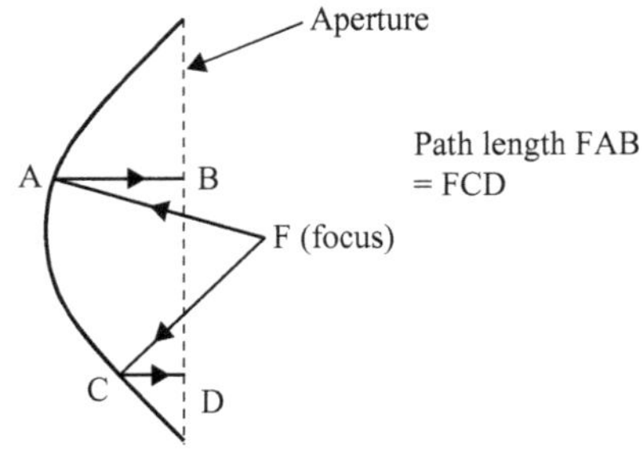

FIGURE 8-42 Parabolic Reflector

8.5.2.2 Parabolic Reflector

If a source is fed from the focus of a parabolic reflector, the aperture is an equiphase surface since all path lengths from source to aperture are equal, as shown in Figure 8-42.

For uniform illumination the beamwidth $\theta_B = \frac{51\lambda}{D}$ and the first sidelobes are 13.2 dB lower than the main beam. In practice, the feed is not uniformly distributed over the aperture since the feed at the focal point is a horn or dipole plus reflector, which results in an approximate sinusoidal aperture illumination. The beamwidth now becomes $\theta_B = \frac{70\lambda}{D}$, and the first sidelobe level is 23 dB below the main beam.

One shortcoming of the parabolic reflector is that the feed presents some *aperture blocking,* which results in backscatter, higher sidelobes, and a lower effective aperture area than the physical area of the aperture. A further shortcoming is that a large reflector requires a long focal length since otherwise the feed at the focal point would not illuminate the parabolic surface adequately. This increases the size and length of the structure. Both these problems can be overcome using a Cassegrain antenna.

8.5.2.3 Cassegrain Antenna

The Cassegrain antenna illustrated in Figure 8-43 (top) is a development of the basic parabolic reflector and stems from an idea used in optical reflecting telescopes. The system uses a main parabolic reflector and a smaller subreflector, which may be a plane surface

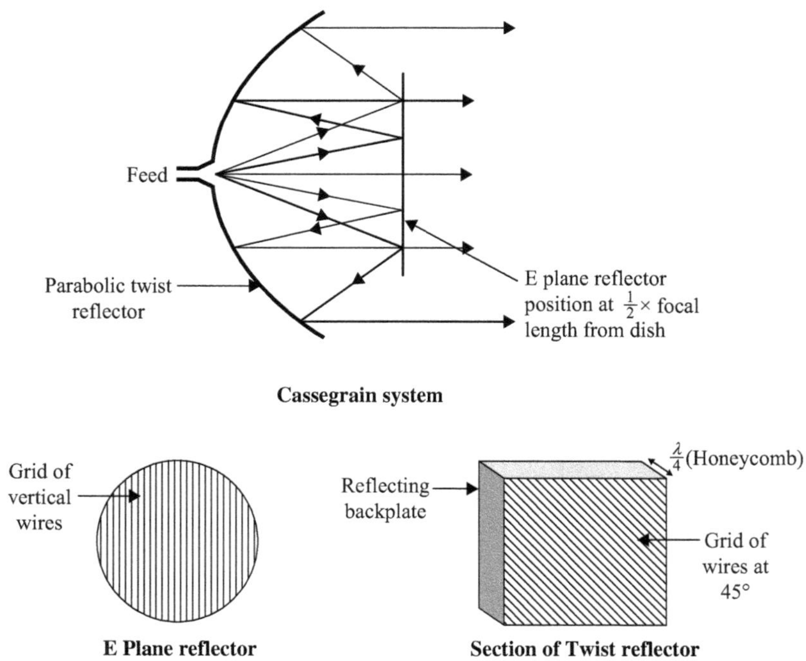

FIGURE 8-43 Cassegrain Antenna

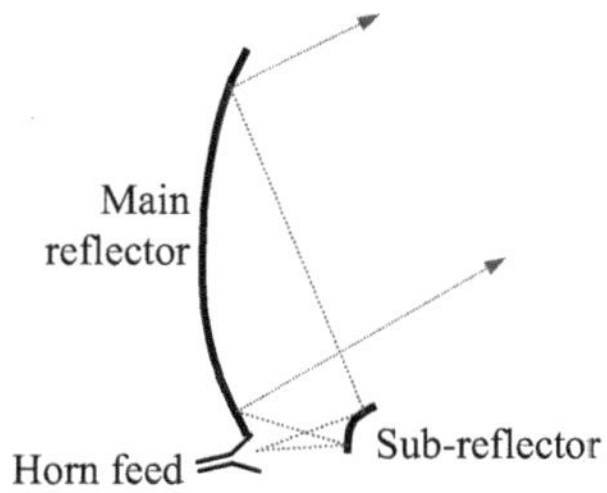

FIGURE 8-44 Offset Cassegrain

or a reversed parabola. Furthermore, if the subreflector is an **E** plane reflector (Figure 8-43, lower left) and the parabolic reflector is a twist reflector (Figure 8-43, lower right), then the aperture blocking effect of the conventional Cassegrain or front-fed reflector system is removed with resulting smaller sidelobes and higher gain. The **E** plane reflector reflects vertically polarized signals from the feed provided that its wire spacing is $\ll \lambda/2$. The twist reflector reflects the vertically polarized incident waves as horizontally polarized waves which then pass straight through the **E** plane reflector as it now appears to be transparent to horizontal polarization. The reverse happens on reception of the horizontally polarized echoes.

Another variation that overcomes aperture blocking is the offset Cassegrain, as shown in Figure 8-44. This arrangement permits rapid scanning by movement of the lightweight subreflector. Larger angular demands can be met by the more sluggish movement of the main reflector. The inverted parabolic subreflector also enables the system to have a long focal length (narrow, high gain beam) without being physically long.

8.5.3 Array Antennas

Many antennas used in today's radars are in fact arrays of radiating/receiving elements. Arrays exist in three basic geometries:

- Linear, in which elements are arranged in a line
- Planar, in which elements are arranged on a plane surface
- Conformal, in which the elements are arranged in three-dimensional space on the surface of the vehicle carrying the radar

8.5.3.1 Two-Element Linear Array

Consider the simplest array of two isotropic elements separated by a distance, d, as illustrated in Figure 8-45. If the two elements radiate signals of equal magnitude and phase,

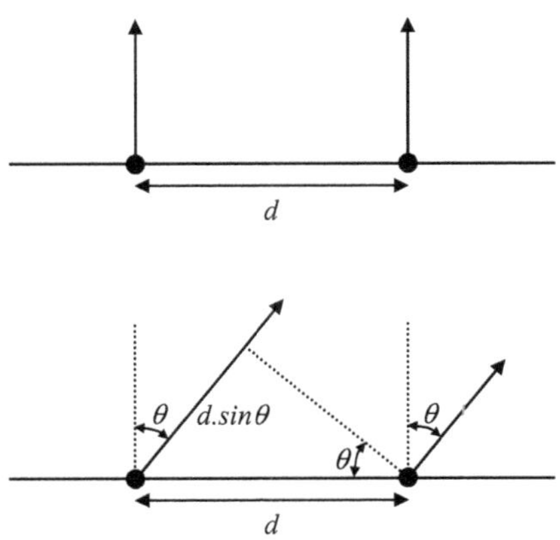

FIGURE 8-45 Two-Element Array

then the resultant E field has a maximum of twice the value from each individual element for a signal propagating at right angles to the array, as shown in the upper diagram of Figure 8-45. (There would also be a peak E field at 180° to this signal by virtue of the symmetry of the array, but in practice the supporting structure would not support this beam.) The line perpendicular to the array is the mechanical boresight of the array, and this coincides with the electrical boresight of the radiation pattern. This is known as a *broadside array*. Signals propagating at an angle, θ, to the mechanical boresight would have a phase difference of

$$\Delta\varphi = \frac{2\pi}{\lambda} d \sin\theta \tag{8-29}$$

and there would be partial cancellation of the fields radiated by the two elements leading to a reduced total E field. If $d = \lambda/2$ then the cancellation of two signals occurs at an angle $\theta = \pm 90^\circ$; in other words, the beam pattern has nulls along the axis of the array. Thus, the shape of the beam is a function of the array geometry and wavelength.

If, however, $d = \lambda/2$ and the elements are fed with signals of equal magnitude but in antiphase, then there would be total cancellation for propagation at $\theta = 0^\circ$, or a broadside null, and the resultant E field has a maximum of twice the value from each individual element for a signal propagating along the line of the array, $\theta = \pm 90^\circ$. This situation is known as an *end-fire array*.

Similarly, signals received by the elements arriving at the angle, θ, result in currents in each element having a phase difference given by equation (8-29). If the currents from both elements are added with equal magnitude and phase weightings, the receiving array has the same beam pattern as the broadside transmitting array. If, however, the currents from both elements are added with equal magnitude and with antiphase weightings, the receiving array would have the same beam pattern as the end-fire transmitting array.

8.5.3.2 Grating Lobes

If $d \geq \lambda$ and the elements are fed with signals of equal magnitude and phase, then the resultant E field has a peak of twice the field from each element at $\theta = 0^\circ$ and also at values of θ, which satisfy

$$\Delta\varphi = \frac{2\pi}{\lambda} d \sin\theta = 2n\pi \tag{8-30}$$

where $n = 1, 2, 3 \ldots$. Multiple solutions to (8-30) are possible only when $d \geq \lambda$. This gives rise to multiple beams, each of the same gain, known as *grating lobes*. A radar using such an array would therefore suffer from angular ambiguity. Consider the limiting case in which $d = \lambda$. For two elements fed in phase there would be a beam centered at $\theta = 0^\circ$ (and 180°) and $\pm 90^\circ$. Grating lobes can be avoided if the element spacing $d < \lambda$.

If each element is truly isotropic, their gains equal unity. Since the peak total E field is twice that arising from each element, a two-element array has a gain of $2^2 = 4$. (Recall that gain is a measure of the peak boresight power density and that power is proportional to the square of the E field magnitude.)

8.5.3.3 *N*-Element Linear Array

The two-element linear array can be extended to any number of elements, N. Simple linear arrays can be constructed from sections of slotted waveguide, as shown in Figure 8-46.

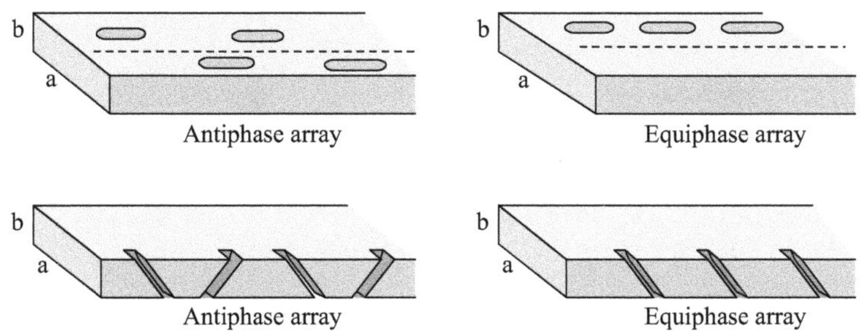

FIGURE 8-46 Slotted Linear arrays

For isotropic elements, the total E field from a linear array of N elements, each of equal magnitude and phase weighting, is given by the *array factor*:

$$E(\theta) = \frac{\sin[N\pi(d/\lambda)\sin\theta]}{N\sin[\pi(d/\lambda)\sin\theta]} \tag{8-31}$$

where θ is the angle with respect to the broadside orientation of the array.

8.5.3.4 Planar Arrays

The idea of linear arrays can be extended in two dimensions to form a planar array in which elements are arranged over a plane surface. The far-field E field pattern is given by the two-dimensional Fourier transform of the surface illumination function. The illumination function is typically a periodic rectangular function across both the horizontal and vertical planes of the array. The width of the "pulse" of the rectangular function is dictated by the aperture width of each element, and the "period" of the rectangular function is set by the spacing between the phase centers of elements.

Elements may be arranged in a rectangular or triangular lattice, as shown in Figure 8-47.

In reality, elements are not isotropic. Elements may be slots in waveguide walls, microstrip patches, dipoles, or even waveguide horns. The elements have their own directional properties, which influence the radiation pattern of any array. In such circumstances it becomes possible to tolerate grating lobes if they can be incurred at angles where the elements have a null, since the overall radiation pattern is the product of the array factor and the elemental radiation pattern. This is illustrated in Figure 8-48.

It is also possible to taper the illumination function of linear or planar phase arrays by feeding each element with an appropriate current and phase. The illumination functions used to control each element are usually sampled forms of the continuous functions mentioned earlier (e.g., Table 8-1) in which the function is sampled at discrete points corresponding to the phase centers of each element in the array. This works well for large

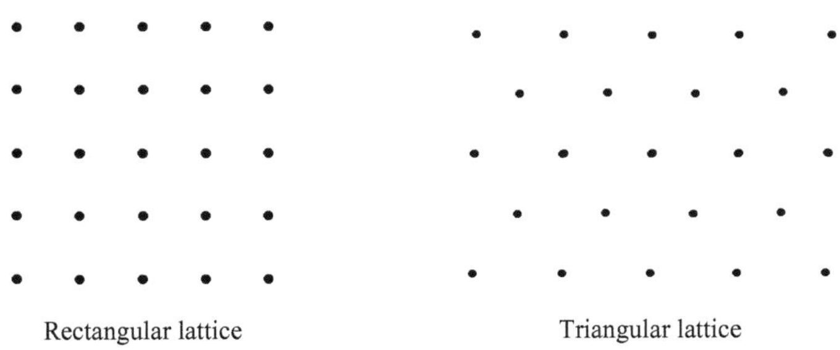

FIGURE 8-47 Planar Array Lattices

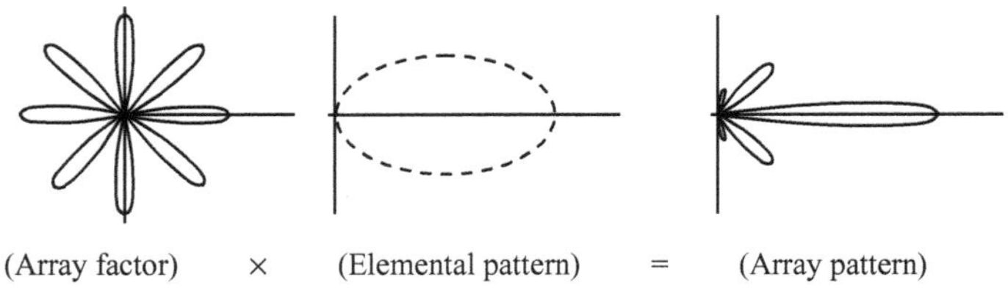

FIGURE 8-48 Array Radiation Pattern

arrays (hundreds of elements or more) but tends to result in degradations to the beam patterns for small arrays of a modest number of elements (e.g., 20) as the illumination function is inadequately sampled by such few elements.

Applying a tapered illumination function to an array results in a broadened beamwidth and consequent loss of gain, which affects both the transmitting and receiving arrays. This reduction is quantified in Table 7-1 and Table 8-1 for various tapered weighting functions. The loss in the gain can be calculated by evaluating the ratio of the coherent to incoherent combination of the weighted signals of each element for the tapered and uniform arrays. This is given by

$$\left(\frac{\sum_{i=1}^{N} |w_i|^2}{\left(\sum_{i=1}^{N} |w_i| \right)^2} \right)_{UNIFORM} \Bigg/ \left(\frac{\sum_{i=1}^{N} |w_i|^2}{\left(\sum_{i=1}^{N} |w_i| \right)^2} \right)_{TAPERED} \tag{8-32}$$

where N is the number of elements in the array, and w_i is the weighting of the ith element.

However, the reduction in element weightings also causes a reduction in the transmitted power. Hence, the transmitting array suffers a double hit as a result of tapering—a reduction in gain and power—whereas the receiving array suffers the reduction only in gain. For example, a uniform weighting function can be applied by setting the magnitude response of all elements to 1. If, however, the magnitude response of all elements is set to 0.5, then the array is still uniformly weighted and its gain is unaffected; however, the transmitted power reduces accordingly.

The array analysis considered so far does not take into account several practical details, such as the mutual coupling between elements and the edge effects encountered by elements near or on the periphery of an array. These can significantly affect the radiation pattern.

8.5.3.5 Conformal Arrays

In conformal arrays, elements are arranged on the surface of the vehicle that carries them. They are distributed in a geometry that conforms to the profile of the vehicle. This may be a non-Euclidean curved surface such as a sphere or a tangent ogive, which is the shape of the nose cone of many (supersonic) airborne vehicles. Such arrays possess the obvious advantage that the array can be accommodated on the outer surface of the vehicle and does not need to occupy space within a radome. For many airborne radars, the shape of the radome is dictated by aerodynamic considerations and the resulting radome shapes are not ideally suited to maintain the best antenna performance, particularly with respect to sidelobe level and the purity of polarization. Conformal arrays can dispense with the radome altogether, but the analysis of these arrays is far from trivial.

8.5.4 Active Electronically Steered Array Antennas

8.5.4.1 Phase Steering

The point was made in the analysis of the two-element array (section 8.5.3) that signals received from an angle θ with respect to the broadside orientation of the array have a phase difference of $\Delta\varphi = \frac{2\pi}{\lambda} d \sin\theta$, which is a repeat of equation (8-29). If, on the other hand, signals are transmitted from each element with a phase difference, $\Delta\varphi$, given by (8-29), then the transmitted beam boresight becomes orientated at the angle, θ. Similarly, if the two elements impart a phase difference of $\Delta\varphi$ given by (8-29) onto the received signals, then the received beam boresight would be orientated at the angle, θ. By adjustment of the phase difference, $\Delta\varphi$, between the elements the transmitted and received beam can be steered to any angle. Practical limitations usually constrain the steering angle, θ_S, to be no greater than $\pm 60°$ from the mechanical boresight of the array.

This can be extended to linear and planar arrays in which a phase difference between neighboring elements is used to steer the beam. By applying a phase gradient across the horizontal and vertical planes of the array, the beam can be steered in both the azimuthal and elevation planes. (Conformal arrays can also be phase steered in much the same way, but here the phase gradients must also account for the position of each element in three-dimensional space, which complicates matters somewhat.)

Some phase shifters have a limited range of settings. For example, a five-bit phase shifter has a phase resolution of $360/2^5 = 11.25°$. A simple phase gradient by which the phase difference between elements, $\Delta\varphi$, must be quantized into $11.25°$ intervals results in a very coarse steering of the beam and would not be very useful. However, it is not necessary to be constrained to simple phase gradients like this, and it is possible to compute more complex (nonuniform) phase weightings, which result in very fine beam steering angle resolution.

8.5.4.2 Phase Shifters

The phase weighting of each element is controlled by a (digital) phase shifter [20]. These have typically used PIN diode switches to switch in lengths of transmission line to impart the necessary phase shift. An example of a phase shifting element is illustrated in Figure 8-49.

A series of PIN diode phase shifters can be used to impart the necessary phase weighting to each element of an array. An example of a five bit phase shifter is given in Figure 8-50.

PIN diode phase shifters are fast (nanosecond switching times) and can be manufactured very small (a few mm^2) and give digital control of discrete phase increments;

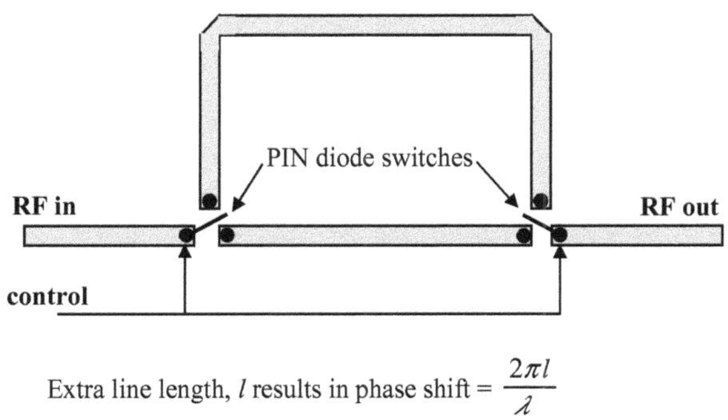

FIGURE 8-49 ■ PIN Diode Phase Shifter

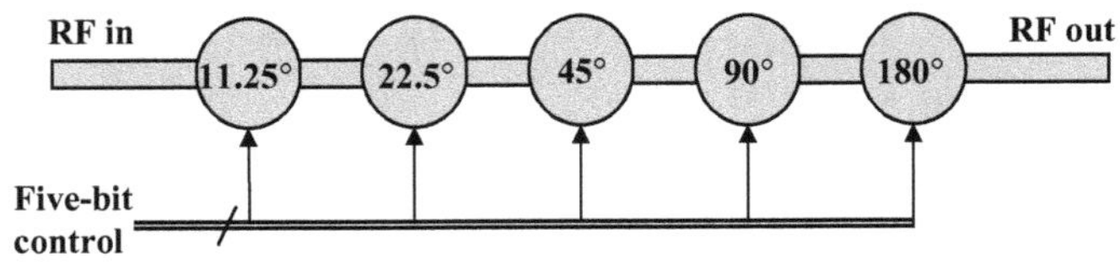

FIGURE 8-50 Five-Bit Phase Shifter

however, they are expensive, moderately lossy, have limited power handling, are hungry on power, and require hermetic packaging.

Alternative phase shifter designs based on lengths of transmission lines incorporating microelectromechanical (MEM) capacitors or thin film ferroelectric phase shifters may soon rival the PIN diode phase shifter.

8.5.4.3 Grating Lobes

In Section 8.5.3, it was noted that for elements of equal magnitude and phase weighting multiple solutions to equation (8-30) would be possible when $d \geq \lambda$. These multiple solutions result in grating lobes and angular ambiguity. However, elements do not have an equal phase weighting when phase steering is applied. If, for two elements, a phase difference of $\Delta\varphi$ is applied, then from (8-29) the beam steering angle, θ_S, is given by

$$\sin\theta_S = \frac{\lambda}{2\pi d}\Delta\varphi \tag{8-33}$$

However, multiple beams could be formed at any solution for θ from

$$\frac{2\pi}{\lambda}d\sin\theta - \Delta\varphi = 2n\pi \tag{8-34}$$

where $n = 0, 1, 2, 3 \ldots.$
The solution for $n = 0$ results in (8-34) becoming

$$\Delta\varphi = \frac{2\pi}{\lambda}d\sin\theta$$

which is a repeat of (8-29) and has the solution $\theta = \theta_S$ given by (8-33).

If the steering angle, θ_S, is limited to $\pm 60°$, then additional solutions to (8-34) could exist. For $n = 1$, equation (8-34) has a solution of $d \geq \lambda/1.866$, and a grating lobe results.

To avoid grating lobes the distance between elements in a linear array and the rectangular lattice array should not exceed

$$\frac{\lambda}{1 + \sin\theta_S} \tag{8-35}$$

or, for the triangular lattice array,

$$\frac{2\lambda}{\sqrt{3}(1 + \sin\theta_S)} \tag{8-36}$$

where θ_S is the steering angle. To accommodate an electronic steering angle of up to 60° off the mechanical boresight, a spacing of $\lambda/2$ is often used.

8.5.4.4 Beam Dilation

One significant consequence of phase steering a beam away from the mechanical boresight of the array is that the dimension of the array resolved across the beam boresight

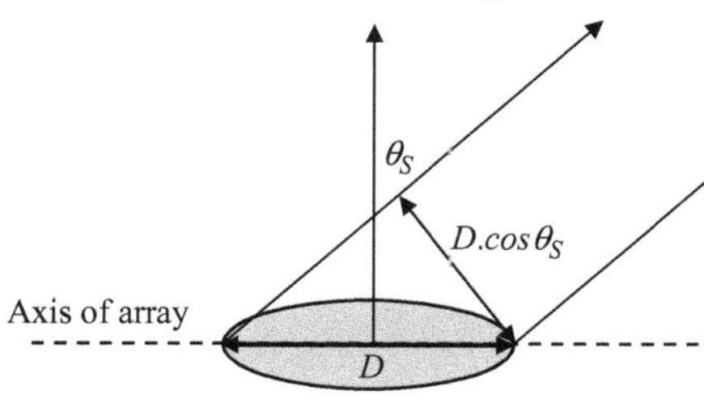

FIGURE 8-51 Beam Dilation

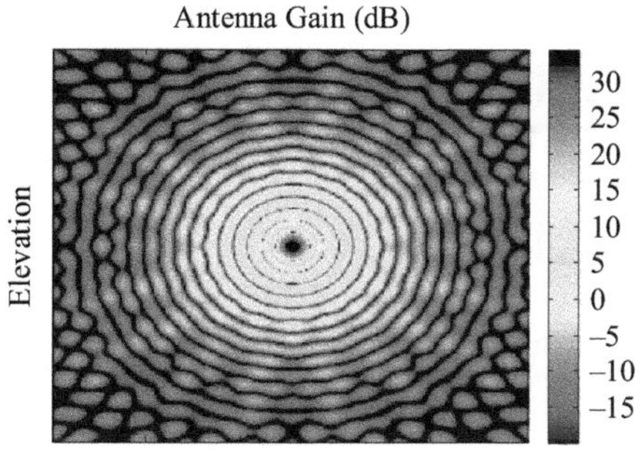

Uniformly Illuminated Circular Array
Azimuth = 0°, Elevation = 0°

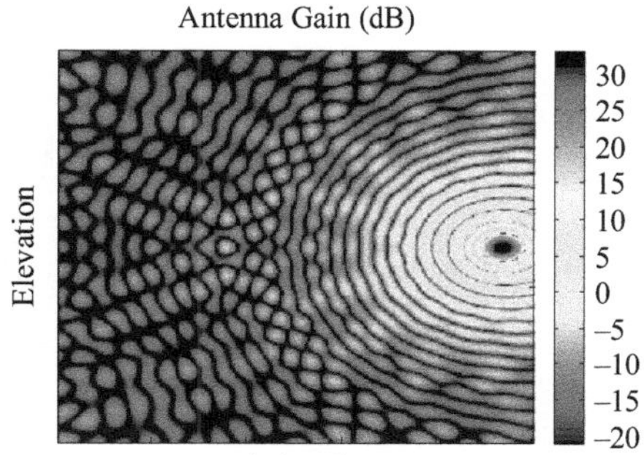

Uniformly Illuminated Circular Array
Azimuth = 40°, Elevation = 0°

FIGURE 8-52 Effects of Beam Dilation

(perpendicular to it) reduces as $\cos\theta_S$, where θ_S is the steering angle. This is illustrated in Figure 8-51. This causes a broadening of the beamwidth and a reduction in main beam gain as $1/\cos\theta_S$. The broadening of the beam means that fewer beam positions are required to search wide areas away from the mechanical boresight (and a corresponding loss of angular resolution/tracking accuracy); however, the reduction in main beam gain causes a reduction in detection performance. The reduced detection performance can be compensated for by increasing the dwell time (increased processing time and gain) when off axis. The effects of fewer beam positions and longer dwell times negate each other.

Figure 8-52 illustrates the effects of beam dilation and the beam pattern of a circular array antenna uniformly illuminated. The left-hand plot shows the beam pattern for zero phase steering in both azimuth and elevation, whereas the right-hand plot shows the beam pattern when steered to 40° in azimuth and 0° in elevation and thus the increase in the azimuth beamwidth.

8.5.4.5 Transmit/Receive Modules

AESA antennas permit control of the phase and amplitude of each element. This requires a certain amount of microwave circuitry and its associated control, power supplies, and cooling to be accommodated within a transmit/receive module on the array face. A typical transmit/receive module is illustrated in Figure 8-53. This design uses the same phase shifter in the transmit and receive paths, which ensures coincidence between the transmit and receive beam boresight. However, different amplitude weighting may be applied in each path via a suitable adjustment of the gain of the amplifiers. This enables any desired tapered illumination function to be applied and even permits a different function to be used on transmit to that used on reception.

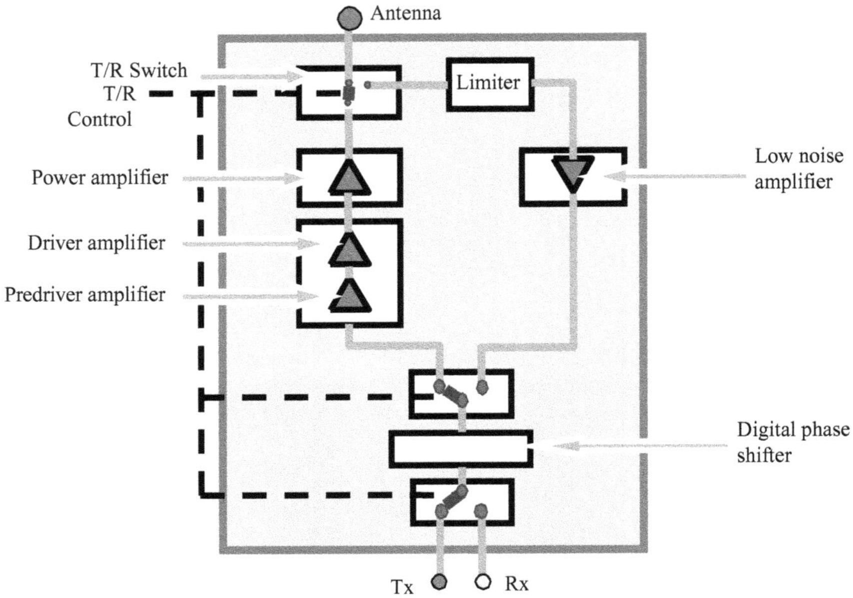

FIGURE 8-53 Transmit/Receive Module

Circuit miniaturization and cooling requirements pose considerable design difficulties. These problems are exacerbated at higher frequencies because of the increased packing density required to maintain an element spacing of $\lambda/2$ coupled with the poorer efficiency in converting DC electrical power to RF.

8.5.4.6 Advantages of AESA Antennas over Conventional Mechanically Steered Antennas

Many radars are now equipped with AESA antennas, which have opened up several new and exciting possibilities. Conventional mechanically steered antennas tend to be limited by the following:

- The aperture illumination is fixed by its design.
- The antenna has a fixed dwell time.
- The data rate is also fixed since a limited range of scan functions are available.

However, a conventional antenna does not suffer from beam dilation and is low cost.

By contrast, an AESA antenna boasts the following features:

- Rapid beam steering. The antenna beam can be redirected in the time it takes to switch the appropriate phase shifters, which is typically of the order of several nanoseconds for PIN diode phase shifters. There is no mechanical inertia to overcome, and the beam can be directed very accurately and very quickly. This means that the radar beam can adopt a series of positions to conduct a search. The search function is no longer constrained to an azimuth scan at a fixed elevation but can, for example, scan an undulating horizon or some more complex profile. The search function may be interrupted to revisit high-priority targets to update track files. Indeed, a multitude of search-and-track functions can be interleaved in rapid succession, and they appear to be conducted simultaneously to a human operator.
- Multiple independent beams. A multi-element AESA antenna can produce multiple beams, each in a different angular position. This also requires multiple receiver

channels, one for each beam, to process the received signals. This is particularly useful in generating multiple stacked beams to permit ground-based air surveillance radars to estimate target elevation. This is also used in the production of the split beam pattern used by monopulse angle tracking.

- Control of radiation pattern. One is no longer constrained to a fixed illumination function and hence a fixed radiation pattern. This could be adapted during operation in response to the local conditions. One area of considerable interest is the ability of an AESA antenna to form a null, or even several nulls, at any desired angles. This is used for the rejection of interference from a known direction, which can also be determined by the antenna [16]. It is then possible to adaptively form and steer nulls toward jamming sources while simultaneously steering the main beam as required. Naturally, the shape of the main beam and the sidelobe levels suffer if a null is steered close to the main beam.
- High output power. Each element may typically contribute only a few watts of RF power; however, the (vector) sum of all contributions from many elements results in a high transmitted power.
- Graceful degradation. A transmitter with a single source of high-power RF also has a single point of failure. However, the loss of a few elements from a large array has an insignificant effect on the radar performance. The effect of an increasing proportion of failed elements is a gradual (or *graceful*) degradation in radar performance. As elements fail, the antenna gain drops and the sidelobe levels increase. The total loss of around 5% is considered tolerable [25]. However, a greater proportion of failures may be tolerated depending upon the failure mode (e.g., loss of transmission only, high receiver noise figure).
- Low radar cross section (RCS) antenna. A conventional radar antenna makes a significant contribution to the RCS of the platform carrying it since it has a large monostatic RCS[5] within its own beamwidth. Of particular concern is the RCS of a military aircraft accommodating a forward (and often downward) looking antenna since any threat to the aircraft can be expected to come from the forward (lower) sector. To counter this, the surface of an AESA antenna may be canted upward slightly such that its specular reflection is directed away from the angle of illumination. It would even be possible to cant the surface at an angle such that there is a null in the scattered radiation back in the direction of an illuminating source directly ahead of the aircraft. Phase steering may be applied to offset the electrical boresight of the beam away from its mechanical boresight.

It is clear that AESA antennas have many advantages. Because of them it has been possible to amalgamate functions previously conducted by several radars into one multi-function radar system. They have also spawned research interest in schemes dictating the task scheduling of such radars [26, 27, 28]. If the full potential of a multi-function radar is to be realized, it is important to establish robust rules dictating the most efficient use of the radar's resources particularly as its workload increases, which generally follows with increasing number of targets.

[5]Monostatic radars are those in which the transmitter and receiver are co-located. The monostatic RCS is therefore the target RCS seen at an angle of reception equal to the angle of illumination.

References

[1] Penley, W. H., "The Early Days of Radar in the UK Notes for Talks." Presentation to various organizations in Dorset, UK, October 1993, 1-26. Full text available at http://www.scribd.com/doc/44642132/Early-Radar

[2] Skolnik, M. I. (Ed.), *Introduction to Radar Systems*, 2d ed., McGraw-Hill, 1980.

[3] Stralka, J. P. and Fedarko, W. G., "Pulse Doppler Radar," Chapter 4 in M. I. Skolnik (Ed.), *Radar Handbook*, 3d ed., McGraw-Hill, 2008.

[4] DiFranco, J. V. and Rubin, W. L., *Radar Detection*, Raleigh, NC: Scitech, 2004.

[5] Nitzberg, R., *Radar Signal Processing and Adaptive Systems*, Artech House, 1999.

[6] Keel, B. M., Pulse Compression Fundamentals," Chapter 20 in M. Richards (Ed.), *Principles of Modern Radar*, vol. 1, Raleigh, NC: Scitech, 2010.

[7] Pace, P. E., *Detecting and Classifying Low Probability of Intercept Radar*, Artech House, 2004.

[8] Long, W. H. and Harriger, K. A., "Medium PRF for the AN/APG-66 Radar," *Proceedings of the IEEE*, **73**(2), February 1985, 301–311.

[9] Long, M. W., *Radar Reflectivity of Land and Sea*, Artech House, 1983.

[10] Alabaster, C. M. and Hughes, E. J., "Clutter Mapping for Airborne Pulse Doppler Radar," Paper presented at the IET Clutter Modelling Seminar, February 19, 2008, Savoy Place, London.

[11] Nathanson, F. E., *Radar Design Principles*, 2d ed., Raleigh, NC: Scitech Publishing Inc., 1999.

[12] Skolnik, M., *Radar Handbook*, 2d ed., McGraw-Hill, 1990.

[13] Nitzberg, Ramon, *Adaptive Signal Processing for Radar*, Artech House, 1991.

[14] Lyons, R. G., *Understanding Digital Signal Processing*, Addison-Wesley, 1997.

[15] Guerchi, J. R., *Space-Time Adaptive Processing for Radar*, Artech House, 2003.

[16] Madksinghe, D., "Performance Analysis of an Airborne High PRF Phased Array Radar in a Jamming Environment," Paper presented at the Proceedings of Radar Conference, Adelaide, September 3–5, 2003.

[17] Cooley, J. W. and Tukey, J. W., "An Algorithm for the Machine Calculation of Complex Fourier Series," *Math. Comput.* **19**, 297–301, 1965.

[18] Wikipedia contributors, "Window Function," *Wikipedia, The Free Encyclopedia*, http://en.wikipedia.org/wiki/Window_function.

[19] Hughes, E. J. and Lewis, M., "Adaptive Spatio-Temporal CFAR and Multi-Hypothesis Tracking System," Paper presented at the Third EM&RS DTC Technical Conference, Edinburgh, 2006.

[20] Pozar, D. M. (Ed.), *Microwave Engineering*, 2d ed., John Wiley & Sons, 1998.

[21] Emery, D., "The Digital Synthesis of HF Surfacewave Radar Waveforms," MSc thesis, Cranfield University, CDS Shrivenham, UK, October 2004.

[22] Nyquist, H., "Thermal Agitation of Electric Charge in Conductors," *Phys. Rev.* **32**, 110, 1928.

[23] Friis, H. T., "Noise Figures of Radio Receivers," *Proc. IRE,* **32**(7), July 1944, 419–422.

[24] Balanis, C., *Antenna Theory: Analysis and Design*, 3d ed., Wiley, 2005.

[25] Alabaster, C. M. and Hughes, E. J., "An Examination of the Effect of Array Weighting Function on Radar Target Detectability," *IEEE Trans. Aerospace and Electronic Systems,* **46**(3), 1364–1375, July 2010.

[26] Martin, S., "Scheduling in Multi-Function Radars," (MESE 15), MSc Thesis, Cranfield University, Shrivenham, Swindon, SN6 8LA, July 2001.

[27] Gillespie, W., "Scheduling of a Multi-function Phased Array Based Airborne Intercept Radar," (MESE 18), MSc Thesis, Cranfield University, Shrivenham, Swindon, SN6 8LA, July 2004.

[28] Gillespie, W., Hughes, E. J., and Lewis, M., "Scan Scheduling of Multi-function Phased Array Radars Using Heuristic Techniques," Paper presented at the IEEE International Radar Conference, May 9–12, 2005, 513–518.

PART IIA

High Pulse Repetition Frequency Pulse Doppler Radar

CHAPTER 9

High Pulse Repetition Frequency Pulse Doppler Radar

Chapter Outline

9.1 INTRODUCTION

9.1.1 A Review of the Problems of Low Pulse Repetition Frequency

Pulsed radars that are not required to be Doppler sensing systems have traditionally employed low pulse repetition frequency (PRF) waveforms. A *low* PRF is defined as one which is sufficiently low so as to avoid range ambiguities; see Chapter 3 for further details. The maximum unambiguous range, R_{mu}, associated with a PRF, f_r, is given by

$$R_{mu} = \frac{c}{2f_r} \tag{9-1}$$

To avoid range ambiguity, the maximum detection range of the radar for all target and clutter returns is required not to exceed R_{mu}. This ensures that all returns are first-trace echoes, that is, that they fall within the first receiving period, and their range may be determined from a simple range delay timing method. Low PRF waveforms are typically low duty ratio waveforms and employ a large number of range cells matched to the range resolution of the radar distributed throughout the receiving period. Target range is therefore readily available from low PRF systems.

Sidelobe clutter (SLC) and main beam clutter (MBC) may also be present depending on the scenario and may be spread across a considerable proportion of the range cells. The low PRF radar response to surface clutter is described in Section 6.6. SLC is strongest at closer ranges; the first few range cells, which are occupied by the strongest SLC, may be attenuated or even gated out to prevent saturation of the receiver. MBC would also be present if the antenna illuminates the surface and occupies range cells corresponding to the footprint of the main beam on the earth's surface. Discrete target returns may therefore be detected in the presence of surface clutter, either SLC or MBC, which could overwhelm the returns from targets of interest. Automatic gain control (AGC) and constant false alarm rate (CFAR) mechanisms within the radar adjust the detection threshold in response to local clutter levels, but even so, targets of interest may go undetected in conditions of high clutter.

For ground-based radars, all SLC and MBC has zero or near-zero Doppler shift and may be rejected using a moving target indicator (MTI) radar clutter cancellation filter (Section 7.2). Similarly, airborne systems employing platform motion compensation (PMC) correct MBC to a centre Doppler frequency of zero Hertz which can subsequently be rejected using an MTI clutter filter. However, SLC is spread over a considerable Doppler band, depending on platform velocity, and cannot be filtered out since targets of interest would also be rejected.

The baseband output of a coherent pulsed radar is a sampled form of the Doppler signal (Section 4.6). The sampling rate is set by the PRF and therefore needs to be at least twice the frequency of the highest expected Doppler frequency to avoid aliasing ambiguities. Low PRF radars seldom employ PRFs which are sufficiently high to avoid such ambiguities. This leads to ambiguous repetition in the frequency domain, and hence velocity ambiguities arise. The spectral spread of SLC from a fast-moving airborne platform results in ambiguous repetition of the SLC in the frequency domain, as described in Section 6.6 and illustrated in Figure 6-21. This compounds the clutter problem since MBC and its ambiguous repetition now occupy much of the Doppler band, and SLC plus multiple ambiguous repeats of SLC occupy the whole Doppler band. In these circumstances (low PRF and high-velocity airborne platform), there are very high levels of clutter across the whole Doppler band and detection performance is severely degraded. If MBC were filtered away it would result in very large bands of blind velocities and the rejection of a high proportion of the targets of interest.

Given their high degree of Doppler ambiguity and the clutter-related problems, particularly those associated with MBC, low PRF waveforms are not a viable proposition for most airborne pulse Doppler radar applications. The solution to these problems is to increase the PRF such that Doppler, and hence velocity, ambiguities are avoided. This leads to *high* PRF waveforms.

9.1.2 A Brief Overview of High PRF Pulse Doppler

High PRF waveforms are those whose PRF is high enough to avoid velocity ambiguity. Pulse Doppler processing based on the fast Fourier transform (FFT) ensures that target velocity data are readily available, so high PRF waveforms lend themselves well to velocity search-and-track modes. The use of high PRFs tends to result in range ambiguities for many radar applications. Should target range data be required, frequency modulation (FM) ranging techniques or multiple PRFs may be used. Main beam clutter can readily be rejected, but multiple ambiguous repeats of sidelobe clutter in the range dimension can be problematic. A low sidelobe antenna is therefore required. For high-velocity (airborne) platforms, sidelobe clutter is spread across a broad band and leads to poor detection performance against slow-moving targets and even those receding more rapidly. However, fast-closing targets are readily detected as they fall above the clutter spectrum and their detection is limited by noise only. High PRF waveforms are also characterized by high duty ratios and hence high-energy waveforms. Furthermore, a large number of pulses are transmitted and received in a given beam dwell time, which offers the potential for high processing gain. Therefore, high PRF has good long-range detection performance against fast-approaching targets when considerable clutter is present. The high duty ratios do, however, result in considerable eclipsing losses. These may be minimized using multiple PRFs such that the eclipsed "blind" ranges of different PRFs are staggered away from each other.

9.2 | PRF SELECTION FOR UNAMBIGUOUS DOPPLER/VELOCITY

9.2.1 The Doppler Band

Figure 9-1 illustrates the spectrum of a periodic train of rectangularly pulsed modulated signal such as the transmitted signal of a coherent pulse Doppler radar of carrier frequency = f_{tx} , PRF = f_r and pulse width = τ (assuming no pulse compression modulation is applied). This spectrum is characterized by a line spectrum having a component at f_{tx} and components at multiples of f_r offset above and below f_{tx} whose powers adhere to a sinc^2 function envelope.

Upon reflection from a moving target each spectral component within this spectrum is subject to a Doppler shift of

$$f_d = \frac{2V_r}{\lambda} \tag{9-2}$$

where V_r is the radial velocity of the target and λ is the wavelength. The frequency of the received central carrier component, f_{rx},,becomes

$$f_{rx} = f_{tx} + f_d \tag{9-3}$$

The Doppler shift is positive for closing targets giving $f_{rx} > f_{tx}$, whereas the Doppler shift is negative for receding targets and gives $f_{rx} < f_{tx}$. If the spectral spread of the transmitted signal is a small proportion of its center frequency, then the variation in wavelength (λ) across the signal bandwidth is negligibly small and each spectral component is subject to approximately the same Doppler shift given by (9-2). If the frequency scale of Figure 9-1 is expanded to encompass the central carrier component and the first harmonic of the PRF lines above and below the carrier, the spectrum of Figure 9-2a is obtained. Figure 9-2b illustrates the received spectrum from a closing target with the transmitted

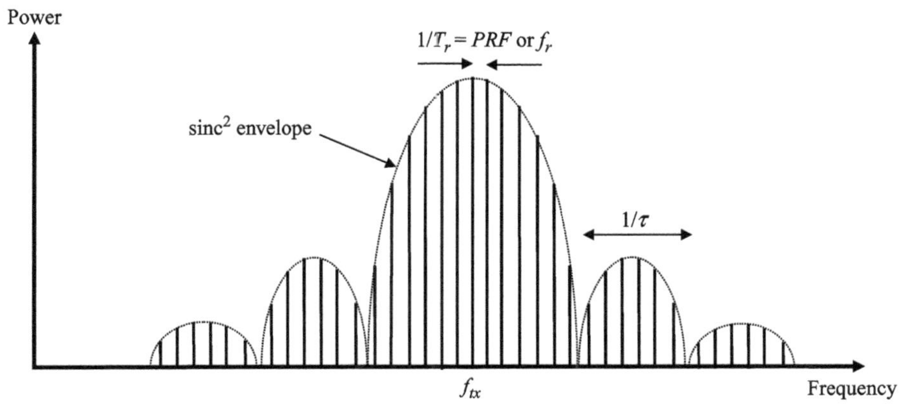

FIGURE 9-1 Pulsed Modulated Carrier Spectrum

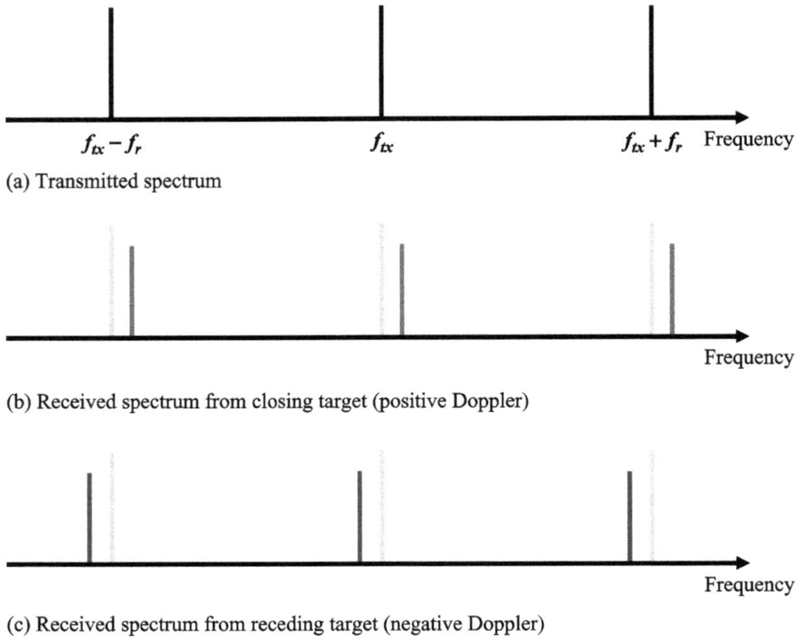

FIGURE 9-2 Central Portion of Pulsed Modulated Carrier Spectrum: (a) Transmitted Spectrum; (b) Received Spectrum from Closing Target (Positive Doppler); (c) Received Spectrum from Receding Target (Negative Doppler)

spectrum shown faintly for reference. Each spectral component has been offset by a positive Doppler shift. Figure 9-2c illustrates the received spectrum from a receding target with the transmitted spectrum shown faintly for reference. This time each spectral component has been offset by a negative Doppler shift. The frequency and sign (positive or negative) of the Doppler shift is obtained from the FFT processing of the complex baseband signal.

FFT Doppler processing must be applied over a suitable band that captures just one spectral component of the received spectrum. Note that within the bands illustrated in Figure 9-2, three spectral components are apparent. However, if the band is narrowed further from $f_{tx} - f_r/2$ to $f_{tx} + f_r/2$ giving a total bandwidth of f_r, then only one spectral component of the received spectrum would be captured. The FFT processing of the complex baseband signal must therefore cover the band $-f_r/2$ to $+f_r/2$ (described in Section 7.3). The FFT processing subdivides this frequency interval into a set of contiguous Doppler filters, as illustrated in Figure 9-3 for an example set of 32 Doppler filters (i.e., 32-point FFT processing).

The Doppler cells illustrated in Figure 9-3 map into the spectrum as in Figure 9-4.

A high-speed closing target whose Doppler shift approaches $f_r/2$ would be detected in the 15th Doppler cell. Should it accelerate any further, its Doppler shift would exceed $f_r/2$, and it would fall above the processed band. However, at the point that this spectral component exits the upper limit of the processed band, the Doppler shifted spectral component originally transmitted at $f_{tx} - f_r/2$ is received at the frequency $f_{tx} - f_r/2 + f_d$ and

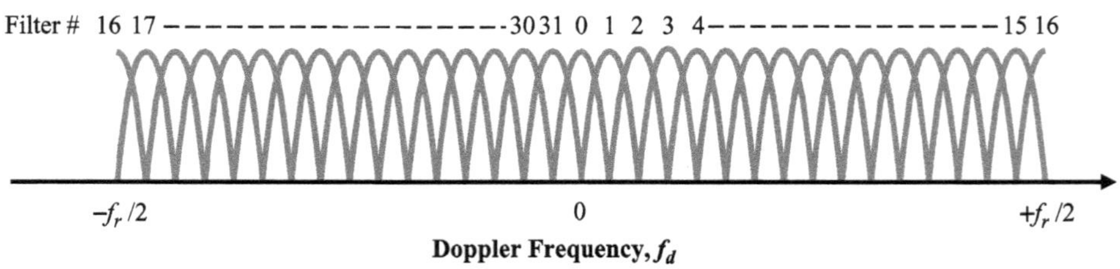

FIGURE 9-3 Subdivision of the Doppler Band into FFT Doppler Cells

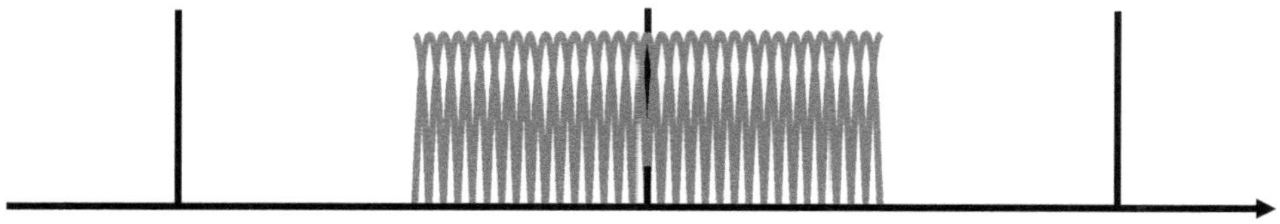

FIGURE 9-4 The Doppler Band

enters the lower end of the processed band and would be seen in the 17th Doppler cell. Note how the 16th Doppler cell straddles the upper and lower limits. This cell corresponds to the case of $|f_d| = f_r/2$ which is the limiting case of obtaining just two pulse samples per period of the Doppler waveform. The Doppler band of $-f_r/2$ to $+f_r/2$ always captures one and only one received spectral component per target. If it were any wider, it would be possible to capture two components pertaining to the same target and two targets would be indicated. If the band were any narrower, it would be possible to miss a closing or receding target whose Doppler shift was equal to, or close to, $f_r/2$. Therefore, the Doppler bandwidth must always equal f_r.

9.2.2 Doppler/Velocity Ambiguity

When the Doppler waveform is undersampled, aliasing ambiguities arise. Doppler and velocity measurements are ambiguous when two or fewer pulse samples of the Doppler waveform are obtained. This occurs when $|f_d| \geq f_r/2$. The aliasing ambiguity is illustrated in Figure 9-5; Doppler shifted received spectral components are shown by heavy lines, and the transmitted spectral components are shown by faint lines.

The target return illustrated in Figure 9-5 exhibits an apparent Doppler shift of $-f_d$. However, the true Doppler shift could be $+(f_r - f_d)$; that is, the PRF line below the carrier is positively shifted into the Doppler band. These possibilities correspond to either a slow receding target or a much faster closing target; clearly it is operationally important to resolve these cases! In fact, the ambiguity problem could be even more complex since the true Doppler shift could be $+(mf_r - f_d)$ or $-(mf_r + f_d)$, where $m = 0, 1, 2, 3 \ldots$. These are the cases of the m-th harmonic PRF lines being shifted into the Doppler band ($m = 0$ corresponds to the central carrier component).

9.2.3 Maximum Unambiguous Doppler/Velocity

For a Doppler band extending from $-f_r/2$ to $+f_r/2$, ambiguity arises if $|f_d| \geq f_r/2$ is possible. This establishes a maximum unambiguous Doppler shift of

$$f_{dmu} = \pm {}^{f_r}/_2 \tag{9-4}$$

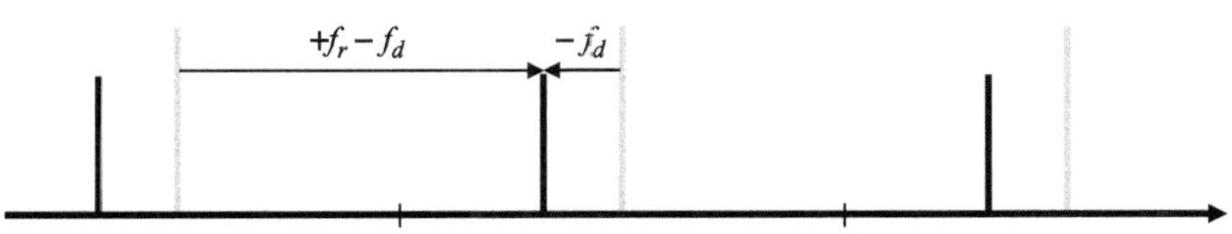

FIGURE 9-5 Doppler Aliasing Ambiguity

Substituting (9-4) into (9-2) and rearranging gives a maximum unambiguous velocity, V_{rmu}, of

$$V_{rmu} = \pm\frac{\lambda f_r}{4} \tag{9-5}$$

Alternatively, one may write that there is an unambiguous velocity bracket that is $\lambda f_r/2$ (m/s) wide.

9.2.4 Definition of High PRF

If the maximum target velocity ever to be encountered can be anticipated, equation (9-5) can be used to set a PRF sufficiently high to avoid velocity ambiguities; this is the definition of *high PRF.* Note that it is not possible to impose a numerical lower limit for high PRF as it is entirely dependent on the target velocities which may be encountered and on the transmitted wavelength.

9.2.5 High PRF/Low PRF Conundrum

Note how high PRF contrasts with low PRF, which is defined as a PRF sufficiently low to avoid range ambiguity (Section 3.4.2). Again, it is not possible to impose a numerical upper limit for low PRF because this is determined by the range bracket of interest. However, the requirements of high and low PRF are often contradictory. Consider an example of a 10 GHz pulse Doppler radar ($\lambda = 0.03$ m), which is required to determine the velocity of targets traveling with a maximum radial velocity of ±340 m/s (approximately Mach 1 at sea level). From (9-2) the maximum Doppler shift is ±22,667 Hz. For unambiguous velocity measurement, a PRF of 45,333 Hz is therefore required. However, from (9-1), this PRF gives a maximum unambiguous range, R_{mu}, of 3309 meters. Such a low value of R_{mu} may give problems because the detection performance of many radars is likely to exceed this, in which case range ambiguities arise. Whilst it is not certain that the use of high PRF definitely results in range ambiguities, they are likely in many applications. Similarly, it is not certain that the use of low PRF definitely results in velocity ambiguities, although they are quite likely in many applications. However, in a few radar applications, the maximum target velocity is quite low and the use of a long wavelength is feasible. These applications also require modest detection ranges for which it is possible to find values of PRF satisfying both the high PRF and low PRF criteria simultaneously.[1] The clutter problems of particularly low PRF values may make this solution undesirable; nevertheless, the possibility exists. Pulse Doppler radar in which the PRF avoids range ambiguity is sometimes dubbed *low PRF pulse Doppler*. Unfortunately, in general, finding PRFs that avoid both range and velocity ambiguities is not possible, and in such cases one may interleave a high PRF mode for velocity measurement with a low PRF mode for range measurement.

9.2.6 High PRF Ambiguity Diagram

An example ambiguity diagram for a pulsed radar with a PRF of 50 kHz (pulse repetition interval [PRI] = 20 μs), pulse width of 10 μs, and a carrier frequency of 10 GHz is given

[1] One such application is short-range battlefield and security surveillance systems. Synthetic aperture radar (SAR) systems also operate at PRFs yielding unambiguous range and Doppler data.

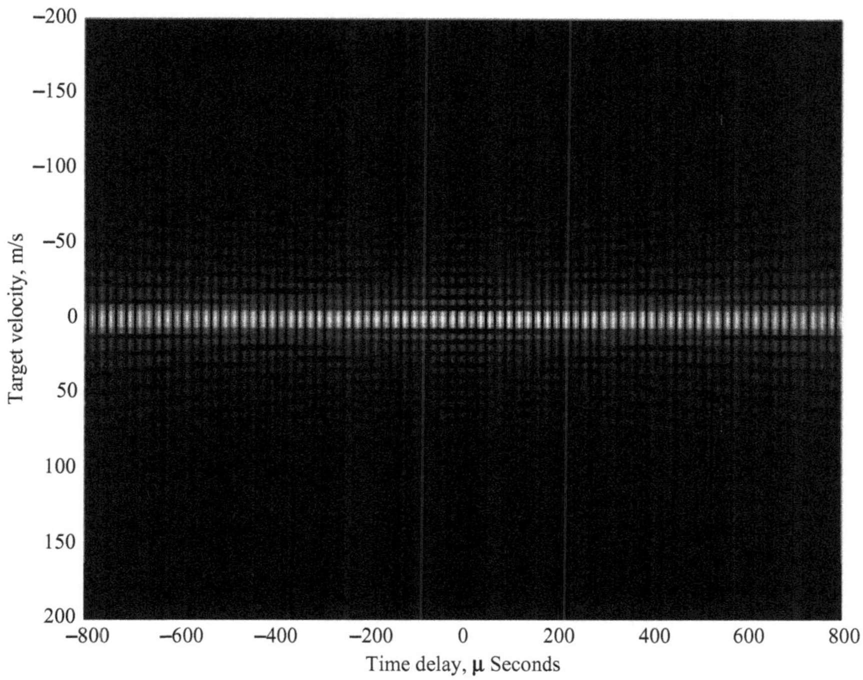

FIGURE 9-6 ■ High PRF Ambiguity Diagram

in Figure 9-6. This PRF results in a maximum unambiguous range, $R_{mu} = 3$ km. The peak response in the ambiguity diagram at 0,0 is repeated along the time axis at intervals of 20 μs (or 3 km of range) and is indicative of the range ambiguity. The 10 μs pulse width gives rise to each response in the ambiguity diagram having a width of 10 μs in the time axis (therefore a range resolution of 1.5 km) and a duty ratio of 0.5. This is also evident in Figure 9-6 since the ambiguous repeats of the response along the time axis at zero velocity have a duty ratio of 0.5. The PRF and carrier frequency combination give rise to an unambiguous velocity space of ± 375 m/s, so there are no repeats of the response within the displayed velocity space of ± 200 m/s. Therefore, for a velocity space of interest of ± 200 m/s, this pulsed waveform is unambiguous in velocity. The ambiguity diagram is plotted for a burst of 64 rectangular pulses. This results in a total waveform duration of $64 \times 20\ \mu s = 1.28$ ms. Each response therefore has a frequency resolution of $1/1.28$ ms $= 781$ Hz or a velocity resolution of 11.7 m/s and is the width of each response in the velocity axis. The rectangular shape of the pulses leads to strong sidelobes in the velocity axis. The main lobe responses are centered along the zero velocity axis in Figure 9-6, and the sidelobes are evident extending above and below each main lobe response in the velocity domain.

9.2.7 Offsetting the Doppler Band

It has previously been stressed that the Doppler band covers a range of Doppler shifts from $-f_r/2$ to $+f_r/2$ and therefore has a total bandwidth of f_r. This gives rise to an unambiguous velocity space of $V_{rmu} = \pm\frac{\lambda f_r}{4}$, which is equation (9-5). Figure 9-4 illustrated the example of the distribution of 32 Doppler cells across the band of $-f_r/2$ to $+f_r/2$. However, an alternative way of presenting this is to illustrate the distribution of Doppler cells from 0 to f_r. Doppler cells 0 to 15 cover the band $f_d = 0$ Hz to $f_d = +f_r/2$ Hz, respectively, and therefore cover the positive Doppler shifts (i.e., closing targets). Doppler

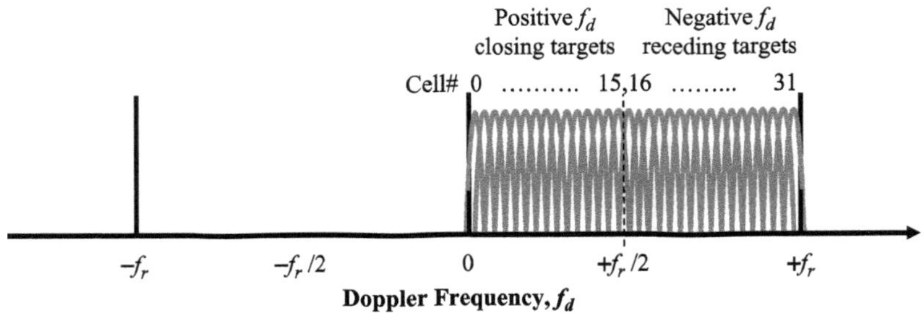

FIGURE 9-7 Alternative Distribution of Doppler Cells

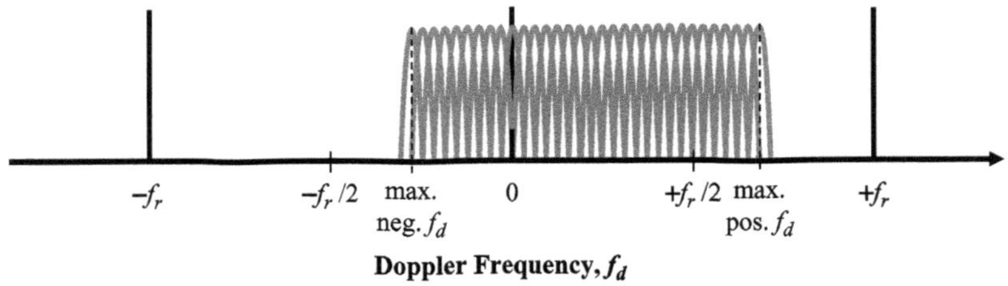

FIGURE 9-8 Offset Doppler Band

cells 31 to 16 cover the band $f_d = 0$ Hz to $f_d = -f_r/2$ Hz, respectively, and therefore cover the negative Doppler shifts (i.e., receding targets). The new distribution of Doppler cells is illustrated in Figure 9-7.

The coverage provided by the Doppler cells, whether their distribution is imagined as in Figure 9-4 or Figure 9-7, processes target returns over a bracket of positive Doppler shifts (closing targets) that is equally wide as the bracket of negative Doppler shifts (receding targets). This is useful when the radar is likely to detect closing targets over the same velocity bracket as receding targets. Surveillance radars, especially those that continually scan 360° in azimuth, are likely to detect closing and receding targets having the same maximum speeds. However, this is not the case for all radars. Notable exceptions to this are the cases of active radar missile seekers and airborne fire control radars (FCRs) aboard fast strike aircraft, both of which scan only a forward looking arc; for the airborne FCR case this is typically ±60° off the axis of the aircraft. The maximum relative velocity of a closing target, such as an inbound missile, is likely to be far greater than the maximum relative velocity of a receding target. This applies when platform motion compensation is not used (see Section 4.3), so all velocities are relative to the platform rather than the ground. In these circumstances, one requires coverage of greater positive Doppler but can tolerate coverage of a reduced negative Doppler. The Doppler band can therefore be offset toward the positive Doppler domain, as shown in Figure 9-8. This is easily imagined from the distribution of Doppler cells as shown in Figure 9-7, since this time cells 0 to 23 can be imagined covering all positive Doppler and cells 31 to 24 covering all negative Doppler. The band is therefore offset into the positive Doppler region but, crucially, still has a total bandwidth of f_r. This can be done only in high PRF pulse Doppler because of the absence of velocity ambiguities.

9.2.8 Dependence on Carrier Frequency

The Doppler shift given by equation (9-2)

$$f_d = \frac{2V_r}{\lambda} = \frac{2v_r f_{tx}}{c} \qquad \text{repeat of (9-2)}$$

is directly proportional to the carrier frequency, f_{tx} (or inversely proportional to its wavelength). Therefore, the higher the carrier frequency the greater the magnitude of the Doppler shift for a given target velocity. The use of higher carrier frequencies increases the sensitivity of the Doppler shift on target velocity. This has several secondary effects on the selection of PRF.

First, a greater spread of Doppler shifts requires higher values of PRF to avoid velocity ambiguity. Conversely, the use of higher carrier frequencies renders velocity ambiguities more likely. Consider an airborne early warning (AEW) pulse Doppler radar operating on a nominal a carrier frequency of 3 GHz and required to measure target velocities between ± 750 m/s (approximately $\pm$Mach 2.3). This results in Doppler shifts of between ± 15 kHz and, therefore, if velocity ambiguities are to be avoided, requires a PRF of $\geq$ 30 kHz. Contrast this with a typical requirement for an airborne FCR operating on a nominal carrier frequency of 10 GHz and required to measure target velocities of $+1500$ m/s to -600 m/s (approximately +Mach 4.7 to −Mach 1.9). This results in Doppler shifts of $+100$ kHz to -40 kHz and this time, if velocity ambiguities are to be avoided, requires a PRF of $\geq$ 140 kHz.

A second ramification of the increased sensitivity of the Doppler shift on target velocity at higher carrier frequencies concerns the relationships between velocity resolution and the integration time, which is described in Section 4.5.2. The Doppler resolution is the inverse of the integration time, t_{int} (or the *coherent processing interval, CPI*):

$$\Delta f_d = \frac{1}{t_{int}} \qquad \text{repeat of (4-15)}$$

This results in a velocity resolution given by

$$\Delta V_r = \frac{\lambda \Delta f_d}{2} = \frac{\lambda}{2.t_{int}} \qquad \text{repeat of (4-16)}$$

This relationship illustrates that velocity resolution is directly proportional to wavelength (or inversely proportional to carrier frequency). Finer velocity resolution is obtainable from a higher frequency radar for a fixed CPI. Conversely, the advantage of the higher frequency is that for a fixed velocity resolution a shorter CPI and therefore a shorter beam dwell time are required, permitting higher scan rates and faster data rates. Naturally, the advantage of a higher frequency may be cashed in as improved velocity resolution, increased data rate, or some measure of both. Another consequence of a higher carrier frequency is the finer beamwidth (finer angular resolution), which would also lead to a shorter beam dwell time for a fixed scan rate. Therefore, the effects of the shorter CPI time required to meet the desired velocity resolution and the shorter beam dwell time resulting from the finer angular resolution are likely to be entirely consistent with each other for a given choice of carrier frequency.

It ought to be stressed at this point that whatever velocity resolution may be required can always be obtained, irrespective of carrier frequency. However, the penalty for the use of a low carrier frequency (yet still demanding fine velocity resolution) is that the CPI duration grows very long.

It is also worth recalling that in the FFT processing of the Doppler signals in the radar baseband (Section 7.3) the Doppler resolution is given by the PRF (f_r) divided by the

FFT point size (n), that is,

$$\Delta f_d = \frac{f_r}{n} \tag{9-6}$$

For a fixed value of Δf_d and therefore, t_{int}, there are multiple solutions of combinations of f_r and n. Higher values of PRF are likely to result in larger numbers of transmitted pulses in a given CPI duration and are processed using larger FFT sizes. Lower values of PRF tend to result in smaller FFT sizes for a given CPI duration. However, if the CPI duration is the same in both cases then so, too, is the Doppler resolution.

Many other aspects of PRF selection are dependent on the choice of carrier frequency and will be considered in greater detail in later sections: blind velocities (where the Doppler shift equals a multiple of the PRF); the spread of MBC and therefore the width of the rejection notch and its associated blind velocity; and the spread of SLC. In fact, it is fair to say that every aspect of radar performance is touched in some way by the carrier frequency selection. As is always the case with radar, every parameter is related to every other parameter.

9.3 TIMING ASPECTS OF HIGH PRF

9.3.1 Duty Ratio

The duty ratio of a pulsed radar is defined as

$$duty\ ratio = \frac{\tau}{T_r} = \tau f_r \qquad \text{repeat of (3-1)}$$

where τ is the pulse width, and T_r is the pulse repetition interval, *PRI* (= 1/PRF).

Typical radar pulse widths are in the order of microseconds or several tens of microseconds. For low PRF waveforms this results in a duty ratio of around 1%, or even less. The receiving period is $T_r - \tau$ and may be as much as 99% of T_r. It is subdivided into many range cells. Pulse compression modulations are often applied across the transmitted pulse to reduce the range resolution to something much finer than would be obtained from the unmodulated pulse. These sorts of pulse widths, when used in conjunction with high PRF waveforms, result in considerably higher duty ratios. It is rare to find a transmitted pulse width from a high-power transmitter of less than a few tens of nanoseconds since it is difficult to modulate high-power RF devices at narrower pulse widths (although this is routinely achieved for laser applications). Furthermore, very narrow pulse widths have small energy content and very large noise bandwidths, both of which mitigate against long-range detection performance.

Considering the examples given in Section 9.2.8 above; the 3 GHz AEW radar requires a PRF $\geq$ 30 kHz (PRI $\leq$ 33.3 μs) to avoid velocity ambiguities, whereas the 10 GHz FCR requires a PRF $\geq$ 140 kHz (PRI $\leq$ 7.143 μs) to avoid velocity ambiguities. Even a modest pulse width of 3 μs incurs a duty ratio of 9% (AEW system) or 42% (FCR system). The duty ratios of high PRF systems are typically very large and may even be as much as 50% in some systems. This means that the average transmitted power may be as much as 50% of its peak transmitted power. High PRF waveforms are typically of high average power and can enjoy large processing gains because of the large number of pulses in each CPI and correspondingly large FFT sizes. High PRF waveforms are therefore very good at long-range detection.

9.3.2 Eclipsing Losses and Blind Ranges

Blindness occurs when an echo is received during the transmitted pulse since the receiver is isolated during the transmitted pulse. The received echo completely overlaps the transmitted pulse resulting in a condition of total eclipse and hence blind ranges at

$$R_{blind} = \frac{cxT_r}{2} \qquad \text{repeat of (3-5)}$$

for $x = 0, 1, 2, 3 \ldots$.

Historically, radars used leading-edge detection, and unless the rising edge of the echo pulse fell within the receiving period a target remained blind. This situation would persist for time delays of between xT_r and $xT_r + \tau$, where τ is the transmitted pulse width. Modern radars tend not to use leading-edge detection and are sensitive to the *energy* within each range/Doppler cell. (Leading-edge detection is sometimes used by some modern range tracking systems and as a means to counter repeater jamming.)

A condition of partial eclipse occurs when the received echo partially coincides with the transmitted pulse and is illustrated in Figure 9-9. The energy in the received echo that is lost due to its time coincidence with the receiver *OFF* time constitutes an *eclipsing loss*. Should half the received echo overlap with the transmitted pulse, 50% of the echo energy is lost. This represents an eclipsing loss of 3 dB. The target may nevertheless be detected if the energy in the remaining 50% exceeds the detection threshold. A radar may be blind to small targets should a small degree of partial eclipse occur but may be able to maintain detections of large targets even with a significant proportion of eclipsing. If the threshold for detection demanded a situation in which the degree of partial eclipse should not exceed 50%, then blindness occurs at ranges between

$$R_{blind} = \frac{c(xT_r - \tau/2)}{2} \quad \text{to} \quad \frac{c(xT_r + \tau/2)}{2} \qquad (9\text{-}7)$$

The high duty ratios of high PRF waveforms can lead to some severe eclipsing problems. Blind ranges occur at range intervals corresponding to the maximum unambiguous range of the PRF; thus, they come thick and fast for high PRFs. Blindness occurs at

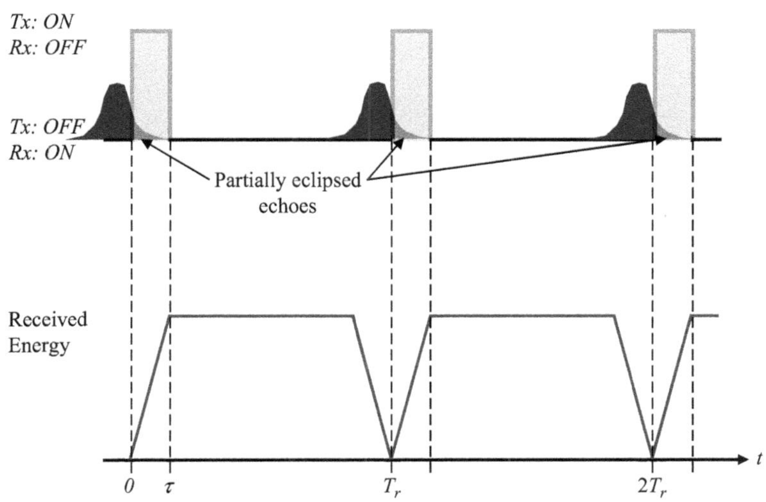

FIGURE 9-9 Eclipsing

intervals of 5 km for a PRF of 30 kHz (AEW example) and at intervals of 1071 m for a PRF of 140 kHz (FCR example). The blind range intervals given by (9-7) can be very broad and result in significant bands of blindness for high duty ratio waveforms. For a duty ratio of 50%, all target returns are totally eclipsed at precisely $\frac{cxT_r}{2}$ and zero eclipsing only occurs at ranges of precisely $\frac{c(xT_r+\tau)}{2}$. At all other ranges the target returns are partially eclipsed, which is the general situation. For duty ratios above 50% some degree of partial eclipsing is guaranteed since the returning pulse width always exceeds the receiver *ON* time. Part of the received energy is always lost; therefore, it was a waste of power for the transmitter to have generated it in the first place. For this reason, it is a pointless waste of energy to use a duty ratio exceeding 50%.

The solution to the eclipsing problem is to use two (or more) suitably chosen PRFs whose blind range intervals are staggered away from each other over a range bracket of interest. The radar must now alternate its operation over the two PRFs to detect targets over the range bracket of interest. It is unlikely that a PRF pair can be selected whereby the blind range brackets of one PRF do not coincide with those of the other PRF over the whole instrumented range of the radar. The radar must be capable of operating on several PRFs such that the appropriate PRF pair may be used over specific range brackets of interest. If the two PRFs are denoted f_{r1} and f_{r2} with PRIs T_{r1} and T_{r2} and pulse widths τ_1 and τ_2, then the condition to ensure target visibility, based on the 50% partial eclipse criterion for detection is that

$$\frac{c(xT_{r1}-\tau_1/2)}{2} \text{ to } \frac{c(xT_{r1}+\tau_1/2)}{2} \neq \frac{c(yT_{r2}-\tau_2/2)}{2} \text{ to } \frac{c(yT_{r2}+\tau_2/2)}{2}$$

over the range interval of R_{lower} to R_{upper} where $x = 0, 1, 2, 3\ldots$ and $y = 0, 1, 2, 3\ldots$ and x need not necessarily equal y. The optimal situation is one of minimal overlap between the blind range brackets of the two PRFs; thus, PRF selection becomes an optimization problem.

The earlier example of the 3 GHz AEW radar concluded that a PRF $\geq$ 30 kHz would be necessary to avoid velocity ambiguities. Thus, PRFs of 30.5 kHz and 31.0 kHz both avoid velocity ambiguities. The eclipsing losses are shown in Figure 9-10 for a duty ratio of 0.1 in both PRFs in which the trace of long dashes represents the eclipsing loss of 30.5 kHz, the trace of short dashes represents the eclipsing loss of 31.0 kHz, and the solid

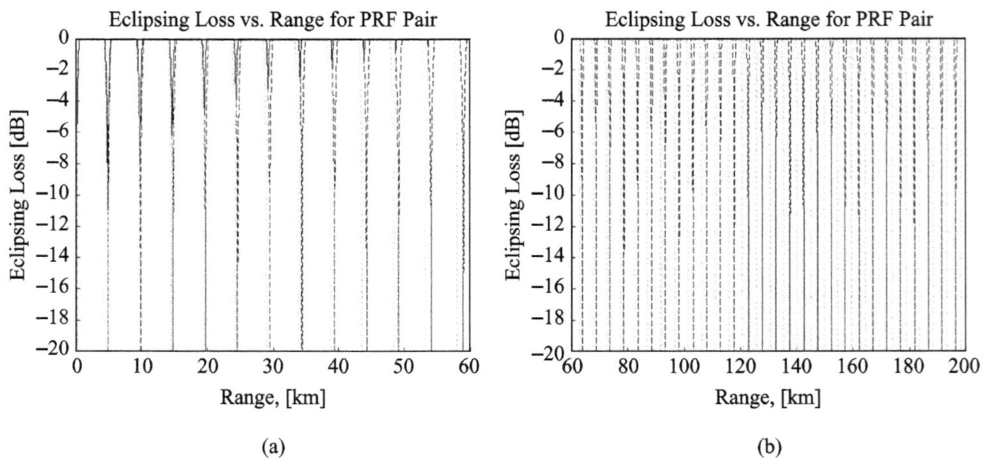

FIGURE 9-10 Eclipsing Losses of PRF Pair: 30.5 kHz (long dashes), 31.0 kHz (short dashes), worst-case (solid) (a) 1–60 km; (b) 60–200 km

trace represents the best case eclipsing loss over the two PRFs. One can see from the plot in Figure 9-10a that at the lower ranges the blind ranges are not sufficiently staggered away from each other and a worst-case eclipsing loss of 10.9 dB is incurred at a range of 4.878 km. A series of other regions of high losses are also incurred at multiples of 4.878 km but with ever reducing loss as range increases and the stagger between the blind ranges of the two PRFs unwinds. By about 55 km, no appreciable eclipsing losses remain. Clearly, this PRF pair is unsuitable for target detection at ranges below 55 km. However, at ranges between 60 km and 200 km (Figure 9-10b) no appreciable eclipsing losses remain for this PRF pair, so it is ideally suited for target detection in this range bracket. This is a significant region of the midrange bracket for a major AEW radar. At ranges beyond 250 km, the blind ranges of the two PRFs start falling in synchronism with each other again, and significant eclipsing losses result.

A radar may alternate its operation over the two PRFs in several ways. First, it may subdivide the beam dwell time into two processing intervals, one for each PRF. The PRF is now being changed in the time the radar takes to scan through a half its beamwidth. Alternatively, the radar may cover the search volume by two scan bars, one on each PRF. The former strategy enjoys half the CPI time as that of the latter and is therefore limited to half the FFT size and processing gain. On the other hand, the latter strategy now requires two scan bars to be completed and thus halves the revisit rate of the former. The selection of which strategy is employed is a balance between detection performance and data rate.

An important aspect of eclipsing also concerns the effects of partial eclipsing on pulse compression waveforms. If the returning echo is partially eclipsed, some portion of the first or last part of the echo is not received, so this portion of the modulation is lost and the remaining echo no longer correlates perfectly in the matched filter. Only a partial correlation can be expected, depending on the degree of eclipsing. This results in a reduced signal to noise ratio and a degradation in detection performance. If linear FM "chirp" is used, then the partially eclipsed echo contains only the linear FM over a reduced bandwidth and the compression ratio is reduced accordingly. A correlation peak is still obtained since the echo still correlates with the appropriate portion of the matched filter; however, the correlation peak is reduced in accordance with the energy lost through the partial eclipsing. The linear FM on the received fragment of the echo does not retain the full bandwidth of the transmitted signal, so the range resolution is also reduced accordingly. For phase coded compression waveforms, partial eclipsing removes some of the code bits, which also results in a loss of signal-to-noise ratio (SNR). The bandwidth of the signal is set by the bit time and is unaffected by the missing bits from the code, so the range resolution is not degraded. However, the correlation sequence of the fragment of the code that remains with the complete code held in the matched filter is likely to exhibit large time (range) sidelobes. It is appropriate to design codes that are robust to partial eclipsing.

9.3.3 Ranging in High PRF

Range gating may be employed alongside high PRF operation in much the same way as implemented in low PRF systems and is known as *range gated high PRF*. For a pulsed waveform of a duty ratio = 0.1 (10%), nine range cells matched to the transmitted pulse width may be accommodated across the receiving period. If pulse compression is used, then the range cell width would be matched to the compressed pulse width in which case the number of range cells across the receiving period increases by a factor equal to the compression ratio. The relatively short transmitted pulse widths typical of high PRF waveforms may not necessitate a high degree of pulse compression, if any. In any case, target echoes and clutter returns are highly ambiguous in range, and recovering the true target range could be awkward. However, range gating has the benefit of subdividing the clutter return over the range cells, whereas the target return is confined to just one range cell (or perhaps it straddles neighboring cells). This minimizes the amount of clutter competing with target echoes in each range cell.

Some applications require very high PRFs and duty ratios of close to 50%, as was the case for the 10 GHz airborne FCR example of Section 9.2.8. Assuming that no pulse compression is applied results in a situation of having just one range cell within the receiving period. This offers no range resolution whatsoever. All target returns, including clutter, fall within the one range cell that is ambiguously repeated on each receiving period. In these circumstances any range measurement and hence any ability to resolve in the range coordinate, including the subdivision of clutter, is impossible.

In many high PRF modes, no attempt is made to recover target range data. The high PRF waveform is well suited to the measurement of velocity and is used in velocity searching and tracking radar modes. Nevertheless, it is necessary to select PRF pairs that, between them, avoid blind ranges and permit detection over the range bracket of interest even when range is not actually being measured. Furthermore, high PRF waveforms are well suited to long-range detection, even though the target range may not be measured.

Nevertheless, it is possible to determine target range in high PRF modes of operation. The two high PRF examples used here—those of moderate duty ratios giving several range cells in each receiving period and those of high duty ratios giving just one range cell per receiving period—have two quite distinct solutions.

9.3.4 Range Gated High PRF

In range gated high PRF, there are several range cells across the receiving period that afford a reasonable degree of range resolution. Targets seen in a particular range cell have an *apparent* range based on the assumption that they are first-trace echoes. However, the *true* range, R_{true}, may differ from the apparent range, R_{app}, by multiples of the maximum unambiguous range, R_{mu} (see also Section 3.4.1). This is expressed as

$$R_{true} = R_{app} + xR_{mu} \quad x = 0, 1, 2, 3 \ldots \qquad \text{repeat of (3-7)}$$

where x is the order of the range ambiguity. Both R_{app} and R_{mu} are PRF dependent, and different values of both are obtained if the radar PRF is changed. The true range is of course fixed at the true location of the target. The introduction of a second PRF leads to

PRF1: $R_{true} = R_{app1} + x_1 R_{mu1}$

PRF2: $R_{true} = R_{app2} + x_2 R_{mu2}$

in which the suffix 1 or 2 refers to the PRF number. This yields two equations with two unknown quantities (x_1 and x_2) that can be solved for R_{true}. For robust operation, particularly when multiple targets are present, a radar may operate on more than two carefully chosen PRFs to solve the range ambiguity. In essence, this is precisely the situation faced by medium PRF systems in which multiple PRF operation is necessary to solve range and velocity ambiguities. For high PRF operation, velocity is unambiguous and only the range ambiguity need be solved. High PRF may therefore be considered a simplification of the medium PRF case. An interesting algorithm proposed by Trunk and Kim [1] is based on processing ambiguous range data over several PRFs to correctly resolve the true ranges of targets in high and medium PRF systems and is described in more detail in Section 12.5.3. This subject will be considered in greater depth in Chapter 12.

9.3.5 FM Ranging

When only one range cell occupies the receiving period, there is no resolution of target range. In these cases, one can use FM ranging techniques to determine the target range. Since the techniques of FM ranging are combined with a pulsed signal, also known as an *interrupted continuous wave*, the technique has become known as frequency modulated interrupted continuous wave (FMICW). FM ranging and FMICW waveforms are considered in Chapter 10.

9.3.6 Other High PRF Ranging Techniques

A host of alternative methods to resolve range ambiguity have been used. Some of these are quite quirky solutions specific to certain radar systems. One rather interesting solution has been implemented for pulsed waveforms of moderately low duty ratios in which there are several range cells within each receiving period and for which only a modest degree of range ambiguity is present. The solution involves tuning the RF carrier frequency by $f_r/4$ and observing the resultant shift in the phase of the returns. If the phase shift on the returns is $< 90^\circ$ the target is a first trace echo, whereas if the phase shift on the returns is between 90° and 180° the target is a second trace echo. Hence, this technique can be used only to resolve whether a target is first or second trace. This technique works as follows.

The phase shift over the two-way journey to a target at range R is

$$= \frac{2\pi}{\lambda}(2R)$$

$$= \frac{2\pi f_{tx}(2R)}{c}$$

where f_{tx} is the original carrier frequency. If the carrier frequency is increased by a quarter of the PRF, then the phase shift over the two-way journey to the target becomes

$$= \frac{2\pi(f_{tx} + f_r/4)(2R)}{c}$$

The difference between these two phase shifts is thus

$$= \frac{2\pi(f_r/4)(2R)}{c}$$

$$= \frac{\pi f_r R}{c}$$

If the echo is first trace, then $R < \frac{c}{2f_r}$. Substituting this into the previous expression for the difference between the two phase shifts gives

$$< \frac{\pi f_r c}{c 2 f_r}$$

which, after cancellation, results in a difference between the two phase shifts of $< \pi/2$.

However, if the echo is second trace $\frac{c}{2f_r} < R < \frac{c}{f_r}$ then the difference between the two phase shifts is between

$$\frac{\pi f_r c}{c 2 f_r} \quad \text{to} \quad \frac{\pi f_r c}{c f_r}$$

which, after cancellation, results in the difference between the two phase shifts of between $\pi/2$ and π.

Another technique to overcome the range ambiguity of high PRF waveforms is to impart a code onto pulses within a burst of high PRF pulses corresponding to the CPI. The maximum unambiguous range of the system is now associated with the time duration of the coded sequence rather than the PRI. Levanon [2] describes a technique of bi-phase modulation of a train of pulses transmitted at a high PRF using Ipatov codes. The coded returns are received into a slightly mismatched receiver that correlates returns with several cycles of the transmitted code. This processing ensures that returns do not exhibit any range sidelobes if the number of cycles of the transmitted code is two more than the number of cycles stored in the correlation receiver. The slightly mismatched correlation receiver, which is responsible for the sidelobe-free performance, comes at the expense of a small processing loss, typically in the order of 0.17 to 1.0 dB for code lengths between 13 and 121 bits. For the 13-bit Ipatov code, one cycle of the code is therefore transmitted over 13 pulses, and its maximum unambiguous range is 13 times the inherent value of the PRF. Long code lengths result in long maximum unambiguous ranges. However, if some pulses are not received, perhaps because of eclipsing or target scintillation, then range sidelobes are incurred. Range sidelobes are undesirable because the presence of a large discrete target or a large clutter return in one range cell can cause range sidelobes to spill over into neighboring range cells and mask small targets in these cells. The elimination of range sidelobes using Ipatov codes (and the mismatched correlation receiver) is therefore highly desirable and can lead to significant improvements in detection performance as well as offering long unambiguous range measurement.

9.4 HIGH PRF RADAR RESPONSE TO CLUTTER

The response of a high PRF system to surface clutter is considered for the case of an airborne radar in a look-down attitude since this presents the most general picture [3, 4, 5, 6]. Ground-based systems can be considered as a specific case in which the platform's ground velocity is zero and the platform altitude is also zero.

9.4.1 Distribution of Clutter in the Doppler Domain

The spectrum of both MBC and SLC is illustrated in Figure 9-11 and was considered in Section 6.4. SLC occupies a portion of the Doppler band from $-2V/\lambda$ to $+2V/\lambda$, where V is the platform's ground velocity. PMC is applied to reference all returns to ground velocity by offsetting all Doppler shifts by the Doppler frequency of the main beam boresight. This has the effect of translating the clutter spectrum in frequency such that MBC is centered at zero Doppler. The bandwidth of SLC can be wide for high platform velocities; for example, for a platform velocity of 250 m/s and $\lambda = 0.03$ m ($f_{tx} = 10$ GHz), SLC extends over a bandwidth of 33.33 kHz.

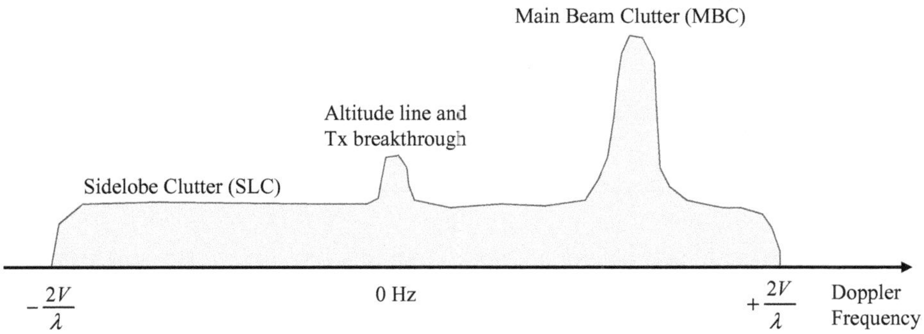

FIGURE 9-11 Surface Clutter from Airborne Radar in the Doppler (Velocity) Domain

From the definition of high PRF, all returns, including those of clutter, are unambiguous in Doppler (velocity). Therefore, the PRF must be greater than the widest spread in SLC ever likely to be encountered. The previous example would require a PRF, $f_r > 33.33$ kHz. Since the ground velocity of targets may exceed that of the platform, high closing speed target returns fall at a Doppler frequency above the band of SLC, so an even higher value of PRF is required to avoid velocity ambiguity.

The spectrum of MBC and SLC is repeated at intervals of the PRF; each spectral component of the transmitted spectrum results in a band of MBC/SLC, as shown in Figure 9-12. The spread in SLC is typically far smaller than the PRF, so SLC does not extend outside the Doppler band. Likewise, the repetition of the clutter spectrum about neighboring PRF lines does not encroach within the Doppler band.

MBC is centered at zero Doppler (after PMC has been applied) and is spread over a small fraction of the Doppler band [7]. MTI filtering may be applied to reject MBC and targets of very low ground velocities. The MTI filter rejects slow ground moving targets (GMTs), which are seldom of any concern in many airborne radar applications. In fact, they are often considered a distraction. However, the MTI filter would also reject the returns from a helicopter in a hover, although returns from the fast-moving rotor blades could still be detected. Choosing the width of the MTI rejection notch is a delicate balance between the rejection of unwanted MBC and GMTs on the one hand and the possible rejection of targets of interest on the other. Typical values of the rejection band are nevertheless a small proportion of the Doppler band, and high PRF waveforms offer good rejection of MBC and low-speed GMTs. MTI filter rejection cannot be extended to cover the band of SLC since too high a proportion of the Doppler band would be rejected and many targets of interest could not be seen.

The ability to detect targets very much depends on the velocity of the targets and the levels of SLC. Three different situations are portrayed in Figure 9-13. Target **A** is a fast closing target and thus has a large positive Doppler that is above the band of SLC. Here it is competing only with background noise for detection, so good detection performance can be expected. A similar situation would exist for very high velocity opening targets. Target

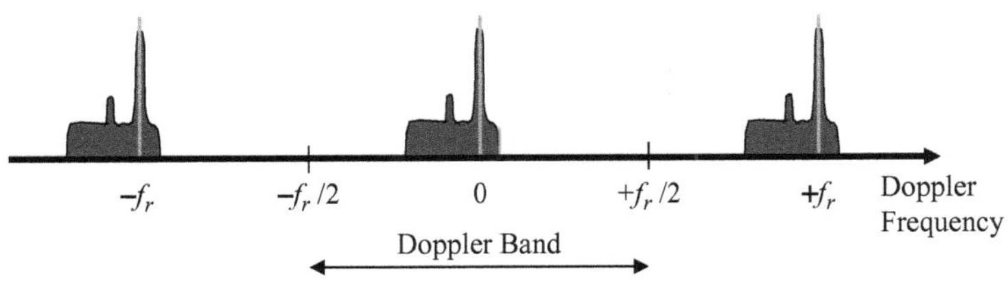

FIGURE 9-12 Repetition of Clutter in Doppler Domain for High PRF

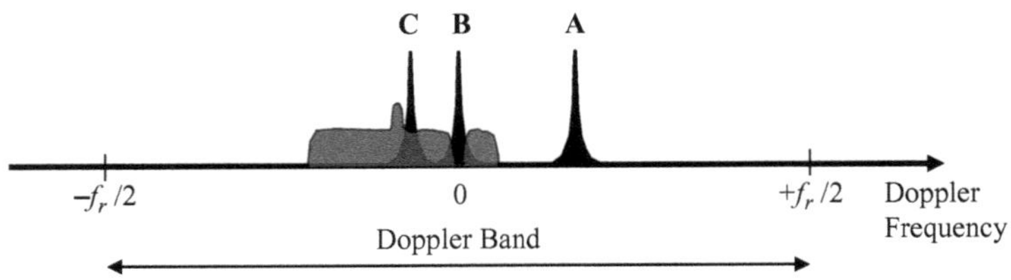

FIGURE 9-13 Target Detection as a Function of Target Velocity

B is stationary (or very low speed) and falls within the rejection width of the MTI filter and thus is rejected by the filter. Even if no MTI filtering were performed, the presence of a large MBC return in a look-down attitude would more than likely overwhelm the target return and so it would not be detected. Target **C** is a slow-moving target that coincides with SLC. In fact, all targets of low closing velocities and moderately fast opening velocities coincide with SLC. The ability of the radar to detect such targets is dictated by the signal-to-clutter ratio (SCR), which is a function of the relative target radar cross section (RCS) and clutter strength. High PRF waveforms have a poor detection performance against these targets. Consequently, in the absence of PMC for air-to-air engagements, side-aspect and tail-chase scenarios would be problematic using high PRF. All-aspect detection is not always possible, particularly at low altitudes [6].

One upshot of FFT processing is that the clutter is spread over several velocity cells, whereas target returns are largely confined to a single velocity cell (although some straddling into neighboring velocity cells does occur). Therefore, moving targets are competing only with the portion of SLC that falls within its velocity cell for detection. This portion of SLC arises from a contour on the surface of constant Doppler. The Doppler shift from the surface is given by

$$f_d = \frac{2V \cos\theta_{az} \cos\theta_{el}}{\lambda} \qquad \text{repeat of (4-11)}$$

where θ_{az} is the look angle with respect to velocity vector in azimuth, and θ_{el} is the depression angle with respect to velocity vector in elevation. A contour of constant Doppler is where a cone of constant angle, θ, offset from the velocity vector intersects the surface and is illustrated in Figure 9-14.

Projecting this cone into a three-dimensional view, as in Figure 9-15, one can see that the cone intersects a flat surface in a hyperbolic locus.

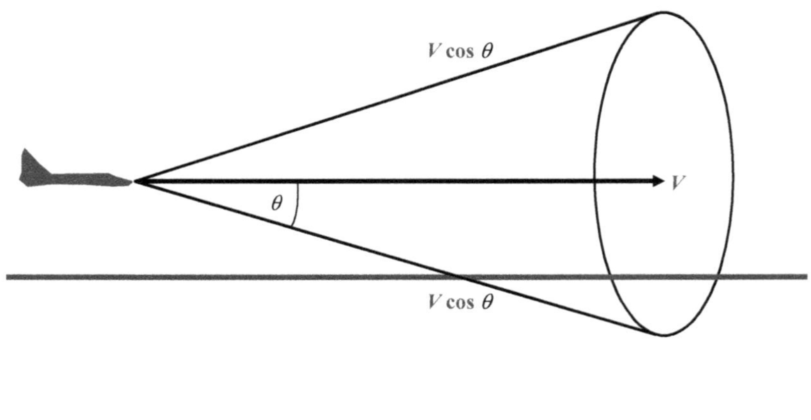

FIGURE 9-14 Two-Dimensional View of Contour of Constant Doppler

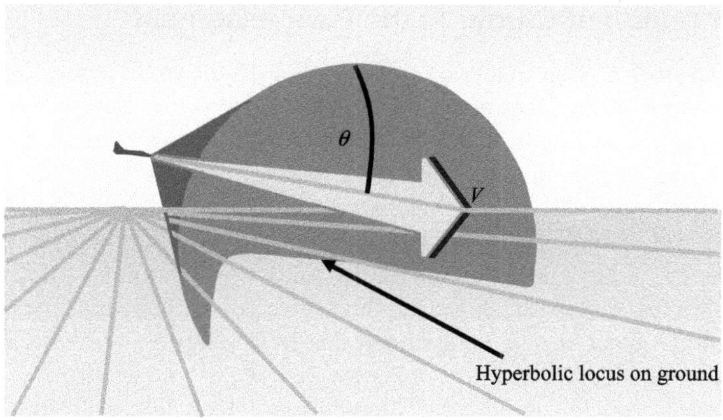

FIGURE 9-15 ■ Three-Dimensional View of Contour of Constant Doppler

Given that Doppler cells have a width of PRF/(FFT_size) or f_r/n then clutter from an area accommodating a small spread in the cone angle, θ, is received into each Doppler cell. This is depicted in plan view in Figure 9-16, in which clutter in three distinct Doppler cells is highlighted. The Doppler cell centered at 0 Hertz (prior to the application of PMC) is centered on a region that is perpendicular to the platform velocity vector. A midrange Doppler cell couples clutter from a divergent hyperbolic locus that is broad and has an apex close to the radar platform. A Doppler cell toward the top end of the Doppler band couples clutter from a divergent hyperbolic locus centered at a smaller angular offset from the velocity vector; this gives a tighter hyperbola whose apex is farther from the radar platform.

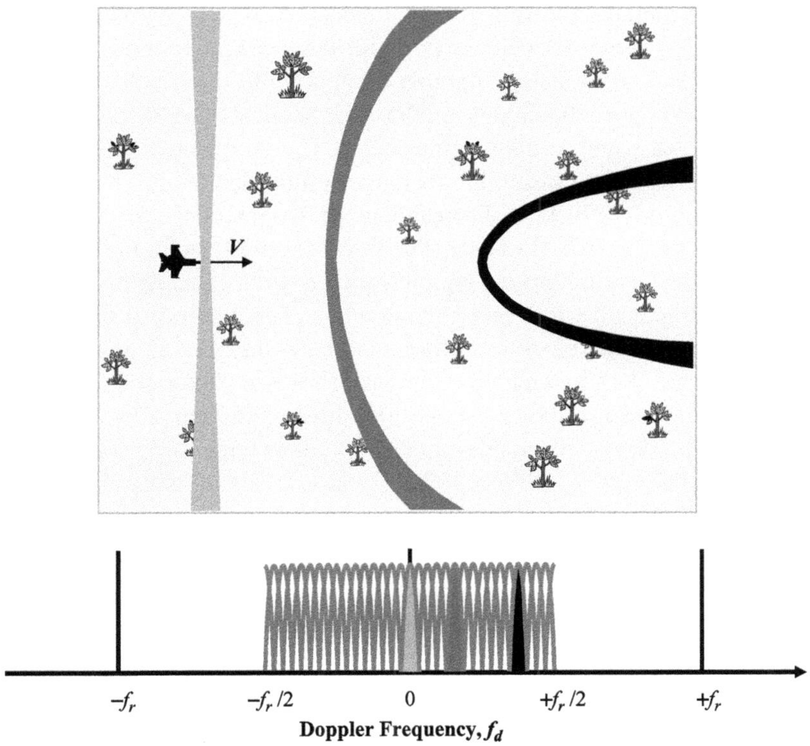

FIGURE 9-16 ■ Regions of Surface Clutter within Doppler Cells in Plan View

9.4.2 Distribution of Clutter in the Range Domain

MBC is spread over a range interval corresponding to the main beam footprint on the ground. MBC starts at the slant range of the lower edge of the intersection of the main beam on the surface and ends at the slant range of the upper edge of the intersection of the main beam on the surface. This range interval may well exceed one maximum unambiguous range interval, R_{mu}, at shallow grazing angles, in which case MBC occupies all range cells and is ambiguously repeated across the receiving periods. At very large grazing angles the radar is looking almost vertically downward and MBC may well be confined to just a few range cells; however, this scenario is unlikely in many pulse Doppler radar applications. Since it may be assumed that MTI filtering is used to reject MBC it needs no further consideration.

Of far greater concern is the SLC distribution, which was also discussed in Section 6.4.2. SLC is always spread over a large range interval that probably exceeds one maximum unambiguous range interval, R_{mu}, so ambiguous repetition across several receiving periods is most likely. The situation is depicted in Figure 9-17. Time gating the SLC out is not an option in high PRF as it is in low PRF, because multiple ambiguous repetitions of SLC occupy all range cells. In any one receiving period there are multiple repeats of SLC; the higher the PRF, the greater the number of ambiguous repetitions and the worse the problem. SLC can be problematic for high PRF systems.

The best, perhaps only, defense against SLC for high PRF systems is not to incur it in the first place. It is incumbent on high PRF systems to use an antenna system of the lowest sidelobe level as possible.

The superposition of multiple ambiguous repeats of SLC throughout the receiving period tends to smear out the clutter to a near-uniform level across all range cells. Each range cell therefore contains a contribution of first trace SLC, second trace SLC, third trace SLC, and so on. Range gating does, however, divide the total SLC return up fairly uniformly across all range cells. The greater the number of range cells, the smaller the proportion of the total within each range cell. Targets, on the other hand, are largely confined to a single range cell, although range cell straddling may occur. Therefore, targets are competing only with the portion of SLC that falls within its range cell for detection. This portion of SLC arises from a contour on the surface of constant range. Contours of equal range correspond to circles centered on the radar location. The finite width of each range cell means that they accept clutter from a ring-shaped region from the surface, the width of the ring being a function of the range cell width and grazing angle. The width of this ring is given by $\Delta R \sec \theta_g$, where ΔR is the range resolution, and θ_g is the grazing angle. The situation is illustrated in Figure 9-18 in which the regions of surface clutter coupling to one range cell (selected at random) are shown for the first three ambiguous repetitions. The rings are offset from each other by the maximum unambiguous range, R_{mu}, associated with the PRF.

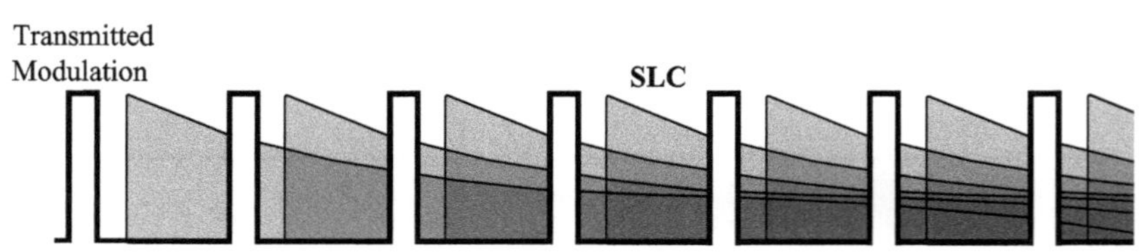

FIGURE 9-17 High PRF Response to SLC in Range Domain

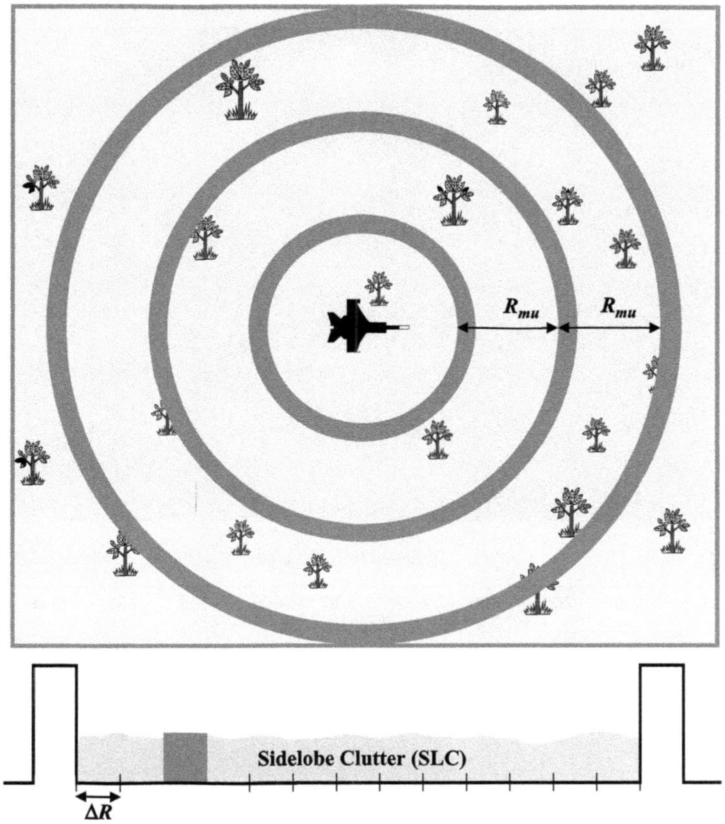

FIGURE 9-18 Regions of Surface Clutter within Range Cells in Plan View

9.4.3 Distribution of Clutter in Range and Doppler

The previous sections have described the typical spread of clutter in the Doppler and range dimensions for high PRF systems. It is now necessary to combine this into a two-dimensional format illustrating the distribution of clutter in range/velocity space.

Section 6.4.3 described an *ideal clutter map* that depicts the distribution of clutter in range/velocity space. An example ideal clutter map is given at Figure 6-17 for a typical airborne radar flying at an altitude of 5000 m and at a velocity of 250 m/s in a 10.8° look-down and 0° azimuth look-ahead attitude. The ideal clutter map simply shows the distribution of the clutter across the range and velocity space of interest and has not been velocity and range gated in accordance with the processing parameters of the radar. From these data one may now derive a *folded clutter map* presenting the same clutter data but "folded" into a region of range/velocity space of one unambiguous range interval by one unambiguous velocity interval. For high PRF systems the total velocity space of interest is accommodated by the unambiguous velocity interval, but there are multiple repetitions of the unambiguous range interval within the range space of interest. Figure 9-19 presents an example of a folded clutter map for a high PRF waveform representative of an airborne FCR operating on a PRF of 140 kHz and employing 1024-point FFT processing. Parameters of the FCR are summarized in Table 9-1. Clutter is plotted over the 1024 velocity cells, arranged across the Doppler band from 0 Hertz up to the PRF (140 kHz, in this case), as in Figure 9-7; each Doppler cell is some 136.7 Hz wide (corresponding to a velocity resolution

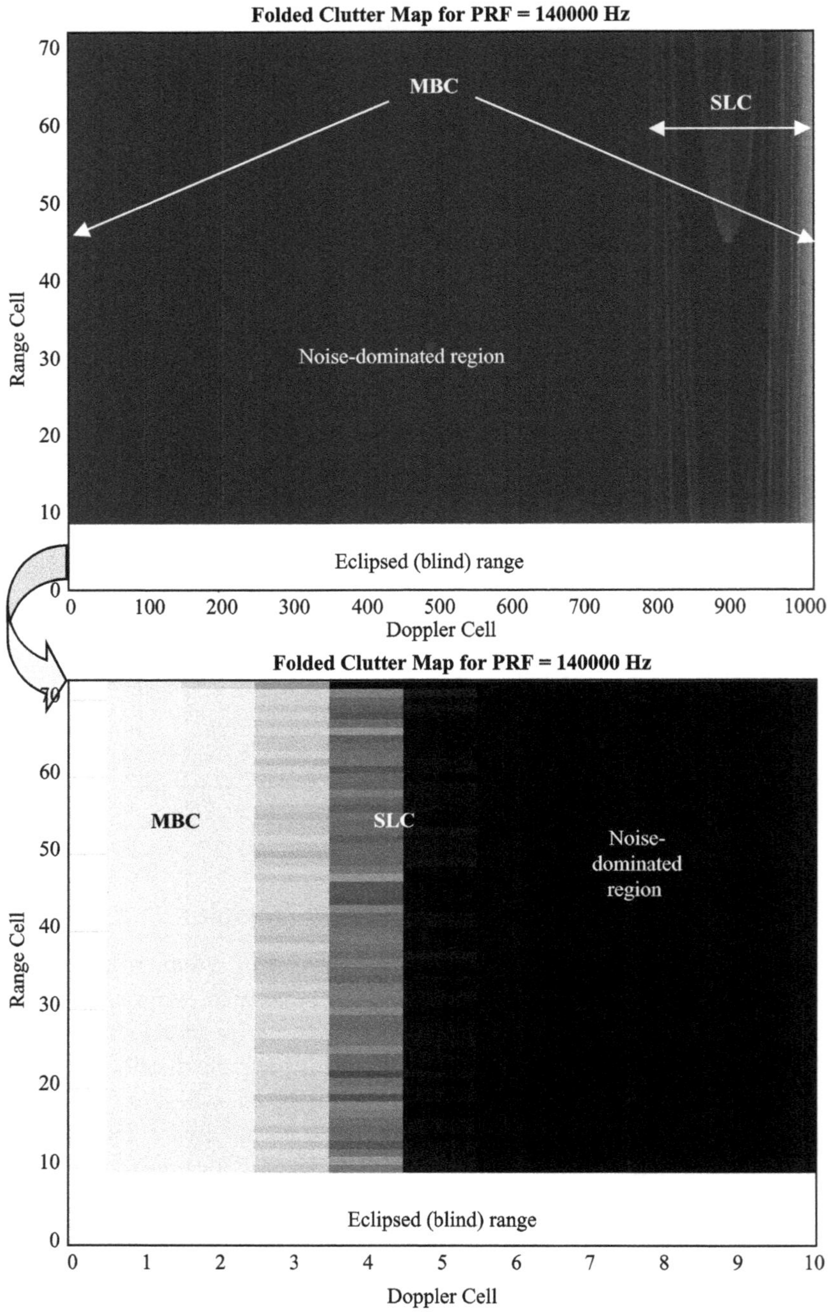

FIGURE 9-19 Folded Clutter Map for a Generic High PRF Airborne FCR Case

of 2 m/s). The number of range cells in one unambiguous range interval is the maximum unambiguous range divided by the range resolution (i.e., $1071/15 = 71.4$), which rounds up to 72 range cells. The folded clutter map therefore contains $1024 \times 72 = 73\,728$ range/velocity cells; each of dimensions 15 m $\times$ 2 m/s. The upper plot of Figure 9-19 shows the distribution of clutter returns within the whole of this interval, whereas the lower plot shows an expanded view of the clutter distributed over the first 10 velocity cells. Platform motion compensation was conducted, as is evident from the strong MBC

TABLE 9-1 ■ Parameters of Example Airborne FCR

Parameter	Value
Carrier Frequency, f_{tx}	10 GHz
PRF, f_r	140 kHz
Maximum Unambiguous Range, R_{mu}	1071 m
Duty ratio, DR	0.1 (10%)
Transmitted Pulse Width, τ	714.3 ns
Range Resolution, ΔR	15 m
FFT Size, n	1024
Velocity Resolution, ΔV_r	2 m/s
FFT Processing Loss	3 dB
CPI Time, t_{int}	7.314 ms
Antenna Beamwidth (both planes), θ_B	3.5°
Antenna Sidelobe Level	−17.5 dB
Azimuth Scan Angle, θaz	0°
Elevation Scan Angle, θ_{el}	−10.8°
Platform Altitude	5000 m
Platform Velocity, V	250 m/s
Receiver Noise Figure, F	5 dB
GMT Velocity	Mean = 0, σ_n = 1.5 m/s (Gaussian)

return centered at velocity cell 1 and its first ambiguous repetition in velocity cell 1024. There is a thin sliver of SLC in the positive Doppler domain extending up to velocity cell 5 (lower plot); however, a much broader band of SLC is faintly visible in the negative Doppler domain extending down to velocity cell 796. Receiver noise was also included, and indeed it is noise that dominates velocity cells 6 to 795. Thus, approximately 77% of the range/velocity space is noise limited, and the remaining 23% of the detection space is dominated by clutter, of which only a very small fraction is MBC.

The minimum range associated with the 714 ns transmitted pulse width, $R_{min} = 107$ m, so the number of eclipsed range cells is $107/15 = 7.14$, which rounds up to 8. The radar model assumes one additional blind range cell due to the transmit to receive switching time and eclipsing losses. Hence, the first nine range cells are blind and are plotted as white in Figure 9-19. Blindness therefore exists across 9 of the 72 range cells, or 12.5% of the detection space.

The folded clutter map illustrates the clutter within one unambiguous range and velocity interval. The PRF of 140 kHz coupled with the carrier frequency of 10 GHz results in an unambiguous velocity space coverage of ±1050 m/s. This encompasses the full velocity space of interest for most airborne FCR applications. However, this PRF results in an unambiguous range of only 1071 meters, which does not cover the full range space of interest. A reasonable range of interest may be in the order of 100 km or more. If one assumes a range of interest of 105 km, then 98 ambiguous repetitions are incurred over this range, requiring 7000 range cells, each of 15 m resolution. The distribution of clutter across the full range/velocity space of interest can be depicted on an *unfolded clutter map* by ambiguously repeating the folded clutter map as many times as necessary to obtain the necessary coverage. In the previous example this requires 98 repetitions of the folded clutter map in the range dimension; no repetitions are required in the velocity dimension since this is a *high* PRF. An example of an unfolded clutter map for a high PRF radar is

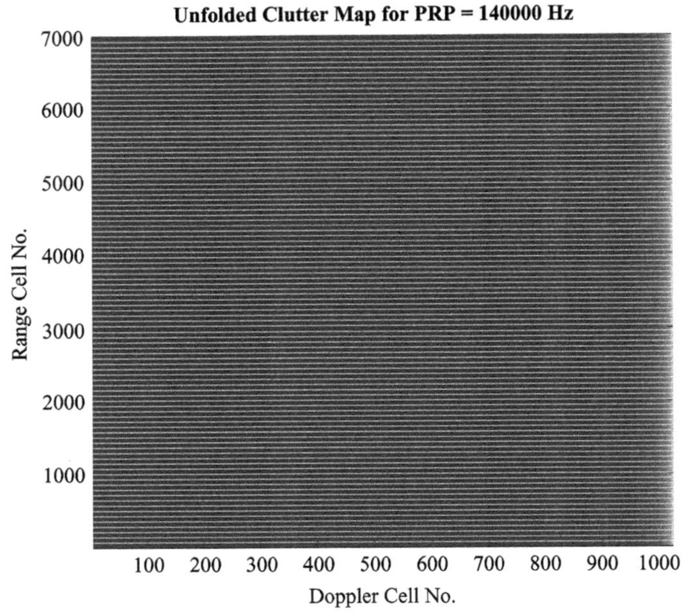

FIGURE 9-20 Unfolded Clutter Map for a Generic High PRF Airborne FCR Case

given in Figure 9-20, which pertains to the same radar model defined in Table 9-1. It is formed by stacking 98 copies of the upper plot in Figure 9-19 on top of each other in the range dimension. MBC and SLC still occupy the same Doppler cells as in Figure 9-19, but now there are 98 repetitions of the eclipsed ranges (horizontal white strips).

The total detection space of interest comprises 7000 range cells, each of 15 meter resolution, spanning a total range of 105 km by 1024 velocity cells, each of 2 m/s resolution, spanning a velocity bracket of -1050 m/s to $+1050$ m/s. This gives a total of 7,168,000 range/velocity cells. Eclipsing renders 12.5% of the total detection space blind.

9.5 SURFACE-BASED HIGH PRF RADARS

The most significant difference between airborne and surface-based high PRF radars is the differing clutter situations. Clutter maps for surface-based radars can be derived by setting the platform velocity to zero and the altitude to zero. For a stationary radar platform all MBC and SLC is centered at zero velocity and subject to a small degree of velocity spreading due to the motion of the clutter surface. There is no velocity spreading on account of the beamwidth of the main beam, as there is for airborne radar, because there is no variation of relative velocity across the beam; the relative velocity is zero at all angles. This tends to confine MBC to a narrower band than would be obtained for the airborne case (assuming identical clutter surfaces). The high speed typical of many airborne platforms causes a considerable spread in SLC. To avoid generating velocity ambiguous clutter returns the PRF must be very high. However, SLC from surface radars is confined to a narrow band, and the avoidance of velocity ambiguous clutter returns does not demand a particularly high PRF. Most ground-based radars operate from a stationary platform; however, naval radars move at the ship's velocity. This gives some Doppler spreading of MBC and SLC, but this is minimal on account of the modest platform velocity.

SLC is received from zero range out to the range of the local horizon. For radars operating over a fairly smooth portion of the earth's surface, such as over the sea surface, the range to the horizon, R_h, in kilometers, is given by

$$R_h \approx 4.1\sqrt{h} \qquad \text{repeat of (6-13)}$$

where h is the height of the phase center of the antenna above the surface in meters. For high altitude airborne systems the range to the horizon can be several hundreds of kilometers and may even exceed the detection range of the radar. However, this is not likely for surface-based systems. Consider, for example, a radar mounted on a 10 meter mast ($h = 10$), the range to the horizon, $R_h = 13$ km. This limits the range extent of clutter to something very much less than the airborne case. In mountainous terrain the range to the horizon may be even less, if in a valley, but could exceed this value if on high ground. If the extent of SLC in range exceeds the maximum unambiguous range of the PRF, then SLC occupies all range cells across the receiving period and clutter is perceived to exist at all ranges.

Surface-based radars designed for the detection of airborne targets, such as air defense radars, may see no MBC when searching above the horizon, although SLC is present from zero range out to the range to the horizon. MBC would be present when searching angles of elevation close to or below the local horizon, as is often the case in air defense search radars. The severity of the clutter situation depends on the elevation of the radar beam with respect to the horizon. This dictates the level of clutter attenuation required. Clutter arises from the surface over a range interval that, in all probability, exceeds the maximum unambiguous range of the PRF and thus would be spread across all range cells. Fine resolution Doppler processing does not succeed in minimizing the clutter problem as it does with airborne systems; there is no subdivision of the clutter across the Doppler cells since all the clutter is at or close to zero Doppler. An MTI clutter canceller filter can suppress both MBC and SLC but renders the radar blind to targets of zero relative velocity and at velocities corresponding to Doppler shifts equal to multiples of the PRF. Blind velocities occur at

$$V_b = \frac{\lambda m f_r}{2} \quad m = 1, 2, 3 \ldots \qquad \text{repeat of (7-8)}$$

The Doppler blindness around zero relative velocity can be problematic for surface-based air defense systems since they would fail to detect a helicopter in a hover (although the rotor blades could be detected) or crossing targets (see Section 4.4.2).

Volume clutter arises from clouds, rain, fog, and mist; dust; insect swarms; or flocks of birds and can occupy large volumes of the sky. Weather-related sources of volumetric clutter drift at the wind speed and direction. Severe sources of volumetric clutter such as intense storms are often highly localized and may not fill the entire beam at long ranges. Targets of interest are not likely to fly into an intense storm cloud, but target detection may still be hampered since the ambiguous range responses of the target and volumetric clutter may coincide, in which case they would not be resolvable. Resolving targets from volumetric clutter is not always possible in range due to the highly ambiguous nature of range data; however, no such ambiguity exists in the Doppler/velocity domain for high PRFs. The volume clutter is offset in the Doppler band by the relative velocity of the wind with respect to the radar look direction. The requirements for the MTI filtering of volume clutter are likely to differ from those for surface clutter. At low elevation angles the clutter is dominated by MBC at or near zero Doppler from the surface, whereas at high elevation angles the clutter return comprises SLC from the surface plus MBC from volumetric clutter having a modest Doppler offset. A tunable clutter filter can be used to reject volumetric clutter. Doppler processing would enable the user to discriminate between volume clutter and targets of interest since airborne targets generally have a relative velocity that exceeds the relative velocity of the wind. High PRF waveforms can achieve excellent rejection of volumetric clutter using optimized MTI cancellation[8]. Indeed, a high level of clutter rejection is required because distant targets give an ambiguous response that may coincide with short-range volumetric clutter. A three-pulse canceller achieves an MTI improvement factor of 3 dB against rain or chaff of 5 m/s spectral standard deviation with a low PRF waveform of 500 Hz at S-band, whereas at a high PRF of 10 kHz the improvement factor would be 52 dB. MTI filters may be optimized to maximize the SCR by optimizing the weightings of a large transversal filter (Section 7.2.6) and can offer an improvement in the probability of detection of between 2 and 4 dB [8].

Surface-based air defense radars are normally required to provide a rapid indication of target velocity; hence, Doppler processing is required. High PRF pulse Doppler techniques could be considered a viable candidate technique. The selection of a suitably high PRF pushes the first blind velocity above the velocities of interest. Low-level aircraft and cruise missile targets rarely exceed Mach 1, so a reasonable velocity bracket of interest may be ± 400 m/s. For a 10 GHz carrier frequency this yields Doppler shifts between ± 26.7 kHz, so for high PRF operation a PRF $\geq$ 53.3 kHz would be required. The maximum unambiguous range of this PRF is $R_{mu} = 2.8$ km. For a 3 GHz carrier frequency, the required PRF $\geq$ 16 kHz, giving $R_{mu} = 9.4$ km. Either way, the range of interest exceeds R_{mu} and range ambiguities and blind (eclipsed) ranges are incurred. This leads to the use of multiple CPIs, each using a different PRF, to minimize the eclipsing losses (Section 9.3.2) and resolve the range ambiguities, which will be discussed in greater depth in Chapters 11 and 12.

Surface-based radars designed to detect surface targets inevitably suffer from high levels of MBC. Battlefield surveillance radars can display a clutter map on a sector plan position indicator (PPI) display, as illustrated in Figure 9-21. This display format modulates the intensity of the display in proportion to the strength of received signals. Data are displayed in polar coordinates in which the radar location is taken to be at the center of the display; the distance from the display center is calibrated in terms of range and the angle around the display corresponds to the azimuth angle. This display format is popular in 360° surveillance radars. A restricted angular range of the PPI display may be presented on a *sector PPI* display, as on the right-hand side of Figure 9-21. In this case the radar's location is at the apex (bottom central) part of the display. Patches of clutter are displayed as the white areas, and discrete targets appear as the white points on the display.

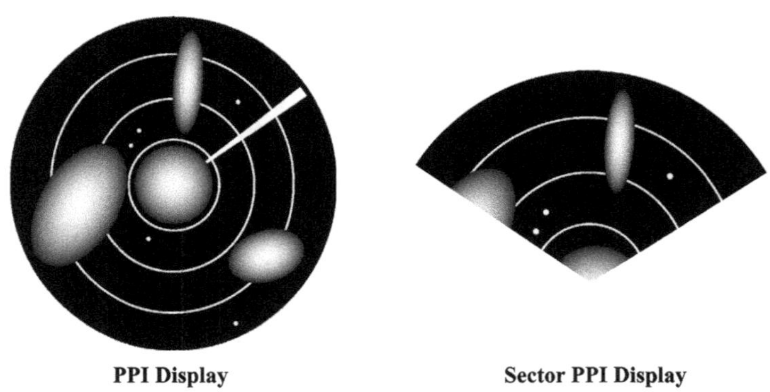

FIGURE 9-21 Plan Position Indicator Display (left) and Sector PPI Display (right)

Land forces operating a battlefield surveillance radar often find it useful to switch out the MTI filter in undulating terrain since the display is now dominated by the clutter returns. This constitutes a *clutter map*, that is, a display of clutter using the polar coordinates of range and azimuth. Clutter maps in undulating terrain may have regions in which no clutter is displayed (only background noise and interference). The lack of clutter is indicative of dead ground such as an area in the shadow of a hill, which could conceal enemy forces. If the radar cannot see into these areas, neither can a human observer, and he approaches them with caution!

In many cases land forces are interested in detecting moving surface targets such as troops and vehicles over modest ranges. Velocities of interest may typically span ± 30 m/s. The radar must be reasonably lightweight, which favors a high carrier frequency. Assuming a frequency of 15 GHz, the spread of target Doppler shifts is ± 3 kHz. High PRF operation therefore demands a PRF ≥ 6 kHz, which in turn has a maximum unambiguous range of 25 km. It is quite probable that the maximum unambiguous range exceeds the range of interest and the range to the horizon and even the maximum detection range of the radar. This ensures that all target returns are first trace, that no range ambiguity exists, and that there are no blind ranges due to eclipsing within the range of interest, except the first minimum range, which is unavoidable. The selection of such a PRF may also be regarded as a low PRF.

There are few radar applications for which it is possible to select a PRF that avoids range ambiguity (i.e., a *low* PRF) and also avoids velocity ambiguity (i.e., a *high* PRF). These come about for radars of short detection ranges and low target velocities and are inevitably ground-based applications. In general, it is not possible to find PRFs that avoid ambiguity in both domains, particularly for airborne applications.

9.6 | SUMMARY OF THE CHARACTERISTICS OF HIGH PRF

- A *high* PRF is sufficiently high to avoid velocity ambiguity. This tends to result in a short unambiguous range, and range ambiguity often, but not always, results.
- Target velocity data are readily available, and high PRF waveforms are well suited to velocity search-and-track modes.
- Target range data are likely to be highly ambiguous. Range ambiguities may be resolved by comparing the ambiguous range data over two or more CPIs, each using a different PRF.
- Additional methods of measuring range are available and are based on the use of long-term phase coding or frequency modulations.
- High PRF offers good rejection of MBC using MTI filtering since MBC occupies a small proportion of the Doppler band. High PRF is therefore a good option in conditions of high MBC. Very few targets of interest would be rejected by the MBC filtering.
- Blind velocities associated with the repetition of the MTI filtering of MBC at Doppler shifts equal to multiples of the PRF fall above the maximum velocity of interest. No blind velocities are incurred.
- For airborne systems, the spread of SLC can occupy an appreciable proportion of the Doppler band and so cannot be filtered away since too many targets of interest would also be rejected.

- The SLC problems are exacerbated due to the multiple ambiguous repeats of SLC in the range dimension, particularly for airborne radars. Slow-moving targets may therefore be difficult to detect since they coincide with the band of SLC.
- The best way to minimize SLC is to use an ultra low sidelobe antenna. In fact, the use of a low sidelobe antenna is imperative for high PRF systems.
- (Fast) closing targets are seen in a region of the Doppler band in which there is no clutter. High PRF therefore offers good long-range detection performance against (fast) closing targets.
- High PRF waveforms typically have high duty ratios, sometimes as much as 50%.
- The high duty ratios of high PRF waveforms lead to significant eclipsing losses and a series of closely spaced blind ranges at multiples of the maximum unambiguous range of the PRF. This may be addressed by alternating the operation of the radar over two or more PRFs and leads to a reduction in either the processing gain (detection performance) or data rate.
- The high duty ratios of high PRF waveforms give rise to high average powers, which enable good long-range detection performance.
- High PRF waveforms illuminate a target with a large number of pulses and offer the scope for large processing gains (large FFT sizes). This also enables good long-range detection performance.

The real strength of high PRF is its long-range detection capability for (fast) closing targets in scenarios of high clutter levels. Its Achilles' heel is its limited detection capability against slow and receding targets due to the high levels of SLC. Whilst target velocity information is readily available, range can normally be determined using only more elaborate waveforms and processing methods, which tend to reduce the detection performance. High PRF lends itself well to airborne applications for long-range detection of approaching intruders in a look-down attitude (high clutter). It is used in airborne early warning radars for long-range searches below the horizon, airborne intercept radars and fire control radars in long range air-to-air velocity search and look-down, shoot-down modes.

CHAPTER 10

Frequency Modulation Ranging in High PRF

Chapter Outline

10.1 | FREQUENCY MODULATION RANGING IN CONTINUOUS WAVE SYSTEMS

10.1.1 Linear Frequency Modulation Ranging

In Chapter 4, continuous wave (CW) radars were presented as systems that sense Doppler shifts and thus provide velocity measurement and are ideal motion detectors. A pure CW signal has an infinitesimally narrow bandwidth and is therefore incapable of any range measurement; all target returns, no matter what their ranges, are detected simultaneously. Pulsed modulation broadens the signal bandwidth and, of course, enables range to be measured with a resolution inversely proportional to the signal bandwidth, B, via the pulse delay ranging method. However, other forms of modulation also increase the signal bandwidth and enable range to be measured. Phase and/or frequency modulation (FM) may be imparted onto a CW signal for the measurement of range. In much the same way as pulsed modulation allows the time delay between the transmission of the signal and the reception of a target return to be established, so, too, does the use of phase or frequency modulation. The most commonly used form of modulation is a linear frequency modulation (LFM), and it gives rise to FM ranging in CW systems and high pulse repetition frequency (PRF) pulsed radars.

An LFM transmitted and received waveform is illustrated in Figure 10-1. The transmitted frequency is increased linearly with time over a bandwidth, B, over the duration of a coherent processing interval (CPI), t_{int}. The received signal preserves the same linear frequency modulation but displaced in time by $t = 2R/c$, on account of the range, R, of the target. This displacement in time gives rise to an instantaneous frequency difference between the transmitted and received signals known as the range delay frequency, f_R (not to be confused with f_r, the PRF). Coherent processing of the received signal using fast Fourier transform (FFT) techniques provides a measurement of f_R and hence range.

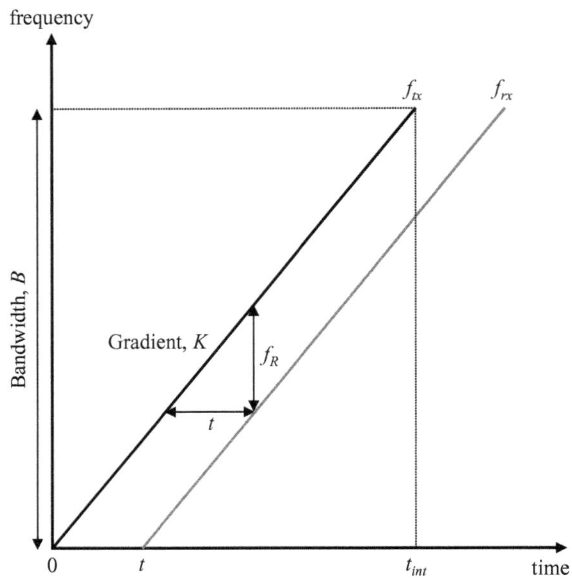

FIGURE 10-1 Linear Frequency Modulation

The gradient of the linear frequency ramp, K, is given by

$$K = \frac{B}{t_{int}} \tag{10-1}$$

The range delay frequency is given by

$$\begin{aligned} f_R &= tK \\ \therefore t &= \frac{f_R}{K} \end{aligned} \tag{10-2}$$

Furthermore, $t = \frac{2R}{c}$, giving

$$R = \frac{cf_R}{2K} = \frac{cf_R t_{int}}{2B} \tag{10-3}$$

The resolution with which the range can be measured is governed by the resolution with which the range delay frequency can be measured. This is given by the inverse of the CPI time; hence,

$$\Delta R = \frac{c\Delta f_R t_{int}}{2B}$$

where

$$\Delta f_R = \frac{1}{t_{int}}$$

giving

$$\Delta R = \frac{c}{2B}$$

which is identical to (3-21), the equation for the range resolution for a simple pulsed waveform with matched filter reception and also a pulse compression system, again using matched filter reception. Indeed, LFM is tantamount to *chirp* pulse compression, except that the latter also includes narrow-band pulsed modulation. The range resolution of a radar

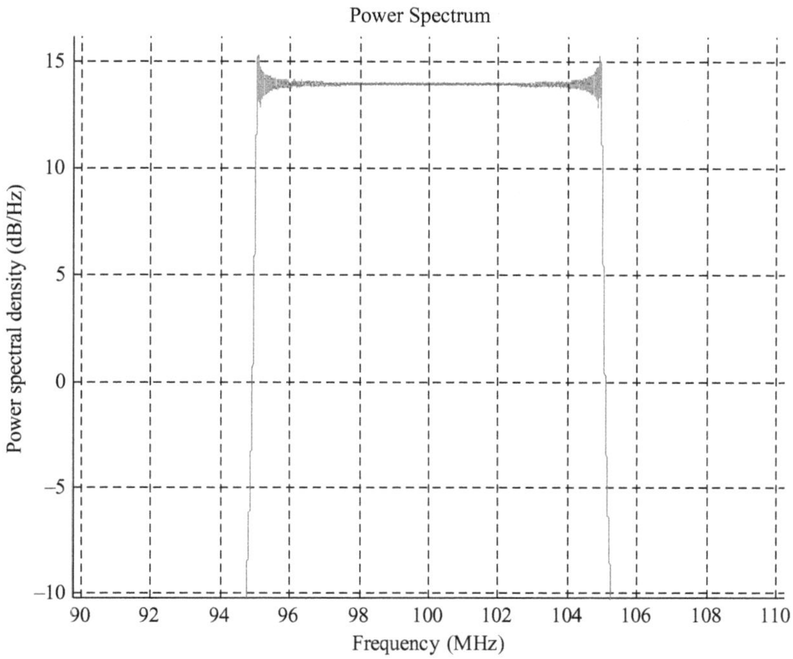

FIGURE 10-2 ■ Spectrum of LFM Ramp Waveform

is inversely proportional to the signal bandwidth, B, as given by (3-21), irrespective of the modulation employed, be it pulsed, phase, or frequency modulation or any combination of these. Unlike any pulsed waveform, however, frequency modulated continuous wave (FMCW) ranging waveforms continue to detect targets down to zero range.

The spectrum of a LFM ramp waveform is given in Figure 10-2 for a nominal carrier frequency of 100 MHz, a peak deviation frequency of 5 MHz (i.e., $B = 10$ MHz) and a modulation frequency of 1 kHz (i.e., $t_{int} = 1$ ms). This spectrum is identical to that of the linear FM *chirp* pulse of Figure 3-14, which is plotted for the same modulation parameters. The spectrum is approximately constant over the signal bandwidth of 95 MHz to 105 MHz, indicating that all frequencies within this band are equally represented.

10.1.2 Stepped Frequency Continuous Wave

The waveforms of Figure 10-1 and Figure 10-2 are derived from a continuous analog frequency sweep. This may be approximated by a set of frequency samples over the appropriate band and is shown in Figure 10-3, known as *stepped frequency continuous wave* (SFCW). The radar tunes its transmitted frequency over a succession of closely spaced frequencies. It must dwell a sufficient time at each frequency to allow returns to be received from the most distant target expected. In this way the radar collects returns at each sample frequency. The returns are processed in an inverse fast Fourier transform (IFFT), which transforms the data from the frequency domain into the time domain and hence range.

Although this is commonly regarded as a CW waveform, most practical implementations actually gate out the radio frequency (RF) during the time it takes to retune the carrier frequency to each new sample frequency plus a small extra time allowance for space charging (see Section 10.1.4) and actually results in a pulsed output.

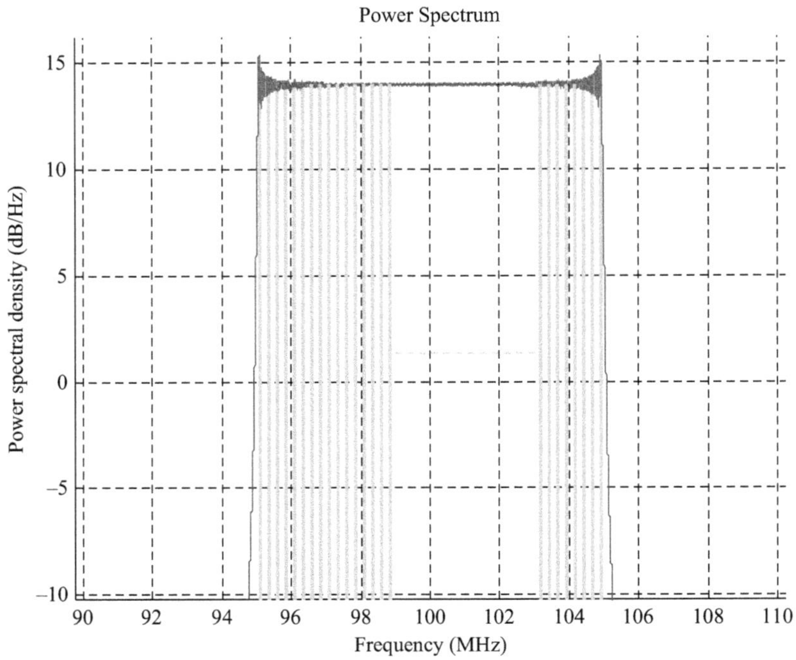

FIGURE 10-3 Stepped Frequency Continuous Wave

Since the frequency data are sampled, its transformation into the range domain via the IFFT results in an aliasing ambiguity. For N frequency samples with a constant spacing frequency of δf over the total bandwidth of B, one has

$$B = (N - 1)\,\delta f \tag{10-4}$$

Ambiguities in time would be incurred for a delay time exceeding $1/\delta f$. This leads to a maximum unambiguous range, R_{mu}, given by

$$R_{mu} = \frac{c}{2\delta f} \tag{10-5}$$

Thus, a long unambiguous range demands a small frequency spacing, δf. If also fine range resolution is required, then a large bandwidth, B, is also required, and the number of frequencies, N, required becomes very high and the time to transmit the entire waveform becomes correspondingly long.

10.1.3 Two-Phase Linear FM Ranging

The FM ramp of Figure 10-1 cannot continue indefinitely since eventually the ever-increasing bandwidth would carry the carrier frequency outside the limits supported by the hardware. The linear ramp waveform could be repeated in each CPI and would appear as a periodic sawtooth function. However, a more practical waveform is a linear "up" ramp with a positive FM gradient followed by a linear "down" ramp with a negative FM gradient, as shown in Figure 10-4. If the *up* and *down* phases of the waveform have equal durations and equal and opposite gradients, then the carrier frequency is restored to its original value after one complete cycle. Such a waveform is depicted in the upper plot of Figure 10-4 and is a triangular waveform of frequency, f_m. The lower plot indicates of the difference between the transmitted and received frequencies during the cycle. Two

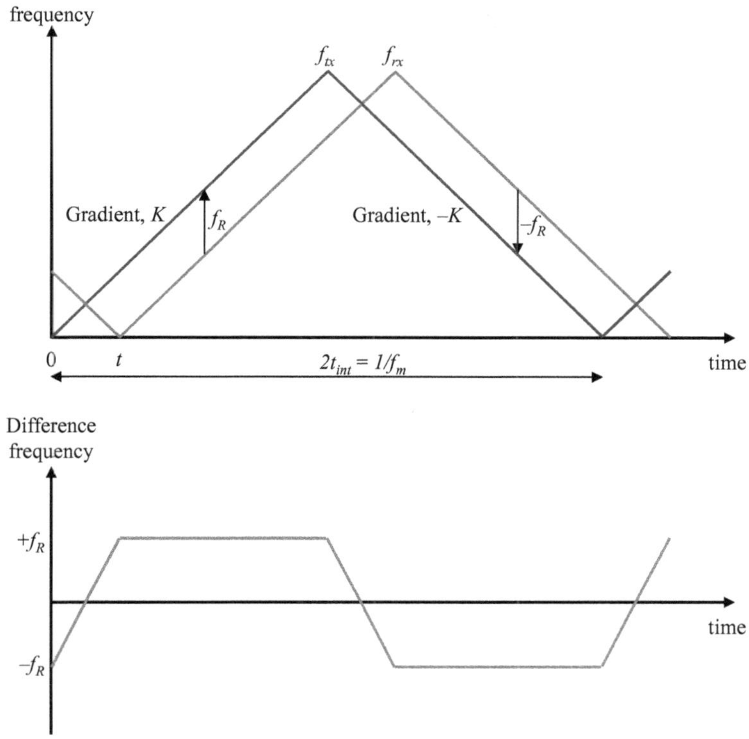

FIGURE 10-4 Two-Phase Triangular LFM Waveform

CPIs are now taken; the first during the *up* phase yields a measurement of $+f_R$, and the second during the *down* phase yields a measurement of $-f_R$. (The system works equally well with positive or negative gradients except that a negative range delay frequency is measured for a negative gradient.)

For the difference frequency to be extracted, the local oscillator (LO) used within the receiver must be frequency modulated over the same bandwidth and in synchronism with the transmitted signal. For homodyne radars (i.e., those that down-convert signals directly to baseband in one down-conversion stage), the LO is a sample of the transmitted signal that clearly has the identical modulation characteristics as the transmitted signal. The difference frequencies are produced in the baseband during each phase of the cycle and may be measured using FFT techniques. For superheterodyne receivers, the first LO frequency is offset from the transmitted frequency by the system's first intermediate frequency (IF) center frequency. Signals in the first IF signal path therefore have a frequency given by the IF center frequency plus the difference frequency. Subsequent stages of down-conversion result in baseband outputs at the difference frequency and may be measured using FFT techniques. In superheterodyne receivers, the LO must be phase locked to the transmitted signal to ensure that they both follow the same modulation and preserve their phase coherency. A phase locked synthesizer system such as that of Figure 8-18 can ensure that these conditions are met. In the design of phase-locked loops a tracking error can be incurred by which the changes on the output of one loop exhibit a small error with respect to changes on its input. The tracking error is determined by the transfer function of the loop response. For the case of two phase-locked loop synthesizers in which one loop generates the LO signal with LFM and the other loop is enslaved to the first and produces the transmitter drive signal, the second loop is expected to produce an output that tracks

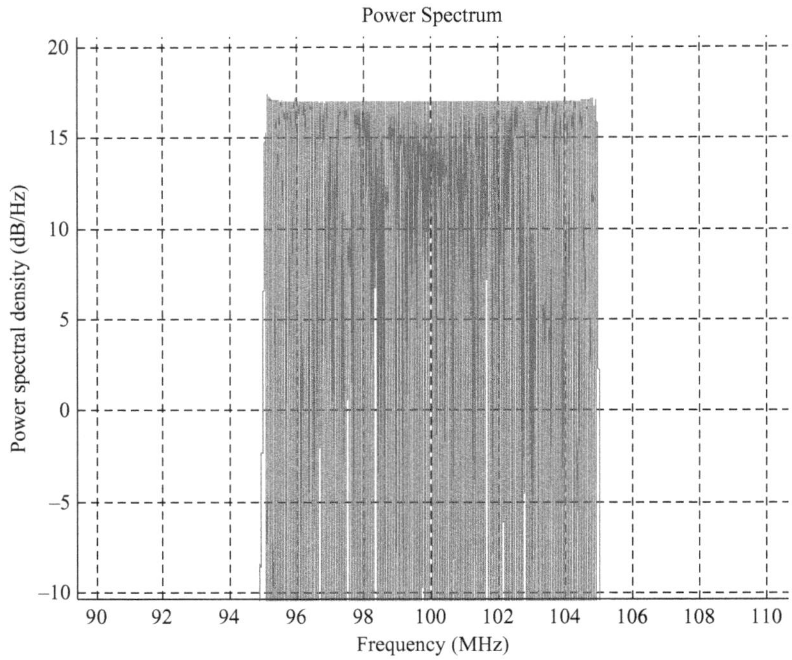

FIGURE 10-5 Spectrum of Two Phase LFM Waveform

the frequency ramp on its input. However, there is inevitably a time displacement between the two. The low-pass filtering function of the loop response often introduces a time delay between the LO ramp and the ramp on the transmitted signal. The transfer function of the loop must be carefully designed to minimize this tracking error. Any time displacement between the FM on the transmitted signal and the FM on the LO introduces a range displacement and is a source of error. Even after good loop design, any remaining error can be calibrated out in the processing stage.

Further sources of error in the ranging function are non-linearities in the transmitted and/or LO FM ramps. Non-linear FM ramps cause variation in the gradient, K, and by reference to equation (10-3) result in a corresponding variation in the indicated range. Modern direct digital synthesizers (DDSs) (Section 8.3.1) can produce reference signals with highly linear FM ramps.

The spectrum of a two phase LFM waveform is plotted in Figure 10-5 for a nominal carrier frequency of 100 MHz, a peak deviation frequency of 5 MHz (i.e., $B = 10$ MHz) and a modulation frequency of 500 Hz (i.e., a phase duration, $t_{int} = 1$ ms). As for the single LFM ramp, the spectrum is approximately constant over the signal bandwidth but has some fine level undulation, dependent on the modulation parameters.

10.1.4 Space Charging

If the modulating frequency is f_m, then the waveform has a period of $1/f_m$. This period is divided into two equal CPIs; however, the CPI duration, t_{int}, must be somewhat less than a half this period. At the start of each phase of the transmitted signal, returns are still being received from the previous phase of the cycle. Hence, as the transmitted signal starts its *up* ramp, target returns from the previous *down* ramp are still being received. Similarly, at the start of the transmitted *down* ramp, target returns from the previous *up* ramp are still being received. During these times the difference frequency is itself ramping between

$+f_R$ and $-f_R$, as is evident in the lower trace of Figure 10-4. These changeover periods could last for as long as the time delay to the maximum detection range of the radar, which must be calculated on the basis of the largest target or clutter return to which the radar is ever likely to be exposed. During this changeover period, the system cannot assume that returns are those of the current transmitted phase, so the processing must be blanked. The early part of the transmission of the new CPI and the subsequent blanking of returns from the processing are known as *space charging*, since it conveys the image that the volume of space surrounding the radar out to its maximum detection range must be filled, or *charged*, with the new waveform before returns can be processed. A space charging period must be allowed at the start of any new CPI, whereby the transmitted waveform used during the previous CPI differs from that used during the new CPI. Space charging must be allowed for in all LFM waveforms and for pulsed waveforms when the PRF is changed.

Let the maximum detection range be R_{max}. The time delay for the maximum range is therefore $t_{max} = \frac{2R_{max}}{c}$. It is often prudent to allow for a 50% margin and set the space charging time at the start of each new CPI, t_{SC}, to

$$t_{SC} = 1.5\frac{2R_{max}}{c} \tag{10-6}$$

The 50% margin caters for the case that a target (or clutter) RCS some five times larger than the largest anticipated could be detected at 1.5 times the maximum range.

Applying space charging to the two phase linear FM waveform of Figure 10-4 blanks the times during which the difference frequency ramps between $+f_R$ and $-f_R$. This reduces the CPI to

$$t_{int} = \frac{1}{2f_m} - t_{SC} \tag{10-7}$$

Thus, each CPI captures the times when the difference frequency is constant at $\pm f_R$ and enables range to be measured using equation (10-3); each CPI returns an independent measurement of range. The allowance made for space charging time means that some of the transmitted FM ramp is not used. This reduces the CPI time and the usable portion of the signal bandwidth, which, in turn, degrades the frequency resolution and hence the range resolution. It is therefore desirable to ensure that the period of the FM ramp is considerably longer than the space charging time.

10.1.5 Two Phase Linear FM with Doppler

Should a moving target be detected the returns are Doppler shifted. As a result, the received waveform is offset up or down by the appropriate Doppler shift. This is illustrated in Figure 10-6, together with the difference frequency plot for a closing target.

Two difference frequencies, f_1 and f_2, are measured: one in each CPI. These are given by

$$f_1 = f_R - f_d$$
$$f_2 = f_R + f_d$$

Adding these two expressions together and dividing by 2 gives

$$f_R = \frac{f_2 + f_1}{2} \tag{10-8}$$

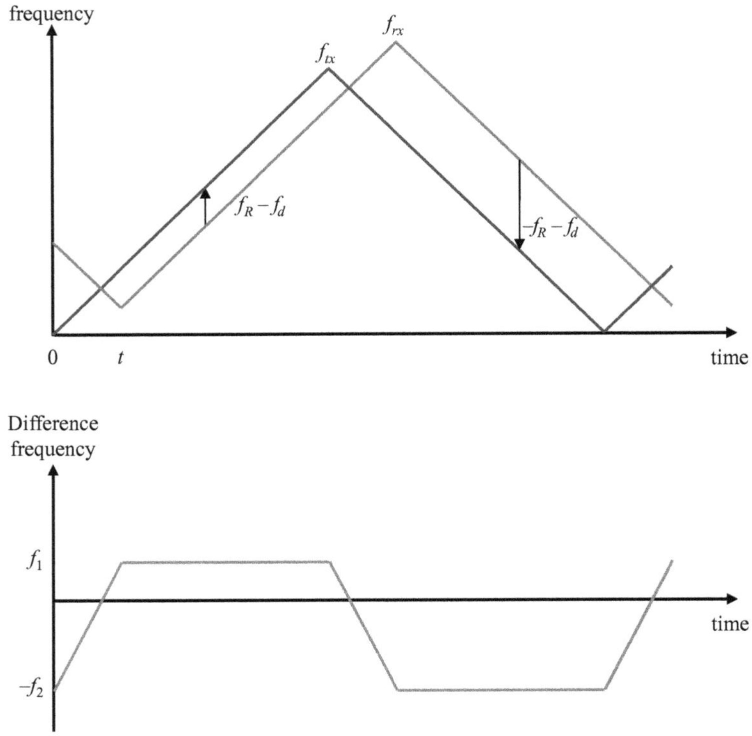

FIGURE 10-6 Two Phase Triangular LFM Waveform with Doppler

Subtracting these two expressions and dividing by 2 gives

$$f_d = \frac{f_2 - f_1}{2} \tag{10-9}$$

Over the two CPIs, both the range and velocity of the target may be extracted. As ever, the resolution with which all frequencies are measured is given by $1/t_{int}$, where t_{int} is given by (10-7). The reduced Doppler frequency resolution is therefore given by

$$\Delta f_d' = 1/t_{int} \tag{10-10}$$

The FM gradient is given by

$$K = 2f_m B$$

The reduced bandwidth, once space charging has been taken into account, is given by

$$\begin{aligned} B' &= K t_{int} \\ &= B2f_m t_{int} \\ &= B(1 - 2f_m t_{SC}) \end{aligned}$$

This leads to a reduced range resolution of

$$\begin{aligned} \Delta R' &= \frac{c}{2B'} = \frac{c}{2B\,(1 - 2f_m t_{SC})} \\ \Delta R' &= \Delta R \frac{1}{(1 - 2f_m t_{SC})} \end{aligned} \tag{10-11}$$

The range resolution is degraded by the factor $\frac{1}{(1-2f_m t_{SC})}$.

If f_m is set such that $t_{SC} = \frac{1}{20} \cdot \frac{1}{f_m}$, then the space charging time amounts to 10% of the duration of each phase and the range resolution is degraded by 1/0.9 or 1.11 (i.e., 11% more coarse). Similarly, the velocity resolution becomes 11% more coarse.

10.1.6 Selecting the Modulation Parameters

The starting point in setting the modulating frequency, f_m, and bandwidth, B, should ideally be the requirements to meet a specified range and velocity resolution. However, this is constrained by the maximum detection range (and hence the space charging time), limitations on the maximum FM ramp gradient and its linearity that the hardware can generate, and the maximum beam dwell time at any given angular position.

The velocity resolution requirement dictates the Doppler resolution requirement, which, from (10-10) in turn fixes the lower limit on the CPI time, t_{int}. The maximum detection range sets the space charging time via (10-6). The sum of $t_{int} + t_{SC}$ sets the minimum phase duration that is, the time to complete each *up* ramp or each *down* ramp. The period of the modulating frequency is then set to be twice the phase duration. This can be expressed by rearranging (10-7) as

$$\frac{1}{f_m} = 2\left(t_{int} + t_{SC}\right)$$

Of course, there is no reason t_{int} should not be set longer than the bare minimum. This leads to a lower modulation frequency. However, at least one complete cycle of the waveform must be completed within the beam dwell time. Increasing t_{int} has several advantages:

- Increased integration gain
- Finer velocity resolution
- Reduced FM ramp gradients that, for a given bandwidth, B, ease the burden on the hardware to produce rapid tuning rates of oscillators and synthesizers

For a fixed FM bandwidth, B, a reduction in f_m means that the space charging time is a smaller proportion of the phase duration, which in turn minimizes the degradations in range and velocity resolution. As an approximate rule-of-thumb, f_m must be set such that $t_{SC} \leq \frac{1}{20} \cdot \frac{1}{f_m}$, which from the previous examples limits the degradation in range and velocity resolution to 11% (compared with the cases for which $t_{SC} = 0$). Thus, an alternative method of setting f_m would be to calculate t_{SC} and set $f_m \leq \frac{1}{20 t_{SC}}$ (subject to beam dwell time constraints).

If very distant targets are detectable, it may be possible that the delay time, t, exceeds $1/f_m$. The received modulation is now delayed by more than one complete cycle of the waveform and range ambiguity results. The modulating frequency, f_m, must be set sufficiently low to avoid range ambiguity. This is directly analogous to the requirement for a low PRF to avoid range ambiguity.

The bandwidth is set on the basis of the range resolution requirement. Again, the reduced bandwidth, B', equation (10-11) dictates the range resolution. Having set both the bandwidth and the modulation frequency, the gradient of the FM ramp, K, is also set. Modern hardware can support tuning rates in the order of several GHz per second.

EXAMPLE

Consider a battlefield surveillance radar operating at a transmitted frequency of 15 GHz using an FMCW waveform comprising two LFM triangular modulation phases. Let the update rate be 100 Hz, the maximum range of interest be 25 km, the maximum velocities of interest be 35 m/s, the range resolution be $\leq$ 10 m, and the velocity resolution be $\leq$ 2.5 m/s. Calculate the parameters of the modulation and FFT processing which would be necessary.

Solution

$f_{tx} = 15$ GHz; therefore, $\lambda = \frac{c}{f_{tx}} = \frac{3e8}{15e9} = 0.02$ m or 2 cm.

The update rate of 100 Hz fixes $f_m = 100$ Hz, which has a period of 10 ms.

This gives a phase duration of $\frac{1}{2f_m} = \frac{1}{2\times 100} = 5$ ms

The space charging time (with 50% margin) is

$$t_{SC} = 1.5\frac{2R_{max}}{c} = \frac{1.5 \times 2 \times 25e3}{3e8} = 250\ \mu\text{s}$$

The space charging time is 1/40 of the period of the modulation frequency and is thus well within the 1/20 rule-of-thumb. In fact, the 1/20 rule would permit a maximum range of interest of 50 km (with the 50% margin applied).

The integration time is $t_{int} = 5\text{ ms} - 250\ \mu\text{s} = 4.75$ ms and is the CPI time.

This yields a frequency resolution of $\Delta f = 1/t_{int} = 1/4.75e-3 = 210.53$ Hz

Since this equates to the Doppler resolution, the velocity resolution is therefore

$$\Delta V_r = \frac{\lambda \Delta f_d}{2} = \frac{0.02 \times 210.53}{2} = 2.1\ \text{m/s}$$

This is within the limit of 2.5 m/s.

Applying the limit of 2.5 m/s would require an integration time, $t_{int} = 4$ ms, and therefore would permit a space charging time of up to 1 ms (1/10 the modulation period, $R_{max} = 100$ km).

The range resolution, ΔR, of 10 m requires a bandwidth, B', of 15 MHz.

From

$$\Delta R' = \frac{c}{2B'} = \frac{c}{2B\,(1 - 2f_m t_{SC})}$$

one has

$$10 = \frac{c}{2B}\frac{1}{(1 - 2 \times 100 \times 250e-6)} = \frac{c}{2B}\frac{1}{0.95}$$

giving

$$B = \frac{3e8}{10 \times 2 \times 0.95} = 15.789_5\ \text{MHz}$$

which rounds up to 16 MHz.

Thus, the FM ramp gradient, K, is 16 MHz/5 ms, or $K = 3.2$ GHz/sec, which is feasible.

Check

A range resolution of 10 meters requires a time resolution of 66.67 ns. Given a gradient of 3.2 GHz/s, the frequency resolution required must be < 66.7 ns × 3.2 GHz/s = 213.44 Hz. The frequency resolution available is 210.53 Hz and so meets this requirement.

To calculate the FFT processing bandwidth, the maximum and minimum difference frequencies must be found. The maximum difference frequency arises for targets at maximum range (maximum f_R) and maximum positive velocity (maximum $+f_d$).

At the maximum range, the delay time, $t = \frac{2R}{c} = \frac{2 \times 25e3}{3e8} = 166.7\ \mu s$

From (10-2) this results in a maximum range delay frequency of

$$f_R = tK = 166.7e - 6 \times 3.2e9$$
$$= 533{,}333 \text{ Hz}$$

The maximum Doppler shift is $f_d = \frac{2V_r}{\lambda} = \frac{2 \times 35}{0.02} = 3500$ Hz

Therefore, the maximum difference frequency is

$$f_2 = f_R + f_d = 533{,}333 + 3500 = 536{,}833 \text{ Hz}$$

The minimum difference frequency arises for targets at zero range (minimum f_R) and maximum negative velocity (maximum $-f_d$), that is, $f_R = 0$ and $f_d = -3500$ Hz. Therefore, the minimum difference frequency, $f_1 = 0 - 3500 = -3500$ Hz. The FFT processing must therefore cover the band -3500 Hz to $+536,833$ Hz, that is, a total bandwidth of 540,833 Hz. Rounding this up to 550 kHz requires a sampling rate of no lower than 1.1 MHz to avoid aliasing ambiguities. Using these numbers, a total of 5225 samples would be available in each CPI. A smaller FFT (fewer samples, lower sampling rate) could be implemented at the expense of a reduced modulation bandwidth, reduced ramp gradient, and ultimately a coarser range resolution.

10.1.7 Problems of CW Systems

The limitations of CW systems were discussed in Section 4.5.4. The main drawback is that inadequate transmitter to receiver isolation leads to an upper limit on the transmitted power which may be used which, in turn, limits the detection range. However, on the plus side, the use of FM ranging restores the measurement of range and the ability to alleviate potential clutter problems through its subdivision in range, which may go some way toward extending the detection range. However, in the context of FM ranging, a new limitation arises: that of confusion when multiple targets are present. Since each target is characterized by a pair of difference frequencies, multiple targets detected within one beam position give rise to several difference frequencies, which could lead to confusion when attempting to correlate them into the appropriate pairs. Figure 10-7 illustrates the eight responses from four targets in the baseband output using two-phase FM ranging. The darker-colored responses are those detected during the *up* phase, and the lighter-colored responses are seen in the *down* phase. However, how should they be paired up? Which response in the *up* phase should be paired with the response in the *down* phase from the same target? It is not at all clear, and the confusion grows only worse as the number of targets increase. The amplitude of the response might be used as a discriminant to resolve the confusion, but this is unreliable due to target scintillation. A narrow beam may be used

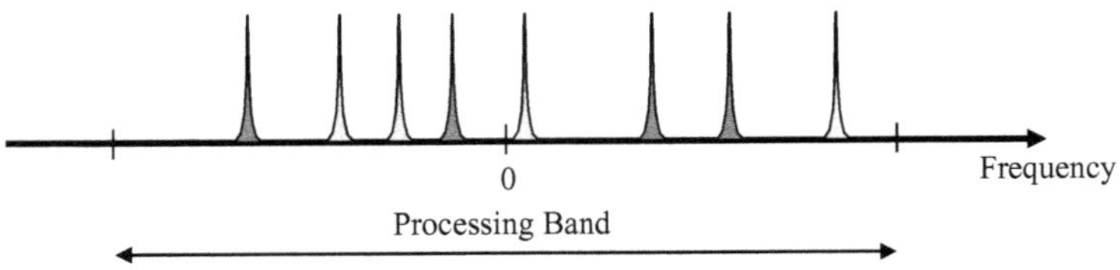

FIGURE 10-7 Response from Four Targets in Baseband Using Two-Phase FM Ranging

so that the number of target responses at any one time is minimal. This may minimize the problem but does not overcome the confusion from close formations of targets. Further, it may be assumed that targets move at a constant velocity such that their range changes linearly with time. Thus, over several scans the tracker would observe a constant rate of change in f_R and no change in f_d, which results in pairs closing together toward a fixed central positive (Doppler) frequency if closing or diverging from a fixed negative (Doppler) frequency if receding. This takes several scans to resolve but eases the task of target correlation *provided* the simplifying assumption holds, which may never be known.

There is no surefire way of resolving the confusion from multiple targets. However, one technique to mitigate the problem is to introduce a third phase in the modulation cycle; this is commonly implemented in high PRF pulsed radars.

10.2 FM RANGING IN PULSED SYSTEMS

10.2.1 Three-Phase FM Interrupted Continuous Wave

The FM ranging techniques described in Section 10.1 may be extended to pulsed systems and are known as *FM interrupted continuous wave* (*FMICW*). The modulation waveform is extended to include a third phase of zero gradient, as in Figure 10-8, and it is commonly implemented in high PRF systems of high duty cycle. Three CPIs are now required—one for each phase. Three difference frequencies are now derived for each target—one in each phase/CPI. The Doppler frequency is readily available in the zero gradient phase, and the range delay frequency, f_R, is found from equation (10-8), as for the two phase waveform. Of course, it is now necessary to ensure that the modulating frequency is chosen to allow at least one complete cycle to be transmitted within the beam dwell time. This causes a reduction in the CPI time allowed for each phase. One also has to suffer three space charging times during each modulation cycle. This becomes particularly damaging since a high PRF waveform is a long-range mode and the duration of each space charging time, given by equation (10-6), becomes correspondingly long. The CPI duration is now curtailed due to the requirement to accommodate three CPIs and three space charging intervals per beam dwell. This limits the size and duration of the FFT and hence frequency resolution and also the scope for integration gain and hence the detection range.

A single CPI equal to the beam dwell time may be used in a high PRF waveform for a velocity search mode. This would obtain the longest detection range possible in a noise-limited environment and boasts the finest velocity resolution. However, it lacks any ranging capability, and sidelobe clutter could be problematic at long ranges and would limit the detection range of slow speed targets. The introduction of FMICW ranging restores the ranging capability and may alleviate the clutter problem to some extent, but this comes at a considerable cost to the (noise-limited) detection range and velocity resolution.

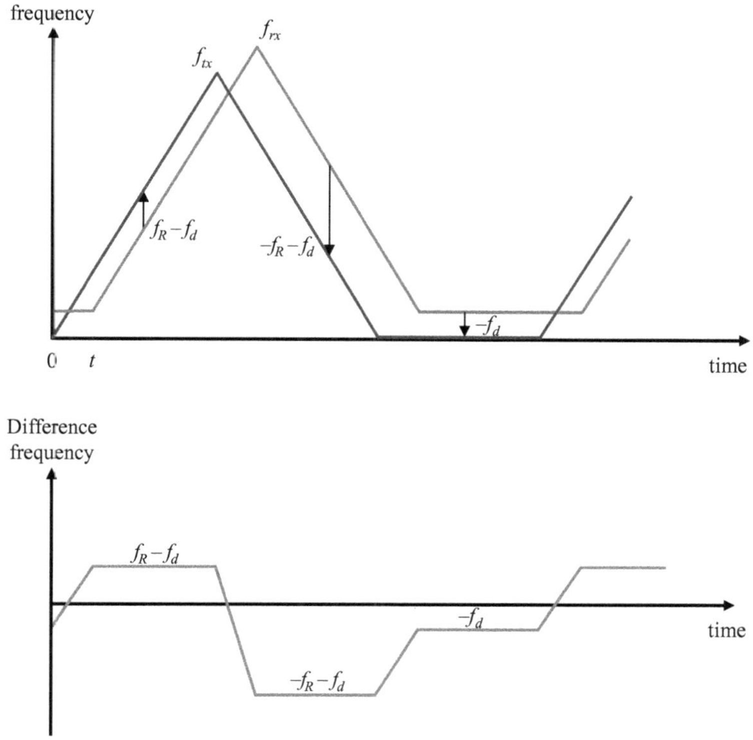

FIGURE 10-8 Three-Phase FMICW

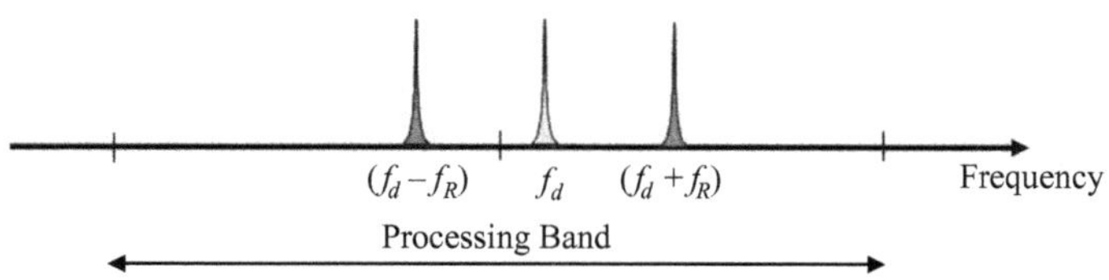

FIGURE 10-9 FMICW Triplets

As a result of the three phases, each target yields three difference frequencies: $-(f_d - f_R)$, $-(f_d + f_R)$, and $-f_d$ (Figure 10-9). These are sometimes referred to as a *triplet*. The central response (light colored) in the triplet falls at the Doppler frequency and hence indicates the velocity of the target. The two outer responses (dark) are symmetrically spaced about the central one, and this spacing is the range delay frequency and hence indicates the range to the target. A slower target results in the triplet center being shifted toward zero frequency, and a longer range target results in a wider triplet. The central component is readily identifiable since this arises during the zero gradient phase.

The responses from multiple targets are more readily correlated since each target yields a symmetrical triplet whose central component is determined in the zero gradient phase. Figure 10-10 illustrates the case of four targets with the light responses indicating

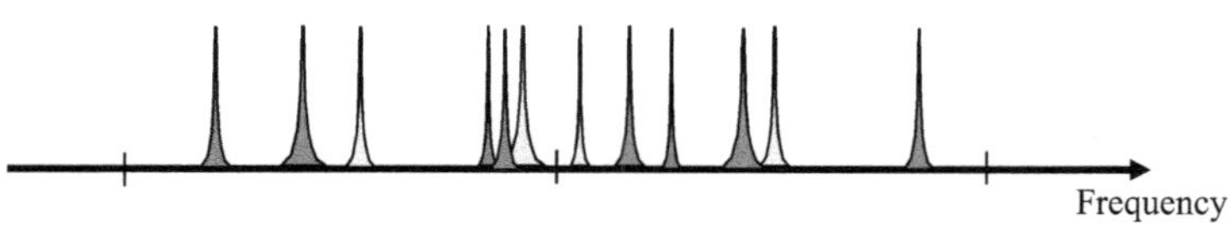

FIGURE 10-10 Multiple FMICW Triplets

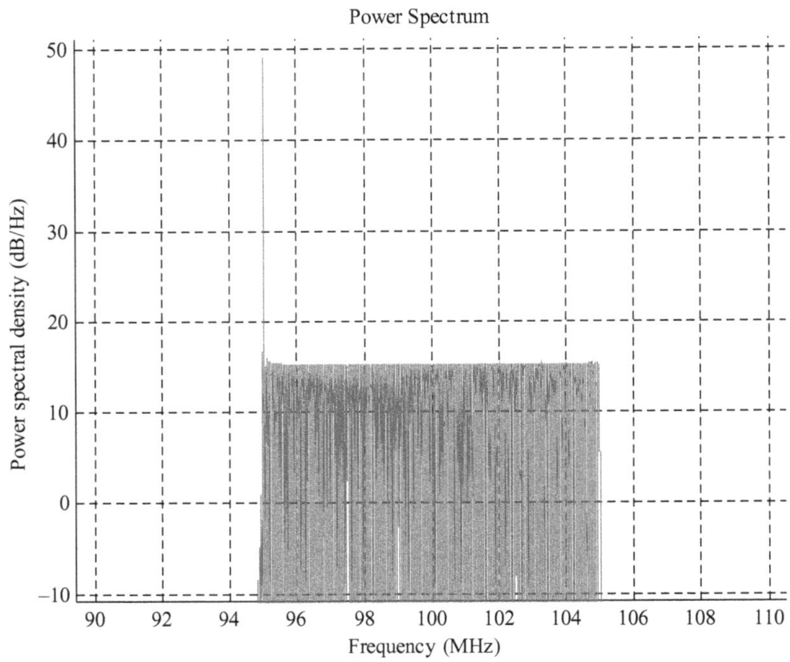

FIGURE 10-11 Spectrum of Three-Phase FMICW Waveform

the central components and the dark responses the outer components. Symmetrical triplets are more easily identifiable, particularly when the central component is known, so target responses are more readily correlated compared with the pairs of the two-phase FM ranging case. However, the scope for confusion still exists and is exacerbated as the number of targets in the beam increases. Noise-generated false alarms may also give rise to false target responses, which would compound the cross-correlation problem. The result of cross-correlation of responses is to declare incorrect target ranges. Cross-correlation is unlikely to persist for very long since a miscorrelated triplet is unlikely to remain symmetrical as the range or velocity of the real targets change (and noise-generated false alarms are unlikely to persist at the same frequency cell for long).

The spectrum of a three-phase FMICW waveform is given in Figure 10-11, for a nominal carrier frequency of 100 MHz, a peak deviation frequency of 5 MHz (i.e., $B = 10$ MHz) and a modulation frequency of 333 Hz (i.e., a phase duration, $t_{int} = 1$ ms).[1] This again exhibits a near constant spectrum like the other linear FM cases except for the large response at 95 MHz due to the lengthy dwell at this frequency during the zero gradient phase.

10.2.2 Sine Wave FM Ranging (Sine FM)

An alternative frequency modulation for extracting range in high PRF pulsed systems is a sinusoidal modulating waveform. Here the transmitted carrier frequency is sine wave modulated; the received signal exhibits a phase shift, ϕ, with respect to the transmitted

[1] Actually, the high PRF pulsing element is omitted from this spectrum but is normally very narrow band and has a negligible effect on the overall spectrum. Typically, the spectrum is dominated by the FM.

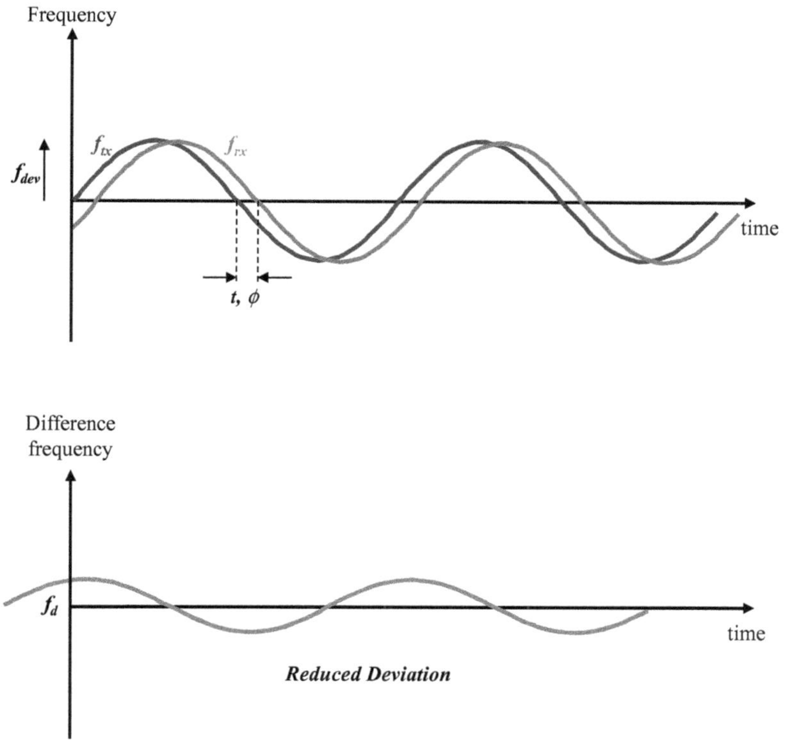

FIGURE 10-12 Sine FM Ranging

signal, which is a function of the target range; the waveforms are illustrate in Figure 10-12 for a stationary target. If a return from a moving target is received, there would be an additional Doppler frequency offset on the received waveform. The difference frequency now undergoes a sinusoidal variation and is known as the *reduced deviation*. The phase and peak deviation of the reduced deviation are functions of ϕ and hence the target range. The reduced deviation varies around a central frequency equal to the target Doppler shift, f_d.

A target at a range R results in a time delay given by $t = 2R/c$, which in turn results in a phase difference between the transmitted and received modulation of ϕ. If the angular frequency of modulation is ω_m then

$$\omega_m = \frac{\phi}{t}$$

and therefore

$$\phi = \omega_m \frac{2R}{c} \tag{10-12a}$$

giving

$$R = \frac{c\phi}{2\omega_m} \tag{10-12b}$$

If the peak (angular) frequency deviation applied to the transmitted signal is ω_{dev}, then the modulation envelope of the transmitted signal can be expressed as

$$\omega_{dev} \sin(\omega_m t)$$

(Here t refers to time, in general, and not the specific time delay of a target return.)

The received modulation envelope, assuming zero Doppler shift, can be expressed as

$$\omega_{dev} \sin(\omega_m t - \phi)$$

The reduced deviation is formed by extracting the difference between the transmitted and received modulations and so has a modulation envelope expressed as

$$\begin{aligned}
&\omega_{dev} \sin(\omega_m t) - \omega_{dev} \sin(\omega_m t - \phi) \\
&= 2\omega_{dev} \sin \tfrac{\phi}{2} \cos\left(\frac{2\omega_m t - \phi}{2}\right) \\
&= 2\omega_{dev} \sin \tfrac{\phi}{2} \cos\left(\omega_m t - \frac{\phi}{2}\right) \\
&= 2\omega_{dev} \sin \frac{\phi}{2} \sin\left[\omega_m t + \left(\frac{\pi}{2} - \frac{\phi}{2}\right)\right] \qquad (10\text{-}13)
\end{aligned}$$

The extent of the reduced deviation is therefore $2\omega_{dev} \sin \frac{\phi}{2}$. The reduced deviation is a sine wave that exhibits a phase advance of $\left(\frac{\pi}{2} - \frac{\phi}{2}\right)$ with respect to the transmitted modulation. FFT processing can be used to measure the extent of the reduced deviation and hence the value of ϕ, which in turn can be used to calculate the range of the target from equation (10-12b). Figure 10-13 plots the extent of the reduced deviation as a function of the phase shift, ϕ, due to the range delay. This plot has a half sine wave variation with a peak of $2\omega_{dev}$ at $\phi = \pi$ radians.

From Figure 10-13 it is apparent that a range ambiguity could exist since values of the extent of the reduced deviation intersect the curve at two points (or more if values of ϕ greater than 2π are considered). The ambiguity is avoided if the modulation parameters are set such that values of ϕ can be guaranteed to be less than π. This sets a maximum unambiguous range, $R_{mu} = \frac{c}{4 f_m}$. In practice, values of ϕ are typically limited to $1/30$ of the period (i.e., $\pi/15$ radians or 12°) and are confined to a region in which the extent of the reduced deviation is approximately a linear function of ϕ.

For example, for a maximum detection range of 100 km, the maximum time delay is 667 μs. If this represents $1/30$ the modulation period, then the full period must be 20 ms, giving a modulating frequency of 50 Hz. If the transmitted frequency deviation is 4 kHz, then the extent of the reduced deviation for a target at 100 km would be 836 Hz.

One reason sine FM is popular is that it is convenient in range tracking systems in which the reduced deviation is tracked within a phase-locked loop. Being sinusoidal, the reduced deviation can be used as an input into a phase-locked loop. The phase and extent of the reduced deviation are both range dependent, so locking to the phase of the reduced deviation provides a continuous tracking function of target range.

FIGURE 10-13 Extent of Reduced Deviation as a Function of ϕ

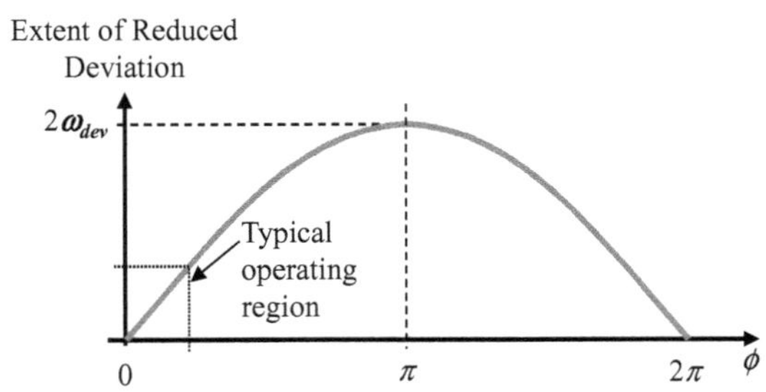

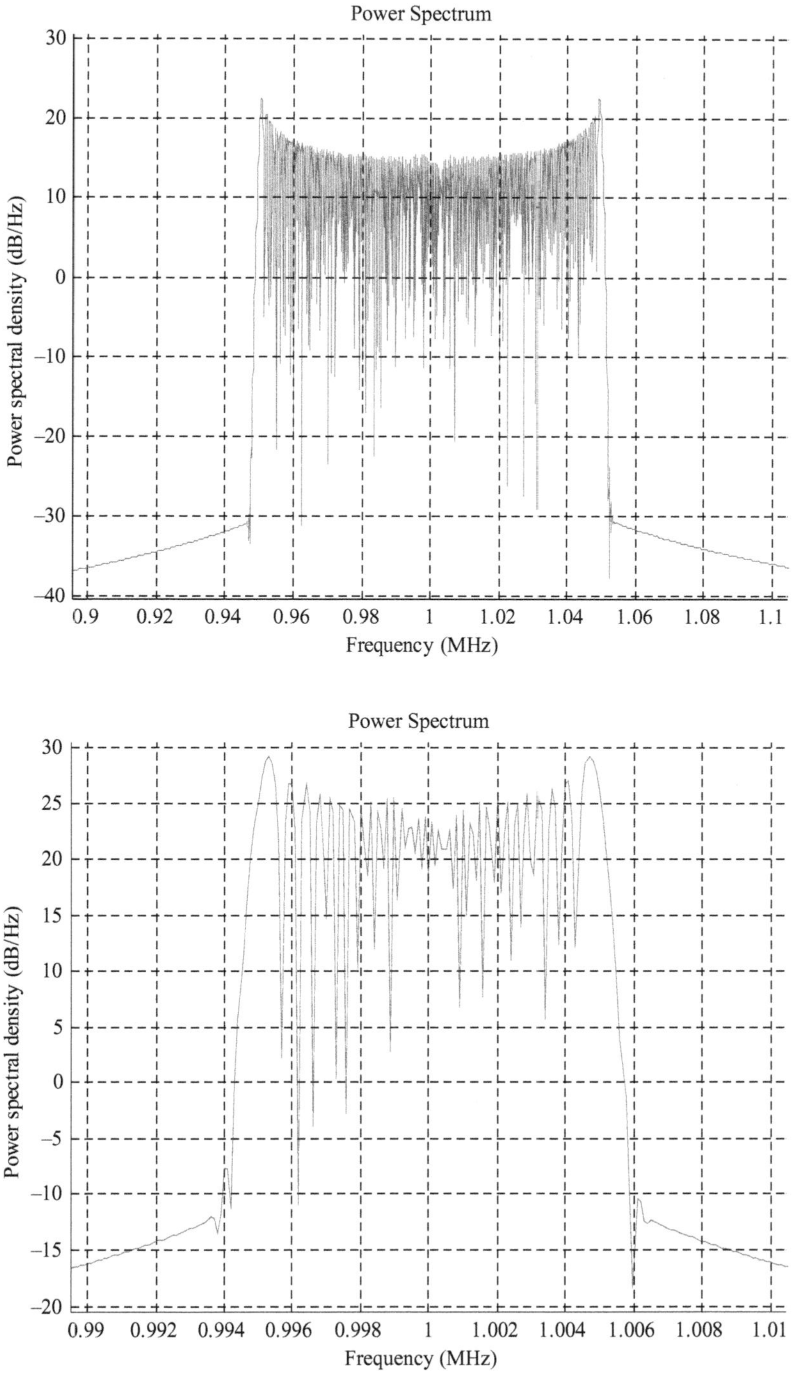

FIGURE 10-14 Spectrum of Sine FM Signal: $m = 500$ (Upper Plot); $m = 50$ (Lower Plot)

The accuracy of any measurement of target range using sine FM is dependent on accurate knowledge of the modulation parameters, specifically f_{dev} and f_m. The voltage of a sine wave frequency modulated carrier signal of (angular) frequency, ω_{tx}, is given by

$$= A\cos\left(\omega_{tx}t + m\sin\omega_m t\right) \tag{10-14}$$

TABLE 10-1 ■ Bessel Tables

Order n	$m = 1$ $J_n(1)$	$m = 2$ $J_n(2)$	$m = 3$ $J_n(3)$	$m = 4$ $J_n(4)$	$m = 5$ $J_n(5)$	$m = 6$ $J_n(6)$	$m = 7$ $J_n(7)$	$m = 8$ $J_n(8)$	$m = 9$ $J_n(9)$	$m = 10$ $J_n(10)$
0	0.765	0.224	–0.260	–0.397	–0.178	0.151	0.300	0.172	–0.090	–0.246
1	0.440	0.577	0.339	–0.066	–0.328	–0.277	–0.005	0.235	0.245	0.043
2	0.115	0.353	0.486	0.364	0.047	–0.243	–0.301	–0.113	0.145	0.255
3	0.019	0.129	0.309	0.430	0.365	0.115	–0.168	–0.291	–0.181	0.058
4	0.002	0.034	0.132	0.281	0.391	0.358	0.158	–0.105	–0.266	–0.220
5		0.007	0.043	0.132	0.261	0.362	0.348	0.186	–0.055	–0.234
6			0.011	0.049	0.131	0.246	0.339	0.338	0.204	–0.014
7				0.015	0.053	0.130	0.234	0.321	0.328	0.217
8				0.004	0.018	0.057	0.128	0.223	0.305	0.318
9					0.005	0.021	0.059	0.126	0.215	0.292
10						0.007	0.024	0.061	0.125	0.208
11							0.008	0.026	0.062	0.123
12								0.010	0.027	0.063
13									0.011	0.029
14										0.012

where m is the modulation index and is given by

$$m = \frac{f_{dev}}{f_m} \tag{10-15}$$

Expansion of this expression gives rise to an infinite series of terms, namely,

$$\begin{aligned} &= A\,[J_0(m)\cos\omega_{tx}t + J_1(m)\cos(\omega_{tx} + \omega_m)\,t - J_1(m)\cos(\omega_{tx} - \omega_m)\,t + \cdots \\ &\quad J_2(m)\cos(\omega_{tx} + 2\omega_m)\,t - J_2(m)\cos(\omega_{tx} - 2\omega_m)\,t + \cdots] \end{aligned} \tag{10-16}$$

where $J_n(m)$ is the Bessel function of the first kind and n-th order.

This series describes a spectrum having an infinite number of frequency components at regular frequency offsets of $n\omega_m$ either side of the carrier component. The Bessel functions determine the amplitudes of the n-th order sidebands. In practice, high-order sidebands become very small, and components with amplitudes of less than 1% of the unmodulated carrier level may be ignored. The spectra of two sine FM signals are shown in Figure 10-14. The upper spectrum has a carrier frequency of 1 MHz, $f_{dev} = 50$ kHz and $f_m = 100$ Hz. Thus, the modulation index $m = 500$, and the modulation is very wide band. The spectrum is not flat, as would be the case of LFM (the dip in the center reflects the more rapid rate of change in frequency here), and individual sidebands are too close to be resolved. The lower plot is for a carrier frequency of 1 MHz, $f_{dev} = 5$ kHz and $f_m = 100$ Hz, giving $m = 50$ and so is not as wide band as the upper plot. In this case individual sidebands spaced at offsets of 100 Hz are visible. As m decreases, the modulation becomes more narrow band and fewer sidebands are evident in the spectrum. Table 10-1 gives the Bessel function values for integer values of modulation index up to 10, and Figure 10-15 plots the sideband level as a continuous function of the modulation index.

This analysis is relatively simple for sine FM waveforms because the modulation envelope is a single sine wave function. It is considerably more difficult for linear FM waveforms because the linear modulation functions comprise an infinite Fourier series of

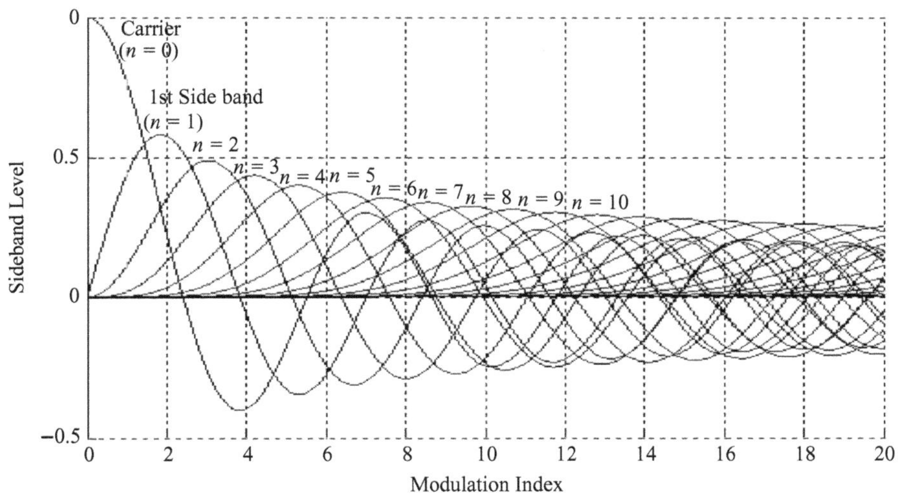

FIGURE 10-15 Sideband Level versus Modulation Index

sine and cosine functions, each having its own modulation indices. The relative simplicity of sine FM signals enables an exact modulation index to be set. For example, if a modulation index of 10 ($m = 10$) is required, then the modulation index can be adjusted until the sixth sideband attains its first null. In this way, precise values of f_{dev} and f_m may be set to ensure an accurate indication of range. The advent of direct digital synthesizers enables high-integrity sine wave reference signals to be produced for the modulation of the carrier.

References

[1] Trunk, G. V. and Kim, M. W., "Multiple Target Ambiguity Resolution," in *Proceedings of the Radar Conference, Record of the 1994 IEEE National,* March 29–31, 1994, 91–94.

[2] Levanon, N., "Mitigating Range Ambiguity in High PRF Radar Using Inter-Pulse Binary Coding," *IEEE Transactions Aerospace & Electronic Systems*, **45**(2), April 2009, 687–697.

[3] Chun-cheng, Z., Jian-jun, G., Jian, L., and Jing-tai, S., "Analysis of Ground Clutter Characteristics of Airborne Forward Looking Pulse Doppler Radar in High PRF," Paper presented at the CIE International Conference on Radar, Shanghai, October 2006.

[4] Alabaster, C. M. And Hughes, E. J., "Clutter Mapping for Airborne Pulse Doppler Radar," Paper presented at the IET Seminar on Radar Clutter Modelling, London, February 2008.

[5] Randig, G. W., "A Simple Method to Determine Airborne Radar Detection Performance in Clutter," in *Proceedings of the Radar Conference*, Lynnfield, MA, April 1993, 192–196.

[6] Long, W. H. And Harriger, K. A., "Medium PRF for the AN/APG-66 Radar," *Proceedings of the IEEE*, **73**(2), February 1985, 301–311.

[7] Madksinghe, D., "Performance Analysis of an Airborne High PRF Phased Array Radar in a Jamming Environment," in Proceedings of the Radar Conference, Adelaide, September 3–5, 2003.

[8] Quayle, R. S. and Earnshaw, N. C., "Using High PRF Burst Waveforms to Resolve Targets in Long Range Clutter of Limited Extent," in *Proceedings of the Radar International Conference,* Brighton, UK, October 1992, 171–174.

PART IIB

Medium Pulse Repetition Frequency Pulse Doppler Radar

CHAPTER 11

Introduction to Medium Pulse Repetition Frequency Radar

Chapter Outline

11.1 | BASIC CONCEPTS

11.1.1 Definition in Terms of Ambiguities

A medium pulse repetition frequency (PRF) gives rise to ambiguities in both range and velocity [1]. Medium PRF is the general case; special simplifying cases arise for low PRF, in which range is unambiguous, and high PRF, in which velocity is unambiguous.

11.1.2 Use of Multiple Coherent Processing Intervals

Radars operating in a medium PRF mode typically cycle through several coherent processing intervals (CPIs), each transmitting a burst of pulses on a different PRF [2-8, 9, 10]. Therefore, within each CPI a fixed PRF is used that differs from the PRFs used in other CPIs. The entire sequence of PRFs is known as a *schedule* and it must be transmitted within the beam dwell time. Within each CPI, the returns are range gated and Doppler (velocity) gated using fast Fourier transform (FFT) techniques. Target and clutter returns in both the range and velocity domains are highly ambiguous.

The operation on several PRFs is crucially important for three main reasons. First, it enables the radar to resolve the range and velocity ambiguities by comparing the ambiguous data in each PRF. Second, the variation in PRF results in a variation in the clutter distributed across the range and velocity dimensions such that range and velocity cells occupied by large clutter returns in one PRF may be occupied by much weaker clutter returns in another PRF. This improves the chances that small targets may be detected in at least an adequate number of PRFs. Third, each PRF has regions of blind ranges due to eclipsing and blind velocities due to the rejection of main beam clutter (MBC) and its repetition in the velocity domain. These blind regions are PRF dependent; therefore, range/velocity cells that are blind in one PRF may be visible in another PRF.

In several modern systems, the carrier frequency is also changed on each CPI (in spite of earlier fears [9]). Frequency agility brings about improvements in tracking due to the minimization of multi-pathing and glint errors as well as forcing the decorrelation of clutter and target responses. However, frequency agility on a CPI-by-CPI basis also introduces an added complication in that the Doppler shift is directly proportional to frequency as well as the radial velocity of the target (or clutter) return. This means that a particular Doppler shift measured at one frequency does not correspond to the same velocity when observed using a different carrier frequency. Consequently, it is often more meaningful to work in *velocity* rather than *Doppler*, since the velocity of a target is the parameter of interest and is independent of the carrier frequency of the radar used to measure it. In most cases the change in the carrier frequency is a small percentage of the nominal central value, so the change in Doppler is correspondingly small. Nevertheless, this can be appreciable when considering ambiguity resolution, clutter mapping, and the regions of blind velocities. While the introduction of differing carrier frequency in each CPI may add a new level of complexity, it offers the designer another dimension that may be exploited. The selection of both the carrier frequency allied to the PRF must now be borne in mind. No such complication exists in range since the translation of the time delay into range is not dependent on the carrier frequency.

11.1.3 Detection Criterion

Let the total number of PRFs used be N. Typically, N may be in the range of 5 to 9; a value of $N = 8$ has often been used. To resolve the range and velocity ambiguities, target data are required in a minimum number of PRFs. This allows a target to be declared if it is detected in this minimum number. Therefore, let the minimum number of PRFs in which target data are required be M. The bare minimum number of PRFs in which target data are required to resolve the ambiguities is, strictly, two ($M = 2$); however, for many practical reasons, data in a third PRF are required ($M = 3$). A target may be declared if it is detected in M of the total of N PRFs and is known as an *M-of-N* schedule. A *3-of-8* schedule is common [11, 12].

Medium PRF radars therefore operate a double detection mechanism. Within each CPI, data in each range/velocity cell are subject to a constant false alarm rate (CFAR) controlled threshold detection process. On completion of the N CPIs, all threshold detections are subject to an *M-of-N* binary integrator and targets are declared on the basis of being detected at a consistent range and velocity in at least M CPIs [13].

A target remains blind if the radar fails to detect it in more than $(N - M)$ PRFs. Failure to detect a target could be due to overwhelming clutter levels or range/velocity blindness as well as a poor probability of detection. Naturally, there is a greater chance of obtaining target data in the requisite M PRFs as N increases. This is because it becomes more likely that the range/velocity cell in which the target resides avoids a large clutter return or does not coincide with blind regions in a higher number of PRFs. However, it would not do to increase N indefinitely since the beam dwell time must be subdivided into N coherent processing intervals (with N space charging times at the start of each new CPI). A large N reduces the CPI duration, which then limits the scope for processing gain and restricts the velocity resolution. For this reason, medium PRF cannot offer the long-range detection performance or fine velocity resolution and accuracy of high PRF.

11.1.4 Why Use Medium PRF?

At first glance the complexity of medium PRF might cause one to wonder if it's worth all the bother. After all, low PRF modes offer a good ranging capability on account of their unambiguous range data, whereas high PRF modes offer easy velocity measurement due to their unambiguous velocity data. They would appear to have complementary strengths. However, for the airborne radar case, both low and high PRF waveforms have limitations largely determined by the clutter. Low PRF is not a viable Doppler mode because the Doppler spectrum is dominated by multiple ambiguous repetitions of MBC. High PRF has problems in detecting low Doppler shift targets due to multiple ambiguous repetitions of SLC in the time domain. It would appear that they have complementary weaknesses, as well [11]. For sure, if the likely target scenario (range, velocity, clutter scene) were known, the waveform could be optimized accordingly and the resulting waveform would likely be either a high or a low PRF mode, since each is optimal in certain specific circumstances.

Medium PRF retains some of the strengths and weaknesses of both low and high PRF; however, none of the weaknesses are too extreme. A midrange value of PRF avoids the worst of the problems associated with the low and high PRF regimes and, as a consequence, offers a decent level of overall performance. When target range and velocity data are required and the target scenario is not known—or is known to be highly varied (medium or long range,[1] short-range, high speed, slow moving, opening or closing targets)—*and* severe clutter is expected, then medium PRF emerges as a good all-around solution. Medium PRF is, in many respects, the optimal solution to a multi-objective problem.

This problem was originally posed by airborne fire control radars (FCRs) in fast strike aircraft, and medium PRF was first developed in response to their requirements. Since then, medium PRF modes have subsequently been employed in a wide variety of airborne and surface-based radars for a varied range of military applications [14]. The remainder of this book attempts to explain medium PRF in the context of the airborne FCR case but diverts to consider other applications from time to time.

11.2 AMBIGUITIES

11.2.1 Maximum Unambiguous Range and Velocity

For any given PRF there is a maximum unambiguous range given by

$$R_{mu} = \frac{c}{2f_r} \qquad \text{repeat of (3-6)}$$

and a maximum unambiguous velocity given by

$$V_{rmu} = \pm\frac{\lambda f_r}{4} \qquad \text{repeat of (9-5)}$$

[1] Medium PRF cannot achieve very long-range detection on account of the modest processing gains which are feasible.

FIGURE 11-1 Unambiguous Detection Space vs. Wavelength

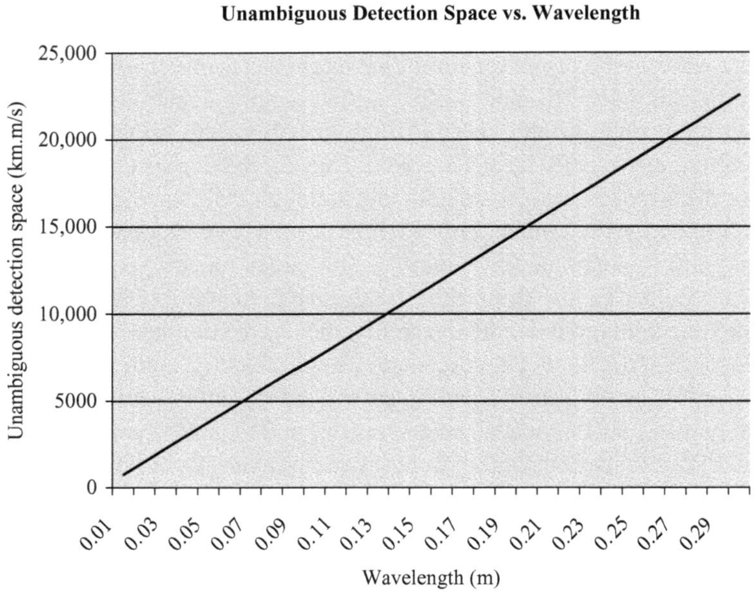

This gives rise to the idea of an unambiguous detection space in the range/velocity plane given by the product $R_{mu}V_{rmu}$.

$$R_{mu}V_{rmu} = \pm\frac{c}{2f_r} \cdot \frac{\lambda f_r}{4} = \pm\frac{c\lambda}{8} \tag{11-1a}$$

This defines an "area" in range/velocity space of

$$R_{mu}V_{rmu} = \frac{c\lambda}{4} \qquad (\mathrm{m^2s^{-1}}) \tag{11-1b}$$

The unambiguous range/velocity detection space is a function of wavelength but is independent of the PRF [14]. This detection space is graphed as a function of wavelength at Figure 11-1. The longer the wavelength (lower frequency), the greater the unambiguous detection space on account of the increase in its velocity dimension.

11.2.2 Range and Velocity Space of Interest

In general, the unambiguous detection space does not encompass the full spread of target ranges and velocities of interest. In other words, the instrumented range and velocity exceed the unambiguous range and velocity. The total detection space of interest is depicted as the light gray rectangle in Figure 11-2. Also shown are three rectangles of equal area (i.e., equal wavelength) representing the unambiguous detection spaces for a low PRF, high PRF, and medium PRF. The low PRF provides adequate range coverage but inadequate velocity coverage, whereas the high PRF provides adequate velocity coverage but inadequate range coverage. The medium PRF provides inadequate coverage in both the range and velocity dimensions. However, the medium PRF rectangle in each axis may be tiled several times to obtain coverage over the whole detection space of interest. This expresses the ambiguous repetition of target data in both the range and velocity coordinates. Since the area of the unambiguous detection space is independent of the PRF, so too is the number of repetitions of this area necessary to cover the detection space of interest [15].

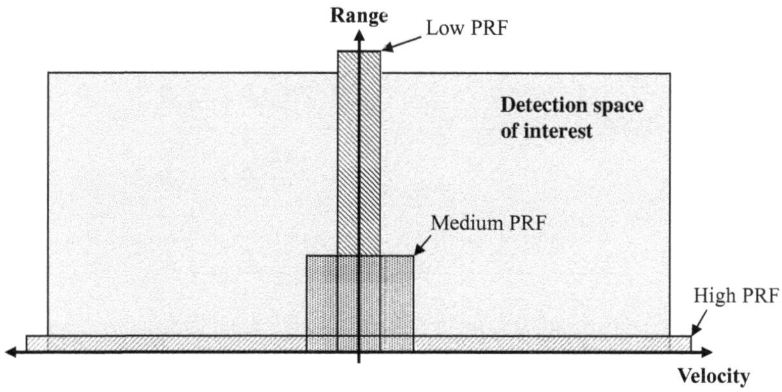

FIGURE 11-2 Detection Space of Interest

11.2.3 Ambiguity Diagram for Medium PRF

The ambiguity diagram for a medium PRF waveform is illustrated in Figure 11-3 for a PRF of 10 kHz and carrier frequency of 10 GHz ($\lambda = 0.03$ m). The response at zero time (range) and zero velocity is repeated at regular intervals in both coordinates. The repetition in time occurs in intervals of $1/f_r = 100\ \mu$s, and the repetition in velocity occurs at intervals of $\lambda f_r/2 = 150$ m/s. The use of a higher PRF would lead to a larger number of repetitions in time (range) coming at closer intervals and fewer repetitions in velocity at wider intervals. Furthermore, the rectangular coverage of the unambiguous detection space (Figure 11-2) would shrink in range and grow in velocity, and the rectangle would tend to that of the high PRF case as the PRF continues to increase. Conversely, the rectangle would distort toward that of the low PRF case as the PRF reduces.

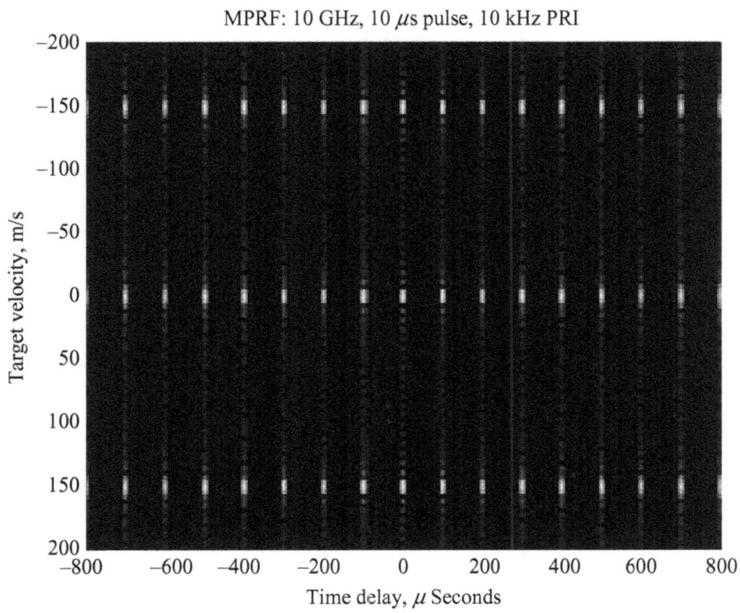

FIGURE 11-3 Ambiguity Diagram for Medium PRF Waveform

11.2.4 Range and Velocity Ambiguities

The ambiguous nature of medium PRF raises doubt over the true range and velocity of all detections. The range ambiguity problem is illustrated in Figure 11-4 which illustrates

FIGURE 11-4 Range Ambiguity

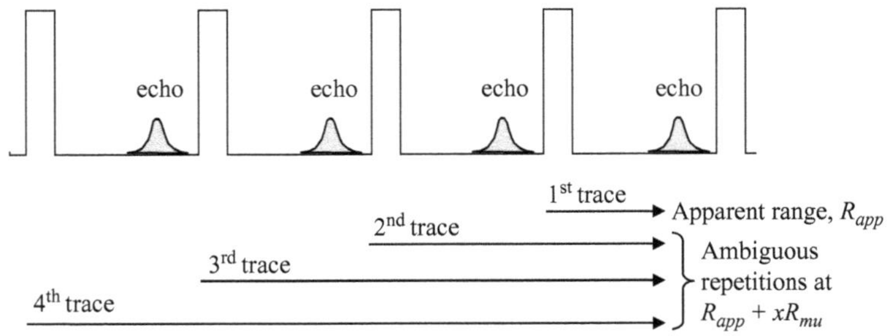

the pulsed modulation of the transmitted signal and the echo returns from a single target. The true range to the target could be its apparent range plus multiples of its maximum unambiguous range, that is,

$$R_{true} = R_{app} + x R_{mu} \tag{11-2}$$

where $x = 1, 2, 3, \ldots$ and represents the order of the range ambiguity.

The Doppler ambiguity problem is illustrated in Figure 11-5 which shows the central components of the transmitted spectrum plus a Doppler shifted return from a single target. The apparent Doppler shift is $+f_d$; however, the true Doppler shift could be its apparent value plus or minus multiples of the PRF, that is,

$$f_{d\_true} = f_d \pm y f_r \tag{11-3}$$

where $y = 1, 2, 3, \ldots$ and represents the order of the Doppler ambiguity.

The ambiguous repetitions in both range and Doppler can be projected into the range/velocity space of interest, as in Figure 11-6. This results in a lattice of possible

FIGURE 11-5 Doppler Ambiguity

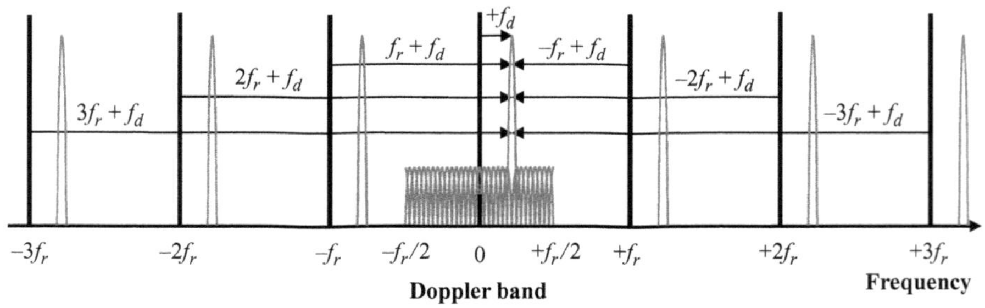

FIGURE 11-6 Lattice of Ambiguous Detection Points

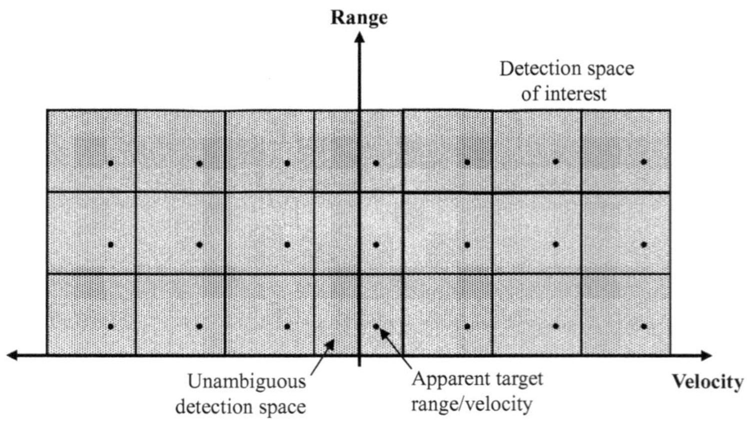

detection points, one of which represents the true range and velocity of the target and the remainder being its ambiguous repetitions. The lattice of ambiguous detection points replicates the multiple peaks of the ambiguity diagram, except that here the detection points are offset by the apparent range and velocity. The ambiguous repetitions of the target detection points all have the same magnitude, and the identity of the true range and velocity is not yet known.

11.3 | MEDIUM PRF RADAR RESPONSE TO SURFACE CLUTTER

As shown in Section 11.2.4, discrete targets are range and velocity ambiguous. Such ambiguities are not confined to discrete target returns but apply also to clutter. The response of a medium PRF system to surface clutter is considered for the case of an airborne radar in a look-down attitude since this presents the most general picture [16, 17, 18]. Ground-based systems can be considered as a specific case in which the platform's ground velocity is zero and the platform altitude is also zero.

11.3.1 Distribution of Clutter in the Doppler Domain

The spectrum of surface clutter, both MBC and SLC, is illustrated in Figure 11-7 and was considered in Section 6.4. SLC occupies a portion of the Doppler band from $-2V/\lambda$ to $+2V/\lambda$, where V is the platform's ground velocity [1]. Platform motion compensation (PMC) is applied to reference all returns to ground velocity by offsetting all Doppler shifts by the Doppler frequency of the main beam boresight. This has the effect of translating the clutter spectrum in frequency such that MBC is centered at zero Doppler. The bandwidth of SLC can be wide for high platform velocities; for example, for a platform velocity of 250 m/s and $\lambda = 0.03$ m ($f_{tx} = 10$ GHz), SLC extends over a bandwidth of 33.33 kHz.

The spectrum of MBC and SLC is repeated at intervals of the PRF; each spectral component of the transmitted spectrum results in a band of MBC/SLC, as shown in Figure 11-8. For a typical medium PRF of 10 kHz, the SLC bandwidth may exceed the PRF, so SLC is spread ambiguously over the Doppler domain. The spread in SLC is typically greater than the PRF, so SLC extends well beyond the Doppler band. Likewise, the repetitions of the clutter spectrum about neighboring PRF lines encroach within the Doppler band.

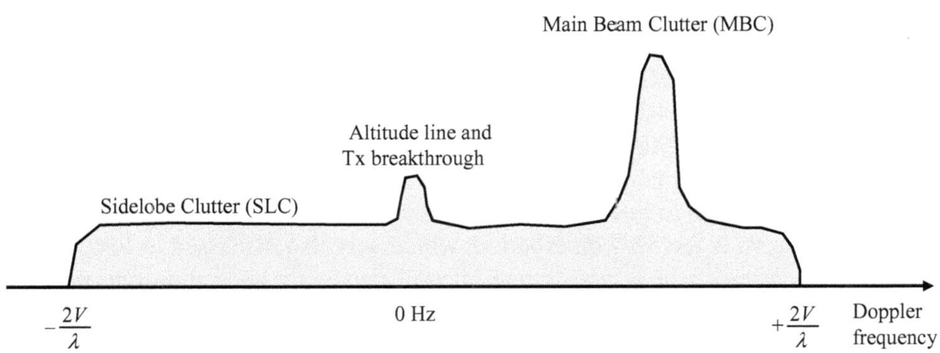

FIGURE 11-7 Surface Clutter from Airborne Radar in the Doppler (Velocity) Domain

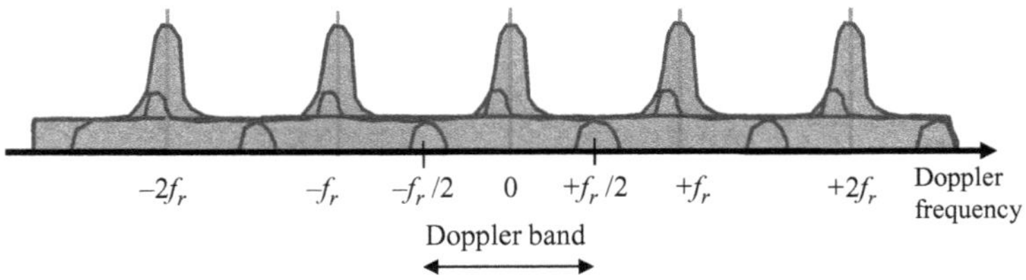

FIGURE 11-8 Repetition of Clutter in Doppler Domain for Medium PRF

MBC is centered at zero Doppler (after PMC has been applied) and occupies the central portion of the Doppler band. For medium PRF radar the portion of the Doppler band occupied by MBC is considerably greater than for high PRF. Moving target indicator (MTI) filtering may be applied to reject MBC and targets of very low ground velocities. The MTI filter rejects slow ground-moving targets (GMTs) and SLC and targets whose Doppler shift is equal to a multiple of the PRF (± the half-width of the MTI filter). Thus, some targets of interest could be rejected. There is considerably greater risk of rejecting targets of interest in medium PRF than for high PRF because in medium PRF (1) the rejection width is proportionally wider and (2) targets whose Doppler shift is close to a multiple of the PRF are also rejected (which never arises in high PRF). Recall that for low PRF, MBC would occupy most, if not all, the Doppler band, and its rejection would result in the rejection of most, if not all, targets. Even without any MTI clutter rejection, discrete targets are unlikely to be seen because of the overwhelming MBC.

In many scenarios, such as low altitude look-down, SLC rises above the receiver noise floor and becomes the factor that limits the detection of targets of interest. Targets now compete with SLC in all Doppler cells, and, unlike high PRF, no target velocities are clear of SLC. Target detection is determined by the signal to noise-plus-clutter ratio (SNCR) in the range/velocity cell occupied by the target. The situation worsens (compared with high PRF) because of multiple ambiguous repeats of SLC falling within the cell of interest. Section 9.4.1 showed that contours of equal Doppler map as hyperbolic loci onto flat terrain. For medium PRF, several such hyperbolic loci may now contribute SLC into a single velocity cell. This situation is depicted in Figure 11-9.

11.3.2 Distribution of Clutter in the Range Domain

As with high PRF, MBC is spread over a range interval corresponding to the main beam footprint on the ground. MBC starts at the slant range of the lower edge of the intersection of the main beam on the surface and ends at the slant range of the upper edge of the intersection of the main beam on the surface. This range interval may well exceed one maximum unambiguous range interval, R_{mu}, at shallow grazing angles, in which case MBC would occupy all range cells and would be ambiguously repeated across the receiving periods. However, there are fewer ambiguous repeats of MBC across range with medium PRF compared with high PRF due to the longer maximum unambiguous range associated with the lower values of PRF. At very large grazing angles the radar is looking almost vertically downward, and MBC may well be confined to just a few range cells. However, this scenario is unlikely in many pulse Doppler radar applications. Since it may be assumed that MTI filtering is used to reject MBC, it needs no further consideration.

Also of concern is the SLC distribution, which was also discussed in Section 6.4.2. SLC is always spread over a large range interval inevitably exceeding one maximum unambiguous range interval, R_{mu}, so ambiguous repetition across several receiving periods is most likely. The situation is depicted in Figure 11-10 (similar to Figure 9-17). Time

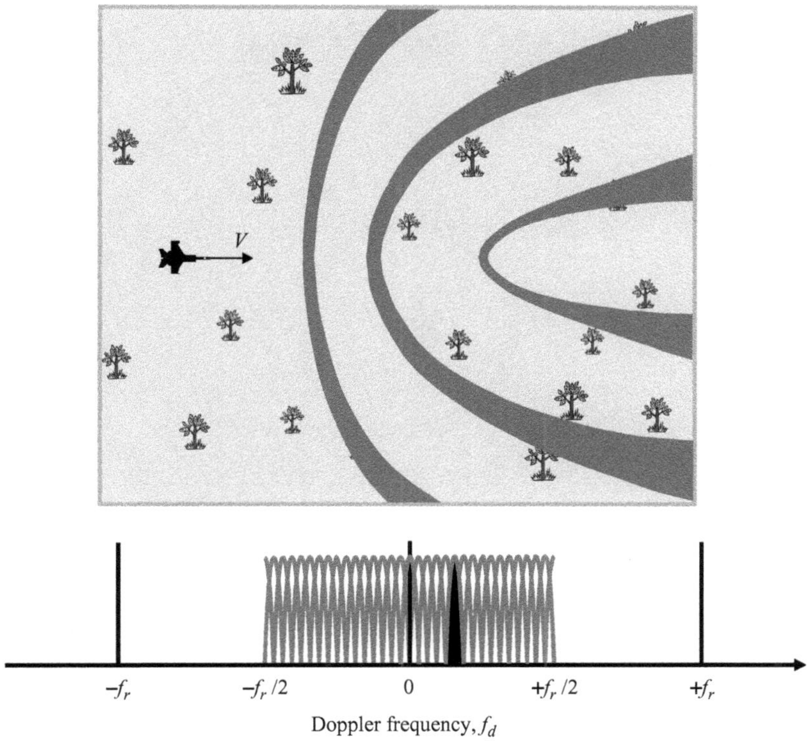

FIGURE 11-9 Regions of Ambiguous SLC Coupling to an Individual Doppler Cell

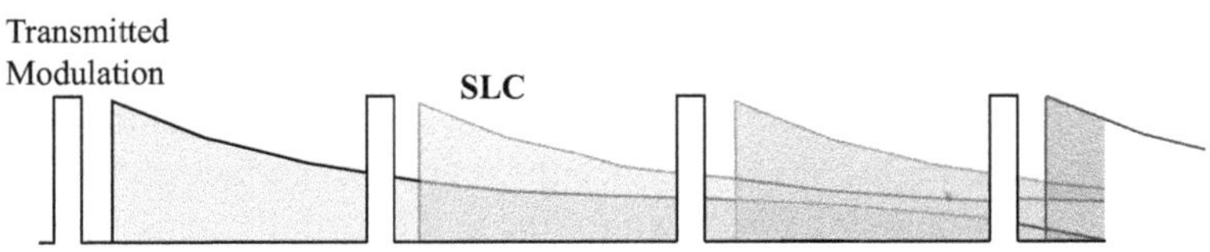

FIGURE 11-10 Medium PRF Response to SLC in Range Domain

gating the SLC out is not an option in medium PRF as it is in low PRF, because multiple ambiguous repetitions of SLC occupy all range cells. In any one receiving period there are multiple repeats of SLC; the higher the PRF, the greater the number of ambiguous repetitions and the worse the problem. SLC can be problematic in medium PRF, but the situation is not nearly as bad as it is for high PRF systems.

The superposition of multiple ambiguous repeats of SLC throughout the receiving period tends to smear out the clutter to a near-uniform level across all range cells. Each range cell therefore contains a contribution of first trace SLC, second trace SLC, and so on. Range gating does, however, divide the total SLC return up fairly uniformly across all range cells. The greater the number of range cells, the smaller the proportion of the total within each range cell. Targets, on the other hand, are largely confined to a single range cell, although range cell straddling may occur. Therefore, targets are competing only with the portion of SLC that falls within its range cell for detection. This portion of SLC arises from a contour on the surface of constant range. Contours of equal range correspond to circles centered on the radar location. The finite width of each range cell means that it accepts clutter from a ring-shaped region from the surface. The width of the ring is a function of the range cell width and grazing angle and is given by $\Delta R \sec\theta_g$,

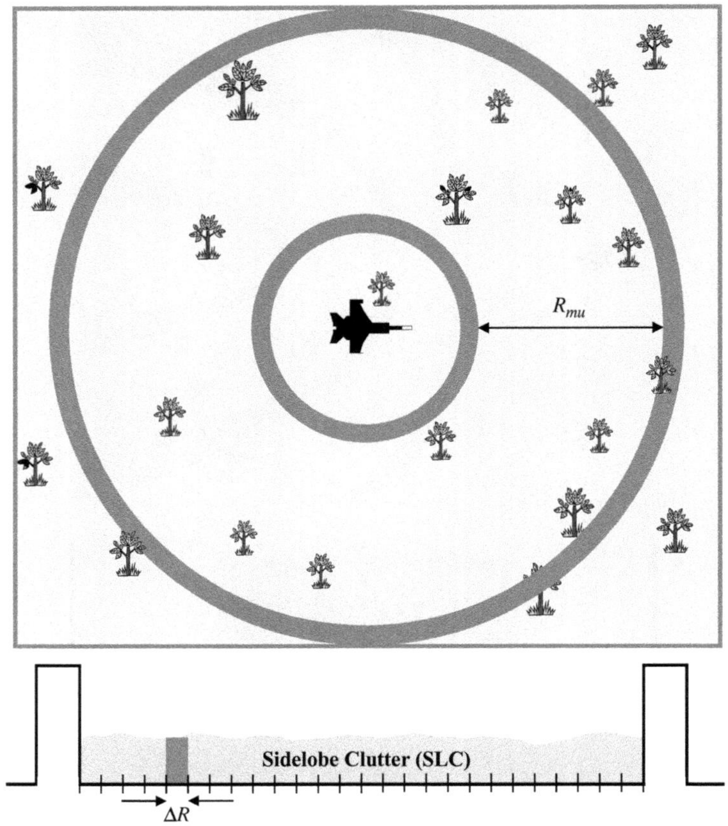

FIGURE 11-11 Regions of Ambiguous SLC Coupling to an Individual Range Cell

where ΔR is the range resolution, and θ_g is the grazing angle. The situation is illustrated in Figure 11-11 (similar to Figure 9-18) in which the regions of surface clutter coupling to one range cell (selected at random) are shown for the first two ambiguous repetitions. The rings are offset from each other by the maximum unambiguous range, R_{mu}, associated with the PRF. In medium PRF the problem is less severe than it is for high PRF since (1) typically more range cells of finer resolution tend to be used (longer PRI), which subdivides the SLC more finely, and (2) fewer ambiguous repeats are incurred so fewer, more widely separated, range rings contribute SLC to any individual range cell.

It is necessary to tightly control sidelobe levels in medium PRF to avoid coupling SLC from distant ranges into the radar. At lower platform altitudes it becomes increasingly more important to minimize sidelobe levels. A sidelobe level of −45 dB may typically be required at altitudes below 2000 meters, whereas sidelobes levels of −35 dB may be acceptable above 10,000 metres [11].

11.3.3 Distribution of Clutter in Range and Doppler

The previous sections described the spread of clutter in the Doppler and range dimensions which is typical for medium PRF systems. It is now necessary to combine this into a two-dimensional format illustrating the distribution of clutter in range/velocity space.

The clutter residing in one range/velocity cell arises from patches on the ground where the hyperbolic loci giving rise to clutter in the velocity cell of interest overlap with the range rings giving rise to clutter in the range cell of interest. Figure 11-12, formed by merging

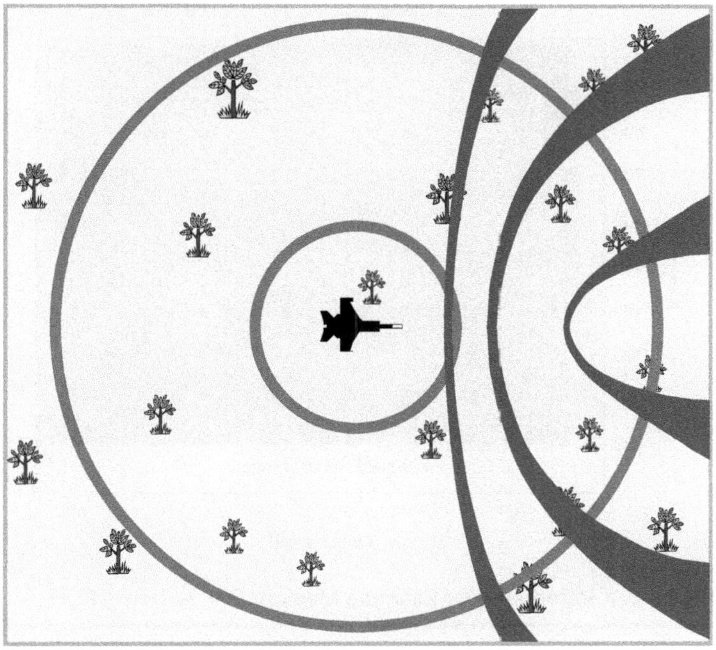

FIGURE 11-12 Regions Giving Rise to Surface Clutter in One Range/Velocity Cell

the plots of Figure 11-9 and Figure 11-11, shows how resolution in both the range and velocity dimensions reduces the area of clutter patches coupling to each range/velocity cell. Furthermore, the strength of the clutter received from each patch of overlap is scaled by the two-way gain of the antenna in the direction of each patch as well as being a function of its range and backscatter coefficient (which is, in turn, a function of grazing angle, terrain type, and polarization).

Section 6.4.3 described an *ideal clutter map,* which depicts the distribution of clutter in range/velocity space. An example ideal clutter map is given in Figure 6-17 for a typical airborne radar flying at an altitude of 5000 m and at a velocity of 250 m/s in a 10.8° look-down and 0° azimuth look-ahead attitude. The ideal clutter map simply shows the distribution of the clutter across the range and velocity space of interest and has not been velocity and range gated in accordance with the processing parameters of the radar. From these data, a *folded clutter map* may be derived that presents the same clutter data but "folded" into a region of range/velocity space of one unambiguous range interval by one unambiguous velocity interval. For medium PRF systems the total range/velocity space of interest is comprised of multiple repetitions of the unambiguous range/velocity interval.

Figure 11-13 presents an example of a folded clutter map for a medium PRF waveform which is representative of an airborne FCR operating on a PRF of 10 kHz and employing 64-point FFT processing. The scenario and radar parameters are very similar to those of Section 9.4.3, except for the changes to the pulsed parameters to a medium PRF mode. Parameters of the FCR are summarized in Table 11-1. Clutter is plotted over the 64 velocity cells, arranged across the Doppler band from 0 Hertz up to the PRF (10 kHz in this case), arranged as in Figure 9-7; each Doppler cell is 156.25 Hz wide (corresponding to a velocity resolution of 2.34 m/s). The number of range cells in one unambiguous range interval is the maximum unambiguous range divided by the range resolution i.e., $15000/15 = 1000$.

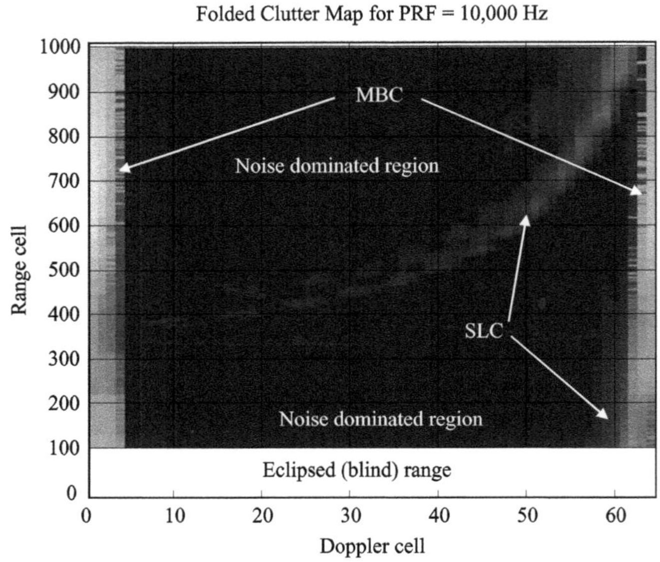

FIGURE 11-13 Folded Clutter Map for a Generic Medium PRF Airborne FCR Case

TABLE 11-1 Parameters of Example Medium PRF Airborne FCR

Parameter	Value
Carrier Frequency, f_{tx}	10 GHz
PRF, f_r	10 kHz
Max. unambiguous range, R_{mu}	15 km
Duty Ratio, DR	0.1 (10%)
Transmitted Pulse Width, τ	10 μs
Range Resolution, ΔR	15 m
FFT Size, n	64
Velocity Resolution, ΔV_r	2.34 m/s
FFT Processing Loss	3 dB
CPI Time, t_{int}	6.4 ms
Antenna Beamwidth (both planes), θ_B	3.5°
Antenna Sidelobe Level	−17.5 dB
Azimuth Scan Angle, θ_{az}	0°
Elevation Scan Angle, θ_{el}	−10.8°
Platform Altitude	5000 m
Platform Velocity, V	250 m/s
Receiver Noise Figure, F	5 dB
GMT Velocity	Mean = 0, σ_n = 1.5 m/s (Gaussian)

The folded clutter map therefore contains $64 \times 1000 = 64\,000$ range/velocity cells, each with dimensions of 15 m × 2.34 m/s.

Figure 11-13 shows the distribution of clutter returns within the whole of this interval. Platform motion compensation was conducted, as is evident from the strong MBC return centered at velocity cell 1 and its first ambiguous repetition in velocity cell 64. Some SLC breaks through the receiver noise, leading to some clutter-dominated and some noise-dominated regions. SLC commences in range cell 333, which corresponds to a range of 5000 m (i.e., the platform altitude). Its Doppler shift increases as its range increases since this corresponds to SLC at shallower grazing angles and closer to the velocity vector of the platform. Second trace SLC re-enters the lower numbered range cells of the folded clutter map close to the first repetition of MBC in the higher numbered velocity cells. In a different scenario (e.g., lower altitude or higher clutter backscatter), SLC would be expected to dominate over a larger portion of the range/velocity cells. The distribution of SLC within the folded clutter map would also vary with PRF.

The minimum range associated with the 10 μs transmitted pulse width, $R_{min} =$ 1500 m, so the number of eclipsed range cells is $1500/15 = 100$. The radar model assumes one additional blind range cell due to the transmit to receive switching time and eclipsing losses. Hence, the first 101 range cells are blind and are plotted as white in Figure 11-13. Blindness therefore exists across 101 of the 1,000 range cells or ~10% of the detection space, in keeping with the 10% duty ratio. MBC is dominant in the first four and last four velocity cells, that is, over 12.5% of the detection space.

The folded clutter map illustrates the clutter within one unambiguous range and velocity interval. However, the PRF of 10 kHz results in an unambiguous range of 15 km and an unambiguous velocity space coverage of ±75 m/s. If a range of interest of 105 km is assumed, then 7 ambiguous repetitions would be incurred over this range, requiring 7000 range cells, each of 15 m resolution. If a velocity space of interest of 1500 m/s (±750 m/s) also is assumed, then 10 ambiguous repetitions would be incurred over this velocity space, requiring 640 velocity cells, each of 2.34 m/s resolution. The distribution of clutter across the full range/velocity space of interest can be depicted on an *unfolded clutter map* by ambiguously repeating the folded clutter map as many times as necessary to obtain the necessary coverage. In the previous example, this requires 7 repetitions of the folded clutter map in the range dimension and 10 repetitions in the velocity dimension. An example of an unfolded clutter map for a medium PRF radar is given in Figure 11-14, which pertains to the same radar model defined in Table 11-1. It is formed by tiling 7 copies of the plot in Figure 11-13 adjoining each other in the range dimension and then 10 copies of that replicated over the velocity dimension. MBC and SLC still occupy the corresponding range/velocity cells as in Figure 11-13, but now there are 7 repetitions of the eclipsed ranges (horizontal white strips) and 10 repetitions of MBC.

The total detection space of interest comprises 7,000 range cells, each of 15 meter resolution, spanning a total range of 105 km by 640 velocity cells, each of 2.34 m/s

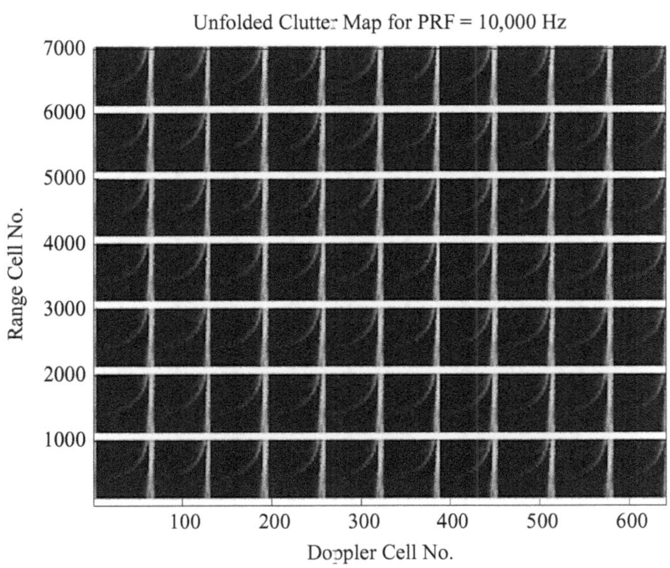

FIGURE 11-14 Unfolded Clutter Map for a Generic Medium PRF Airborne FCR Case

resolution, spanning a velocity bracket of −750 m/s to +750 m/s. This gives a total of 4,480,000 range/velocity cells. Eclipsing renders 10% of the total detection space blind, and about 12.5% of the detection space is dominated by MBC.

The most significant difference for surface-based medium PRF radars is the simplified clutter situation. The situation for surface-based medium PRF radars is much the same as that for high PRF systems previously described in Section 9.5.

11.4 BLINDNESS IN MEDIUM PRF RADAR

The unfolded clutter map of Figure 11-14 can be used to deduce regions in which the radar would be blind. Blindness arises from two main reasons: (1) due to eclipsing losses; and (2) due to the rejection of MBC (or levels of MBC that overwhelm discrete targets). Figure 11-15 shows the detection space for the medium PRF radar described in Section 11.3 in which blind range/velocity cells are shown in black. Approximately 21.25% of the detection space is blind in this example, so for random target placement there is a 21.25% probability that the radar remains blind to it.

A third phenomenon, overwhelming SLC, can also contribute to missed targets. The returning echo strength from a target and from SLC is shown in Figure 11-16 as a function of range. The profile of the SLC repeats at intervals of R_{mu} due to the ambiguous repetition in range. However, the target echo has a power that reduces as $1/R^4$ in keeping with the radar range equation with the general level of the target echo profile being set by its radar cross section (RCS). As range increases, there comes a point when the SLC is more powerful than the target echo. Initially, this occurs for a narrow range interval. However, this repeats with ever broadening range intervals. The SCR varies with range and gives rise to ever broader bands of poor SCR with increasing range. The problem is worse for lower RCS targets. Strictly, the radar is not blind to such targets since the target echo is still received; however, the SCR is too low to declare a detection.

When this effect is considered, the blind zone map can be modified for a fixed RCS target to include the bands of "blind" ranges due to overwhelming SLC, as in Figure 11-17.

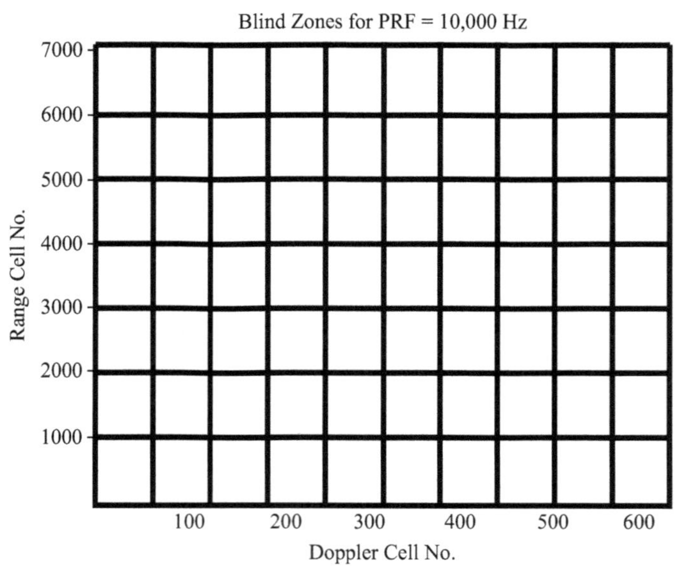

FIGURE 11-15 Blind Zone Map for PRF of 10 kHz

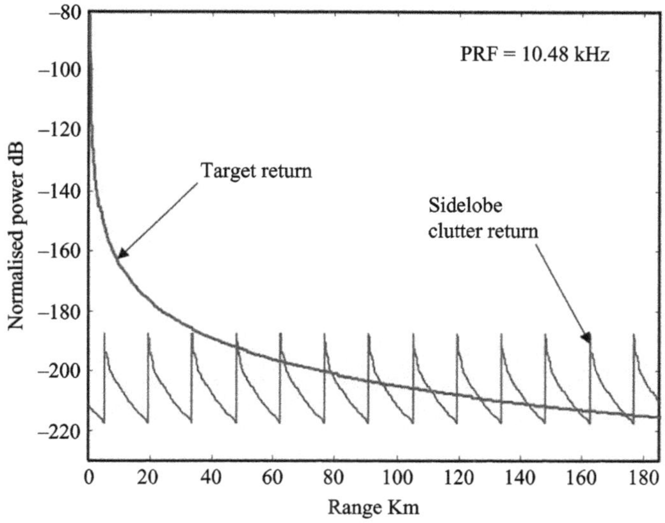

FIGURE 11-16 SLC and Target Return Power versus Range

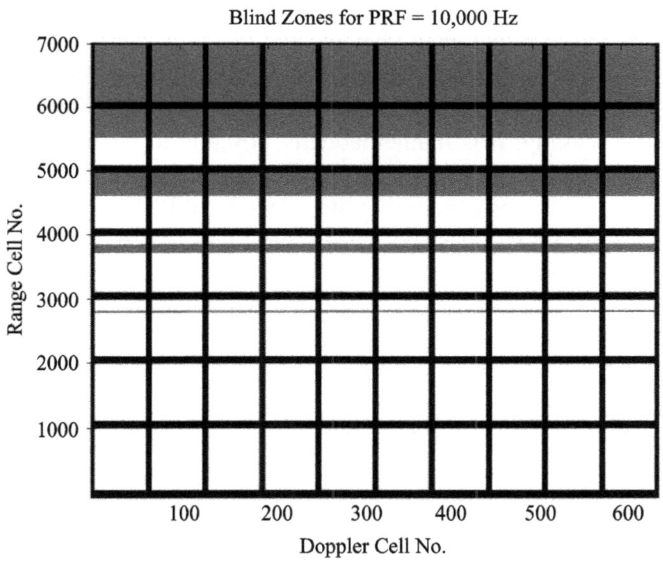

FIGURE 11-17 Blind Zone Map with SLC (Fixed Target RCS)

A similar, though subtly different, blind zone map would result from operation on a different PRF. The exact locations of all the blind regions are PRF dependent, except for the first eclipsed range along the bottom of the map and the first case of MBC rejection up the left-hand edge of the map, which remain fixed in all PRFs. However, in any single PRF there is a significant probability that a target goes undetected, because of either the blindness inherent in the PRF or overwhelming SLC.

11.5 | MULTIPLE PRF SCHEDULES

Medium PRF modes require the radar to cycle its operation through several CPIs, each on a different PRF. In the context of medium PRF it is common to refer to the pulse repetition intervals (PRIs) rather than the PRF. All CPIs must be transmitted and all returns processed within the beam dwell time. This must also accommodate the transmission of space charging pulses at the beginning of each new CPI (see Section 10.1.4). This places a limit on the duration of each CPI, and hence the velocity resolution, which in turn is controlled by setting an upper limit on the mean PRI and FFT size. FFT sizes of 16, 32, or 64 are typical of currently fielded equipment. Thus, the processing gains are far more modest than high PRF waveforms, and, consequently, medium PRF struggles to attain the detection ranges of high PRF modes.

Cycling the operation over several PRFs is done for two main reasons. First, target data received in each PRF are ambiguous in range and velocity. The true range and velocity can be decoded by comparing the ambiguous target data in each PRF, provided that the PRFs adhere to certain requirements. The decodability requirements are set by the target extraction algorithm, which performs the range and velocity decoding, and, to some extent, on the nature of the expected targets. Most modern implementations resolve both range and velocity ambiguities, but some, often older, systems resolve only the target range; velocity may then be resolved through recourse to the tracker files. Second, some targets may fall into a range/velocity cell that is blind in some PRFs but visible in others. Furthermore, a target may fall in cells occupied by high levels of SLC in some PRFs but much lower levels of SLC in other PRFs. The greater the number of PRFs used, then, the greater the probability that any given target would be detected in a larger number of PRFs.

Let the total number of PRFs (CPIs) used be N, and let the number of PRFs in which target data are required to decode the true range/velocity be M. This defines an *M-of-N* schedule. Strictly speaking, it is mathematically possible to decode the true range and velocity of targets on the basis of ambiguous data received in just two PRFs. However, in many practical cases, issues concerning the selection of PRFs and robust target extraction algorithms require target data in at least three PRFs ($M = 3$). Since blindness and overwhelming SLC can occupy as much as 50% of the range/velocity detection space of interest, ensuring data in three PRFs then requires that the schedule includes a total of six PRFs since, on balance, half of these fail to detect the target. However, the lower limit is to have target data in three PRFs; should some target scintillation reduce its RCS during one of these PRFs the target may go undetected, in which case data are available only in two PRFs. Target range and velocity now cannot be decoded, and the target extraction algorithm does not declare a target detection. Target detection would be far more robust if there were detection in four PRFs [2]. This therefore requires a schedule of eight PRFs ($N = 8$). Typically, $N = 8$ and $M = 3$, giving rise to a 3-*of*-8 schedule. Increasing the schedule to nine PRFs ($N = 9$) [19] may seem an attractive option, particularly in conditions of high SLC, since there is an even greater likelihood of obtaining target data in the requisite three. However, the beam dwell time constraint may well push the mean PRI limit to a value that is too low to be practical or limit the FFT size to such an extent that there is insufficient processing gain to achieve the required detection performance.

It would be far better to design shorter schedules (fewer PRFs, lower N) by carefully selecting PRFs that permit robust target extraction while also minimizing blindness even in the presence of strong clutter. These are the criteria by which the quality of a medium PRF schedule may be judged. It must be stressed that the quality of any schedule depends on the number and precise values of PRFs used; no individual PRF can be considered good or bad. Rather, the combination of PRFs coupled with the target extraction algorithm dictate the quality of the schedule. These issues are considered in greater depth in the ensuing chapters of this book.

CHAPTER 12

Factors Affecting the Choice of PRFs

Chapter Outline

Many factors [2, 11] influence the choice of exact values of pulse repetition frequencies (PRFs), the total number of PRFs used within a medium PRF schedule, N, and the minimum number of PRFs in which target data are required, M. Some factors are more important than others, and in some cases PRFs have been selected on the basis of one or two of the most important factors with little or no regard to factors of lower priority. It is fair to say that the design of the PRF schedule (selection of precise values of PRFs and the numbers N and M) is fundamental to many aspects of the performance of the radar. It is worth noting that the performance of a radar is dictated by the *combination* of the PRFs used; no individual PRF can be regarded as good or bad in its own right. The quality of any medium PRF schedule is an attribute of how all the PRFs work together. Some key factors influencing schedule design are discussed in the following sections of this chapter. These factors tend to place constraints and limitations that guide the selection of PRF values; they do not dictate precise values to be selected. Indeed, the selection of precise values of PRF is an area shrouded in mystery and is commonly perceived as something of a *black art*. Recent work, however, has brought solid engineering principles to bear on this problem.

12.1 DECODABILITY

12.1.1 Introduction

A *medium PRF* is defined as one which results in ambiguous range and Doppler (velocity) data. As presented in Section 11.2.4, the range ambiguity can be expressed as

$$R_{true} = R_{app} + x R_{mu} \qquad \text{repeat of (11-2)}$$

where $x = 1, 2, 3, \ldots$ and represents the order of the range ambiguity, R_{true} is the true range, R_{mu} is the maximum unambiguous range of the PRF, and R_{app} is the apparent range as seen in the first range interval. Likewise, the Doppler ambiguity can be expressed as

$$f_{d\_true} = f_d \pm y f_r \qquad \text{repeat of (11-3)}$$

where $y = 1, 2, 3, \ldots$ and represents the order of the Doppler ambiguity, f_{d_true} is the true Doppler shift, f_r is the PRF (also the Doppler bandwidth), and f_d is the apparent Doppler shift as seen in the Doppler band.

When the target is observed using several PRFs a set of these equations is obtained:

$$R_{true} = R_{app1} + x_1 R_{mu1} = R_{app2} + x_2 R_{mu2} = R_{app3} + x_3 R_{mu3} = \ldots \tag{12-1}$$

and for Doppler

$$f_{d\_true} = f_{d1} \pm y_1 f_{r1} = f_{d2} \pm y_2 f_{r2} = f_{d3} \pm y_3 f_{r3} = \ldots \tag{12-2}$$

where the subscript 1, 2, 3 ... is an index referring to each PRF.

This indicates that the true range and Doppler are independent of the PRF in which the target is seen but the apparent range and Doppler shift, the order of the range, and Doppler ambiguities and the maximum unambiguous range and Doppler vary with each PRF.

Solving equations (12-1) and (12-2) to obtain the true range and Doppler (velocity) is known as *decoding* the range and velocity. The ability to decode the true range and Doppler over the full range/Doppler space of interest imposes constraints on the combinations of exact PRF values chosen, known as *decodability constraints.*

In decoding range, the apparent ranges, R_{app1}, R_{app2}, $R_{app3} \ldots$ are measured in each PRF. Furthermore, the maximum unambiguous ranges, R_{mu1}, R_{mu2}, $R_{mu3} \ldots$ are functions of the PRF and are easily computed by the system. This leaves the only unknowns as the order of ambiguity in each PRF: $x_1, x_2, x_3 \ldots$ and the true range. Likewise, for Doppler, the apparent Doppler shifts are measured in each PRF, and the PRFs are known to the system, leaving the only unknowns as the order of the Doppler ambiguity in each PRF: $y_1, y_2, y_3 \ldots$ and the true Doppler. For each coordinate (range, Doppler) this always leaves one more unknown than the number of equations. Nevertheless, this apparent impasse can be solved by imposing constraints on the PRFs and by making some assumptions. Strictly speaking, it is mathematically possible to decode the true range and Doppler using only two PRFs by selecting values of the order of ambiguities (x_1 and x_2 for range and y_1 and y_2 for Doppler) using trial and error until a solution is found. However, it is more common to require data in a minimum of three PRFs ($M = 3$) for robust decoding whilst meeting the decodability constraints on PRF selection.

12.1.2 Decodability Constraints

There are two decodability constraints: one applicable to range, and the other to Doppler. These constraints can be summarized as

$$LCM(PRI_1, \quad PRI_2, \quad \ldots \quad PRI_M) \geq \frac{2R_{max}}{c} \tag{12-3}$$

$$LCM(PRF_1, \quad PRF_2, \quad \ldots \quad PRF_M) \geq f_{D_{max}} \tag{12-4}$$

The first of these can be expressed as "the lowest common multiple (LCM) of the M pulse repetition intervals in which the target has been detected must be greater or equal to the time delay to the maximum range of interest." Similarly, the second may be expressed as "the lowest common multiple of the M pulse repetition frequencies in which the target has been detected must be greater or equal to the maximum Doppler of interest." It is worth stressing that these decodability constraints apply to all combinations of M PRFs/pulse repetition intervals (PRIs) from the total number of N used in the schedule since the system cannot anticipate in which combination of three PRFs target data may be available. The number of combinations of *M-of-N* is given by

$$_N C_M = \binom{N}{M} = \frac{N!}{M!\,(N-M)!} \tag{12-5}$$

For $N = 8$ and $M = 3$, this gives $_8C_3 = 56$ combinations. The decodability constraints of equations (12-3) and (12-4) must apply to all these combinations. Finding PRF values for which all combinations meet these constraints can be awkward and can restrict compatible combinations of PRFs. If a ninth PRF is introduced, $_9C_3 = 84$, and the difficulty of PRF selection deepens. Increasing N compounds the difficulty of decodability.

12.1.3 Decodable Space

There are $_N C_M$ values for the LCM $(PRI_1, PRI_2, \ldots PRI_M)$, all of which must exceed the time delay of the maximum range of interest. The minimum value of all these combinations defines a time, and hence a range, known as the *decodable range*. Similarly, the minimum value of all $_N C_M$ values of LCM $(PRF_1, PRF_2, \ldots PRF_M)$ defines a Doppler frequency, and hence a velocity, which is the *decodable velocity*. The product of these two quantities defines the *decodable space*, which is an area in range/velocity space having units of m · m/s. The PRF schedule facilitates the true range and velocity of targets confined to the decodable space to be determined, whereas targets whose true range and/or velocity places them outside the decodable space give rise to ambiguous responses within the decodable space that would be declared. It is therefore important to ensure that the decodable space envelopes all possible target ranges and velocities since, otherwise, the range and velocity ambiguities would not be fully resolved.

12.1.4 Decodability Using Two and Three PRFs—Decodability Margins

Figure 12-1 shows a timing diagram for two PRFs with maximum unambiguous ranges, R_{mu1} and R_{mu2}. The two pulse timings are initially synchronized, but the position of the pulses of the two PRFs becomes increasingly staggered as time increases. Eventually, the pulses merge back toward synchronism, and the ninth pulse of PRF1 falls into time coincidence with the eighth pulse of PRF2. This results in a new unambiguous range for this PRF pair of $R_{mu1,2} = 8R_{mu1} = 7R_{mu2}$. If this combined unambiguous range exceeds the maximum range of interest, then this PRF pair ensures range decodability over the ranges of interest.

To maximize the unambiguous range for this PRF pair, care must be taken to select suitable values of PRF. In most radars, PRF timings are derived from a highly stable clock such as a quartz crystal oscillator. The PRIs must therefore be a multiple of the clock period. For example, if an oscillator of 10 MHz is used, then all PRIs must be a

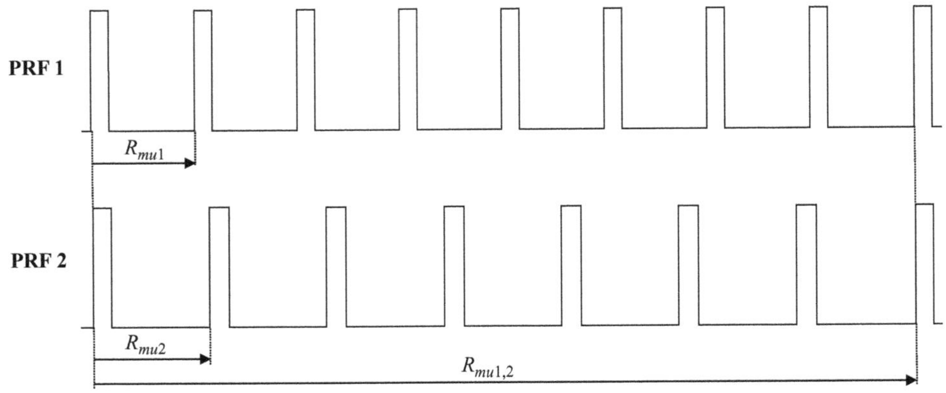

FIGURE 12-1 Timing Diagram for Two PRFs

multiple of 100 ns (and all PRFs have a common multiple of 10 MHz). This means that the PRFs are subharmonically related. If PRFs too close to each other are selected, the pulse stagger would be too small over the first few pulses and again as the pulses merge back toward synchronism. This is seen as the near coincidence of the second pulses and the penultimate pulses in both PRFs in Figure 12-1. The near coincidence of these pulses means that range decoding is not robust at these ranges. On the other hand, if the PRFs differ by too much, the PRIs tend to realign quite quickly, and therefore the unambiguous range of the PRF pair may not exceed the maximum range of interest. It could be very awkward to select PRFs that result in robust range decodability over the full range of interest on the basis of requiring target data in just two PRFs. However, if PRFs can be found, fewer decodability constraints must be met since ${}_8C_2 = 28$. Reducing M may go some way toward offsetting the difficulty in finding decodable PRFs. The problems of finding decodable PRFs can be alleviated (but not entirely dismissed) by introducing a third PRF, as illustrated in Figure 12-2. The first coincidence between all three now occurs

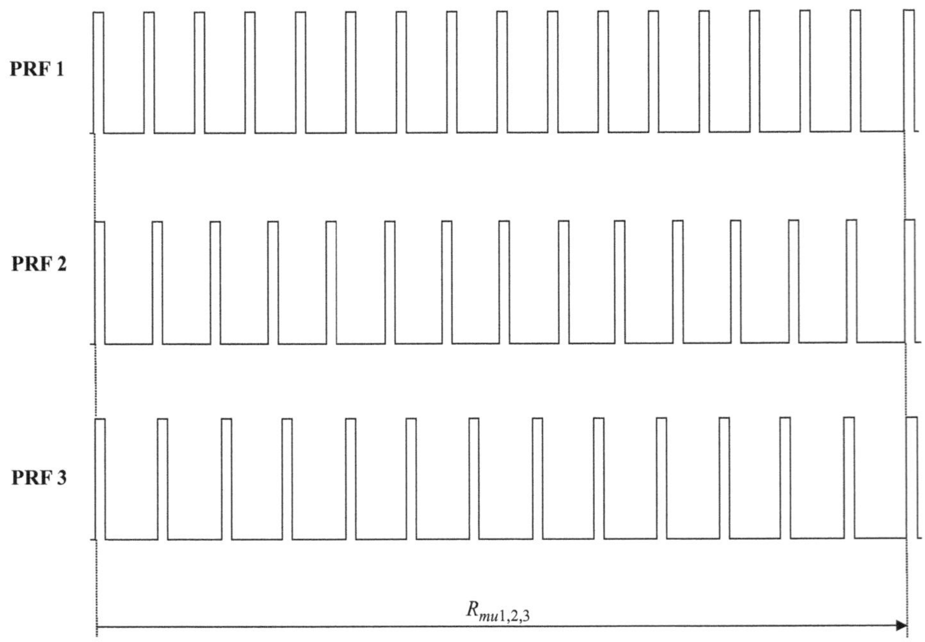

FIGURE 12-2 Timing Diagram for Three PRFs

on the 17th pulse of PRF1, the 15th pulse of PRF2, and the 14th pulse of PRF3. The new unambiguous range of this PRF triplet is $R_{mu1,2,3} = 16R_{mu1} = 14R_{mu2} = 13R_{mu3}$. It is now very much more likely that this would exceed the maximum range of interest. The addition of a third PRF means that a far greater decodable range can be achieved. Furthermore, while there may be near coincidence between pulses of PRF pairs, there are no cases of near coincidence between all three until $R_{mu1,2,3}$. Hence, the requirement for data in three PRFs also results in an improvement in the robustness of range decodability.

For an M-of-N schedule, range decodability is ensured if multiples of any M PRIs do not align to within one range cell of each other over the range of interest. In reality, target responses may be range extended and may straddle neighboring range cells, in which case it becomes necessary to modify the previous statement as follows:

> *For an M-of-N schedule, range decodability is ensured if multiples of any M PRIs do not align to within one range cell* **plus a suitable extra margin** *of each other over the range of interest.*

The "suitable extra margin" is known as the *range decodability margin*. Increasing the range decodability margin improves the robustness of the range decoding but constrains the selection of PRF values more tightly, thereby making it more difficult to find decodable PRF combinations. When it comes to PRF selection, good solutions are the most difficult to find.

Much the same argument can be developed for Doppler decodability as for range, except here one works in the frequency domain. Figure 12-3 shows the central part of the spectra of two pulsed signals of differing PRFs having a common central carrier frequency, f_{TX}. The harmonics of the PRFs lines become increasingly staggered from each other as the offset frequency from f_{TX} increases, but then they merge back into synchronism. It can be seen that the sixth harmonic of PRF1 coincides with the fifth harmonic of PRF2. The unambiguous Doppler bandwidth of this PRF pair is thus $6f_{r1} = 5f_{r2}$. If this value exceeds the maximum Doppler shift expected, then this PRF pair offers Doppler decodability.

Again, careful selection of PRFs derived from a common clock can ensure Doppler decodability over the band of expected Doppler shifts. The problem of near coincidence between certain harmonics of the two PRFs raises doubts over the robustness of the Doppler decodability at certain velocities. Furthermore, the unambiguous Doppler band for this PRF pair may be insufficient.

Once again, the problems of finding PRFs that offer Doppler decodability over a wide spread of Doppler shifts and the robustness of Doppler decodability can be alleviated (but not fully dismissed) by introducing a third PRF, as in Figure 12-4. The first coincidence between all three now occurs on the 12th harmonic of PRF1, the 11th harmonic of PRF2,

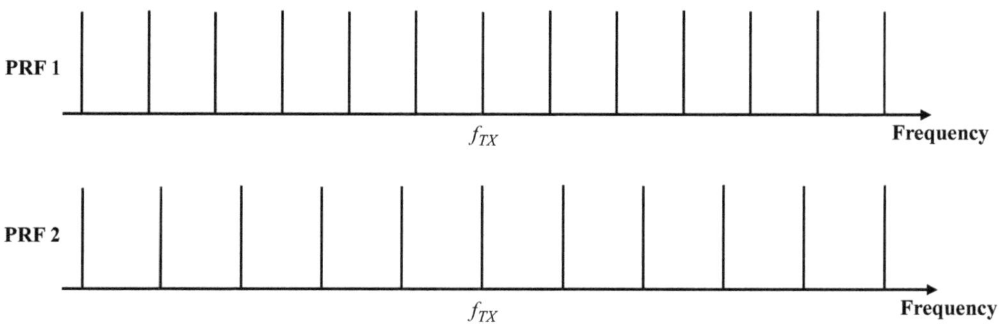

FIGURE 12-3 Spectrum for Two PRFs

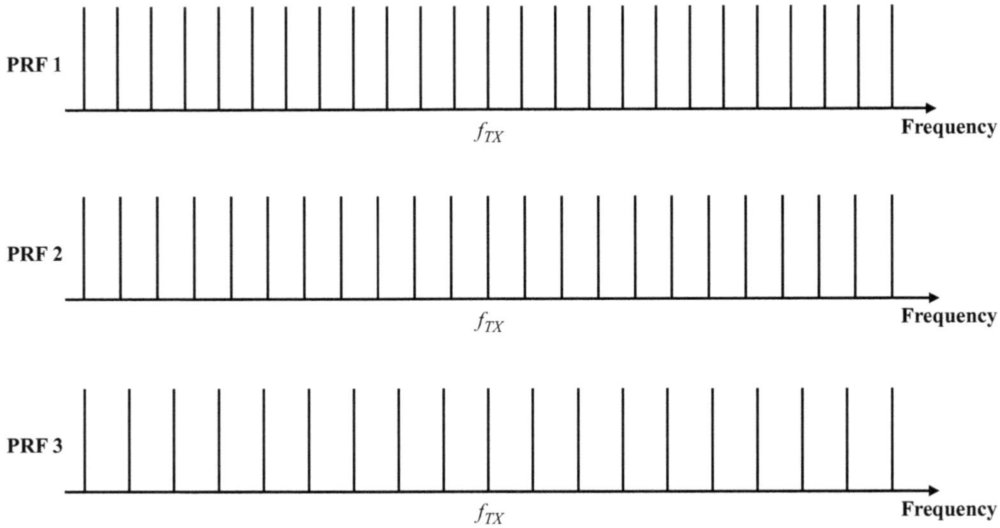

FIGURE 12-4 Spectrum for Three PRFs

and the 9th harmonic of PRF3. The new unambiguous Doppler band of this PRF triplet is: $12f_{r1} = 11f_{r2} = 9f_{r3}$. It is now very much more likely that this would exceed the maximum Doppler of interest. The addition of a third PRF means that a far wider decodable Doppler band can be achieved. Furthermore, although there may be near coincidence between harmonics of PRF pairs, there are no cases of near coincidence between harmonics of all three. Hence, the requirement for data in three PRFs also results in an improvement in the robustness of Doppler decodability.

For an M-of-N schedule, Doppler decodability is ensured if multiples of any M PRFs do not align to within one Doppler cell of each other over the band of interest. In reality, target responses may be extended in Doppler and may straddle neighboring Doppler cells, in which case it becomes necessary to modify the previous statement as follows:

> *For an M-of-N schedule, Doppler decodability is ensured if multiples of any M PRFs do not align to within one Doppler cell* ***plus a suitable extra margin*** *of each other over the band of interest.*

The "suitable extra margin" is known as *Doppler (velocity) decodability margin*. Increasing the Doppler decodability margin improves the robustness of the velocity decoding but constrains the selection of PRF values more tightly, thereby making it more difficult to find decodable PRF combinations. In fact, it is particularly difficult to find good solutions in both range and velocity.

12.1.5 The Coincidence Algorithm

The *coincidence algorithm* (also known as the *unfolding algorithm*) is a highly popular method of decoding range and velocity ambiguities [1, 20, 21]. For each PRF in which a target is detected, all possible ambiguous ranges and velocities are computed out to the maximum range and velocity of interest. As was seen in Section 11.2.4 (Figure 11-6) this results in a regular lattice of possible detection points over the range/velocity space of interest.

For many systems, the *PRIs* are fixed to an integer multiple of range cells, in which case it is convenient to refer to the range cell number in which a target is seen. The coincidence algorithm does not actually require the PRI to be an integer number of range cells; however, it is convenient to describe the algorithm in these terms for now. In fact, the PRI may be quantized into smaller time/range intervals, such as the period of the fundamental clock controlling all system timings, in which case one may refer to the index of the actual time/range intervals used in place of range resolution cells.

The general expression for the ambiguous target range cells is

$$A_{ux} = A_a + xm \quad x = 0, 1, 2, 3 \ldots \tag{12-6}$$

where x is the order of range ambiguity, m is the number of range cells in the PRI, A_a is the apparent range cell in which a target is detected, and A_{ux} is the xth potential ambiguous range cell number.

For example, if a target is detected in the 47th range cell of a PRI of 70 range cells, the 12th range cell of a PRI of 85 range cells and the 25th range cell of a PRI of 97 range cells, then the sets of possible unambiguous ranges, in units of range cells, are

$\{47, 117, 187, 257, 327, 397, 467, 537, \underline{607}, 677, \ldots\ldots, 1237\}$.

$\{12, 97, 182, 267, 352, 437, 522, \underline{607}, 692, \ldots\ldots, 1202\}$

$\{25, 122, 219, 316, 413, 510, \underline{607}, 704, \ldots\ldots, 1189\}$

The sets of potential unambiguous ranges are scanned for any common ranges. If a potential unambiguous range occurs in each of M sets, corresponding to M-of-NPRFs, then a target is reported at that range. In the previous example, a target would be reported to be at range cell 607. Typically, a tolerance of ± 1 or 2 range cells is allowed.

Some medium PRF radars use a fixed range cell size for all PRFs; however, some do not. For those using a fixed range cell size, the range resolution is constant for all PRFs (i.e., for all coherent processing intervals [CPIs]). It is often desirable to maintain a constant duty cycle for all CPIs so that the transmitter load does not vary rapidly and cause problems for the power supply regulation circuits. If a constant duty cycle is to be maintained, then the transmitted pulse width must change in sympathy with the PRI. Furthermore, if a constant range resolution is to be maintained over all CPIs, the compression ratio must vary from CPI to CPI. This adds complication to the requirements of the frequency synthesizer. If, on the other hand, the compression ratio is fixed for all CPIs, the range resolution and range cell width would vary from one CPI to the next. Alternatively, one might hold a constant transmitted pulse width, compression ratio, and range resolution and incur the variation in duty cycle, assuming the transmitter power supply can tolerate the variation in load. There is a simple trade between the convenience of a fixed duty ratio (and the inconvenience of either variable compression ratio or range resolution) and a fixed range resolution (and the inconvenience of either a variable duty ratio or compression ratio). A summary of these trade-offs is presented in the flowcharts of Figure 12-5. Both solutions are represented in fielded systems.

For systems whose range resolution varies with PRF, then all range data must be normalized to a reference range resolution that can be no finer than the coarsest range resolution used. The expression for unfolding the ambiguous target range cells becomes

$$A_{ux} = (A_a + xm)\frac{\Delta R}{\Delta R_{ref}} \qquad x = 0, 1, 2, 3 \ldots \tag{12-7}$$

where ΔR is the range resolution appropriate to the PRF under consideration, and ΔR_{ref} is the reference range resolution.

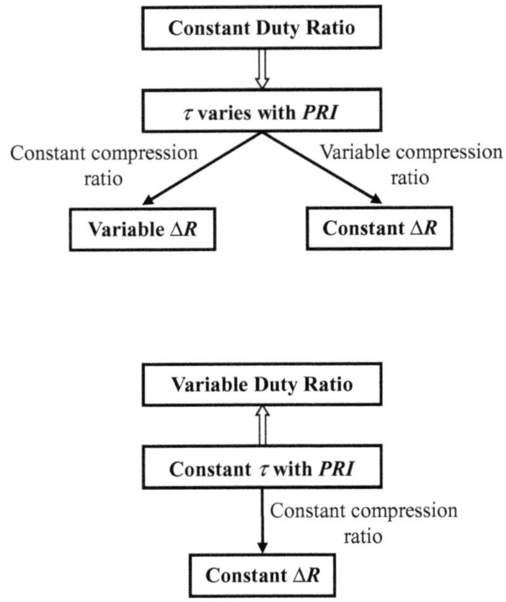

FIGURE 12-5 Flowcharts Concerning Pulse Timings

In the Doppler domain, the Doppler bandwidth is always equal to the PRF, and this is subdivided into a whole number of Doppler cells equaling the fast Fourier transform (FFT) size. In some radars the Doppler cell width is fixed and in others it varies with the PRF. Each CPI uses a different PRF, so, if the FFT size remains fixed in all CPIs, the CPI duration and hence the Doppler resolution vary from one CPI to the next. Conversely, to obtain a constant Doppler cell resolution across all CPIs, the CPI duration must remain fixed and the FFT size must vary in relation with the PRF. Again, there exists a simple trade between the convenience of constant FFT size (but the inconvenience of variable Doppler resolution) and the convenience of constant Doppler resolution (but the inconvenience of variable FFT size). The former was the preferred solution in early pulse Doppler radars, but both solutions are represented in modern fielded systems. When combined with the various options on the pulse timings summarized in Figure 12-5, the detection performance may vary over successive CPIs due to the possibilities of differing pulse widths, duty ratios, noise bandwidths, and FFT processing gains used.

The expression for unfolding the ambiguous target Doppler cells becomes

$$F_{uy} = (F_a + yn)\,\frac{\Delta f_d}{\Delta f_{d\_ref}} \qquad y = 0, 1, 2, 3 \ldots \tag{12-8}$$

where y is the order of the Doppler ambiguity, n is the FFT size, F_a is the apparent Doppler cell number, F_{uy} is the yth potential ambiguous Doppler cell number, Δf_d is the Doppler cell resolution appropriate to the PRF under consideration, and Δf_{d_ref} is the reference Doppler cell resolution.

Unfolding the ambiguous target data in both range and Doppler creates a lattice work of potential detection points over the range/Doppler space of interest. Figure 12-6 presents three such lattices arising from a single target observed in three PRFs. The ambiguous target points are plotted as closed circles (•), open circles (∘), and crosses (×) in PRFs 1, 2, and 3, respectively. The true target range and Doppler are identified as the only range/Doppler cell in which an ambiguous target detection point coincides in all three PRFs. For a 3-of-N scheme, a target would be declared at this point.

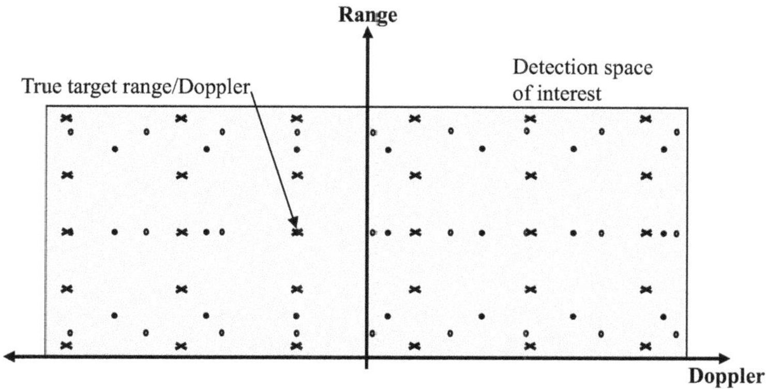

FIGURE 12-6 Ambiguous Target Data for a Single Target Observed in Three PRFs

If the decodable space of the PRF set used here fails to exceed the detection space of interest, more than one point of coincidence would be obtained and ambiguity would still be present. If the maximum decodable range is less than the range of interest, then further points of coincidence would occur at other ranges, whereas if the maximum decodable Doppler does not exceed the Doppler space of interest there would be points of coincidence at other Doppler shifts. A medium PRF schedule may be decodable in range but not in Doppler, or vice versa. Usually, the priority is given to range decodability and some systems seek to correlate returns from the various PRFs in the range coordinate only. Doppler/velocity ambiguity may be resolved through recourse to the tracker. Doppler processing is still beneficial, even when the PRF set is not fully decodable in Doppler, because the action of FFT processing reduces the clutter in each Doppler cell and improves target detection in clutter limited scenarios.

Lower values of PRF (but still within the medium PRF regime) tend to favor range decodability since the maximum order of range ambiguity reduces as the PRF reduces. Conversely, higher values of PRF (within the medium PRF regime) tend to favor velocity decodability since the maximum order of Doppler ambiguity reduces as the PRF increases. Higher and lower values of medium PRFs exhibit tendencies toward the strengths and weaknesses of high and low PRF, respectively. Full decodability in range and Doppler requires a fair spread of PRF values, typically over an octave bandwidth. It may be desirable to use one set of PRFs to resolve range and a different set of PRFs to resolve velocity [1].

12.1.6 The Chinese Remainder Algorithm

An alternative method of decoding is an algorithm based on the *Chinese remainder theorem* (CRT) [2, 15, 22]. This method is applied to decoding range ambiguity, although there is no reason it could not be applied to decode velocity ambiguity. The algorithm applies additional constraints on PRF selection that make it virtually impossible to find PRF sets satisfying both the conditions for range and velocity decodability. Therefore, algorithms based on the CRT are normally implemented in range only, as presented here.

In addition to the range decodability requirement of (12-3), the CRT algorithm requires all PRIs to be subdivided into an integer number of contiguous range resolution cells of fixed width. Furthermore, the numbers of range cells in all combinations of M PRIs from the total number of N PRIs in the schedule must be coprime.[1] For a 3-of-8 schedule all 56 combinations ($= {}_8C_3$) of PRIs must have coprime numbers of range cells.

[1] Coprime (also known as *relatively prime*) numbers have no common positive factor (divisor) except 1.

In a three-PRI system, each PRI, expressed in numbers of range cells, is designated m_1, m_2, and m_3 $(m_1 > m_2 > m_3)$. From the corresponding ambiguous range cell numbers, A_1, A_2, and A_3, in which the target is detected, the unambiguous target range cell number, R_c, is given by

$$R_c = (C_1 A_1 + C_2 A_2 + C_3 A_3) \bmod (m_1 m_2 m_3) \tag{12-9}$$

where

$$\begin{aligned} C_1 &= b_1 m_2 m_3 \equiv (1) \bmod (m_1) \\ C_2 &= b_2 m_1 m_3 \equiv (1) \bmod (m_2) \\ C_3 &= b_3 m_1 m_2 \equiv (1) \bmod (m_3) \end{aligned} \tag{12-10}$$

and b_1 is the smallest positive integer that, when multiplied by $m_2 m_3$ and divided by m_1 as many times as possible, leaves unity as the remainder (b_2 and b_3 are determined similarly).

Thus, the unambiguous target range, R_t, is given by

$$R_t = \frac{R_c}{m_1 m_2 m_3} R_{mu1,2,3} \tag{12-11}$$

where $R_{mu1,2,3}$ is the unambiguous range of the three-PRF combination.

An example of the algorithm is presented in Figure 12-7 in which three PRIs are arranged in order of longest to shortest. In this example the numbers of range cells per PRI are: $m_1 = 11$, $m_2 = 9$, and $m_3 = 7$. A single target is detected in the ambiguous range cells: $A_1 = 10$, $A_2 = 3$, and $A_3 = 7$. Start with the equation set (12-10) to find C_1, C_2, and C_3. Starting with $b_1 = 1$, the equality is tested for in the first line of (12-10). If the equality does not hold, increase b_1 by 1 and retest the first line of (12-10) for equality. Repeat the operation as many times as is necessary, increasing b_1 by 1 on each successive attempt until the equality in the first line of (12-10) is obtained and calculate C_1. Repeat

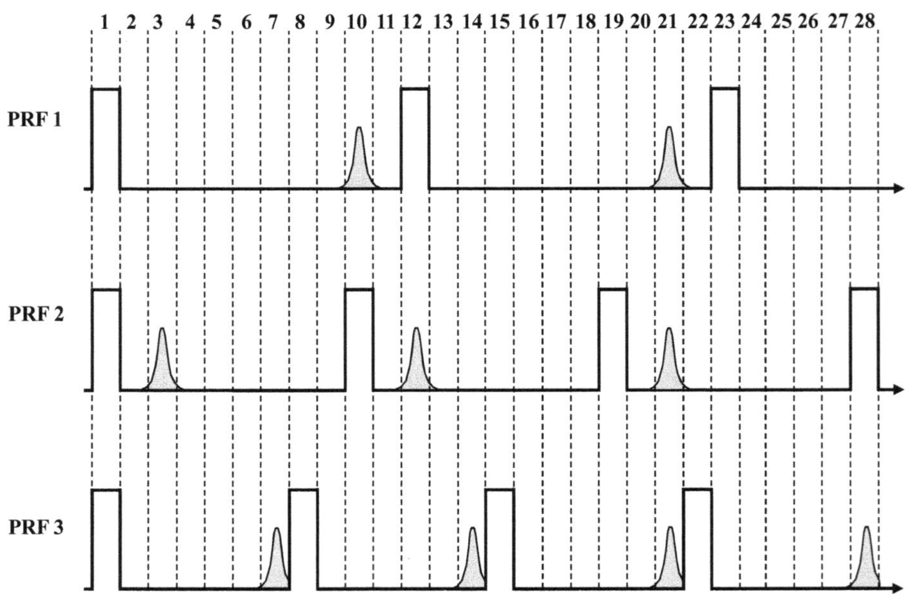

FIGURE 12-7 3PRF Solution Using the Chinese Remainder Theorem

the procedure for C_2 and C_3. The solution in this example is

$$\begin{aligned} C_1 &= 441 \qquad (b_1 = 7) \\ C_2 &= 154 \qquad (b_2 = 2) \\ C_3 &= 99 \qquad (b_3 = 1) \end{aligned}$$

Substituting these values into (12-9) gives

$$\begin{aligned} R_C &= ((441 \times 10) + (154 \times 3) + (99 \times 7)) \text{ modulo } (11 \times 9 \times 7) \\ &= (5565)\text{modulo}(693) \\ &= 21 \end{aligned}$$

Finally, solve (12-11). For the PRIs in question, their lowest common multiple is their product: $m_1 \cdot m_2 \cdot m_3 = 11 \times 9 \times 7 = 693$. Therefore, $R_{mu1,2,3} = m_1 \cdot m_2 \cdot m_3 = 693$.

This last stage of the algorithm accounts for the eventuality that the PRIs may have a common multiple that is less than their product. This does not arise in this case; therefore, $R_t = R_C = 21$.

The result can be verified by eye since it is obvious from inspection of Figure 12-7 that there are coincident target returns in range cell 21 in all three PRFs. This inspection is, in fact, an application of the coincidence algorithm.

The Chinese remainder theorem is a well-proven technique that is simple to implement in hardware [2]. However, the theorem works only for very few combinations of PRFs. First, in its conventional application, PRIs are required to be divisible into integer numbers of range cells of fixed width. Second, for a 3-of-8 schedule, all 56 combinations of three PRIs must satisfy equations (12-3) and (12-10) if a target's range is to be decoded from detections in any three PRFs. (This problem is exacerbated in a 3-of-9 schedule in which there are 84 combinations of three PRIs). These limitations have serious implications for the choice of PRF schedule. Moreover, it would be reasonable to assume that so few fully decodable PRF sets exist that there is little to no scope to optimize other aspects of the radar's performance through PRF selection. A further drawback of the CRT is that small errors in the measurement of one of the ambiguous ranges, $A_1, A_2, A_3 \ldots$ result in a large error in the resolved range, R_C, with no confidence factor available to indicate the quality of the solution [15, 23, 24].

To implement the CRT in Doppler, the Doppler band would need to be subdivided into an integer number of contiguous Doppler cells of uniform bandwidth. Since the Doppler bandwidth is equal to the PRF, f_r, and the number of Doppler cells is given by the FFT size, n, then Doppler bandwidth $= f_r/n$. Furthermore, the FFT size must change in sympathy with the PRF to maintain a fixed Doppler bandwidth across both all PRFs ***and*** FFT sizes because all combinations of M PRFs must be coprime. These conditions constrain the selection of PRFs yet further, and it is very unlikely that any PRF sets can be found that are decodable in range and Doppler when using the Chinese remainder theorem [25]. It may be possible, however, to use one set of PRFs to resolve range and a different set of PRFs to resolve velocity [1].

Coincidence algorithm and CRT are mainstream techniques used in fielded systems; however, a variety of alternative methods of ambiguity resolution have been reported in the literature [9, 10, 23]. Many of these are described in simplified circumstances, like single target detection or velocity ambiguity resolution in low PRF only, and it remains to be seen how such algorithms may cope in practice.

12.2 | BLINDNESS

12.2.1 Blind Zone Maps

Any particular value of medium PRF results in blindness in range due to eclipsing and in velocity due to the blanking of main beam clutter (MBC) and its repetition in the Doppler domain. Even if MBC is not blanked, it can sometimes be so powerful that it would overwhelm discrete targets, so for all intents and purposes the radar would be blinded. Similarly, sidelobe clutter (SLC) results in bands of range blindness that worsen at increasing ranges. As a result, a considerable proportion of the range/velocity space of interest would be blind. Blindness issues in a medium PRF were discussed in Section 11.4. The blind range and velocity cells can be depicted on a *blind zone map*, as in Figure 11-15 (without SLC) or Figure 11-17 (with SLC). Given a random distribution of targets throughout the range/velocity space of interest, a high proportion of targets would not be detected, either through blindness or overwhelming SLC. The location of all blind regions in the range/velocity space are dependent on the precise value of PRF. Different values of PRF have similar looking, although subtly different, blind zone maps since regions of blindness differ, except for the first eclipsed range along the bottom of the map and the first case of MBC rejection up the left-hand edge of the map, which remain fixed in all PRFs. By selecting several PRFs, the regions of blindness are dispersed, so targets that may be blind in one PRF become visible in others. Over a sufficient number of PRFs, the hope is to be able to detect any given target in at least M of them. Normally, detection is required in three PRFs ($M = 3$); however, Simpson [2] showed that, against scintillating targets, the probability of detection is improved substantially if the number of clear PRFs is increased to four. Target detection is far more robust if the aim is for target detection in $M + 1$ PRFs (i.e., one more than the lower limit), since a missed detection in one CPI can be tolerated and the target is still detected in the requisite M PRFs. Blindness in an individual PRF may cover some 20% of the detection space of interest. The additional problems of SLC very much depend on the relative strength of the clutter and of the targets of interest. In severe cases, overwhelming SLC can extend blindness to as much as 50% of the space of interest. Thus, if the aim is for detection in four PRFs, then as many as eight PRFs would be required in the total schedule ($N = 8$). The addition of a ninth PRF would further improve target detectability (if a decodable solution can be found). A composite blind zone map can be generated for the entire schedule by overlaying and aligning the blind zone maps of each individual PRF. Looking down through the N layers of the blind zone maps, it may be observed that any given range/velocity cell may be blind in some PRFs and clear in others. This has been done to generate the blind zone map of Figure 12-8 for a 3-of-8 schedule based on the hypothetical radar model of Table 12-1. Range/velocity cells that are visible in fewer than three PRFs are colored black, so the radar remains blind to targets falling in these cells. The gray cells indicate target visibility in precisely three PRFs, which is the marginal case, and white cells indicate target visibility in four or more PRFs, where target detection should be good. Figure 12-8 illustrates the bands of "blind ranges" due to overwhelming SLC that grow ever wider at longer ranges. A 5 m^2 target is visible in three or fewer PRFs in 57.4% of the detection space of interest, with most of the blindness being incurred at ranges greater than 100 km.

If SLC is removed from the scene, the blind zone map for this schedule appears as in Figure 12-9. In the absence of SLC the total blindness is now 6.82% (based on visibility in fewer than four PRFs). However, this still includes the blindness due to the first blind

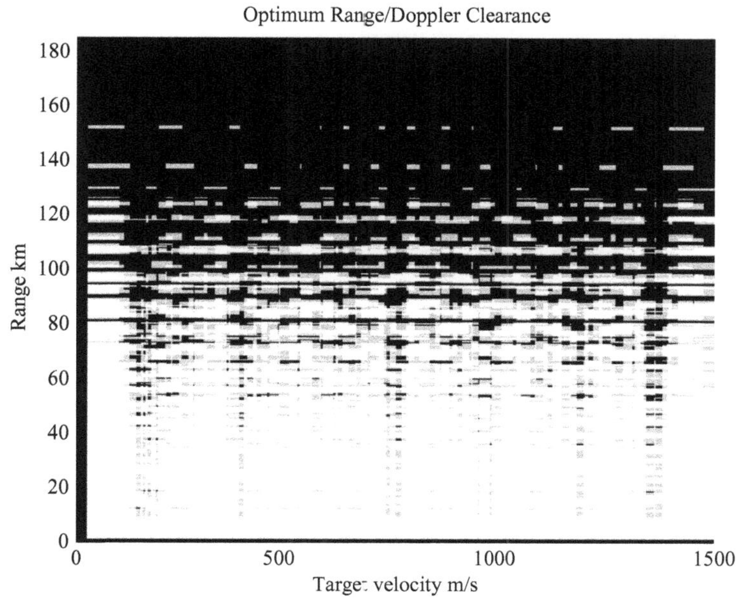

FIGURE 12-8 Blind Zone Map for an Eight-PRF Schedule (with SLC)

TABLE 12-1 Airborne FCR Model

Parameter	Value
Carrier Frequency, f_{tx}	10 GHz
PRIs (μs)	63.11, 69.97, 77.07, 81.31, 90.06, 99.90, 109.75, 119.00
Ambiguity Resolution	Coincidence Algorithm (Range and Doppler)
Transmitted Pulse Width	7 μs
Duty Ratio	Variable: mean = 7.89%, peak = 11.09%
Pulse Compression	2 MHz linear FM, compression ratio = 14
Range Resolution	75 m (0.5 μs)
Blindness Due to Eclipsing	15 range cells
FFT Size	64
Receiver Noise Figure, F	5 dB
MBC/GMT Rejection Notch Width	±1.67 kHz (±25 m/s)
Maximum Range of Interest	185.2 km (100 nautical miles)
Maximum Velocity of Interest	±1500 m/s (±100 kHz)
Antenna Beamwidth (Both Planes)	3.9°
Antenna Sidelobe Level	−30 dB
Antenna Scan Rate	60°/s
Azimuth Scan Angle	0°
Elevation Scan Angle	−6°
Platform Altitude	5000 m
Platform Velocity, V	300 m/s
Clutter Backscatter Coefficient	0.01 m^2/m^2
Target RCS	5 m^2

range (along the bottom of the map) and the first blind velocity (up the left-hand side of the map), which are unavoidable. These unavoidably blind regions account for 2.26% of the total blind zone map and leave 4.56% (= 6.82 − 2.26) avoidably blind or 95.44% (= 100 − 4.56) avoidably clear in at least four PRFs. Careful PRF selection should aim to minimize blindness/maximize target visibility [13]. Target visibility should be maintained

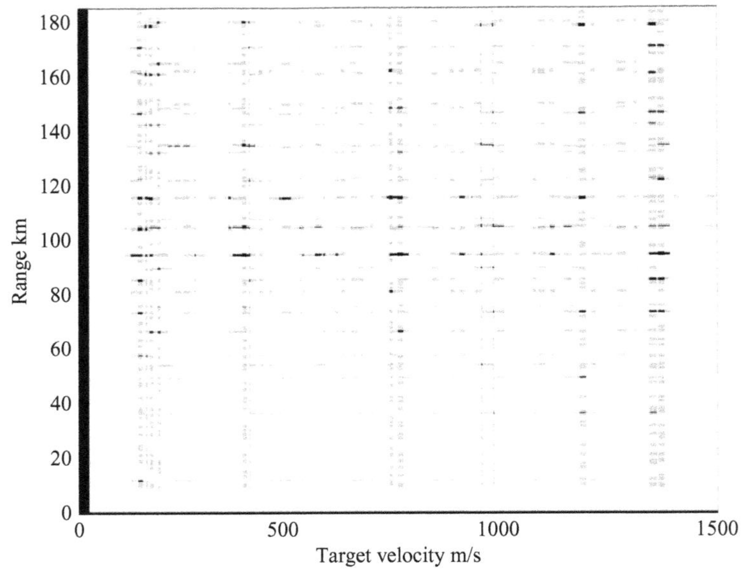

FIGURE 12-9 Blind Zone Map for Eight-PRF Schedule (without SLC)

in 96% or more of the detection space of interest [26]; recent work [3, 4, 5, 7, 27] suggests that this is perfectly achievable.

The greater the number of PRFs used within a schedule, the lower the percentage of blindness. This is because there is greater variation in the positions of blind ranges, blind velocities, and strong clutter patches across a large number of CPIs, and therefore targets are more likely to be detected in clear regions in more PRFs. Simply put, as N increases, there is a greater chance of avoiding blindness and overwhelming clutter in M of these PRFs. Visibility problems can arise for short schedules of low N.

12.2.2 Blind Velocities

A dangerous situation could potentially arise in which the first few repetitions of the clutter notch filter coincide in too many PRFs, leaving fewer than M PRFs clear at these Doppler frequencies. At specific velocities, this results in radar blindness, which persists over all ranges and gives rise to a *blind velocity*. The blind zone map of a poorly designed schedule is given in Figure 12-10, in which blind velocities are incurred at 145 m/s and 290 m/s, plus a marginal one at 430 m/s. Blind velocities often fall at typical aircraft velocities and mean that a target could close toward the radar at these velocities all the way to zero range and go undetected. Clearly, the utmost care should be taken to select precise PRF values to ensure this does not occur.

12.2.3 Blindness Margins

With regard to Figure 12-8, the dominant cause of blindness is not the eclipsing loss or MBC rejection but the clutter. Blindness exists above 100 km because overwhelming clutter coincides in too many PRFs. This gives rise to the notion of a blindness margin.

The blindness margin is the maximum size a clutter patch can attain in both its range extent and velocity extent before it begins to form aliases in the detection space that reduce the detection performance, causing blind ranges or blind velocities. The blindness arises because there are multiple PRFs where the clutter is not staggered sufficiently and an overlap is occurring. The range/velocity cells corresponding to this overlap region remain clear of the clutter return in too few PRFs; hence, the probability of detection,

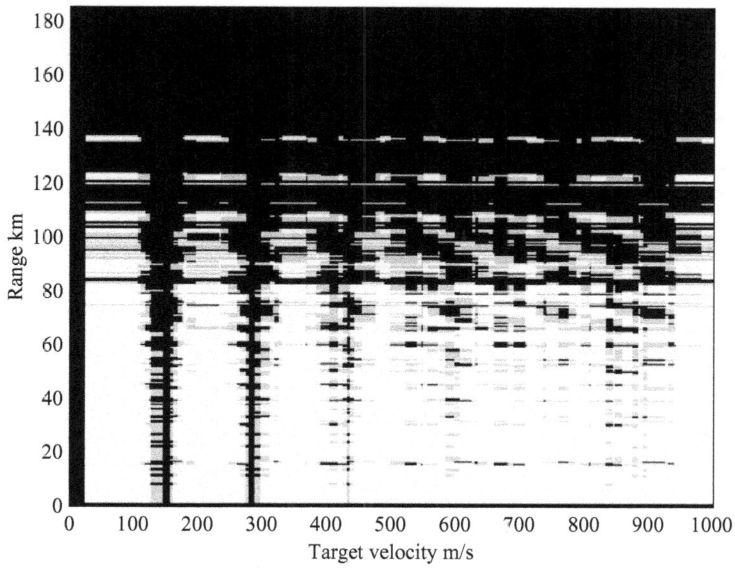

FIGURE 12-10 Blind Velocities of a Poorly Designed Schedule

P_D, is reduced. The blindness margin has a lower limit in the range dimension which is governed by the eclipsing caused by the minimum range, R_{min}, (set by the pulse width and the transmit to receive recovery time) and the ambiguous repetition of the eclipsed ranges. In the velocity domain, the lower limit of the blindness margin is determined by the notch filter used to reject the MBC and its ambiguous repetition in Doppler. If the constraints on the lower limits are observed, then there would be no totally blind ranges or velocities; however, there may still be isolated regions of blindness in the range/velocity space of interest. These isolated regions occur where the PRFs eclipsed in range are not the same PRFs overwhelmed by the MBC returns in velocity; hence, there are insufficient clear PRFs to allow the ambiguities in the system to be resolved and declare a target detection. Blindness results when the offending clutter patch causes blindness that overlaps in $N - M + 1$ PRFs, such as overlap in 6 PRFs for a 3-of-8 schedule, since this leaves only 2 clear PRFs, which is insufficient to declare a target detection.

A clutter patch could be centered at any range and, in the absence of platform motion compensation, centered at any velocity. To explore the blindness margins, one must imagine that the clutter patch can grow in its range and velocity extents until ambiguous repetitions of the clutter patch in the range and velocity dimensions overlap in too many PRFs and a condition of blindness occurs. The range blindness margin is shown in the timing diagram of Figure 12-11 for an 8-PRF schedule. This shows a clutter patch and its ambiguous repetitions in range for all 8 PRFs. The range blindness margin is the range extent of the clutter patch over and above the minimum value of R_{min} caused by eclipsing necessary to produce blindness in $N - M + 1$ PRFs—that is, an overlap of the ambiguous repetitions of the clutter in 6 PRFs for a 3-of-8 schedule. Such an overlap does occur at a value of range offset below the start range of the clutter patch. Should the timing diagram have been extended to the right, it is quite possible that further blind ranges may be incurred at greater ranges.

Should the true start range to the clutter patch differ from that shown in Figure 12-11, a different blindness margin would be applicable. Furthermore, once the clutter patch exceeds the blindness margin appropriate for that range, the offset range to the new blind range would also differ from that shown in Figure 12-11. The range offset can be positive (a later range) or negative (an earlier range). Hence, the blindness margin and the range

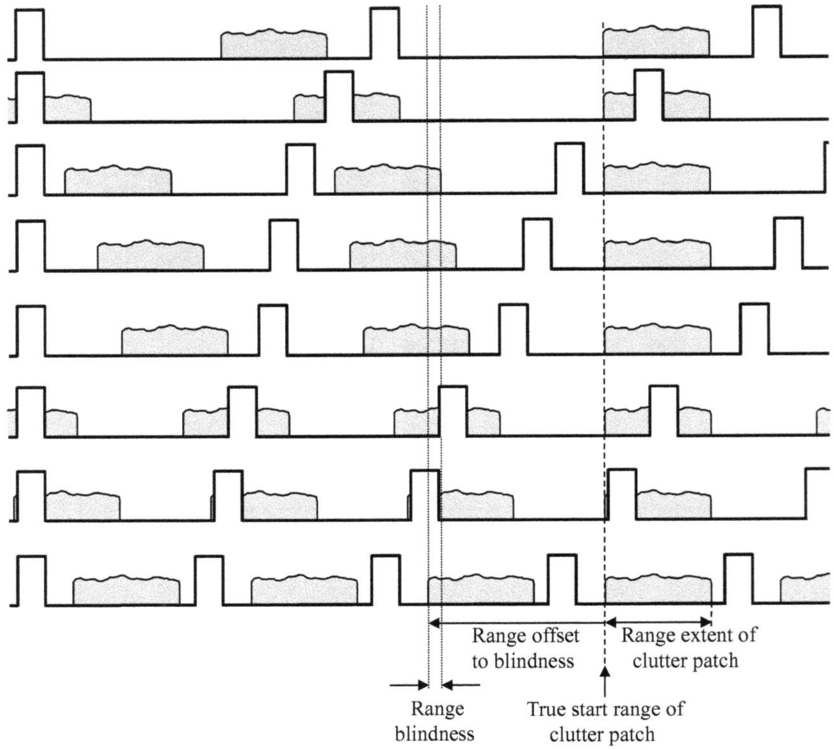

FIGURE 12-11 Range Blindness Margin

offsets to blind ranges vary with the actual range to the clutter patch for any given PRF schedule.

A very similar situation can also be conceived in the velocity dimension and is shown in Figure 12-12. Here the central part of the spectrum is shown for signals of eight differing PRFs and a common carrier frequency, f_{TX}; harmonics of the PRF lines are also shown in the spectra. A patch of clutter is shown centered at some Doppler shift and having a certain extent in Doppler, as indicated. The ambiguous repetitions of the clutter patches in the Doppler domain are also shown. The velocity blindness margin for the given PRF schedule would equate to the extra velocity extent of the clutter patch over and above the minimum value given by the width of the MBC rejection filter necessary to produce blindness in $N - M + 1$ PRFs—that is, an overlap of the ambiguous repetitions of the clutter in 6 PRFs for a 3-of-8 schedule. The Doppler extent of the clutter patch shown in Figure 12-12 far exceeds the velocity blindness margin because its ambiguous repetitions overlap in 6 PRFs in three locations, resulting in three bands of blind velocities at three different Doppler offsets away from the true center of the clutter patch. If the spectra of Figure 12-12 were to be extended across a wider frequency band, then it is quite possible that further blind velocities would become apparent. The blind velocities may have negative or positive Doppler offsets with respect to the central Doppler of the clutter patch. As with range, the velocity blindness margin and Doppler offsets to the blind velocities vary with the actual center Doppler of the clutter patch, for any given PRF schedule.

Figure 12-12 depicts the true clutter spectra centered at a large positive Doppler. If platform motion compensation (PMC) is applied, the clutter is centered at zero Doppler shift and so appears either side of the central carrier component. The ambiguous repetitions of the clutter are likewise centered on the harmonics of the PRF lines above and below the carrier frequency. This situation is illustrated in Figure 12-13. The spectra, including

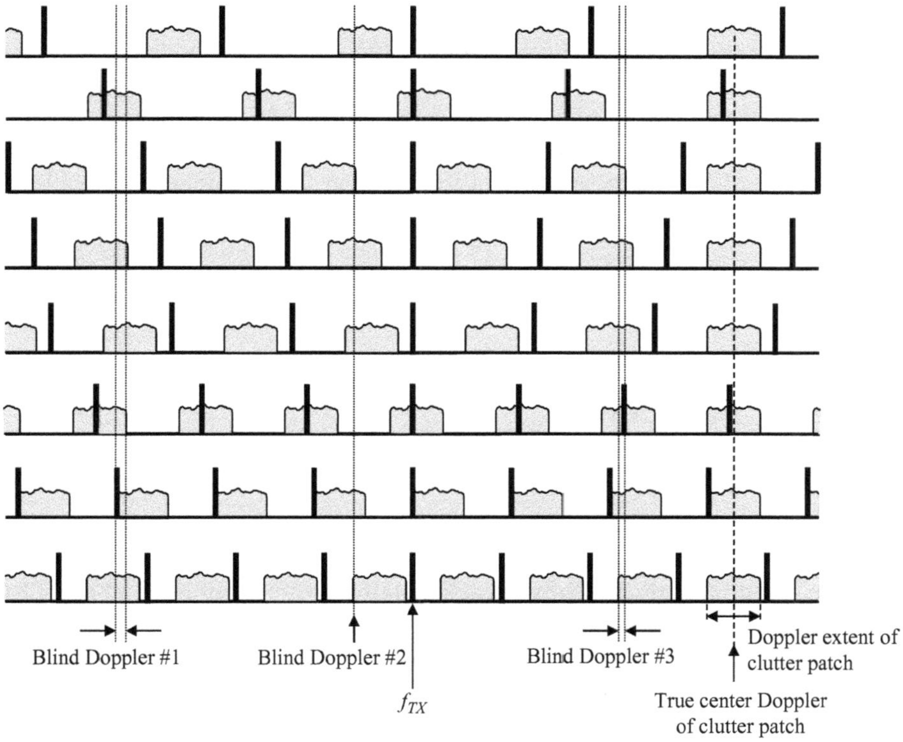

FIGURE 12-12 Velocity (Doppler) Blindness Margin (without PMC)

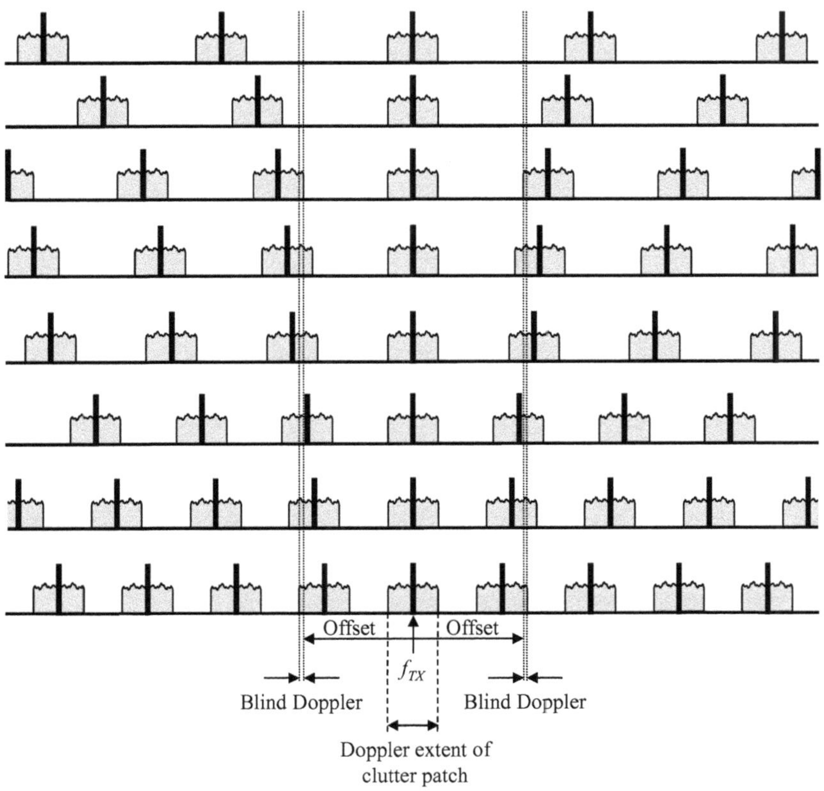

FIGURE 12-13 Velocity (Doppler) Blindness Margin (with PMC)

the clutter and its ambiguous repetitions, from all eight PRFs are symmetrical about the central carrier frequency, giving rise to blind Doppler offsets symmetrically above and below f_{TX}. This is the classic blind velocity problem discussed in Section 12.2.2.

The range and velocity blindness margins, together with the corresponding offset ranges and velocities, may be computed for all values of range and velocity within the detection space of interest for any given PRF schedule. These parameters are all functions of the precise values of PRF selected. PRFs should ideally be selected to maximize the range and velocity blindness margins. PRF schedules that minimize blindness maximize target visibility.

12.3 | PRF LIMITS

12.3.1 Maximum PRF

As PRFs increase, the radar tends toward the characteristics of high PRF. The repetition of SLC in the time domain becomes more problematic and the eclipsing losses become ever more severe and blind ranges occur more regularly, making it difficult to find PRFs that are free of blind ranges that are range decodable. Higher PRFs tend to accentuate range-related problems. There is no upper PRF limit that may conveniently be quoted except for the obvious fact that eventually, at a sufficiently high PRF, no velocity ambiguity is incurred and, by definition, one has reverted to a *high PRF*. In practice, one is likely to keep well below this extreme since the advantages of medium PRF become diluted and the disadvantages of high PRF become accentuated as one selects higher values of PRF. A few specific considerations may guide selection of a maximum PRF.

12.3.1.1 Maximum Duty Cycle

The duty cycle of a pulsed signal increases with increasing PRF, for a fixed transmitted pulse width. Recall that

$$duty\ ratio = \frac{\tau}{T_r} = \tau f_r \qquad \text{repeat of (3-1)}$$

It is quite likely that the transmitter is designed to run either at a fixed duty cycle or that it is rated for a maximum value. A maximum value limits the PRF (f_r) from (3-1), accordingly. For example, if an upper limit on the duty cycle of 20% is imposed, then the PRF is limited to $f_r \leq 0.2/\tau$. In the example of the radar model described by Table 12-1, the pulse width, $\tau = 7\ \mu s$; therefore, the minimum PRI $= 7\ \mu s/0.2 = 35\ \mu s$, or a maximum PRF of 28,571 Hz.

12.3.1.2 Eclipsing Losses

For any single PRF, the proportion of the range/velocity space of interest that is blind due to eclipsing losses is equal to the duty cycle (assuming the *transmit/receive switching delay* to be negligibly small). As the duty cycle increases, so too does range blindness due to eclipsing. This makes it increasingly more difficult to select PRFs that offer a large range blindness margin. Again, a duty ratio of 20% is a reasonable upper limit.

12.3.1.3 Repetition of SLC

The Achilles' heel of high PRF is the multiple repetitions of SLC in the range dimension, and this can become the problem once again as higher values of medium PRFs are selected. For airborne systems the SLC is spread over a wide band, as determined by the ground velocity of the radar platform. The selection of higher values of PRF within a medium PRF schedule can result in strong levels of SLC filling the detection space. In high clutter scenarios, this can lead to broad range intervals in which small discrete targets are overwhelmed by SLC; this, in turn, exacerbates the SLC related blind ranges for long-range targets, described in Section 11.4.

12.3.1.4 Maximum FFT Size versus Doppler Resolution

For a fixed FFT size, n, the Doppler cell width increases with the PRF. This is consistent with a shortening of the CPI time since it takes less time to recover n samples as the PRF increases. Recall that the Doppler cell width (or Doppler resolution) is given by

$$\Delta f_d = \frac{1}{t_{int}} \qquad \text{repeat of (4-15)}$$

where the CPI time is $t_{int} = n/f_r$, giving $\Delta f_d = f_r/n$. The Doppler resolution may be restored if the FFT size, n, is increased along with the PRF, since this restores the CPI time. Limits imposed on the Doppler resolution fix the minimum CPI time, which may then be obtained through a combination of selecting the appropriate PRF and FFT size.

12.3.2 Minimum PRF

As lower values of PRF are selected, the waveform tends toward the attributes of *low PRF*, which is characterized by having MBC occupying a considerable proportion of the Doppler band (Section 6.6.2). Lower PRFs accentuate Doppler-related problems, and some of the useful features of medium PRF are eroded as ever lower PRFs are selected. As with the maximum PRF, there is no hard and fast lower limit on the PRF, except to say that as the PRF continues to decrease one ends up avoiding all range ambiguities and, by definition, operates on a *low PRF*. In practice, MBC considerations guide the selection of a suitable lower limit.

One mechanism to derive a lower PRF limit is to stipulate that the MBC/GMT rejection filter should not exceed a certain proportion of the Doppler bandwidth ($= f_r$). Some have suggested a limit of 50% [2, 5] and others have suggested a 25% limit [28]. Applying these limits to the radar model of Table 12-1, in which the MBC/GMT rejection width is 3333 Hz, results in a minimum PRF of 6,667 Hz (maximum PRI $= 150\ \mu$s) for the 50% limit or 13,333 Hz (maximum PRI $= 75\ \mu$s) for the 25% limit. At a PRF of 6,667 Hz, half the Doppler band is obliterated by the MBC/GMT rejection filter, and there would be broad bands of blind velocities repeated at only 6,667 Hz intervals, thus rendering half the detection space blind. The case of the 50% limit is shown in Figure 12-14 in which MBC is shown with a bandwidth of half the PRF. The inclusion of too many extremely low values of PRF compound the difficulty of finding precise PRF values offering full velocity decodability and reasonable velocity blindness margins.

12.3.3 Mean PRF

The mean value of the PRI must be controlled to allow the transmission of the entire schedule within the beam dwell time. This time is fixed in a mechanically scanning antenna

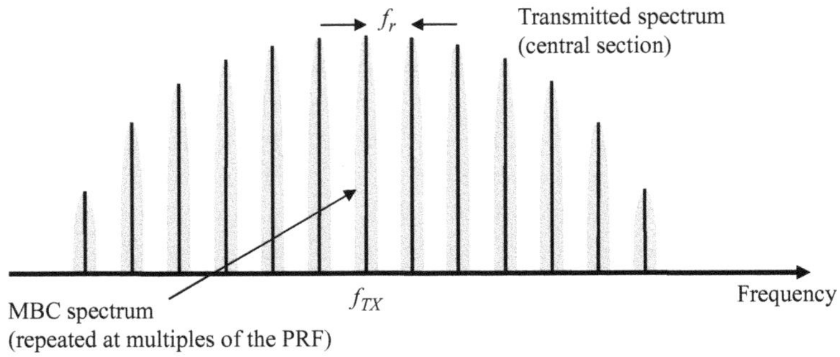

FIGURE 12-14 MBC Occupation of 50% of the Doppler Band

by the duration it takes for the antenna to scan through its own beamwidth. The beam dwell time need not be fixed for an active electronically scanned array (AESA) antenna; nevertheless, there are still limits on the beam dwell time imposed by the workload and prioritization of tasks, as dictated by the task scheduler.

Recall that each CPI must be separated by a *space charging time* (Section 10.1.4) at the beginning of each new CPI, during which time returns must be blanked from the processing. The blanking is necessary to ensure that returns processed under the new CPI do not arise from distant targets that have been illuminated by the waveform of the previous CPI. In the context of medium PRF waveforms, a radar transmits N CPIs within the beam dwell time; each on its own specific PRF. When the radar switches from one PRF to the next, pulse returns from distant targets may continue to be received on the previous PRF for some while into the new CPI. Having switched into the new CPI, the FFT processing assumes that returns relate to the new PRF and not the previous one. The blanking time must be sufficiently long to allow for the most distant target expected. During this time, the transmitter continues to emit pulses at the new PRF, since these will give rise to future echoes, but the processing must be blanked. The situation is depicted in Figure 12-15.

The space charging time, t_{SC}, has previously been given as

$$t_{SC} = 1.5\frac{2R_{max}}{c} \qquad \text{repeat of (10-6)}$$

in which the factor of 1.5 makes an additional allowance of 50% to cater for the possible detection of very large targets beyond the normal maximum detection range of the radar. For a medium PRF system, the space charging time can be very damaging since it represents a time in which no returns may be processed. For a medium PRF schedule comprising N CPIs, then N space charging times must be incurred. These N space charging times can consume a considerable proportion of the beam dwell time. Longer schedules of larger N

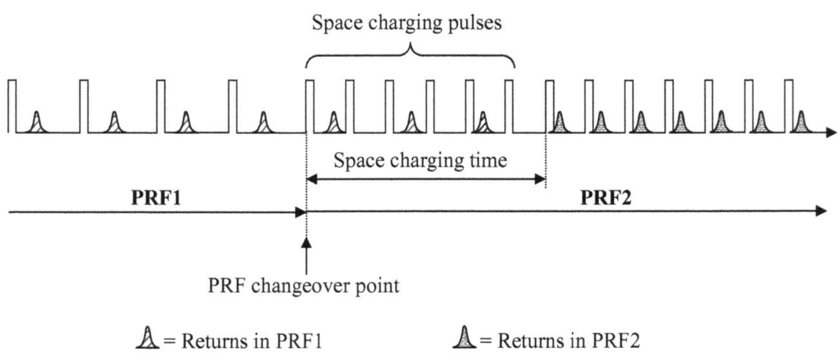

FIGURE 12-15 Space Charging Pulses in Medium PRF

impact on the radar's time budget with a double hit since not only must each CPI duration be shorter (less processing gain, coarser velocity resolution) but also more time is lost to space charging.

For a schedule of N CPIs, with a uniform FFT size of n, the mean PRI is constrained by

$$\textit{Beam dwell time} \leq N\left[\left(n \times \overline{PRI}\right) + t_{SC}\right] \qquad (12\text{-}12)$$

Considering the radar model of Table 12-1, one has $N = 8, n = 64$, and $t_{SC} = 1.7$ ms. For a beamwidth of 3.9° and a scan rate of 60°/s one obtains a beam dwell time of $3.9/60 = 0.065$ or 65 ms. Inputting these parameters into (12-12) yields a maximum value of the mean PRI to be 100.4 μs (mean PRF should not be less than 9960 Hz). The mean CPI duration cannot exceed 6.4256 ms, and the eight space charging times amount to 13.6 ms, which constitutes 21% of the beam dwell time. While this does not impose an upper limit on any PRI that may be selected, there is insufficient time to tolerate too many long PRIs within the schedule. The actual mean value of the eight PRIs used in this radar model is 88.77 μs and is considerably less than the maximum allowed here. In fact a mean PRI of 88.77 μs would permit a scan rate of 66°/s and accommodate a beam dwell time of 59.05 ms and a CPI duration of ~5.7 ms.

Blindness in range resulting from eclipsing and blindness in velocity due to the rejection of MBC are both functions of the PRF. Therefore, target visibility in range/velocity space may be maximized through PRF selection in noise-limited cases and can be used as a means of selecting the mean PRF [12, 13]. This is dealt with in more detail in Section 14.1.2 and Chapter 15.

12.4 | GHOSTING

12.4.1 Introduction

Ghosting is the phenomenon whereby ambiguous target detections are incorrectly correlated across M PRFs, resulting in the indication of additional targets at false ranges or velocities. Such a target is known as a *ghost,* and it appears equally strong as genuine targets. Ghosting is a problem that worsens with increasing number of genuine targets; it is also dependent on the precise method of target extraction and the exact values of the PRFs used in the schedule. No radar design can guarantee the elimination of all ghosts, however, much can be done to minimize their occurrence and the impact they have on the radar. Ghosting can arise from a variety of mechanisms, each of which is described in the ensuing sections. Techniques to combat the ghosting problem are also discussed, some of which are related to PRF selection and others are not, but they are all dealt with in this section for the sake of completeness.

12.4.2 Cross-Correlation of Multiple Targets

If several targets are present in the beam, the ambiguous returns of one target could coincide with the ambiguous returns of other targets in M PRFs. A ghost target would be seen at the point of coincidence as well as the genuine target locations. Coincidences like this occur in both the range and velocity dimensions. If a radar correlates targets in range only (and some do), then alignment of the ambiguous returns from multiple targets in the same range cell in M PRFs is required to give rise to a ghost. If a radar correlates targets in both

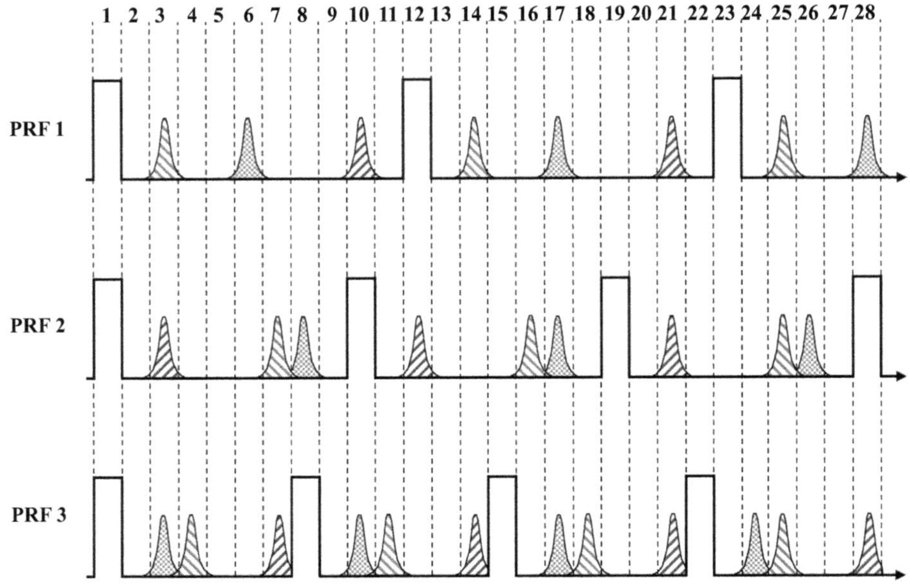

FIGURE 12-16 Range Correlation of Ghosts

range and velocity, then alignment is required in both one range cell and one velocity cell in M PRFs. Target extraction algorithms that correlate in range only are more susceptible to ghosting than those that correlate in both range and velocity (see Section 12.5.2).

The ambiguous returns in the range axis from three targets in three PRFs are shown in Figure 12-16. The targets are shaded differently so that the reader might easily distinguish them, but to a radar there is no such distinction. Let it be assumed that the target extraction requires range coincidence in three PRFs ($M = 3$). The three targets can be seen to align in range cells 17, 21, and 25, which are the true ranges of the targets. However, coincident detections in three PRFs also take place in range cell 3 due to the cross-correlation of all three targets. A ghost would be declared at range cell 3, indicating the presence of a total of four targets. In fact, further cross-correlations are also incurred beyond range cell 28 but within the maximum unambiguous range for this three-PRF combination. The system cannot distinguish between genuine correlations and cross-correlations. Had the criterion for detection been coincidence in just two PRFs ($M = 2$), then ghosts would also have been incurred in range cells 7, 10, 14, and 28.

The ghost targets observed in Figure 12-16 would also correlate in velocity at some point within the decodable velocity space for the three PRFs used in the correlation process. Each target gives rise to a regular lattice of ambiguous detection points in the range/velocity space, as discussed in Section 11.2.4 and illustrated in Figure 11-6. When N_T targets are seen in N PRFs, there are $N_T \times N$ such lattices that can be superimposed in the range/velocity detection space. A target is declared at any range and velocity cell in which there is coincidence in M ambiguous detection points. Figure 12-17 shows the ambiguous detection points of three targets detected in three PRFs. Detection points in PRF1 are denoted by the circles, detection points in PRF2 are denoted by the crosses, and detection points in PRF3 are denoted by the squares. Figure 12-17 plots *all* detection points; all circles, crosses, and squares are assumed to avoid the blind ranges and velocities and regions of overwhelming clutter associated with their specific PRF. There are four points at which there are coincident target detections in all three PRFs: three of these are the correct correlation points denoted by the white squares, and one of these is a ghost,

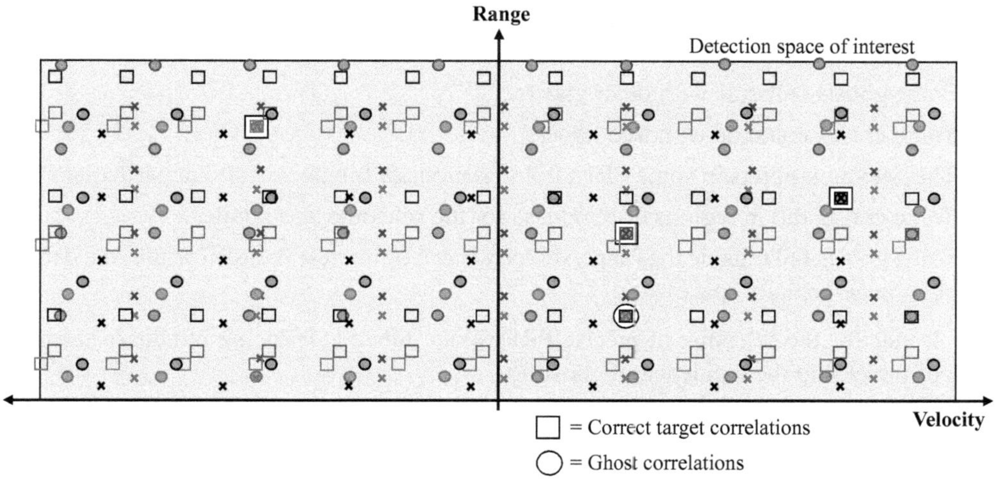

FIGURE 12-17 Correlations in Range and Velocity

denoted by the white circle. One can imagine the plethora of detection points that would arise over an eight-PRF schedule and the increased scope for coincident detection points in three PRFs. One can also imagine the consequences of correlation in range only by collapsing all the detection points onto the range axis.

From Figure 12-17, it may be seen that there are a number of cases of very near coincidence in detection points across all three PRFs. If target responses were slightly extended in range or velocity or if realistic tolerances were applied to represent the range and velocity measurement accuracies, then it is probable that more ghost targets would be declared. When correlating target returns it is necessary to apply some sort of margin since even genuine targets might not align perfectly to give a precise coincidence at one range/velocity cell. The allowable range and velocity margins that must be applied when using the coincidence algorithm not only affect the ability of the processing to extract the true targets but also its susceptibility to ghosting. Hence, the incidence of ghosting is a function of the target extraction algorithm used (i.e., range only versus range and velocity correlation and the allowable margins), and this is discussed in Section 12.5.2.

Ghosting is also a function of the number of targets present, N_T, the total number of PRFs used in the schedule, N, and the minimum number of PRFs in which target data are required to declare a target, M. Assuming that N_T targets are seen in M PRFs, the number of correlation possibilities is given by the following [29, 30]:

Number of correlation possibilities $= N_T^M$

Of these, the number of correct target correlations $= N_T$

leaving the number of ghost correlations as $N_G = N_T^M - N_T$.

Therefore, the number of ghosts is $N_G = N_T\left(N_T^{M-1} - 1\right)$ (12-13)

All these correlation possibilities are guaranteed to occur within the decodable range/velocity space of the M PRF combination involved in the correlation process.[2] However,

[2]Whilst the analysis in [29] was presented in the context of decodability in range only using algorithms based on the Chinese remainder theorem, the analysis on the correlation of multiple targets is, in fact, independent of the decodability algorithm, and it remains the case that all correlation possibilities do occur somewhere within the decodable range/velocity space.

in practice, fewer ghosts are actually seen. This reduction in apparent number of ghosts is due to several reasons:

- Some ghosts coincide with other ghosts.
- Some ghosts coincide with true targets.
- The detection points in some PRFs fall in regions of blindness for that particular PRF.
- Some ghosts fall in regions of blindness for the schedule as a whole.
- Some ghosts fall outside the range/velocity space of interest (but still within the decodable range/velocity space).

Ironically, the selection of precise PRF values aimed at reducing blindness carries a risk of increasing the incidence of ghosting.

If the proportion of the range/velocity space in which the radar has target visibility is given by P_{RV} (see Section 14.1.2)—that is, the proportion not blind due to eclipsing or MBC rejection—then the probability that a randomly placed target response could be detected is also given by P_{RV}. This proportion varies with the precise PRF value; therefore, let $\overline{P_{RV}}$ refer to the proportion of the range/velocity space in which the radar has target visibility in the *mean* PRF used within the schedule. Therefore, on average, the probability of a target response being visible in M PRFs is given by $\overline{P_{RV}}^M$. Since also the number of combinations of drawing M PRFs from the total of N is given by ${}_NC_M = \binom{N}{M}$, the number of ghosts can be estimated at

$$N_G = \left(N_T^M - N_T\right)\binom{N}{M}\overline{P_{RV}}^M \tag{12-14}$$

Finally, the effects of range and velocity gating must be considered.

Given a set of PRIs [29], a reasonable estimate for the expected number of ghosts in the region of interest can be calculated, ignoring the possibility of some ghost targets coinciding with others. If the PRI set results in an unambiguous range/Doppler region with, on average K range/Doppler cells, and the region of interest has, on average, Q repeats of the unambiguous region, then the probability of ghosts in an M-of-N system can be approximated. Assuming N_T targets are placed in the unambiguous region, an approximation for the expected number of ghosts in the region of interest, given a probability of detection of 100%, is[3]

$$N_G \approx QK\left(1+\sum_{i=0}^{M-1}\binom{N_T}{K}\left(\frac{N_T}{K}\right)^i\left(1-\frac{N_T}{K}\right)^{N-i}\right) \tag{12-15}$$

The approximate number of ghosts due to cross-correlations predicted by (12-15) is plotted in Figure 12-18 for $N = 8$, $M = 3$ and between 0 and 60 real targets of random placement ($N_T = 0$ to 60). The accuracy of the approximation improves as the average values of K and Q more closely represent the precise values used in each PRF. Equation (12-15) has been verified by repetitive simulations against a medium PRF radar model operating a 3-of-8 schedule and decoding target range and velocity using the coincidence algorithm. Thirty trials of randomly placed targets have also been conducted for numbers of real targets ranging from 0 to 60. The mean number of ghost correlations within the

[3]This is a corrected form of the equation given in [29] supplied by Hughes.

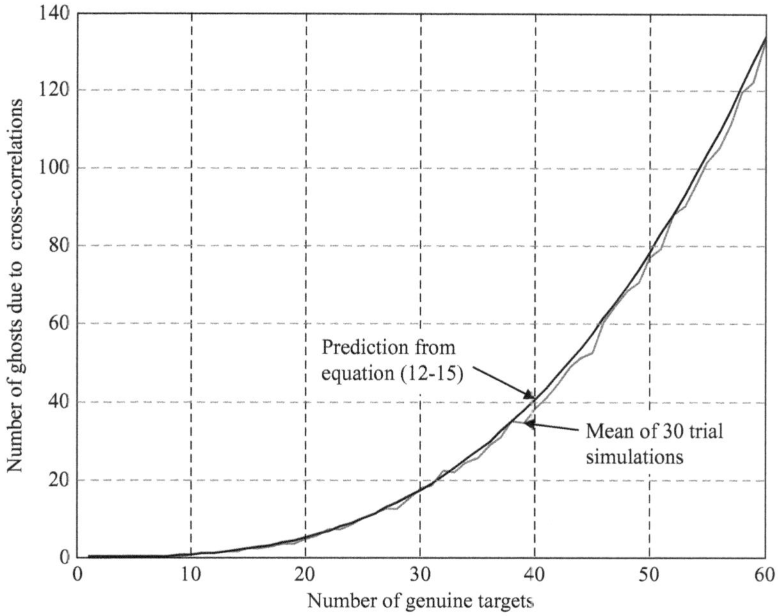

FIGURE 12-18 Number of Ghosts versus Number of Genuine Targets (Smooth curve = Prediction of (12-15), Irregular line = Mean of 30 Trials Simulations)

range/velocity space of interest has been quantified and is also plotted in Figure 12-18. The number of cross-correlation ghosts from (12-15) is in close agreement with the mean number of ghosts seen in the 30 trial simulations. It is clear that the number of ghosts likely to be declared increases very rapidly as the number of real targets present exceeds 40. This has major implications for the havoc that could be brought about through the use of digital radio frequency memory (DRFM) based repeater jammers!

12.4.3 Noise-Generated Ghosts

Noise spikes can trigger the detection of a false alarm that may correlate with ambiguous target responses and result in a ghost detection [12]. Figure 12-16 demonstrated that there were coincident target detections over two PRFs in range cells 7, 10, 14, and 28. Ghosts would not be declared at these range cells for a 3-of-N schedule. However, a noise generated false alarm in range cell 7 incurred in PRF1 would result in a ghost target. Similarly, a noise generated ghost would result from a false alarm in range cell 14 in PRF2 (no false alarm is possible in range cells 10 and 28 in PRF2 because these cells are eclipsed). Likewise, for correlation in range and velocity, a noise-generated false alarm at a range/velocity cell already occupied by ambiguous target detections in $M - 1$ PRFs would trigger the declaration of a ghost target. Inspecting Figure 12-17 reveals numerous instances of coincident target detections in two PRFs.

Similarly, there is the possibility that a single target response in one PRF could correlate with two noise-generated false alarms occurring in the same range/velocity cell in two more PRFs. This, too, would indicate a ghost in a 3-of-N detection system. Naturally, this can be extended for even more noise-generated false alarms, ultimately resulting in the declaration of a (ghost) target due entirely to coincident noise generated false alarms in all M PRFs. This is considered further in Section 14.1.

Noise-generated ghosts are very much like the cross-correlation ghosts described in Section 12.4.2 in that the number of ghosts tends to increase with an increasing number of genuine targets and the P_{FA}. Unlike cross-correlations, however, noise-generated false alarms pop up in random cells and are a sporadic and short-term effect, whereas ghosts generated via the cross-correlation of returns can persist for some considerable time.

12.4.4 Self-Ghosting Due to Extended Target Responses

A single target may ghost with itself if it is extended in range and/or velocity. Section 12.2.3 considered the blindness margins associated with extended returns such as large clutter patches. However, a more modest extension in the range or velocity dimension can cause self-ghosting long before blindness would become a problem. Targets may be extended in range due to the dimension of the target in the range direction, tolerances in the measurement of range, and range cell straddling effects. Some degree of velocity extension is possible for targets having a band of Doppler shifts, such as the variation in the instantaneous velocity of points along a rotating helicopter blade as a function of their distance from the center of rotation. Doppler smearing due to the measurement accuracy and velocity straddling effects are also possible and are synonymous with their counterparts in range.

Figure 12-19 illustrates the marginal case whereby ghosts are just about to form due to the range extension of a single target, assuming $M = 3$. The transmitted modulation is shown in the heavy black line, and the ambiguous returns from a single, extended target are shown in the lightly shaded boxes. As it happens, Figure 12-19 depicts the threshold whereby two ghosts are incurred. One occurs at a small offset range and arises because the stagger between the PRFs is inadequate to prevent a coincidence between the first ambiguous repetition of the target return away from the true range in the lower three PRFs. The second ghost occurs due to a cross-correlation of returns of differing orders of ambiguity over the third, fourth and eighth PRFs. These ghosts are generated for a target

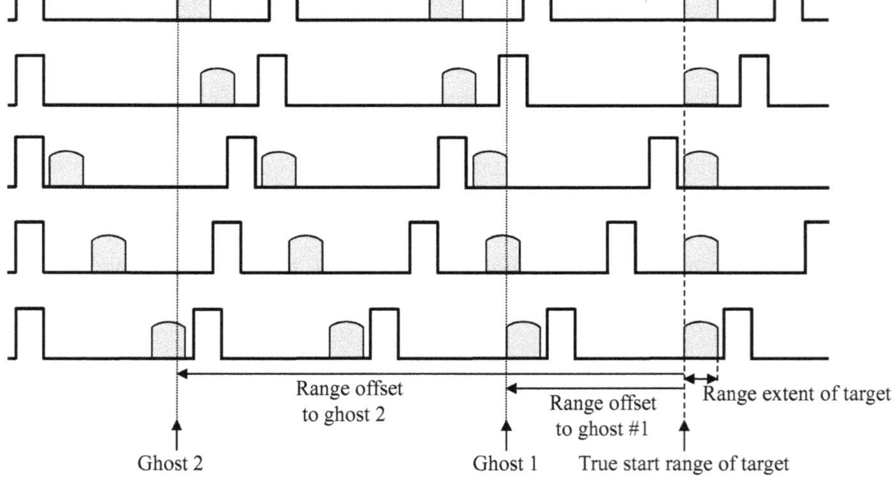

FIGURE 12-19 Ghosting Due to Range Extension of Target

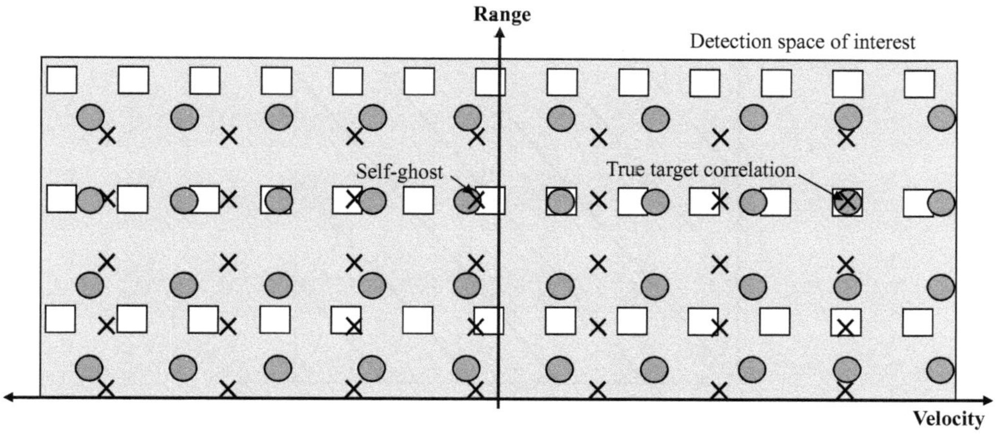

FIGURE 12-20 Self-Ghosting in Range/Velocity Space

response that is only marginally wider than the transmitted pulse width, meaning that just a small degree of range extension would be necessary to form a ghost. In other words, the range decodability margin of the PRF set shown in Figure 12-19 is very small, and the PRF schedule is not particularly resilient to self-ghosting.

A similar problem can occur in the velocity domain and for radars that correlate returns in range and velocity. Self-ghosting occurs only when ambiguous target responses overlap in both dimensions in M PRFs. This is illustrated in Figure 12-20 in which a single target is detected in three PRFs. The detection points are indicated by the circles (PRF1), crosses (PRF2), and squares (PRF3), as in Figure 12-17, except here the circles, crosses, and squares are considerably larger denoting the extension of the target detections in range and velocity. Detections in three PRFs coincide at the true target position, and there is a partial alignment in the three PRFs at a point that would also be declared as a target (but is, in fact, a ghost). As ever, target correlation in both range and velocity is more resilient to the declaration of self-ghosts than correlation in range only.

12.4.5 Decodability Margins and their Depiction Using Skyline Diagrams

As discussed in Section 12.1.4, the decodability requirements demand that multiples of M PRIs (or PRFs) do not coincide with each other to within one range (or velocity) cell to ensure range (or Doppler) decodability. Furthermore, the point was made that decodability is more robust if an extra margin in range (and velocity) is allowed for, over and above the single range (and velocity) cell requirement. The minimum value of these extra margins over all combinations of M PRFs from the total of N PRFs used in the schedule defines the range and velocity decodability margins. These margins dictate the maximum extensions in range and velocity by which a single target could grow before the onset of self-ghosting is incurred.

A convenient way of visualizing the decodability margins is with a *skyline diagram* [31]. Consider the range measurement of a single target in two PRFs. In the absence of any range errors

$$R_{true} = R_{app1} + x_1 R_{mu1} = R_{app2} + x_2 R_{mu2}$$

which is a repeat of equation (12-1) for a two-PRF system. Here, R_{true} represents the true target range, R_{app1} and R_{app2} represent the apparent ranges of the target seen in PRFs 1

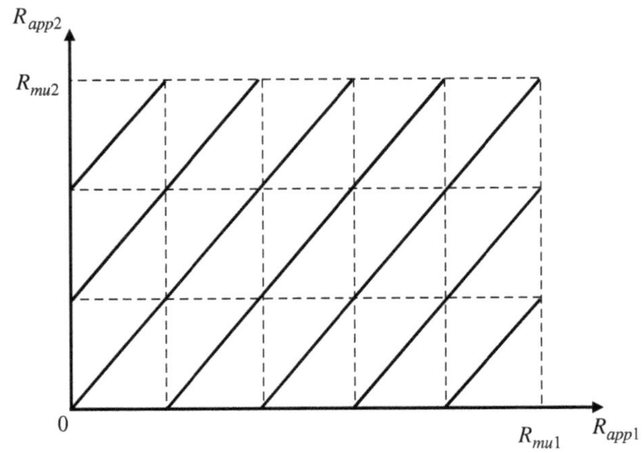

FIGURE 12-21 Range Ambiguity Plots for Two PRFs

and 2, R_{mu1} and R_{mu2} represent the maximum unambiguous ranges of PRFs 1 and 2, and x_1 and x_2 represent the order of range ambiguity of the target return in PRFs 1 and 2, respectively. Rearranging this equation gives

$$\begin{aligned} R_{app1} &= R_{true} - x_1 R_{mu1} \\ R_{app2} &= R_{true} - x_2 R_{mu2} \end{aligned} \tag{12-16}$$

Combining these equations produces

$$R_{app2} = R_{app1} + x_1 R_{mu1} - x_2 R_{mu2} \tag{12-17}$$

Figure 12-21 shows a plot of R_{app2} as a function of R_{app1}. The various diagonal lines relate to successive orders of ambiguity: x_1 and x_2. The measurement of the apparent range in the two PRFs in the absence of any errors would correspond to a point on one of these lines.

The *x-axis* intercept is $x_2 R_{mu2} - x_1 R_{mu1}$, and the *y-axis* intercept is $x_1 R_{mu1} - x_2 R_{mu2}$. The absolute value of the intercept on either axis is therefore $|x_1 R_{mu1} - x_2 R_{mu2}|$ and is the range decodability margin for the two-PRF combination. This represents the magnitude of the difference between multiples of the maximum unambiguous ranges of the two PRFs.

When measurement error is taken into consideration, an apparent target range of R'_{app1} in PRF1 and an apparent target range of R'_{app2} in PRF2 is observed. The error corrupted apparent ranges differ from the true apparent ranges by $\pm r_1$ and $\pm r_2$ in PRFs 1 and 2, respectively. Hence

$$\begin{aligned} \left|R'_{app1} - R_{app1}\right| &\le r_1 \\ \left|R'_{app2} - R_{app2}\right| &\le r_2 \end{aligned} \tag{12-18}$$

The measurement point now lies within a rectangle error bound of dimensions $2r_1$ by $2r_2$, as illustrated in Figure 12-22. Any overlap between the two rectangular error bounds causes a problem since neighboring orders of range ambiguity could be confused and the true target range could not be solved unambiguously. This problem is avoided if

$$|x_1 R_{mu1} - x_2 R_{mu2}| > 2\,(r_1 + r_2) \tag{12-19}$$

In other words, the decodability margin must exceed the sum of the range errors.

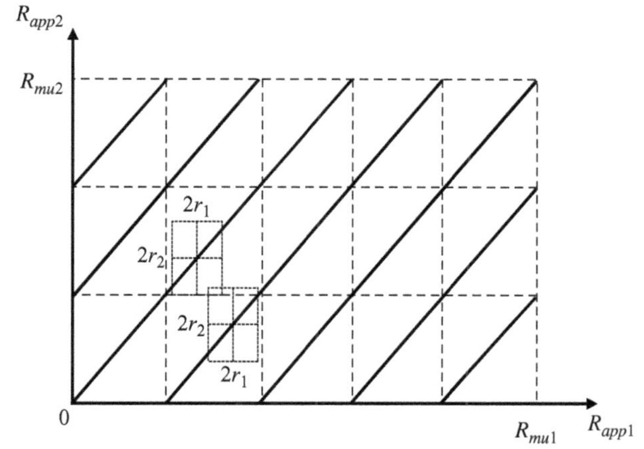

FIGURE 12-22 Range Ambiguity Plots for Two PRFs with Error Bounds

For extended range targets one may set $r_1 = r_2 = r$, in which r represents the range extension of the target (which is common in all PRFs). In this case, the target is located within a rectangular error bound of dimensions r by r. Decodability is assured if the decodability margin exceeds a value of twice the range extension. This theory can be extended to detections in three PRFs and ultimately results in the previous remark that range decodability is assured if the multiples of any three PRFs do not align to within twice the sum of the range errors (or twice the target range extension). Similarly, the same analysis may be applied in the velocity domain to describe the velocity decodability margin. A plot of the range decodability margin as a function of the true range of the target yields a plot known as a *skyline diagram*. A skyline diagram can also be generated in the velocity domain by plotting the decodable velocity margin as a function of the true target velocity. An example of range and velocity skyline diagrams is presented in Figure 12-23 for the 3-of-8 PRF schedule and radar model given in Table 12-1. In this example, the

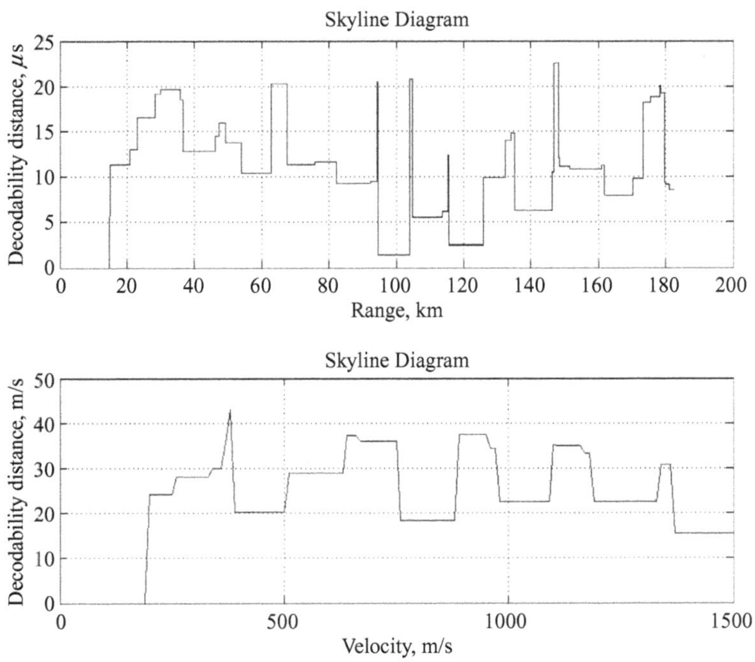

FIGURE 12-23 Rangeand Velocity Skyline Diagrams: Range (Upper Plot); Velocity (Lower Plot)

minimum value on the range skyline of 1.37 μs indicates that the PRF schedule would be admissible for a ghosting check employing an extended pulse width of up to 1.37 μs (= 0.5 μs pulse width +0.87 μs extra tolerance). The minimum velocity decodability margin is 15.4 m/s.

12.4.6 PRF Selection to Minimize Ghosting

The first and most obvious expedient to take to minimize the incidence of ghosting is to select PRF values that maximize the minimum value of the range and velocity decodability margins over all combinations of M-of-N PRFs. A minimum range decodability margin of one range cell is required to ensure that the schedule is decodable in range; however, an extra margin to allow for the measurement accuracy and the range extension expected from targets is a wise addition. Similarly, a minimum velocity decodability margin of one velocity cell ensures velocity decodability and, again, it is prudent to allow an additional margin to cater for the measurement accuracy and any Doppler spreading of target responses. Targets that are extended in range (e.g., large clutter patches) or in velocity (e.g., helicopter blade flashes) may exceed the decodability margins and cause ghosting. The decodability margins vary with target range and velocity, and the *minimum* values of the decodability margins across the detection space of the radar govern the limits on self-ghosting. That said, the minimum values are incurred at a specific range and velocity; elsewhere the margins are greater and there is a greater resilience to self-ghosting. So although the minimum decodability margins drive the PRF selection process, it would also be worth selecting PRFs that boost the mean (or median) decodability margins.

Maintaining large decodability margins guards against self-ghosting, but this strategy also helps minimize the incidence of other ghosting mechanisms. Enforcing large decodability margins ensures that the PRF schedule is robust against all forms of ghosting but imposes constraints that make it more difficult to find compatible schedules. Those schedules which may be found may suffer from poor blindness. Finding precise PRF values that maintain both large decodability margins and large blindness margins for all combinations of M-of-N PRFs whilst also adhering to the maximum, minimum, and mean PRF constraints becomes exceedingly difficult.

The second expedient to observe when selecting PRF values is to design for as large a decodable range/velocity space as possible [22]. Here it is important to stress the distinction between the *decodable space* and the *detection space of interest*. The decodable space is the region bounded by the minimum value of the decodable range by the minimum value of the decodable velocity, where the minimum values are taken over all combinations of M-of-N PRFs, each combination of M-of-N PRFs with its own unique decodable range/velocity space. The detection space of interest is the instrumented range and velocity space and encompasses the spread of ranges and velocities of targets that the radar may be expected to encounter. Naturally, the decodable space should exceed the detection space of interest. The incidence of ghosting is minimized if the margin by which the decodable space exceeds the detection space of interest is maximized. Section 12.4.2 pointed out that all N_T^M correlation possibilities arise somewhere within the decodable space for the M PRFs involved in the correlation process[29]. However, this does not necessarily mean that all ghost correlations fall within the detection space of interest. A proportion of the ghosts falls within the decodable space but outside the detection space of interest. A greater proportion of ghosts falls in this region as the margin between the two increases. The situation is depicted in Figure 12-24 for two different PRF schedules. The upper diagram

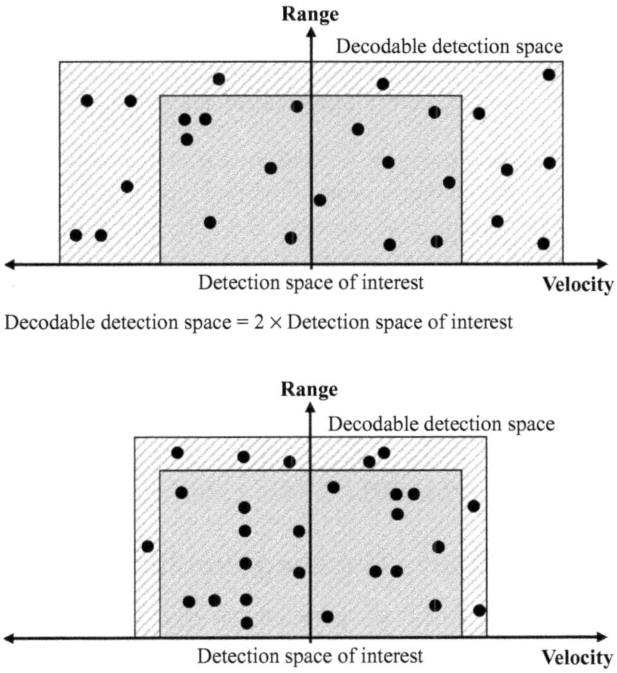

(a) Decodable detection space = 2 × Detection space of interest

(b) Decodable detection space = 1.4 × Detection space of interest

FIGURE 12-24 Ghosts Falling within/outside the Detection Space of Interest

employs PRFs that result in a decodable space twice the area of the detection space of interest. There are 27 correlation possibilities (3 targets seen in 3 PRFs) plotted as solid black circles, of which only 14 fall within the detection space of interest—approximately half the total, in keeping with the ratio of the areas. The lower plot uses PRFs that give a decodable space merely $\times$ 1.4 the area of the detection space of interest. Again, there are 27 correlation possibilities, of which 19 (70%) fall within and 8 (30%) fall outside the detection space of interest, consistent with the ratio of the areas. Given that in both cases the three true target correlations do fall within the detection space of interest this leaves 11 ghosts within the detection space of interest in upper plot and 16 ghosts in the lower plot.

12.4.7 Orbits of Ghosts

The range and velocity at which a ghost appears depends on the ranges and velocities of the genuine targets whose returns have been cross-correlated to form the ghost and on the values of the PRFs in which the returns were detected [29, 22]. The path taken by a ghost as the range and velocity of the real targets change and as the PRF changes is known as the *orbit* of the ghost. Orbits of ghosts can resemble the movement of real targets and may even form tracks for some considerable time. Eventually, the orbit of a ghost suddenly jumps in range and/or velocity as the ambiguous target returns jump through a range and/or velocity ambiguity, and the old track dies and a new one will be formed.

Consider the following (simple) examples of two targets being observed using a 2-of-2 PRF schedule. This gives rise to $2^2 = 4$ correlation possibilities, two of which correspond to the true targets and the other two being ghosts. If both targets move $+L$ range cells in the same direction, then both ghosts also move by $+L$ range cells, both in the same direction as each other and in the same direction as the true targets. In fact, the same result is obtained should the two true targets move through $+L$ range cells where

$$L = (\textit{actual range shift}) \bmod (\textit{number of range cells in any one of the PRIs})$$

Therefore, multiple targets moving in formation (e.g., on a radial course) ghost heavily, and the resulting ghosts match the motion of the real targets.

As the target range continues to increase, the echo in one PRF approaches the end of the receiving period, fades into an eclipsed range, and then reappears early on in the next receiving interval. (This inevitably happens in one PRF before any other.) When this occurs the ghost suddenly jumps in range by the maximum unambiguous range of that PRF. Similar effects happen in Doppler as a target accelerates or decelerates from one ambiguous Doppler interval into its neighboring one.

If now, just one of the real targets shifts by $+L$ range cells, then one ghost shifts by $+L$ range cells and the other shifts by $-2L$ range cells. Both ghosts move but in opposite directions. A change in one of the PRIs also causes both ghosts to shift.

12.4.8 Scenarios that Give Ghosting Problems

Multiple targets seen in the same beam position can give rise to ghosting. If these targets have a similar range, then their returns already correlate in range and correlation only in velocity is required to give a ghost. Similarly, if multiple targets have a similar velocity, then any correlations in range result in ghosts. These situations can arise in the following scenarios.

12.4.8.1 Crossing Targets

These produce returns within the same azimuth and range, and the result is multiple ghost targets. Since these are crossing targets, the ghosting effect is short-term; that is, it persists while two targets are within the radar beamwidth and is not necessarily significant.

12.4.8.2 Formations of Multiple Targets with Similar Velocities Moving Radially

This is seen to be the most significant problem. The individual target returns remain within the radar beamwidth for an extended period and persist on the same azimuth over many antenna scans. Under these circumstances the true targets and ghosts form tracks, and there is the potential problem that the tracker would become overloaded. The *multiple-targets-with-similar-velocities* formation is the worst-case scenario and may be presented to an airborne radar by unwanted ground-moving targets and targets of interest with low closing rates. Figure 12-25 is a result of a simulation to plot target correlation points arising from a single scan of an airborne radar similar to that of Table 12-1 operating a 3-of-8 PRF schedule. The correlation points are displayed over the range/velocity space of interest. No MBC notch filter is implemented so that detections very close to, or even within, the MBC are not removed. MBC is seen as the thick horizontal band of returns centered at zero velocity. Note also that the range and velocity axes of Figure 12-25 are at right angles to the axes of Figure 12-24. The radar has been presented with 20 targets; some are within the MBC response (and so might ghost with the clutter return), some are in close formation (which might be expected to cause heavy ghosting), and others are widely dispersed. The number of correlation possibilities (excluding those due to clutter and false alarms) is therefore $20^3 = 8000$. However, due to the effects described earlier

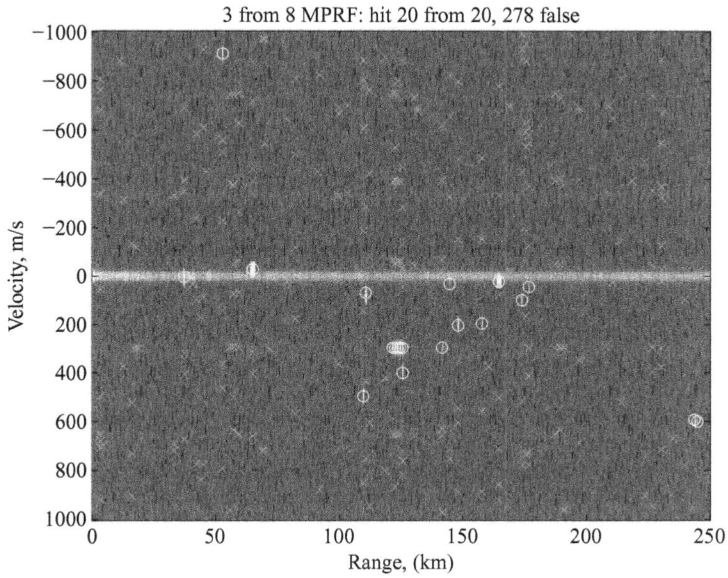

FIGURE 12-25 Simulation of Real and Ghost Target Detections

(Section 12.4.2), only 298 distinct target correlation points are obtained. The 20 correct target correlations are indicated by the circles, and the 278 ghosts are indicated by the crosses.

12.5 | ADDITIONAL TECHNIQUES TO COMBAT GHOSTING

12.5.1 Introduction

While this chapter is ostensibly about PRF selection in medium PRF radars, it seems appropriate at this point to include all techniques to combat the ghosting problem. The following sections cover techniques not concerned directly with PRF selection.

The precise implementation of an algorithm to decode target parameters and declare a detection not only determines the ability of the radar to correlate targets correctly but also its propensity to declare a ghost target. After all, deciding between a true and a ghost correlation are two facets of the same task. The algorithm must declare targets that are correctly correlated and not declare incorrect correlations.

12.5.2 Combating Ghosting via the Correlation in Range and Velocity

Most decodability algorithms are based on the coincidence algorithm since this permits decodability in range and velocity and offers fewer constraints on PRF selection. (Algorithms based on the Chinese remainder theorem are used and normally support only range decodability.) Target correlation in range and velocity is far more resilient to ghosting than correlation in range only [12]. The ghosting phenomenon for radars that correlate in range only may be depicted by collapsing all ambiguous target responses onto the range axis. If all target responses in Figure 12-17 were collapsed onto the range axis, many

FIGURE 12-26 Ghost Correlations in Range versus Rangeand Velocity

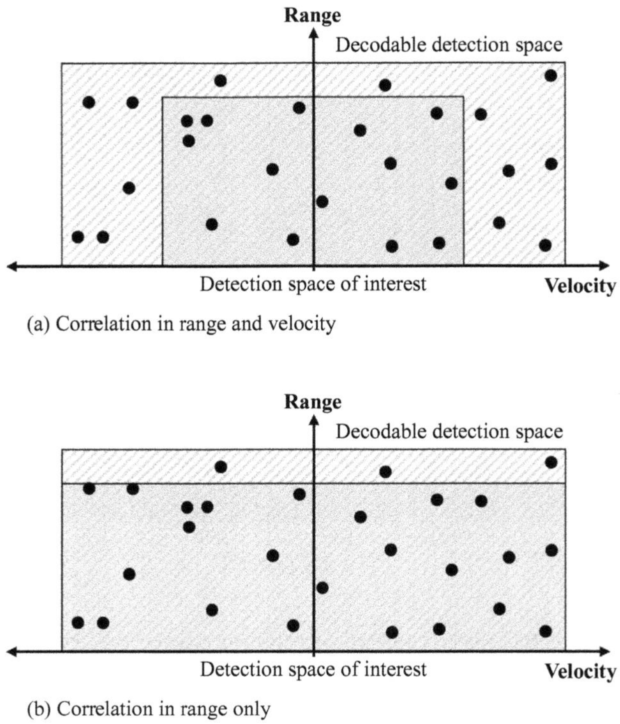

detection points would be crowded along the range axis, and there would inevitably be cases of coincident detection points in $M = 3$ PRFs giving rise to ghosts. It may seem that this would generate many more ghosts than would be incurred by distributing the detection points in the range and velocity space, but this is incorrect; there are always N_T^M correlation possibilities within the decodable space, although some of these may coincide. If all the detection points falling at the ghost ranges were once again strung out along the velocity dimension then there would come a point somewhere within the decodable velocity interval where a ghost would be generated. If the velocity of this ghost falls outside the velocity of interest, then a system in which correlation in range and velocity takes place would not declare a ghost. However, if there is no correlation in velocity then all ghosts within the decodable velocity space would be declared. The situation is shown in Figure 12-26 (based on Figure 12-24) for 27 correlation possibilities indicated by the solid circles. Correlation in range and velocity is depicted in the upper plot for the situation whereby the decodable space is twice the area of the detection space of interest. With a lack of any correlation in velocity, the system acts as if its maximum velocity of interest equals the maximum decodable velocity, and this is depicted in the lower plot. This now captures more ghost targets than previously.

12.5.3 Target Clustering and Extraction Algorithms Using a Maximum Likelihood Method

Target detection points tend not to coincide precisely when detected in several PRFs due to small measurement errors, range and velocity cell straddling, and inaccuracies in compensating for range walk. As a result, clusters of ambiguous detection points tend to form. Correct ambiguity resolution is more robust if the target extraction algorithm first groups nearby detection points into *clusters*. The successful declaration of true target correlations in preference to ghosts is somewhat sensitive to the clustering algorithm and the allowable size of clusters [6, 15, 32, 33]. Clustering of ambiguous detection points in range has been used to decode range ambiguities in systems that alternate their operation over several high PRFs [32, 33]. The algorithm uses a maximum likelihood method and is summarized here. Targets are highly ambiguous in range, and the criterion for declaring a target is that it must be seen in at least three PRFs ($M = 3$); therefore, clusters of three detection points from three different PRFs represent potential targets. This algorithm logs clusters of potential target detections, giving the highest priority to clusters containing detection points in four or more PRFs. The second priority is given to a fixed number of clusters of the three most tightly confined detection points; a maximum of 10 such clusters was permitted and was thought to be a reasonable compromise between correctly resolving range ambiguities and the speed of the algorithm. A maximum likelihood calculation is then performed to test the hypothesis that the potential targets arise from the cross-correlation of noise-generated false alarms or the cross-correlation of one, two, three, or more genuine targets. All cross-correlation possibilities are considered, and the most likely combinations are declared as the targets (those less likely are dismissed as ghosts). The ability of this method to correctly resolve the range ambiguity of multiple targets in preference to declaring ghosts was explored as a function of the error in the measured ambiguous ranges. Gaussian range errors of zero mean and a standard deviation ranging from 0 to 0.3 range cells were imparted onto the measured ranges of two test targets. For zero range errors, the probability of declaring the two targets correctly was marginally less than 1.0, whereas the probability of declaring a ghost was only slightly greater than 0. For a known error of 0.3 range cells, the probability of declaring the two targets correctly reduced to 0.9 and the probability of declaring a ghost increased to 0.1. In practice, the actual range error would not be known. When the error was unknown (but was in fact 0.3 range cells) but was assumed to be 0.2 range cells, then the probability of correct resolution of the two targets was 0.8 and the probability of declaring a ghost became 0.2. The range resolution method remained robust for range errors whose standard deviation was up to 0.2 range cells when two targets were present but the performance degraded as the number of targets increased to three. The algorithm has also been assessed for its ability to resolve the ranges of four targets, all at a common velocity, seen in a medium PRF schedule of five PRFs. Large range errors promoted the declaration of ghosts, and some targets were missed due to blindness; furthermore, noise-generated false alarms tend to result in a ghost. Nevertheless, the algorithm resolved target ranges robustly and in a computationally efficient manner. These algorithms would be equally applicable in solving velocity ambiguities in medium PRF radars. (There was no suggestion that the five PRFs were selected for optimum blindness and decodability margins.)

12.5.4 Combating Ghosting via the Target Extraction Algorithm

An implementation of the coincidence algorithm that was successful in preferentially declaring correct target correlations over ghosts is described by Wiley et al [6]. This study sought to quantify the ghosting performance of a novel target extraction algorithm for two different target scenarios and several optimum PRF schedules. The performance was tested against a model of an airborne FCR similar to that described by Table 12-1. Two types of multiple target scenarios were used; the first placed between one and five targets at random values of range and Doppler, and the second placed 4, 6, 8, or 10 targets at

150 m range intervals, each with the same Doppler. The former gave a random placement of targets, which is perhaps an unlikely situation in reality, whereas the latter represented a close formation, whose range/Doppler centroid was randomly placed on each trial, and is a more likely occurrence. Zero, one, or two noise-generated false alarms of random range/Doppler were also added to permit the possibility of noise-generated ghosts. A small random variation on target range/Doppler was also imparted over successive PRFs to represent random measurement error and had the effect of spreading the clusters of detection points slightly.

The target extraction algorithm considered the *base* targets initially (i.e., target detection points within the first unambiguous range and Doppler intervals) and repeated these detection points into the lattices of ambiguous detection points over the detection space of interest by the addition of multiples of the R_{mu} in range and multiples of the PRF in Doppler. Clusters were then formed through the proximity of detection points in different PRFs. The target extraction algorithm was based on the concept that genuine targets are characterized by clusters of a large number of detection points (i.e., visible in a large number of PRFs) in a small region of range/Doppler space, whereas ghost targets are more likely to be observed in just a few PRFs. It also discounted any clusters containing detection points already attributed to the clusters considered genuine. Therefore, potential ghost target clusters containing the ambiguous detections of genuine targets that were repeated in the range and velocity domains were dismissed. The target extraction algorithm progressed in the following stages:

- The (base) target detection points were expanded into the full lattices of ambiguous returns over the detection space of interest. A range offset was made to allow for the range walk of high Doppler shifted targets.
- Target detection points were sorted in ascending range order.
- Closely spaced targets were formed into pairs.
- Closely spaced pairs were formed into clusters.
- Clusters were checked and possibly subdivided if they contained more than one target detection in any one PRF. In this way, large clusters corresponding to several closely spaced targets generated subclusters, one for each target.
- Clusters and subclusters were ranked in order of the number of detection points.
- The process continued by declaring the cluster or subcluster having the greatest number of detection points to be a real target and tagging the detection points and their repetition in the time and Doppler domains as having already been declared, potentially reducing the size of some other clusters.
- The algorithm progressed from largest to smallest clusters and either declared them as real targets and tagged their detection points or dismissed them as ghosts.

The proximity of detection points that formed pairs and hence clusters was an important variable in the success of the algorithm. Rectangles in range/velocity space of the following dimensions were trialed: 90 m $\times$ 1.2 m/s; 90 m $\times$ 0.75 m/s; 45 m $\times$ 0.6 m/s; and 45 $\times$ 0.375 m/s. These rectangles were based on the range and Doppler resolution and defined the dimensions of the maximum allowable cluster sizes.

Multiple tests were conducted that explored the various combinations of variables for the randomly distributed targets and close formation targets, respectively. These trials resulted in 240 combinations for randomly distributed targets and 192 combinations for close formation targets. A total of 500 experiments of each combination were run to generate statistics on the correctly reported targets, additional targets (i.e., ghosts), genuine targets not reported, and blind targets.

When one to five random targets were applied, approximately 95% of them were correctly reported, irrespective of the schedule, number of false alarms, or allowable cluster size. Additional ghost targets were generally lower than 0.5% of the total applied targets. There was no significant increase in ghosts when one false alarm was applied over the case of zero false alarms. However, a small increase in ghosts (depending on allowable cluster size) for some schedules was seen when two false alarms and only one or two targets were applied and was attributable to the correlation of one false alarm with the other. Generally, the incidence of ghost targets increased with increasing allowable cluster size but reduced with increasing number of real targets. This was perhaps surprising but was due to the fact that in identifying several genuine targets the algorithm tagged most of the detection points, and in so doing the smaller clusters typical of ghosts were more readily discounted.

When 4, 6, 8, or 10 targets in a close formation were applied, about 95% were correctly reported, and that percentage was independent of allowable cluster size and numbers of false alarms but reduced with increasing numbers of targets. The number of ghosts was somewhat greater for the close formation targets than those of random placement. This was due to the fact that all targets had the same Doppler, so alignment in range only was required to form a ghost. The worst-case ghosting of around 7.5% of the total number of applied targets was incurred for the case of 2 false alarms, 10 formation targets, and an allowable cluster size of 90 m $\times$ 1.2 m/s (2-of-6 schedule). Under similar conditions, the best-case ghosting performance of 1.3% of the total number of applied target was obtained for the optimum 3-of-8 PRF schedule. Longer schedules tended to result in fewer ghosts as they gave rise to clusters having a greater number of detection points that were more readily identified as targets in preference to the smaller clusters of ghosts. In many cases no ghosts were reported at all. The numbers of targets not reported followed the trend in ghosting performance since the reporting of a ghost was usually done in preference to the reporting of a target.

12.5.5 Target Extraction Algorithms Based on Non-Coherent Integration

An algorithm similar to that described in Section 12.5.4 that successfully declares true target correlations in preference to ghosts is based on the non-coherent integration of the powers of ambiguous detection points in each range/velocity cell acquired over all N CPIs within the schedule [34]. However, due to slight misalignment of targets in range and velocity over the N CPIs, it is necessary to form clusters of detection points [35], and this results in an algorithm very similar to that of Section 12.5.4 [6]. Clusters of three range cells by three velocity cells are formed, and the clusters are rank-ordered on the basis of the total power within each cluster [35]. This algorithm is based on the premise that genuine target correlations tend to have a higher sum of non-coherently integrated powers than ghosts. (The algorithm of Section 12.5.4 assumes that genuine targets are characterized by a greater number of detection points, so it stands to reason that if all detection points have the same SNR then the rank-order for both algorithms would be identical. If, however, some targets are stronger than others, the rank-order of clusters between the two algorithms may differ.) The strongest target is declared as a genuine target. Ambiguous repetitions of detection points from this target are removed from other clusters and replaced with a sample of the ambient noise level. Having done this, the clustering process is repeated, the sum of the non-coherently integrated power in each cluster is computed, and clusters are rank-ordered according to this sum. The strongest target cluster is again declared as a genuine target. The algorithm repeats this cycle, declaring the strongest target cluster on each iteration and eroding the remaining clusters until no more are left. A (simpler) variant of the deghosting algorithm has been investigated in a noise-limited background in which 15 targets were placed in the detection space of interest and detected in three PRFs [36]. The algorithm correctly declared all 15 targets plus only one ghost target.

The detection performance of the non-coherent integration has also been compared with M-of-N binary integration method [34, 35, 36]. In the former case, the non-coherent sum of powers over all N CPIs is compared with a detection threshold. In the latter case, the data in each cell within each individual CPI are compared with a threshold, and, subsequently, detections are combined in an M-of-N binary integrator. The non-coherent integration offers an improvement over M-of-N binary integration equivalent to a processing gain of around 2 dB in a noise-limited environment.

A hybrid approach may also be adopted based on a method of data selection that simplifies the deghosting at the expense of about 0.5 dB in detection performance with respect to the straightforward non-coherent integration approach [34]. The hybrid approach sets a low detection threshold within each range/velocity cell for data acquired over each individual CPI. Non-coherent integration is then applied across the N CPIs on all cells that registered a target detection. A tally is maintained of the number of target detection points falling within these cells; this records the number of CPIs (or PRFs) in which target data were actually observed above the low initial threshold. The non-coherently integrated sum for all detections in which the target was observed in $M \geq 2$ PRFs is then selected and compared with a second threshold. This second threshold is set as a function of the number of CPIs in which the target data were observed. The ghosting rate is shown to be a function of the signal-to-noise ratio (SNR) and P_{FA} and peaks for SNR $= 7.5$ to 8.5 dB ($P_{FA} = 10^{-4}$ to 10^{-6}). Use of the data selection method reduces the peak ghosting rates by some 20 to 30%.

12.5.6 Combating Ghosting via Target Data Consistency

The orbits of ghosts can resemble the movements of genuine targets quite convincingly, and ghost tracks could be formed. However, in many cases the rate at which a ghost migrates through range cells is inconsistent with its radial velocity. This inconsistency should be noticed by the tracker and the tracker should not initiate a track, if indeed the tracker does correlate the changes in range with the radial velocity. For example, if the scanning of a radar antenna is such that the revisit rate is once per 6 seconds and the target is seen with a radial velocity of 240 m/s (closing), then the range of the target should reduce by 1440 meters on each successive detection. If the target range does not change in sympathy with the radial velocity, the tracker should not initiate a track and should suspect that the target is a ghost or the result of some interference/jamming.

Not all trackers correlate the changes in range seen over successive scans with the radial velocity of a target; however, this is a relatively simple and effective precaution that would minimize the occurrence of ghost tracks. This strategy also has implications for effective jamming and target generation test equipment. Jammers must produce false target signals whose Doppler shift must be tied to the changes in range to be accepted by the tracker of the victim radar. Similarly, target generators used as first-line test equipment to provide an end to end confidence test of the radar must do likewise.

12.5.7 Decorrelating Ghost Tracks Using Multiple PRF Schedules

In spite of all the measures described in the previous sections, ghost tracks may still be formed, particularly for close formations of targets on a radial course. Any given PRF schedule can give rise to ghost tracks whose orbits depend on the precise PRF values used in that schedule, among other factors. If, however, the radar switches to a second PRF schedule using different PRF values, the genuine target tracks remain consistent across the two schedules but the ghost tracks do not [29]. Ghost tracks correlate only for the duration over which the radar operates on any particular PRF schedule. When the radar changes its PRF schedule, the positions (range and velocity) and hence the orbits of ghosts also change, but the true target correlations remain consistent. If the radar alternates its operation over several different PRF schedules, each schedule produces a similar number of ghosts but only small fragments of ghost tracks are formed. When a sufficient number of PRF schedules are used in rapid succession (e.g., a schedule change is made for each new scan), then the orbits of the ghosts decorrelate and ghost tracks are not formed.

Figure 12-27 shows the range/time plot of 10 targets viewed with a single schedule of two PRFs, as follows: two closing targets with equal velocities; four opening targets with equal velocities; three closing targets with differing velocities; and one stationary target. All moving targets are on a radial course. The PRFs are such that there are approximately five ambiguities in range within the range space of interest. Severe ambiguity can clearly be observed, and, since all the ghosts are strong, it is very difficult to determine which tracks are from the real targets and which are ghost tracks. As the probability of detection is set at 100%, some of the ghost tracks can be identified as they have brief breaks in them as ghosts jump through ambiguities and can be dismissed, but this is a special case.

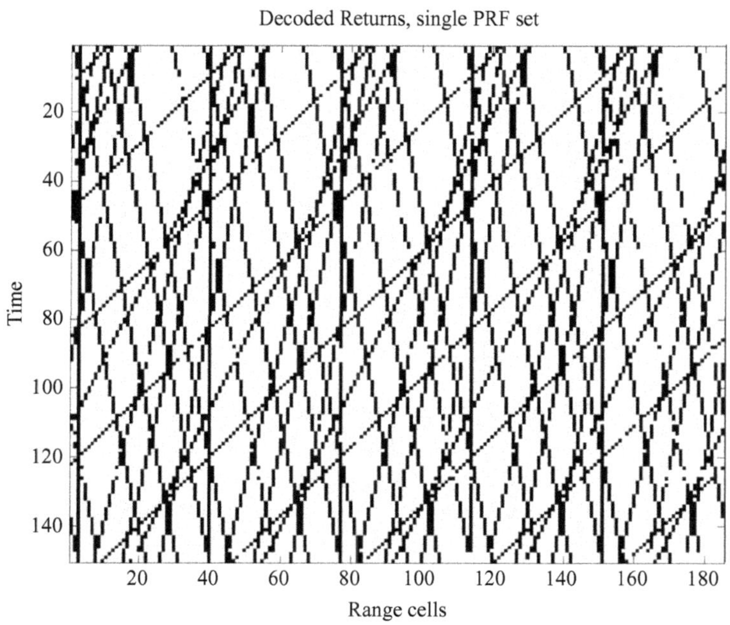

FIGURE 12-27 ■ Tracks for 10 Targets Seen in a Single Schedule of 2 PRFs [29]

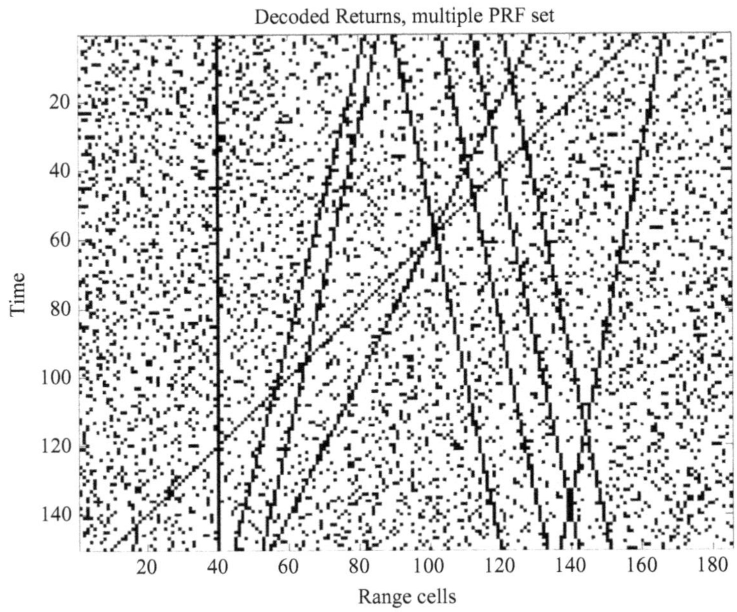

FIGURE 12-28 Tracks for 10 Targets Seen over 5 Schedules of 2 PRFs

Figure 12-28 illustrates the effect of changing the PRF schedule. The same target set is now observed using five PRF schedules, each of two PRFs. The system cycles through the five PRF schedules by changing the schedule on a scan-by-scan basis. The true tracks are clearly visible against a background of track fragments caused by ghosting. Although approximately the same number of ghosts is present, they occur in different locations for each PRF schedule; therefore, the effect of the schedule changes has been to decorrelate the ghost tracks. The five schedules used here are sufficient to cause the ghosts to be completely decorrelated into a random speckle-like background. Unfortunately, the ghost target returns must still be handled by the tracker, so a tracking system that can handle a very high false alarm rate must be used.

This strategy is based on the premise that multiple PRF schedules can be derived, each of which offers robust decodability with large decodability margins, low blindness, and adherence to maximum, minimum, and mean PRF constraints. Some recent methods of PRF selection have been successful in finding several near-optimum PRF schedules, so this is a realistic proposition [37]. In fact, if one is using several different PRF schedules, one may also take advantage of this by ensuring that the small degree of blindness in one schedule does not coincide with the blindness in other schedules. This ensures that, over several schedules, the blindness statistics can be exceedingly low.

12.5.8 Summary of Ghosting

There are several mechanisms by which ambiguous target returns may be miscorrelated and ghosts may be declared. Hybrid mechanisms exist as well, such as a discrete target being miscorrelated with an extended clutter patch.

Targets may be resolved in angle, range, and velocity. For ghosts to occur, multiple targets (including extended targets) must occur that are unresolved in angle; that is, they must be present in the same beam position. If targets are also unresolvable in one of the other dimensions (range or velocity), then ghosts form more readily. This tends to arise momentarily for crossing targets and is more prolonged for close formations of targets, particularly those on a radial path. The severity of the ghosting problem worsens with increasing number of targets. Ghosts can resemble real targets, and ghost tracks may be formed.

Ultimately, no strategy can guarantee the elimination of ghosts since the ghosting phenomena depend on the number of targets and the extension of targets in range and velocity, both of which are factors outside the control of the designer. However, much can be done to combat the ghosting problem. The various strategies may form several layers of defense that the radar can deploy to counter the ghosting problem, summarized as follows:

(i) Don't incur ghosts in the first place by (a) selecting PRFs to maximize the decodability margins and (b) designing a clustering/target extraction algorithm that promotes the declaration of genuine targets while suppressing ghost correlations.

(ii) If ghosts are generated, then minimize the proportion of those seen by (a) selecting PRFs to maximize the decodable space and (b) correlate target returns in range and velocity.

(iii) If ghosts still get through, then don't form a track on them by ensuring the tracker can recognize realistic target motion from unrealistic or inconsistent ghost data.

(iv) If ghost tracks are inclined to form, then break these up by using multiple PRF schedules over successive scans.

It may be very difficult to select PRF values that combat the ghosting robustly while also minimizing blindness and adhering to other PRF constraints. These constraints may not be mutually compatible, so a trade exists between ghosting (decodability), blindness, and other issues. It is even more difficult to find suitable PRF values for several schedules, but recent methods of selecting optimum PRF values have achieved this.

Coping with the ghosting problem is a high priority for the designers of military radar systems. It has profound implications for the radar's ability to deal with digital radio frequency memory (DRFM) jammers and, conversely, for the requirements on DRFM jammers to generate effective false targets.

CHAPTER 13

Medium Pulse Repetition Frequency Schedule Design

Chapter Outline

13.1 SELECTION OF *M* AND *N*

13.1.1 Introduction

Hitherto, ambiguity resolution in medium pulse repetition frequency (PRF) systems can be achieved by comparing ambiguous target data in a minimum of two PRFs ($M = 2$), although practical constraints often dictate that targets must be detected in a minimum number of three PRFs ($M = 3$). This minimum number of PRFs is also the threshold for the declaration of a target in a binary integrator system. The probability of detection increases considerably, especially against scintillating targets, if target data are available in a fourth PRF, that is, one more than the bare minimum necessary to declare a detection. In any individual PRF there exists considerable blindness over the range/velocity space of interest due to eclipsing, the rejection of main beam clutter (MBC), and its repetition in the Doppler domain and due to overwhelming levels of sidelobe clutter (SLC). Therefore, to be sure of obtaining the target data in a sufficient number of PRFs, many more PRFs (N) have to be used within the schedule, all of which must be transmitted within the beam dwell time. The binary integrator sets the threshold requiring target data in at least M-of-N PRFs, hence the reference to an M-of-N schedule. A typical M-of-N schedule used in current systems is 3-of-8.

This chapter addresses the selection of the numbers M and N and considers the ramifications of using schedules that depart from the norm. Longer schedules, $N > 8$, have been perceived as advantageous in heavy clutter situations but also cause complications and compromises in other aspects of the radar performance. Recent work has examined the performance of shorter schedules ($M < 3$, $N < 8$), and these have been found to have several advantages over longer schedules. The question of how best to use the limited beam dwell time arises and is also examined here.

13.1.2 Selection of *M*

Pulse timings in most radars are derived from a fundamental clock. All timings are therefore quantized into intervals that can be no finer than the period of the clock. For example, if a 10 MHz quartz crystal oscillator clock is used, then the timing resolution is 100 ns. The range resolution, ΔR, of the radar is governed by the signal bandwidth, B, via

$$\Delta R = \frac{c}{2B} \qquad \text{repeat of (3-21)}$$

and it is common to match the range cell width to the range resolution. For convenience, it is desirable to make the range cell width a multiple of the clock period. Hence, a signal bandwidth of 2 MHz yields a range cell width of 500 ns (5 clock cycles) and a range resolution of 75 meters. The pulse repetition interval (PRI = 1/PRF) is then usually fixed at an integer number of range cells, again as a matter of convenience, to ensure all timings can be derived from the common clock. In this example, the PRI resolution would be 0.5 μs, and for an allowable PRI range of 35 μs up to 150 μs there are 231 possible PRIs to choose from. Typical PRI resolutions have been given as 0.5 μs [3, 5] and 1 μs [2, 4, 11]. This is a relatively coarse resolution and results in a limited number of allowable PRIs. If a finer PRI resolution is used, then there are very many more allowable PRIs. For example, a PRI resolution of 10 ns (clock frequency = 100 MHz) [6, 7, 8] yields 11,501 possible PRI values over the range of 35 μs up to 150 μs.

All combinations of M-of-N PRFs should be chosen to meet the decodability constraints, with suitable decodability margins, and to minimize blindness and ghosting. When the PRI resolution is coarse, finding suitable PRIs that offer full decodability on the basis of $M = 2$ PRFs is difficult and grows ever more difficult as the range/velocity space of interest increases, as discussed in Section 12.1.4. Full decodability is often possible only on the basis of $M = 3$ PRFs. Recall also that decodability algorithms based on the Chinese remainder theorem (CRT) further constrains all combinations of M-of-N PRIs to be comprised of coprime numbers of range cells; there is no such constraint when using the coincidence algorithm. In practice, 3-of-N schedules offer larger decodable range and velocity spaces than 2-of-N schedules. Although finding 2-of-N PRF schedules that are decodable over a large range/velocity detection space of interest may be more difficult than finding 3-of-N decodable schedules, this is offset by the fewer number of PRF combinations that must meet the decodability requirements. For example, there are 56 combinations of 3-of-8 (${}_8C_3 = 56$) all of which must be decodable; however, there are only 28 combinations of 2-of-8 (${}_8C_2 = 28$) which have to adhere to the decodability requirements.

Irrespective of whether a fine or coarse PRI resolution is used, any PRI must be maintained accurately over the operational use of the radar. For example, an error of ± 10 ns on a nominal PRI of 100 μs represents a percentage error of $\pm 0.01\%$. Inverting these figures results in a PRF of 10 kHz ± 1 Hz. This is about the limit to which the accuracy of the frequency of an oscillator may be maintained over the range of environmental conditions specified for military airborne radars. In particular, master oscillators are often temperature stabilized to protect them from the full temperature range to which the system is exposed. The tenth harmonic of this PRF falls at 100 kHz $\pm$ 10 Hz, so the 10th ambiguous repeat of MBC, assuming a 10 GHz carrier frequency, falls at a velocity of 1500 $\pm$ 0.15 m/s. For a MBC rejection notch width of ± 25 m/s, the error on the 10th ambiguous repetition amounts to 0.6% of its width. These rough calculations demonstrate that a PRI accuracy of ± 10 ns is perfectly acceptable and, conversely, that it is pointless attempting to define PRIs any finer than this.

If it is possible to select PRIs with a fine resolution (assuming ambiguity resolution is via the coincidence algorithm), then PRF solutions can be found offering good decodability and blindness margins on the basis of requiring target data in $M = 2$ PRFs [6, 7, 8]. Furthermore, 2-of-N schedules are viable for applications requiring a large range/velocity detection space of interest such as airborne fire control radars (FCRs). If target data are only required in 2 PRFs then this suggests that fewer than eight PRFs may be required in the schedule. However, the use of 2-of-N schedules raises concerns over the ghosting performance since now coincident target detections are required in only two PRFs (rather than three) for a target to be declared.

13.1.3 Selection of *N*

Increasing the number of PRFs used in a schedule gives rise to a greater decorrelation in the eclipsed ranges, blind velocities due to MBC rejection and its repetition in the frequency domain and regions of strong SLC. The more widely these are dispersed across the N PRFs used, then the greater the probability that targets may be seen in more of the PRFs. This reduces the blindness of the overall schedule and improves the probability of detection in clear regions of the range/velocity detection space. On the downside, increasing N increases the pressure on the time budget since an increased number of space charging periods are incurred and the CPI duration must be reduced, leading to lower processing gains and a loss of velocity resolution.

Recall that the mean PRI is constrained by

$$Beam\ dwell\ time \leq N[(n \times \overline{PRI}) + t_{SC}] \qquad \text{repeat of (12-12)}$$

in which n is the FFT size, $\overline{PRI}$ is the mean PRI, and t_{SC} is the space charging time and is given by

$$t_{SC} = 1.5\frac{2R_{max}}{c} \qquad \text{repeat of (10-6)}$$

where R_{max} is the maximum range of interest, and the factor of 1.5 applies an extra 50% margin to the maximum range to cater for the possible detection of very large but very distant targets and clutter; the 50% margin should be taken only as a rough *rule-of-thumb*. A space charging time needs to be included whenever the radar switches from one coherent processing interval (CPI) to the next. During this time, the radar must transmit pulses on the new PRF (also known as *fill* pulses) but must blank the processing of any returns. The blanking period must be sufficient to allow all returns from the previous PRF to clear the system before processing returns on the new PRF may begin. In essence, the space charging pulses must be generated, but this time cannot be used by the receiver and processor; it is dead time.

Using the hypothetical, but realistic, model of the airborne FCR described in Table 12-1, the beamwidth and scan rate give a beam dwell time of 65 ms and $R_{max} = 185.2$ km, giving $t_{SC} = 1.85$ ms. For a schedule of eight PRFs ($N = 8$) there are eight space charging times totaling 14.8 ms or 22.8% of the total beam dwell time. The inclusion of a ninth PRF ($N = 9$) incurs a total space charging time of 16.65 ms or 25.6% of the beam dwell time. Furthermore, the mean CPI duration ($n \times \overline{PRI}$) is constrained to ≤ 6.275 ms for the eight PRF case and ≤ 5.372 ms for the nine PRF case. Conversely, reducing the schedule to just six PRFs ($N = 6$) incurs just 11.1 ms total space charging time (17.1% of beam dwell time) and permits a CPI duration of up to 9 ms. Increasing N incurs a larger number of space charging times, which is particularly damaging to the radar in so many ways:

(i) More energy has to be put into the transmission of pulses that play no part in the detection process.

(ii) The processing has to be blanked for a higher proportion of the time.

(iii) The mean PRI is constrained to a lower value, which further limits PRF choice and the quality of the schedules which can be found.

(iv) The mean CPI is reduced, leading to reduced processing gain and coarser velocity resolution.

Performance could be recovered by increasing the beam dwell time, which leads to a reduced data rate. Variable scan rate is not feasible for mechanically scanned antennas, so the whole scan pattern needs to be slowed down. This is unacceptable. For active electronically scanned arrays, a variable beam dwell time can be implemented where it is needed and thus is far less damaging.

Naturally, shorter schedules of fewer PRFs (lower N) would be desirable provided that adequate target visibility can be maintained, especially in conditions of high clutter. This is facilitated by setting $M = 2$ (in favor of $M = 3$), which, in turn, is rendered a far more realistic proposition if the PRI resolution is fine. Simpson [2] described a method of PRF selection for a spread of allowable PRIs of between 50 μs and 96 μs with a resolution of 1 μs, giving 47 possible values. An exhaustive search is conducted for every combination of eight PRFs; this amounts to ${}_{47}C_8 = 314$ million combinations.[1] For larger numbers of PRIs, the vast number of possible combinations precludes an exhaustive search. In the numerical examples previously given, a PRI resolution of 0.5 μs gives 231 possible PRIs within the limits of 35 μs to 150 μs, whereas a PRI resolution of 10 ns gives 11,501 possible PRIs within the same range. The task is to select the optimum combination of N of these PRIs. The number of combinations can be astronomical; for example, the number of combinations of 8 from 231 is ${}_{231}C_8 = 1.78 \times 10^{14}$, whereas the number of combinations of 8 from 11,501 is ${}_{11501}C_8 = 7.57 \times 10^{27}$! Increasing the number of allowable PRIs by using fine PRI resolution increases the complexity of selecting the precise values of optimal PRIs, but the quality of the solutions is superior to that of solutions drawn from fewer PRIs of coarser resolution. It is better to invest effort at the design stage to find optimal solutions of fewer PRFs than to accept suboptimal solutions of longer schedules simply because they are easier to find.

13.2 | COMPARISON BETWEEN SCHEDULES

13.2.1 Introduction

A few recent studies have examined the viability of 2-of-N schedules and short schedules of $N < 8$ for airborne radar applications [6, 7, 8, 27]. All these studies have applied evolutionary algorithms (EAs) of some sort to select precise PRF values that offer (near-) optimal

[1] Simpson [2] mistakenly gave the number of possible schedules as 12.6 trillion, which is actually the number of *permutations* of 8 from 47 (${}_{47}P_8 = 12.6 \times 10^{12}$). Since each PRF is used only once, it is appropriate to take the number of *combinations* rather than *permutations*.

performance. These studies are reviewed here. Additional studies have been conducted within the defense industry, but these are not openly published.

13.2.2 Decodability

A variety of PRF schedules have been derived and compared for their suitability in an airborne FCR application [6, 7, 8], using an EA-based optimization procedure for the selection of precise values of the PRFs. During the optimization process, the range/velocity blindness of each trial solution is assessed against a model of a FCR based on Table 12-1 and associated clutter model. The FCR model assumes a fixed carrier frequency of 10 GHz and a spread of allowable PRIs from 35 μs to 150 μs, with a PRI resolution of 10 ns giving 11,501 possible PRIs. Checks are introduced into the process to ensure that each trial PRF set is fully decodable, with a range decodability margin of at least 0.7 μs (i.e., the 0.5 μs range cell width plus 0.2 μs) and is free of blind velocities. The detection space of interest measures 185.2 km in range by $\pm$1500 m/s in velocity. Solutions were sought for the following schedules: 2-of-5, 2-of-6, 2-of-7, 2-of-8, 3-of-8, and 3-of-9. Many high-quality schedules were found, thereby confirming that it was perfectly feasible to find several 2-of-*N* decodable PRF schedules for what is a very large range/velocity space of interest. The skyline diagram depicting the range and velocity decodability margins of the best 3-of-8 schedule is shown in Figure 13-1, and the best 2-of-6 schedule is given in Figure 13-2. Whilst no attempt was made to maximize the decodability margins, it is worth noting that the worst-case decodability margin for the best 3-of-8 schedule was 205.5 meters in range (1.37 μs) and 15.4 m/s in velocity, whereas the worst-case decodability margin for the best 2-of-6 schedule was 121.5 meters in range (0.81 μs) and 1.98 m/s in velocity. The decodable space of the 3-of-*N* schedules was generally far greater than that of the 2-of-*N* schedules.

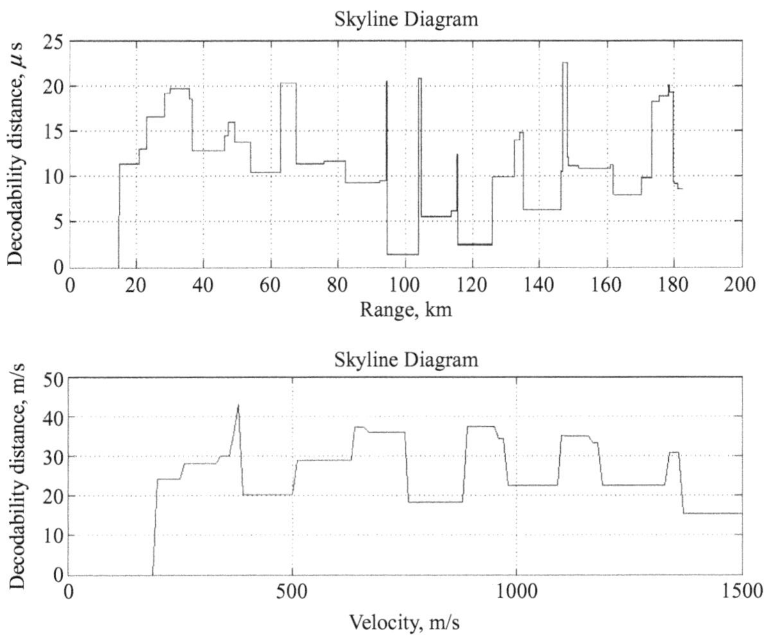

FIGURE 13-1 Skyline Diagram for Best 3-of-8 Schedule

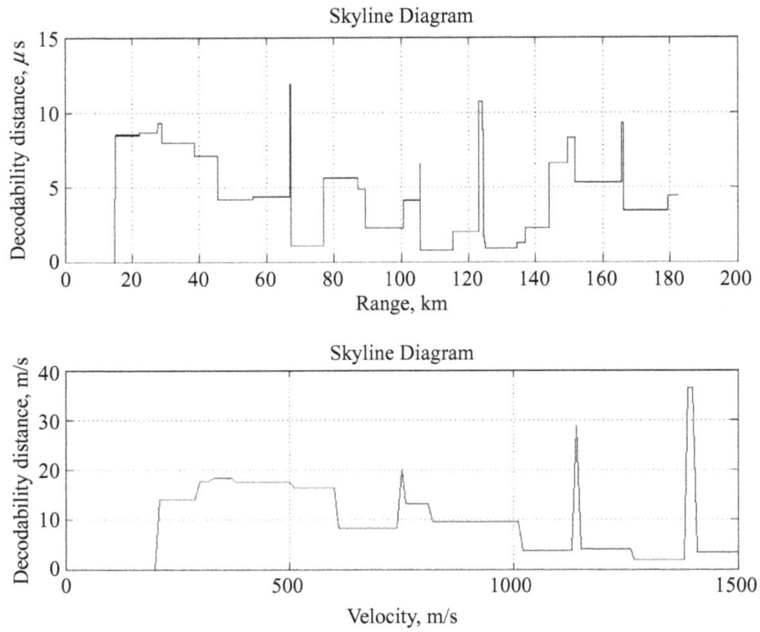

FIGURE 13-2 Skyline Diagram for Best 2-of-6 Schedule

For 2-of-N schedules, the most difficult problem is to find 2-of-8 schedules since this requires the greatest number of combinations (${}_8C_2 = 28$, whereas ${}_7C_2 = 21$, ${}_6C_2 = 15$ and ${}_5C_2 = 10$), all of which must be decodable. Clearly, decodability is easiest for the 2-of-5 case, but finding 2-of-5 schedules is severely restricted by the check to avoid blind velocities. While the 2-of-N schedules were fully decodable over the space of interest, their lower decodability margins and smaller decodable spaces have implications for their ghosting performance.

13.2.3 Blindness

The studies described in Section 13.2.2 set the minimization of blindness as the optimization goal [6, 7, 8]. This resulted in the following near-optimum solutions[2]:

Best 2-of-5, PRIs = 59.90, 68.20, 75.50, 81.50, 93.50 μs
Best 2-of-6, PRIs = 64.04, 74.53, 83.03, 92.07, 100.75, 118.80 μs
Best 2-of-7, PRIs = 73.55, 81.03, 89.76, 99.42, 109.50, 116.46, 125.17 μs
Best 2-of-8, PRIs = 78.92, 81.56, 86.66, 90.46, 99.81, 111.81, 117.09, 128.56 μs
Best 3-of-8, PRIs = 63.11, 69.97, 77.07, 81.31, 90.06, 99.90, 109.75, 119.00 μs
Best 3-of-9, PRIs = 65.00, 65.62, 72.41, 79.96, 83.92, 88.44, 93.30, 102.84, 112.41 μs

The optimization process was driven by attempts to minimize the total blindness in $M + 1$ PRFs over the space of interest in the presence of strong ground clutter and for a 5 m^2 target. Each schedule was assessed to determine the percentage of the range/velocity space of interest which was blind, including blindness due to overwhelming SLC, in $M + 1$ PRFs. A total of 100 repeated runs of the optimization process were conducted to

[2]These solutions are termed "near-optimum" since an exhaustive search was not possible and therefore it was impossible to know whether these solutions were indeed the global optimum.

TABLE 13-1 ■ Blindness Statistics for Near-Optimum PRF Schedules (based on the percentage of the range/velocity detection space of interest blind in $M+1$ PRFs in the presence of strong ground clutter).

M-of-*N*	Min %	Max %	Mean %	Median %	σ %
2-of-5	66.10	66.73	66.43	66.44	0.1434
3-of-8	58.37	59.91	59.01	59.02	0.2803
2-of-6	56.35	57.70	57.12	57.18	0.3316
3-of-9	53.74	55.02	54.46	54.51	0.2656
2-of-7	48.90	50.24	49.46	49.54	0.3437
2-of-8	44.13	45.21	44.59	44.57	0.2296

generate blindness statistics of the near-optimum solutions on each trial. These statistics are presented in Table 13-1 in rank-order from most blindness to least.

It may be noted from these statistics that blindness reduces as N increases and as M reduces, but of the two the reduction in M is the most significant. The blind zone maps for the best 3-of-8, 2-of-8, and 2-of-6 schedules are presented in Figure 13-3, Figure 13-4, and Figure 13-5, respectively. The majority of the blindness is caused by clutter that overwhelms the detection of the 5 m^2 target at long ranges. The schedule having least blindness is the 2-of-8, which has some 14% less blindness than the 3-of-8 schedule. The most noticeable improvement occurs at ranges around 60 to 150 km, beyond which high SLC levels form the dominant cause of blindness. Of particular interest is the fact that the best 2-of-6 schedule has slightly superior blindness statistics to the best 3-of-8 schedule. In the absence of clutter, the total blindness for the best 3-of-8 schedule is 6.82% (based on visibility in fewer than **4** PRFs), but this still includes the blindness due to the first blind range (along the bottom of the map) and the first blind velocity (up the left-hand side of the map), which are unavoidable. These unavoidably blind regions account for 2.26% of the total blind zone map and leave 4.56% (= 6.82 − 2.26) avoidably blind or

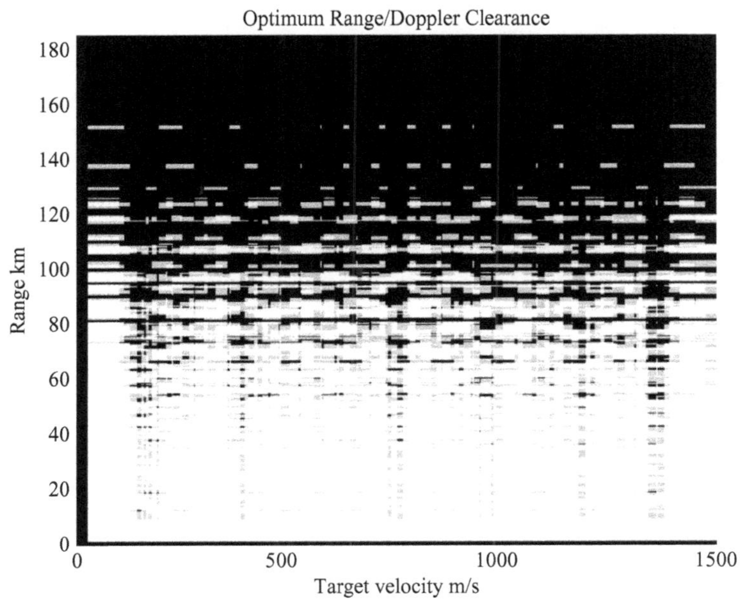

FIGURE 13-3 ■ Blind Zone Map for Best 3-of-8 Schedule (with clutter) (*Black = visibility in fewer than* 3 *PRFs, Gray = visibility in exactly* 3 *PRFs, White = visibility in more than* 3 *PRFs*)

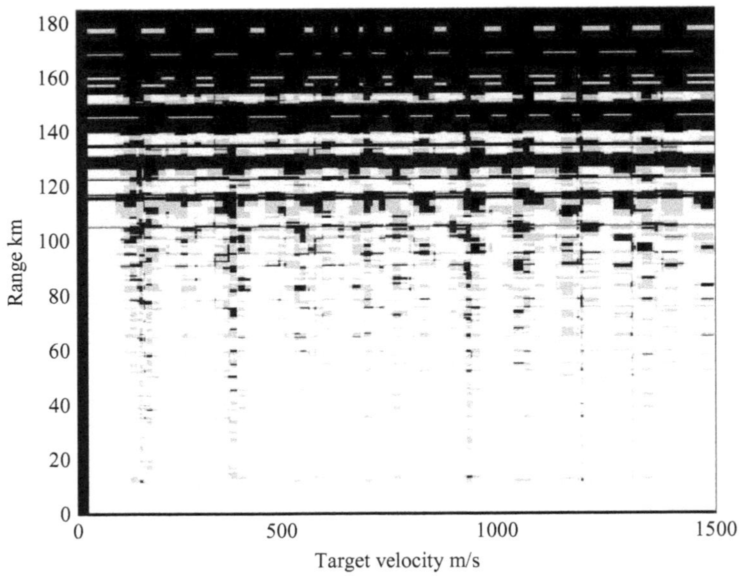

FIGURE 13-4 Blind Zone Map for Best 2-of-8 Schedule (with clutter) (*Black = visibility in fewer than 2 PRFs, Gray = visibility in exactly 2 PRFs, White = visibility in more than 2 PRFs*)

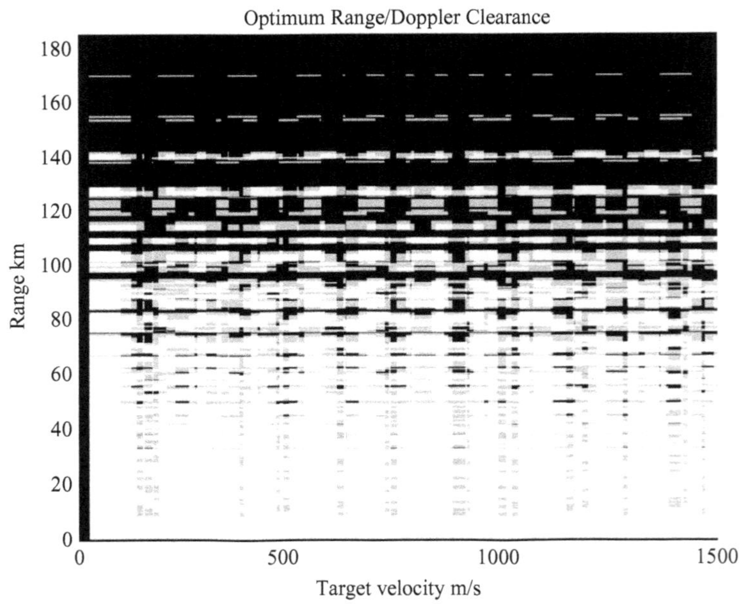

FIGURE 13-5 Blind Zone Map for Best 2-of-6 Schedule (with clutter) (*Black = visibility in fewer than 2 PRFs, Gray = visibility in exactly 2 PRFs, White = visibility in more than 2 PRFs*)

95.44% (= 100 − 4.56) avoidably clear. Similar blindness statistics apply to the best 2-of-6 schedule in the absence of clutter, and, in this respect, a 2-of-6 schedule offers marginally superior performance to the more common 3-of-8 schedules and has the advantage of being a shorter schedule.

There is a suspicion that the evolutionary optimization process acts to position eclipsed ranges and blindness in velocity at positions already occupied by overwhelming clutter returns. Since some small degree of blindness in range and velocity is inevitable, the EA selects PRFs that sneak this blindness into cells already blinded by clutter, and hence no additional blindness is incurred. If true, the PRF schedules have been optimized to the clutter environment and target radar cross section (RCS) used in the models. When the optimization is performed for differing clutter conditions, or no clutter, different solutions are obtained. This suggests that dynamic optimization would be capable of providing near-optimum PRF sets to maximize target visibility in response to the clutter conditions and target sizes prevalent at any one time or beam position. PRF schedules optimized for one set of conditions may be suboptimal when used in a different set of conditions.

Similarly, an optimization procedure based on a genetic algorithm (GA)[3] was used to select precise PRF values for an airborne radar with an AESA antenna in which the beam is electronically steered through $\pm 60°$ [27]. It is therefore necessary to increase the number of pulses in each CPI (and hence the fast Fourier transform [FFT] size and processing gain) to compensate for the beam dilation effects when steered off axis. The detection space of interest measures 400 km in range by ± 40 kHz in Doppler (equates to ± 600 m/s in velocity for a carrier frequency of 10 GHz). This is approximately the same detection space of interest as the FCR study but is some 2.16 $\times$ the range space and 0.4 $\times$ the velocity space of the FCR model. The much greater range space of interest coupled with the much lower velocity space of interest suggests that lower PRFs (longer PRIs) would be favored. The PRF selection procedure includes various constraints on the PRFs to ensure that they adhere to maximum and minimum limits and to ensure that no blind velocities are incurred and that range blindness is avoided at near ranges. The PRI resolution is set at 1 μs. The evolutionary process optimizes PRF selection to minimize range/velocity blindness for 3-of-8 and 3-of-6 schedules. The number of pulses per CPI is adjusted in accordance with the beam position.

The best 3-of-8 schedule has PRIs of 213, 192, 180, 167, 134, 126, 115, and 105 μs. In the absence of any clutter, this schedule is blind over 2.31% of the detection space of interest based on visibility in 3 PRFs. The best 3-of-6 schedule has PRIs of 272, 254, 215, 133, 121, and 117 μs. Again, in the absence of clutter, this schedule is blind over 3.06% of the detection space of interest. However, neither PRF schedule is fully range or velocity decodable over the detection space of interest.

Comparing the two studies, one may observe the favoritism of lower PRFs (longer PRIs) in systems requiring longer decodable ranges and lower decodable velocities. Shorter decodable ranges and higher decodable velocities would favor the selection of higher PRFs (shorter PRIs). The two studies consider quite differing decodable ranges and velocities, albeit with similar areas. However, both obtain similar levels of blindness, but an exact comparison is not possible due to the differing ways blindness is evaluated across the two studies. Both studies affirm the improved blindness resulting from increasing the number of PRFs used in the schedule; both also affirm that careful selection of PRFs ensures that schedules of just 6 PRFs yield acceptable levels of blindness. The latter study achieves this using a much coarser PRI resolution than the former (1 μs, as opposed to 10 ns) but sacrifices decodability in doing so.

13.2.4 Ghosting

The ghosting performance of various near-optimum PRF schedules when used in conjunction with the target extraction algorithm described in Section 12.5.4 has been evaluated using repeated trials [6, 8]. During the course of the ghosting tests the incidence of ghosts

[3]A GA is a specific form of an evolutionary algorithm.

was found to be determined by the inherent inclination of the PRF schedule to cross-correlate target returns or noise-generated false alarms together with the margin of the inequality of (12-3) and (12-4).

$$LCM(PRI_1, \quad PRI_2, \quad \ldots \quad PRI_M) \geq \frac{2R_{max}}{c} \qquad \text{repeat of (12-3)}$$

$$LCM(PRF_1, \quad PRF_2, \quad \ldots \quad PRF_M) \geq f_{D_{max}} \qquad \text{repeat of (12-4)}$$

This inequality is the margin by which the extent of the unambiguous detection range/Doppler space, as defined by the left-hand sides of (12-3) and (12-4), exceeds the range/Doppler space of interest, as defined by the radar model and therefore by the right-hand sides of (12-3) and (12-4). As previously noted, it is difficult to find PRFs that offer a large decodable range/velocity space on the basis of decodability in just two PRFs ($M = 2$), although having a fine PRI resolution helps achieve this. In general, for a fixed PRI resolution, optimum values of PRF for 3-of-N decodable schedules boast considerably larger decodable spaces than optimum 2-of-N schedules. Thus, 3-of-N schedules increase the margin of the inequalities of equations (12-3) and (12-4) over those of 2-of-N schedules. As a consequence, proportionally more of the ghosts fall outside the detection space of interest (but still within the decodable space) for 3-of-N schedules.

The inherent inclination of the PRF schedule to form ghosts depends on the number of PRFs in which target responses are visible and the decodability margins between the combinations of those PRFs. Recall that for N_T targets seen in M PRFs, the number of ghost correlations is given by

$$N_G = N_T(N_T^{M-1} - 1) \qquad \text{repeat of (12-13)}$$

All these ghost correlations exist somewhere within the decodable detection space of the M PRFs involved in the detection process. A target is declared if detected in at least M PRFs, but if targets are seen in more PRFs then more ghosts arise. As blindness reduces, targets are more likely to be seen in a greater number of PRFs, so the incidence of ghosting increases. For targets detected in just 2 PRFs ($M = 2$), then, from (12-13), there are fewer ghosting possibilities than the case of targets detected in 3 PRFs ($M = 3$), all other factors being equal. Furthermore, for any fixed number of PRFs in the schedule greater than 5, there are fewer combinations of 2-of-N than 3-of-N, that is, ${}_NC_2 < {}_NC_3$, for $N > 5$, (${}_5C_2 = {}_5C_3 = 10$). So for schedules of more than 5 PRFs, there are fewer ways target data across the PRFs could combine to give a ghost in 2-of-N schedules than 3-of-N schedules. Shorter schedules of lower N also give rise to fewer PRF combinations, and hence fewer ghosts, and are distinctly more feasible in 2-of-N schedules than 3-of-N schedules.

On one hand, 2-of-N schedules tend to give rise to fewer cross-correlation possibilities compared with 3-of-N schedules. On the other hand, though, 2-of-N schedules have a smaller decodable space, so proportionally more of the cross-correlation possibilities are captured within the detection space of interest compared with 3-of-N schedules. These two effects cancel each other out to some extent. The net result is that 3-of-N schedules have somewhat superior ghosting statistics to 2-of-N schedules, but in both cases the numbers of ghosts can be kept relatively low [6, 8].

Wiley et al [6] trialed an experimental target extraction algorithm based on clustering using several near optimum PRF schedules, as described in Section 12.5.4. The incidence of ghost targets were generally lower than 0.5% of total applied targets for 1 to 5 randomly placed point targets plus 0, 1, and 2 false alarms. The 3-of-8 schedule was consistently the best; there was no consistently worst schedule. There was no significant increase in ghosts when one false alarm was applied over the case of zero false alarms. However, an increase in ghosts to around 1–4% (depending on allowable cluster size) for the 2-of-N schedules was seen when two false alarms and only one or two targets were applied and was attributable to the correlation of one false alarm with the other. Generally, ghost target percentages increased with increasing allowable cluster size but reduced with increasing number of real targets. This is perhaps surprising but is due to the fact that genuine targets are more readily identifiable due to the larger number of detection points in the cluster, whereas the ghost clusters typically contain fewer detection points and are more readily discounted. When tested against a formation of 4, 6, 8, and 10 point targets plus 0, 1, and 2 false alarms, the incidence of ghosting increased for all the PRF schedules since the formation targets all have the same velocity and only correlation in range is required to declare a ghost. The 2-of-6 schedule reports the highest incidence of ghosts that rises from 1 to 6% as the number of targets increases from 4 to 10. The number of ghosts also rises as the allowable cluster size increases and rises very slightly as the number of false alarms increases; it peaked at a value of around 7.5% for the 2-of-6 schedule. The 3-of-8 schedule consistently had the best ghosting performance with a peak value of 1.3% under similar conditions. The best-case ghosting was 0% for all schedules and was obtained under a variety of test conditions. The numbers of ghost targets remained very low for the 3-of-8 schedule but were degraded in the 2-of-N schedules; the peak worst cases were some 2.3 to 5.8 times higher.

2-of-N schedules generally tend to have smaller decodability margins than 3-of-N schedules, so self-ghosting due to range or velocity extended target responses is more likely in the former.

In summary, each schedule type has areas of relative strength and weakness [6], however, optimal schedules for various values of M and N do not differ appreciably. 2-of-N schedules can be considered viable and even advantageous with respect to the more conventional 3-of-N schedules, not only in radar applications such as airborne fire control but especially those of a modest range/Doppler detection space and limited dwell time. In particular, the detection performance of an optimal 2-of-6 schedule is very similar to that of an optimal 3-of-8 schedule but enjoys several benefits of being a shorter schedule.

13.2.5 Clutter-Limited versus Noise-Limited Cases

One conundrum facing a designer is how best to use the available beam dwell time. In short, is it better to use many PRFs each with a relatively short CPI duration or use just a few PRFs but with a longer CPI duration? On the one hand, fewer PRFs, smaller N, has traditionally been seen as leading to increased blindness due to SLC, although recent work has shown that optimum 2-of-N schedules are just as good, if not better, than optimum 3-of-8 schedules in this respect [6, 7, 8]. The use of fewer PRFs also suffers from modest increases in the incidence of ghosting. On the other hand, a long CPI duration permits a greater FFT size, increased processing gain, and improved noise-limited detection performance. A long CPI duration also results in finer velocity resolution, which can also benefit detection performance in clutter. Clearly, the alternative solution of more PRFs/shorter CPI reverses the argument.

The solution to this conundrum may be heavily influenced by whether the radar is operating in clutter limited conditions or noise-limited conditions. In clutter-limited conditions, blindness reduces as N increases and as M reduces [6], so for a fixed M it pays to increase N. The increase in N also reduces ghosting due to the cross-correlation of returns and thus would also reduce ghosting due to extended target responses such as large clutter patches. Furthermore, in clutter-limited conditions, there is little advantage to be had from large FFTs since both targets and clutter exhibit similar processing gains and the overall signal-to-clutter ratio (SCR) shows little improvement. The option of a large number of PRFs of modest CPI duration would seem to be favored in clutter-limited cases, but this has to be tempered with the adverse effects of increased space charging time. In noise-limited cases, it is advantageous to benefit from high processing gains and hence to use fewer PRFs and incur fewer space charging times. The noise-limited case therefore favors the solution of fewer PRFs of longer CPI duration. Again, this must be tempered by maintaining adequate target visibility and ghosting considerations.

Most current medium PRF radars use a single PRF schedule designed for an average range of conditions, some of which are clutter limited and others noise limited. Unless dynamic optimization of the schedule in response to changing conditions is possible, the schedule becomes a compromise at some midpoint between the extremes. Strong SLC (low altitude, look-down) results in an optimum schedule erring toward larger N, whereas weak SLC (high altitude, look-up) encourages the optimum to a lower N. In all cases, there remains no substitute for careful PRF selection and target extraction to achieve the desired performance with as few PRFs as is reasonably possible.

13.3 | OTHER WAVEFORM DESIGN ISSUES

13.3.1 Implications for Single Target Tracking Radars

When tracking a single target, the radar is cued with approximate target data from a surveillance asset—often another radar or even the same radar operating in a different mode. The tracking radar must first search a small area of range/velocity and azimuth/elevation space around the target coordinates to acquire the target, and, having acquired the target, the subsequent range/velocity space of interest may become very small. The range/velocity detection space of interest at any instant is a small fraction of the total detection space of the tracking radar. The target's range may then be tracked using split gate range locking techniques or by monitoring its progression through range cells over time. Similarly, the target's Doppler may be tracked using phase locking techniques or by monitoring its progression through the Doppler cells in time. This is the case whether the tracking radar uses a conventional mechanically steered antenna and monopulse techniques to track the target continuously in angle or an active electronically steered array (AESA) antenna to periodically position the beam on the target. Either way, the tracking window defines a very small range/velocity space of interest. It may be perfectly acceptable to use a single PRF whose maximum unambiguous range and velocity are adequately large to encompass the range/velocity space of interest, as shown in Figure 13-6. This PRF must be selected so that the target response does not fall within the band of MBC rejection or eclipsed ranges and their repetitions in the velocity and range domains. For tracking purposes it may suffice to lock to an ambiguous range and velocity response to isolate the desired target from other returns, and it may not be necessary to resolve the range and velocity ambiguities. If, however, one does want to decode the true range and velocity, then the ambiguous tracking data may be compared with target data from the

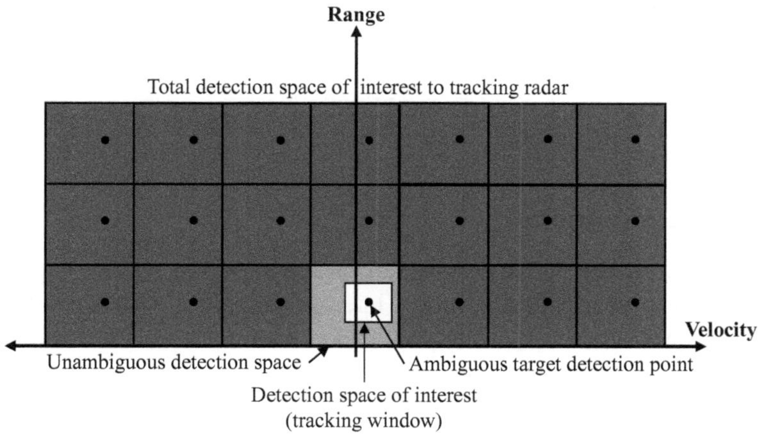

FIGURE 13-6 Single Target Tracking in Range/Velocity Space

surveillance asset to help resolve the issue; alternatively a medium PRF approach may be taken.

The dimensions of the detection space of interest is initially determined by the resolution of the target data received from the surveillance system plus additional allowances to cater for possible target motion in the time between the most recent measurement made by the surveillance system and the acquisition of the target by the tracking radar. This region defines the *search basket* of the tracking radar. (Conventionally, the search basket refers to the azimuth and elevation scan angles necessary but can equally apply to the range/velocity limits.) Once the target has been acquired, the detection space of interest can be shrunk around the target of interest. It is quite probable that the unambiguous detection space of a PRF exceeds the detection space of interest, in which case a single PRF can do the job.

In many scenarios, the ambiguous target detection point may be obscured by SLC. This can be problematic in airborne look-down cases. In these circumstances a PRF must now be selected which gives the most favorable SCR in the vicinity of the target. In many military applications, it may also be necessary to decode the true range and velocity of the target, and of other targets in proximity to the target of interest to ensure that the tracking function is not seduced away from the desired target through the action of any countermeasures, such as from towed decoys. This now necessitates a multiple PRF approach and an M-of-N detection strategy. However, the detection space of interest still needs to be only a small region about the (ambiguous) target detection point. A designed schedule needs only to minimize blindness, ensure decodability (if necessary), and minimize ghosting within a small region of range/velocity space (bearing in mind that targets outside the space of interest could generate ghosts within it). Yi and Byun [38] described a fast method of PRF set selection using a neural network that is optimal for the prevailing clutter conditions; platform parameters such as altitude, velocity, and look angle; and target range and velocity for a tracking radar. As the target migrates to a new location in range/velocity space, a different PRF set is selected in real time to optimize the detection performance within the new tracking window. This study sought to select 3-of-8 PRF sets that ensure decodability and that are optimized for the best target detectability by minimizing range/velocity blindness in the presence of clutter. The PRI resolution was 4 μs, which is very coarse but is sufficient to derive PRF sets with very little blindness over small range/velocity regions. Many PRF sets are used, each covering a small proportion of the total range/velocity space of interest to the tracking radar. As a result, the composite blind zone map using multiple PRF sets of coarse PRI resolution exhibits considerably less blindness than that of a single PRF set selected with a much finer PRI resolution. Since the tracking window is very small, it is quite probable that shorter schedules of moderately fine PRI resolution would achieve similar, or even superior, blindness statistics.

13.3.2 Frequency Agility

Modern military radar systems are designed to operate on several regularly spaced channel frequencies. This is primarily done to ensure interoperability between like radars deployed within range of each other; a typical specification may demand that there is negligible mutual interference between radars offset by four channels when separated by 1 km in a main beam to sidelobe scenario. The channel frequencies would typically occupy a bandwidth of up to 5% of the center frequency. The 5% bandwidth is limited by the microwave hardware on the front-end receiver and perhaps also by the antenna system.

In many cases, a radar may go frequency agile, in which its channel frequency is changed on a CPI-by-CPI basis. For medium PRF radars this means that each PRF burst would be transmitted on a different carrier frequency. Frequency agility can bring about several benefits:

- Frequency agility forces a more rapid decorrelation of clutter than would happen at a fixed frequency (see Section 6.5).
- Angle tracking errors due to glint and multipath [39] are minimized.
- Target detection against rapidly scintillating targets is improved.
- The radar emissions are less likely to be intercepted by enemy receivers such as anti-radiation missile seekers, radar warning receivers (RWRs), tactical electronic support measures (ESMs) systems, and electronic intelligence (ELINT) equipment.
- Enemy countermeasures systems must also be able to track the radar's frequency agility to be effective or to apply wideband noise jamming, which is less effective than narrowband noise.

None of these reasons for frequency agility dictate the precise choice of frequency; rather, the benefits are enhanced the wider the bandwidth of the frequency agility.

For Doppler sensing radars the translation of Doppler frequency into velocity via equation (4-2) is sensitive to the carrier frequency.

$$f_d = \frac{2V_r}{\lambda} = \frac{2V_r f_{TX}}{c} \qquad \text{repeat of (4-2)}$$

A 5% change in the carrier frequency, f_{TX}, brings about a 5% change in the Doppler shift, f_d, for the same radial velocity, V_r. Although this may seem like a relatively small change, it can have significant consequences.

Most (if not *all*) pulse Doppler radars maintain a fixed carrier frequency during each CPI since this ensures a coherent pulse train over the CPI. The carrier frequency may be changed from one CPI to the next. For medium PRF radars, each PRF would then be transmitted on a different carrier frequency, and Doppler shifts seen in one PRF would not align with Doppler shifts seen in another. This has ramifications for the blindness, velocity decodability and ghosting of the schedule but does not influence range decodability since the translation of the time delay into range is independent of frequency. It is desirable to quantify the PRF schedule in *velocity* rather than *Doppler frequency*, since velocity is invariant. It is far better to deal with *velocity* decodability margins and blind *velocity* margins and to construct blind zone maps in range/*velocity* space rather than their equivalents in Doppler. Even small frequency changes can have significant consequences on these parameters, and it is not sufficient to characterize the waveform at a single, nominal center frequency.

The precise choice of frequencies introduces a further degree of complication when designing a medium PRF waveform. The designer is now presented with a total of four degrees of freedom:[4] (1) the total number of PRFs to be used (N); (2) the minimum number of PRFs in which target data are required to declare a detection (M); (3) the precise values of the PRFs; and (4) the precise values of frequencies (or, at least, which channel to use from those available). All these factors are interrelated and cannot be set independently of one another. In much the same way that the quality of a PRF schedule is determined not by the values of individual PRFs but by the manner in which they all work together, so it is with the selection of frequencies. In fact, now PRF selection and frequency selection cannot be considered as separate issues; they must be selected in harmony with each other. One set of PRFs or frequencies cannot be claimed to be "good" or "bad" since the quality of the waveform depends on how PRFs are matched to frequencies. A poor medium PRF waveform could be improved simply by pairing its PRFs to its carrier frequencies differently; neither the precise values of the PRFs nor frequencies need be changed, just the paired combinations.

Section 13.1.3) mentioned that a typical PRI resolution of 1 μs results in several tens of possible PRFs to choose from between the limits of the maximum and minimum allowable. The number of combinations of drawing N PRFs from all those available can be enormous—so large that an exhaustive search to explore all the possibilities is not always feasible. Similarly, the number of frequencies is typically several tens. When the combinations of PRFs and frequencies are now considered, the numbers are astronomical. The introduction of frequency as another degree of freedom at the designer's disposal certainly does make for added complication. However, it actually presents an opportunity to design better waveforms or design several high-quality waveforms. In the discussion on ghosting (Section 12.4.2), the multitude of ghosts that could be brought about by using digital radio frequency memory (DRFM) based repeater jammers was mentioned. In Section 12.5.7, ghost tracks were noted to be decorrelated using multiple medium PRF schedules. Multiple near-optimum medium PRF schedules can be designed if both a sufficiently fine PRI resolution and a sufficient number of channel frequencies are available. The inclusion of frequency agility does indeed enable superior medium PRF waveforms to be designed to enhance the operation of the radar while minimizing the impact of enemy EW systems.

[4] Actually, one might also add the pulse width (or duty ratio) to this list; see Section 14.1.6.

CHAPTER 14

Detection Performance

Chapter Outline

14.1 NOISE-LIMITED DETECTION

14.1.1 Classical Detection Theory

Conventional medium pulse repetition frequency (PRF) processing applies a double thresholding system of detection. First, within each coherent processing interval (CPI), signals in each range/velocity cell within the unambiguous range and unambiguous velocity space are compared with a threshold level to declare whether a target is present. These target detections are ambiguously repeated across the range/velocity space of interest. This process is repeated for all N CPIs; each CPI operates on a different PRF. Second, each range/velocity cell is subjected to a binary integration process, which is looking for target detections in the same range/velocity cell in M out of the N CPIs. A target is declared if detected in a consistent range/velocity cell in at least M of the total of N CPIs. If P_D represents the probability of detection in each CPI, then the probability of detection in exactly M of N CPIs is

$$\frac{N!}{M!(N-M)} P_D^M (1-P_D)^{(N-M)} \qquad \text{repeat of (7-34)}$$

The probability of detection in at least M-of-N CPIs is therefore given by

$$\sum_{m=M}^{N} \frac{N!}{m!(N-m)} P_D^m (1-P_D)^{(N-m)} \qquad \text{repeat of (7-35)}$$

Similarly, if P_{FA} represents the probability of a false alarm in each CPI, then the probability of a false alarm in at least M-of-N CPIs is given by

$$\sum_{m=M}^{N} \frac{N!}{m!(N-m)} P_{FA}^m (1-P_{FA})^{(N-m)} \qquad \text{repeat of (7-36)}$$

Medium PRF and its M-of-N detection criterion are largely designed for operation in clutter-limited conditions, yet for a significant proportion of the operational use of the radar clutter levels are low and the detection performance is noise limited.

Equations (7-35) and (7-36) are overly simplistic as they stand. There may well be variations in the noise level in each CPI, leading to variations in P_D or P_{FA} per cell. Noise levels may vary due to differing values of the CPI duration, fast Fourier transform (FFT) size, and noise bandwidth used in each CPI. The myriad different combinations of parameters are discussed in Section 12.1.5.

Additionally, other considerations must be taken into account, including range/velocity blindness and target scintillation.

14.1.2 Effect of Blindness

Even in homogenous noise, P_D does not remain fixed over all range/velocity cells in each CPI, since different CPIs operate on different PRFs and range/velocity cells in which the target detected in one PRF may be blind in another. Range blindness due to eclipsing and velocity blindness due to the rejection of main beam clutter (MBC) and its repetition in the velocity domain are both functions of the PRF. The variable P_D in each PRF needs to be adjusted down in accordance with the proportion of the range/velocity space, which is not blind.

The proportion of the range space in which targets are visible, P_R, is given by

$$P_R = \frac{r_m - r_b}{r_m} = \frac{R_{mu} - R_{min}}{R_{mu}} \tag{14-1}$$

where r_m is the number of range cells within the pulse repetition interval (PRI), r_b is the number of blind range cells due to eclipsing, R_{mu} is the maximum unambiguous range of the PRF, and R_{min} is the minimum range due to the first eclipsed zone.

Similarly, the proportion of the velocity space in which targets are visible, P_V, is given by

$$P_V = \frac{n - v_b}{n} = \frac{f_r - 2f_{dMBC}}{f_r} \tag{14-2}$$

where n is the number of velocity cells (FFT size) in the Doppler band, v_b is the number of velocity cells rejected to eliminate MBC, f_r is the PRF, and MBC rejection takes place over the band $\pm f_{dMBC}$.

The proportion of the range/velocity space that remains visible, P_{RV}, is given by the product of the two expressions

$$P_{RV} = P_R P_V = \frac{R_{mu} - R_{min}}{R_{mu}} \frac{f_r - 2f_{dMBC}}{f_r} \tag{14-3a}$$

Recalling that $R_{mu} = c/2f_r$ and $R_{min} \approx \frac{c\tau}{2}$ gives

$$P_{RV} = P_R P_V \approx \frac{c/2f_r - c\tau/2}{c/2f_r} \frac{f_r - 2f_{dMBC}}{f_r} \tag{14-3b}$$

which reduces further to

$$P_{RV} \approx (1 - \tau f_r) \frac{f_r - 2f_{dMBC}}{f_r} \tag{14-3c}$$

P_{RV} is the factor by which P_D (and P_{FA}) in equations (7-34) to (7-36) must be reduced to accommodate the blindness in each PRF. A good approximation can be made by using a single value of P_{RV} calculated on the basis of the mean PRF, that is, $\overline{P_{RV}}$. The probability of detection and probability of false alarm in at least M-of-N CPIs can now be adjusted to

$$P_D(M\text{-of-}N) = \sum_{m=M}^{N} \frac{N!}{m!(N-m)!} \left(\overline{P_{RV}} P_D\right)^m \left(1 - \overline{P_{RV}} P_D\right)^{(N-m)} \quad (14\text{-}4)$$

$$P_{FA}(M\text{-of-}N) = \sum_{m=M}^{N} \frac{N!}{m!(N-m)!} \left(\overline{P_{RV}} P_{FA}\right)^m \left(1 - \overline{P_{RV}} P_{FA}\right)^{(N-m)} \quad (14\text{-}5)$$

From this, it can be deduced that P_{FA} in each Doppler/velocity cell required to produce a noise-generated false alarm in the same range/velocity cell in M-of-N CPIs is given by: [12, 40]

$$P_{FA/cell} = \left[\frac{T_d FAR}{N_D N_R \binom{N}{M} \overline{P_{RV}}^{-M} N_{DW}^{M-1} N_{RW}^{M-1}} \right]^{1/M} \quad (14\text{-}6)$$

where

T_d = the total dwell time for the whole schedule of all N CPIs (including the space charging times)

FAR = the (mean) false alarm rate

N_D = the total number of Doppler filters in the whole Doppler space of interest

N_R = the total number of range cells in the whole range space of interest

$\overline{P_{RV}}$ = the proportion of the range/velocity space of interest that is visible in the mean PRF

N_{DW} = the number of Doppler cells within the correlation window

N_{RW} = the number of range cells within the correlation window

This assumes the radar correlates returns in range and velocity. The correlation window of N_{DW} by N_{RW} defines the maximum allowable size of a cluster of detection points in cell numbers. Equation (14-6) allows $P_{FA/cell}$ to be fixed for any desired false alarm rate and thus the detection threshold in each cell to be set appropriately.

Recalling (2-8) $\quad FAR = P_{FA}.B_N$

and (7-23) $\quad \dfrac{1}{FAR} = \dfrac{1}{B_N} \exp\left(\dfrac{v_T^2}{2v_N^2}\right)$

one can write $\quad P_{FA/cell} = \exp\left(\dfrac{-v_T^2}{2v_N^2}\right) \quad$ (14-7)

or $\quad \dfrac{v_T}{v_N} = \sqrt{-2.\ln\left(P_{FA/cell}\right)} \quad$ (14-8)

in which v_T is the threshold voltage, and v_N is the root mean square (RMS) noise voltage in each cell. This enables the appropriate threshold to be set relative to the RMS level of Gaussian white noise to achieve the desired $P_{FA/cell}$ and hence the specified false alarm rate.

The limited range/velocity visibility represents a loss that is PRF dependent. P_{RV} and its inverse, the blanking loss $(= 1/P_{RV})$, are plotted as a function of PRF in Figure 14-1

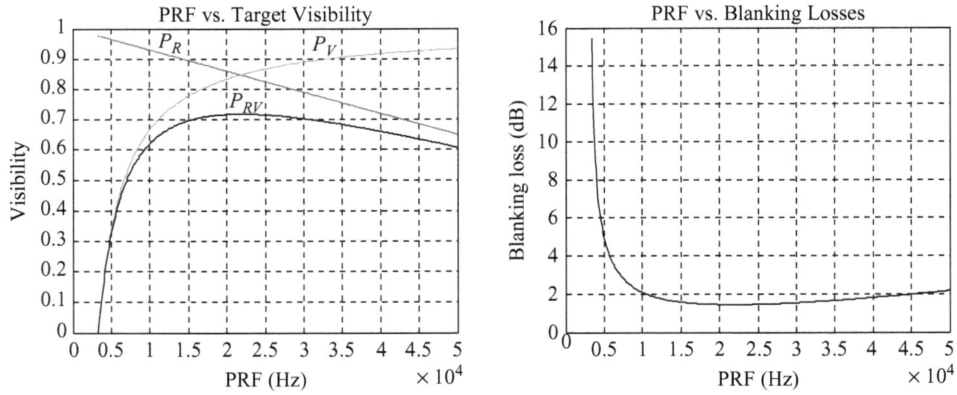

FIGURE 14-1 Range/Velocity Visibility (Left) and Loss (Right) for Fixed Pulse Width

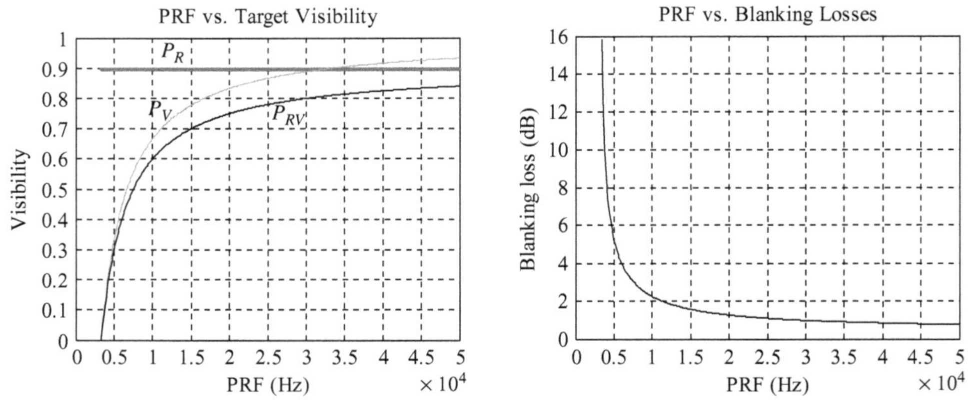

FIGURE 14-2 Range/Velocity Visibility (Left) and Loss (Right) for Fixed Duty Ratio

and Figure 14-2 for two cases. Figure 14-1 assumes a constant transmitted pulse width of 7 μs (and hence a variable duty ratio) with PRF, whereas Figure 14-2 assumes a constant duty ratio of 0.1 (and hence a variable pulse width) with PRF; both cases assume MBC rejection over $\pm$1.67 kHz ($\pm$25 m/s at 10 GHz). P_{RV} becomes zero at a PRF of $2f_{dMBC}$ because of the lack of any velocity visibility at this point, and, as a result, the blanking losses take off markedly as the PRF reduces down toward $2f_{dMBC}$. The minimum PRF for a medium PRF schedule is often set at the point when $P_V = 0.5$.

For the fixed pulse width, the range visibility $= 1$ for PRF $= 0$ Hz and reduces linearly with increasing PRF. Once the PRF has increased to $1/\tau$, the range visibility and hence P_{RV} go to zero and the blanking losses become infinitely high. In between there is an optimum point of maximum P_{RV} and minimum blanking loss that falls at the point of intersection of P_R and P_V. Setting $P_R = P_V$ and solving gives an optimum PRF given by

$$f_{rOPT} = \sqrt{\frac{2f_{dMBC}}{\tau}} \tag{14-9}$$

which is at around 21.8 kHz in this example.

For the fixed duty ratio, the range visibility remains fixed at 1 – *duty ratio*, and P_{RV} asymptotically approaches this value as the PRF increases. Thus, the losses continue to reduce as the PRF increases. A practical limit on the maximum PRF is often determined by the minimum acceptable pulse width (which, in turn, may be governed by consideration of the energy content in the signal, detection performance, bandwidth, and resolution).

For a single look with N_{NCI} CPIs non-coherently integrated and a given P_{FA} per range/velocity cell, the P_D for a non-fluctuating (Swerling case 0) target is given by [40, 41, 42, 43]:

$$P_D = \frac{1}{2} erfc\left(\sqrt{-0.8\ln\left[4P_{FA}\left(1-P_{FA}\right)\right]} + \sqrt{\frac{N_{NCI}}{2}-\frac{1}{2}} - \sqrt{N_{NCI}.SNR + \frac{N_{NCI}}{2}-\frac{1}{2}}\right) \quad (14\text{-}10)$$

where *erfc*() denotes the complementary error function.

The signal-to-noise ratio (SNR) is given approximately by

$$SNR = \frac{\eta^2}{N_{NCI}} + \frac{2\eta}{N_{NCI}}\sqrt{\frac{N_{NCI}}{2}-\frac{1}{4}} \quad (14\text{-}11)$$

where

$$\eta = \sqrt{-0.8\ln\left[4P_{FA}\left(1-P_{FA}\right)\right]} - sign\left(0.5-P_D\right)\sqrt{-0.8\ln\left[4P_D\left(1-P_D\right)\right]} \quad (14\text{-}12)$$

14.1.3 Effects of Target Scintillation

The Swerling cases 1 and 3 model the target return as one that varies slowly compared with the beam dwell time. The SNR of these targets remains constant across all N CPIs within the beam dwell. The overall probability of detection is relatively insensitive to missed detections in just a few PRFs due to blindness [12].

Rapid target scintillation such as Swerling case 2 and 4 type targets results in a lack of correlation in the returns from one CPI to the next. This now causes P_D to vary across N CPIs. If frequency agility is used from one CPI to the next, then this forces the decorrelation of target returns such that Swerling cases 2 and 4 apply. For the cases of radar cross section (RCS) fluctuation, P_D and SNR can be approximated as [40, 41, 42, 43]:

$$P_D = K_m\left[\frac{K_m^{-1}\left(P_{FA}, 2N_{NCI}\right) - 2\left(N_{NCI}-n_e\right)}{\frac{N_{NCI}}{n_e}SNR+1}, 2n_e\right] \quad (14\text{-}13)$$

$$SNR = \left[\frac{K_m^{-1}\left(P_D, 2n_e\right) - 2\left(N_{NCI}-n_e\right)}{K_m^{-1}\left(P_D, 2n_e\right)} - 1\right]\frac{n_e}{N_{NCI}} \quad (14\text{-}14)$$

where

$$n_e = \begin{cases} 1, & \text{for Swerling case 1 target} \\ N_{NCI}, & \text{for Swerling case 2 target} \\ 2, & \text{for Swerling case 3 target} \\ 2N_{NCI}, & \text{for Swerling case 4 target} \end{cases}$$

$$K_m(x, d) = 1 - P\left(\frac{d}{2}, \frac{x}{2}\right) = \text{chi-squared distribution survival function [44]}$$

$$K_m^{-1}(p, d) = \text{inverse chi-squared distribution survival function}$$

$$P(\alpha, x) = \frac{\gamma(\alpha, x)}{\Gamma(\alpha)} = \frac{\int_0^x t^{\alpha-1}e^{-1}dt}{\int_0^\infty t^{\alpha-1}e^{-1}dt} = \text{regularized lower incomplete gamma function}$$

If the probability of detection in an individual CPI is given by P_D, then the probability of detection over the whole dwell, P_{DD}, is given by the M-of-N binary integration system as

$$P_{DD} = \sum_{m=M}^{N} \frac{N!}{m!(N-m)!} \left(\overline{P_{RV}} P_D\right)^m \left(1 - \overline{P_{RV}} P_D\right)^{(N-m)} \tag{14-15}$$

14.1.4 Cumulative Probability of Detection

The detection performance of a radar is often evaluated against a target of a defined RCS that closes from a great range in a head-on aspect down to a range at which a specified cumulative probability of detection is achieved [12]. For example, R_{85} would be the range at which there is an 85% probability of having detected the target once over repeated scans. R_{85} is often similar to the range at which the single scan probability of detection of 50% is achieved. The cumulative probability of detection is very sensitive to the radar parameters over the regime where the single scan probability of detection is 20 to 50% [12].

The cumulative probability of detection for the kth scan, $P_{Dcum}(k)$, for a closing target may then be given by

$$\begin{aligned} P_{Dcum}(k) &= 1 - \prod_{i=1}^{k} [1 - P_{DD}(i)] \\ &= P_{Dcum}(k-1) + P_{DD}(k)(1 - P_{Dcum}(k)) \end{aligned} \tag{14-16}$$

where $P_{DD}(k)$ represents the probability of detection on the kth scan.

14.1.5 Effect of Target Extraction Algorithm

In practice, target responses may not align perfectly in the exact same range/velocity cells over all the PRFs. The declaration of true targets in preference to ghosts has been shown to be far more robust if detection points are first clustered and rank-ordered on the basis of the total integrated power or number of detection points within each cluster (Sections 12.5.3, 12.5.4, and 12.5.5). A simple M-of-N binary integration process is no longer valid for these more complex target extraction algorithms. In fact, it is virtually impossible to be thoroughly analytical about such algorithms, and it is interesting to note that all studies on these algorithms have quantified their performances using the statistics gleaned over a large number of (simulated) trials.

Medium PRF radars that correlate target detections in range only increase the likelihood of false alarms and noise-generated ghosts over those that correlate in both range and velocity. The threshold must therefore be increased to maintain a desired P_D. An increase in the SNR of around 1 dB is required to maintain the same P_D as would be obtained using both range and velocity correlation and can be considered as the loss associated with the lack of velocity ambiguity resolution [12].

14.1.6 Optimum Duty Ratio

As the duty ratio of a pulsed waveform increases, more energy is incident on the target and the noise-limited detection performance improves. Increasing the duty ratio would therefore enable a lower peak power to be transmitted while maintaining the noise-limited detection performance. A lower peak power is advantageous as lower powers are more readily available from solid-state devices and the signal has a lower probability of intercept. Increasing the duty ratio of a medium PRF waveform impacts on the following:

- The energy on target is increased, improving SNR.
- The extent of fully eclipsed regions is increased per PRI, potentially reducing the overall probability of detection and also constraining PRI selection further.
- The first blind range of the radar is increased, unless shorter PRIs are employed.
- The pulse-compression range-time sidelobes occupy a larger proportion of the detection space, increasing the apparent local noise floor (especially if pulses are partially eclipsed).

The initial expectation is that a higher duty may improve detection performance. However, in clutter-limited environments, an increase in pulse energy does not influence the signal-to-clutter ratio. The performance in the velocity dimension is unaffected by a change in duty ratio.

Recall that the duty ratio, d, is given by

$$d = \frac{\tau}{T_r} \qquad \text{repeat of (3-1)}$$

For a target return that approaches an eclipsing situation, the return is initially partially eclipsed and some of the echo energy is gated out. If pulse compression is used, then partial eclipsing also increases the range sidelobes. As the target range continues to increase, the degree of partial eclipse increases and more energy is lost, as depicted in Figure 14-3. Eventually, the partial eclipse is so severe that the target can no longer be detected. For all intents and purposes the radar has become blind to such a target. Total eclipsing occurs when the degree of partial eclipsing becomes so severe that the remaining energy in the target return fails to exceed the detection threshold. The point at which blindness occurs depends on the strength of the target return; however, the radar would always be blind to any target, regardless of its size, when fully eclipsed. It is also useful to define the quantity,

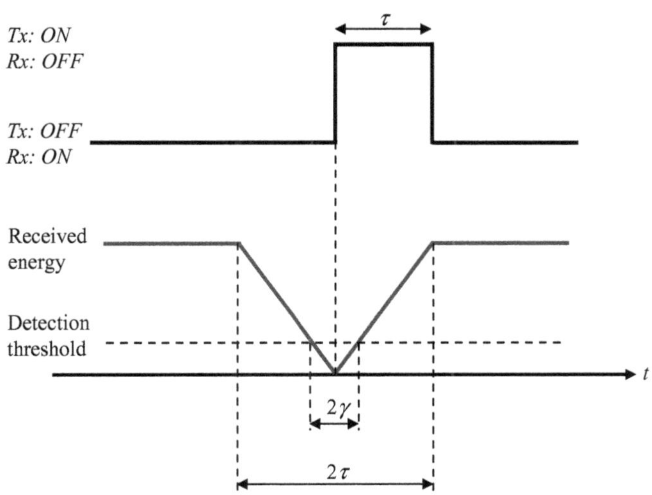

FIGURE 14-3 Eclipsing Timing Diagram

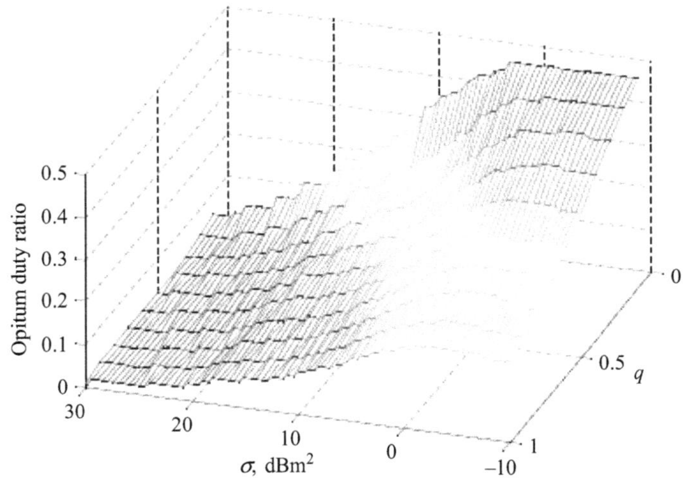

FIGURE 14-4 Optimal Surface of Best Detection Performance as a Function of Duty Ratio (*d*) and Total Eclipsing Factor (*q*)

q, as the proportion of the time that the radar is blind on account of eclipsing losses to the time for which some degree of eclipsing exists. With reference to Figure 14-3

$$q = \frac{\gamma}{\tau} \tag{14-17}$$

For modest sized targets ($\sigma = 10\ \text{m}^2$) and high eclipsing ($q = 0.5$), there is a broad peak in the optimum duty ratio at around 15%, which maximizes P_D aggregated across all range/velocity cells. For smaller targets and greater eclipsing (increasing q), the optimum duty ratio increases toward 20 to 50%. The optimum duty ratio never exceeds 50% since a transmitted pulse width exceeding the receiving time always results in eclipsing losses; exceeding a 50% duty ratio represents a pointless waste of energy. Figure 14-4 presents a surface plot of the optimum duty ratio, d, for a range of target sizes and proportions of total eclipsing, q. The optimum duty ratio results in the best P_D on aggregate over all the range/velocity cells. The right of the figure shows that with small targets a high duty ratio is best; the left of the figure shows that with large targets low duty ratios are superior. As the amount of eclipsing is increased (higher q value) the optimal duty ratio also reduces. The explanation is that as the target size increases, P_D improvements due to integration against the noise are less significant so the losses due to eclipsing are dominating.

In general, if the SNR is expected to be low, higher duty ratios are better as more energy is placed on the target, improving the final SNR. With mean target signals that are of a similar order to the detection threshold level, a small increase in SNR provides a useful increase in P_D. Conversely, with mean target levels much higher than the detection threshold, a small change in SNR produces a negligible change in the probability of detection. The higher the proportion of total eclipsing within the pulse (i.e., higher values of q) then the greater the losses due to eclipsing and the lower the overall probability of detection. Often, the optimum duty ratio needs to be lowered to limit the actual size of the eclipsed regions. When integration is providing little benefit to P_D, as with high SNR returns, the optimum duty ratio is dominated almost entirely by the drive to minimize the actual amount of totally eclipsed signal, and therefore very low duty ratios reign.

The optimum duty ratio is only weakly dependent on clutter levels since, in general, the number of range cells dominated by sidelobe clutter is a small fraction of the total. Significant increases in the clutter levels do not lead to corresponding increases in the number of cells dominated by clutter. There is little scope for optimization of the duty ratio in conditions of high clutter, and, in fact, the optimization of PRF values is far more significant in determining the detection performance of the radar.

14.2 | CLUTTER-LIMITED DETECTION

14.2.1 Introduction

The justification for the medium PRF waveform is its good all-around performance in conditions of high clutter, particularly for airborne radar applications. No discussion on medium PRF pulse Doppler can therefore be complete without some consideration of its clutter-limited detection performance. This topic and several related issues drawn from recent research on the matter are considered here.

14.2.2 Optimal Constant False Alarm Rate Design

For conventional M-of-N processing, detections are made in the unambiguous range/velocity detection space at each PRF. Constant false alarm rate (CFAR) detection is used to declare target detections by deriving a threshold on the basis of the noise-plus-clutter values at cells within a CFAR window. The threshold is applied to a central test cell and is intended to be representative of the local clutter statistics. Each range/velocity cell is tested in turn. The process and some of its potential pitfalls are illustrated in Figure 14-5, which shows the CFAR window superimposed onto a folded clutter map with a target in the central test cell. The two immediate problems are that the clutter varies considerably within the CFAR window and that the target is not confined to just the central test cell because it has associated range sidelobes and Doppler (velocity) sidelobe responses spilling into the CFAR window.

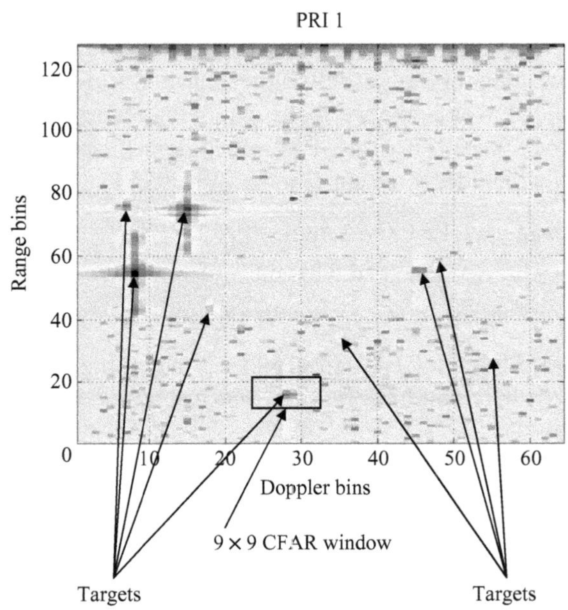

FIGURE 14-5 CFAR Detection in Medium PRF

Conventional radar CFAR detector design is based on the following fundamental assumptions:

- The noise/clutter is locally homogeneous, allowing statistical moments to be generated spatially.
- The statistics of the noise/clutter are stationary allowing accurate moments to be generated temporally.
- The shape of the noise/clutter probability density function is known.
- A low number of samples (typically 30) provide a sufficiently accurate estimate of the moments.

Unfortunately these assumptions do not hold, except for a limited range of scenarios [45]. One scenario where none of the assumptions is likely to be valid is the littoral environment where, in general, the homogeneity and stationarity of the clutter (including SLC) is poor [46].

This raises a number of questions:

- What is the effect on the CFAR estimate of the mean clutter level when non-uniform cells are used across different PRFs?
- What are the optimum CFAR window dimensions?
- Can conventional CFAR detection theory be applied to realistic data?
- How does CFAR performance vary with $M = 3, 4$ for realistic data?
- Which is better, cell averaging CFAR (CACFAR) or ordered-statistic CFAR (OSCFAR)?

These questions have been investigated supported by complex simulation of an airborne fire control radar (FCR), based on the parameters of Table 12-1, including range (pulse compression) and velocity FFT processing sidelobe effects, inhomogeneous, non-stationary clutter models, and the use of actual radar data.

14.2.2.1 Effect of Non-Uniform Cells on CFAR

The range or velocity resolution typically varies with PRF. Conventional cell averaging CFAR samples a fixed number of (range) cells either side of the test cell. Therefore, clutter statistics could be sampled over a range/velocity extent that varies with PRF. Trials using cell averaging CFAR over 30 cells, for an 8-PRF schedule in the presence of homogenous noise have been conducted under three conditions:

(i) Uniform cell sizes in each PRF (baseline)
(ii) Fixed number of cells/variable spatial extent
(iii) Fixed spatial extent/variable number of cells (approximately 30)

This study concluded that all CFAR strategies estimated the same cumulative mean values. Therefore, CFAR over non-uniform cells was just as good as CFAR over uniform cells.

14.2.2.2 Optimum CFAR Window

A scene with nine Swerling case 2 targets, noise, and ground clutter was created, as shown in Figure 14-5. To provide consistent automated results, the threshold multiplier was adjusted until satisfactory detection of all true targets was just obtained; this provides a consistent P_D for all methods, given the same target set. The full detection space of interest comprised 24,750 range cells of 75 m resolution by 515 velocity cells of 5.85 m/s resolution giving 1,272,050 range/velocity cells in total detection space. CFAR widows of: 1, 5, 9, 13, 17, 21, 25 range cells and 1, 3, 5, 7, 9, 11 velocity cells were trialed with a theoretical prediction of $P_{FA} = 10^{-6}$.

Two CFAR strategies were implemented: cell averaging CFAR (CACFAR) and OS-CFAR. The CACFAR detection threshold was set by the following function (assuming a square law detector and noise with an exponential power distribution):

$$P_{FA} = (1 + k)^{-n} \tag{14-18}$$

where n is number of cells in the CACFAR window, and k is the scale factor times the sum of the n squared amplitudes. Similarly, for the OSCFAR, the threshold is set by the following function (making the same assumptions as for the CACFAR case):

$$P_{FA} = \frac{n!\Gamma\left(\frac{n}{2} + k + 1\right)}{\Gamma\left(\frac{n}{2} + 1\right)\Gamma(n + k + 1)} \tag{14-19}$$

where the threshold is k times the median value. In both cases the factor k is derived as the value required to just enable all nine targets to be detected and thus provide a constant P_D.

Repeated trials were conducted to derive statistics on the actual P_{FA} values obtained for all combinations of CFAR window dimensions and strategies. The CACFAR differs from the OSCFAR results and presents an inferior predicted false alarm probability. For cell averaging, the window size of 7×7 gave the lowest P_{FA} of 10^{-5} (median of repeated trials), whereas for the OSCFAR the lowest median P_{FA} of 10^{-16} was obtained for a CFAR window of 13 range cells $\times$ 9 velocity cells. In conclusion, the P_{FA} in M-of-N CACFAR matches theoretical predictions, but that of OSCFAR is somewhat lower than theoretical predictions. The hypothesis that *the theoretical relationship between CFAR threshold levels, number of cells in CFAR window and* P_{FA} *still holds true with realistic data* is reasonably true.

14.2.2.3 CFAR Performance Variation with $M = 3$, 4 for Realistic Data

For this trial, a scene was generated with the number of targets set at 4, 5, 6, 7, 8, 9, 10, 15, and 20. 3-of-8 and 4-of-8 schedules were used together with OSCFAR using a window of 9 range cells by 7 velocity cells. The threshold multiplier, k, was adjusted until all targets were just detected. A total of 30 repeated trials were conducted to generate some test statistics. The results are shown graphically in Figure 14-6, in which the (logarithm of the) number of false alarms, N_{fa}, is plotted against the cumulative distribution function index for the case of nine genuine targets. With nine targets, the cumulative distribution of the number of false alarms can be seen to be highly long-tailed. The 3-of-8 schedule is the worst performer with nine targets.

In general, 4-of-8 schedules gave fewer false alarms than 3-of-8 schedules. Small CFAR windows lead to overestimates of local clutter/noise, which leads to CFAR detection requiring a lower threshold for a given $P_D$but also leads to a higher P_{FA}. Theory predicts that for a given SNR and P_{FA} 3-of-8 schedules should out-perform 4-of-8, whereas the opposite was observed in trials with realistic data. 4-of-8 was better (or 3-of-8 was worse) than theoretical predictions. This is probably due to the assumption made in the theory that the noise/clutter statistics are homogenous, which isn't true of real data.

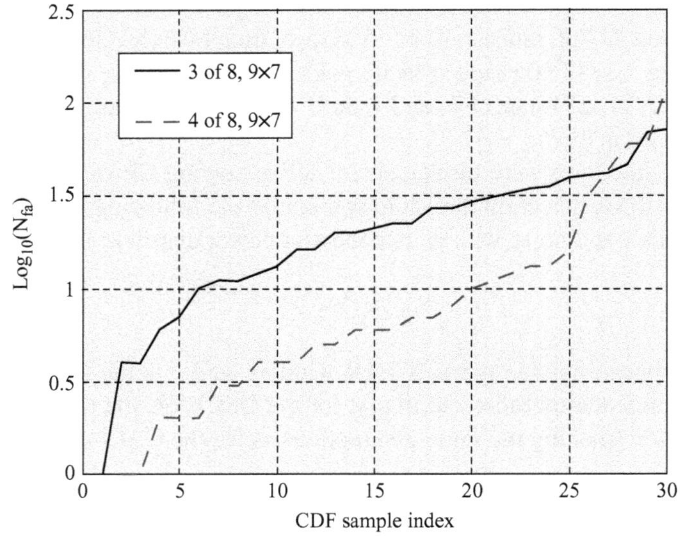

FIGURE 14-6 OSCFAR Results for 3-of-8 and 4-of-8 PRF Schedules for Nine Targets

The number of false alarms increases with the number of real targets. 4-of-8 always has fewer false alarms than 3-of-8. 4-of-8 is particularly good for ≤ 4 real targets. The scene becomes less homogenous as the number of targets increases. The threshold multiplier, k, decreases and it becomes more difficult to estimate true noise/clutter level.

14.2.2.4 CACFAR versus OSCFAR

Linear CACFAR and linear OSCFAR were trialed, both with a CFAR window of 9 range cells × 7 velocity cells and with the number of targets set at 4, 5, 6, 7, 8, 9, 10, 15, and 20. An optimum 3-of-8 PRF schedule was used for all trials. The number of false alarms, N_{fa}, was ascertained for all these combinations, and the results are summarized in Figure 14-7. This shows that for low number of targets, < 7, OSCFAR gives fewer false alarms than

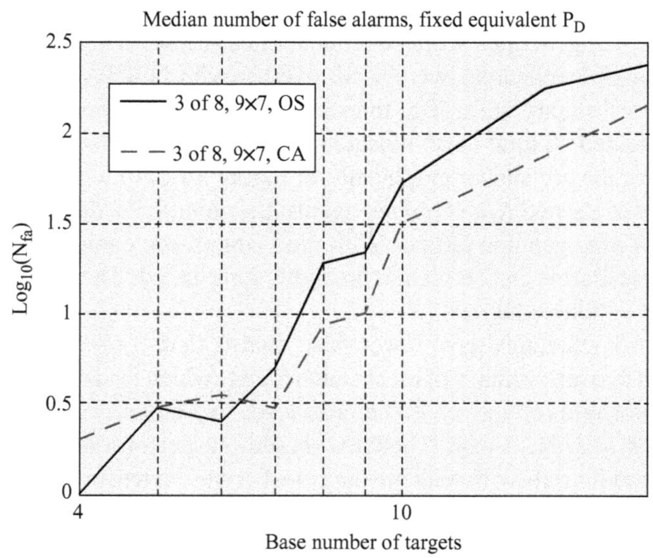

FIGURE 14-7 CACFAR versus OSCFAR False Alarm Performance

CACFAR, whereas a for large number of targets, ≥ 7, CACFAR gives fewer false alarms than OSCFAR.

14.2.3 Detectability Maps

A very useful tool in quantifying the performance of a schedule in clutter is the *detectability map* [16, 47, 48, 49]. A detectability map denotes the minimum target RCS required for detection at each range and velocity cell over the full detection space of interest in a minimum number of PRFs ($= M$).

A detectability map may be derived from the unfolded clutter maps appropriate to each PRF of the schedule. Each PRF has a similar-looking, though different, unfolded clutter map that displays the clutter levels across each range/velocity cell, which is ambiguously repeated to cover the entire detection space of interest (see Section 11.3.3). Range blindness due to eclipsing and velocity blindness due to MBC rejection may also be applied to each unfolded clutter map. In generating a detectability map the unfolded clutter maps of all PRFs in the schedule need to be overlaid and read at a common resolution that can be no finer than the coarsest resolution of any of the unfolded clutter maps. The clutter amplitudes at each range/velocity cell need to be assessed to judge the necessary target RCS required to exceed the noise-plus-clutter with the required SNCR. A detectability map can therefore be derived over the full range and Doppler detection space of the radar and denotes the minimum target RCS required for detection at each range and Doppler cell in an appropriate number of PRFs (often three). For reference purposes it is often convenient to assume a signal-to-noise-plus-clutter ratio (SNCR) of 0 dB leading to a detectability map in which the required target RCS results in a return whose power equals the noise-plus-clutter power. Should a more realistic SNCR be required, for example an SCR $= +20$ dB, then the RCS values in each cell need only to be offset by 20 dB. The detectability map may be thresholded at any given fixed RCS to indicate regions where a target of the given RCS would be visible or not visible. This thresholding forms the classic blind zone map for a medium PRF schedule.

An example detectability map is given in Figure 14-8 for the case of a 3-of-8 schedule. The grayscale coding represents the required target signal level giving a SNCR $= 0$ dB in each cell (on an arbitrary decibel scale). The first blind range running horizontally along the base of the detectability map is colored white. No MBC blanking is applied, and its effects can be seen as the requirement for a very powerful target return to be seen above the MBC in the first few velocity cells running in a vertical strip up the left-hand edge of the map. In general, there is a requirement for target returns to become larger in order to be detected as the range increases. There is some fine variation in target detectability evident in the map due to the variation in clutter strength and number of PRFs in which target visibility is maintained across the range/velocity detection space of interest.

Detectability maps are a useful means of characterizing relative performances in clutter. Each detectability map is valid given the assumptions made in the radar, antenna, and clutter models, and therefore comparisons between detectability maps are also valid. Furthermore, comparisons remain valid irrespective of any offsets that may be applied to the detectability maps so long as a constant offset is applied to all detectability maps. In many design applications it is necessary to compare two detectability maps to judge relative performances. Typically, a detectability map contains a total of around 10^5 to 10^6 range/velocity cells (pixels) and is stored as a two-dimensional array in which each element corresponds to a range/velocity cell. Since each pixel (element) refers to a consistent

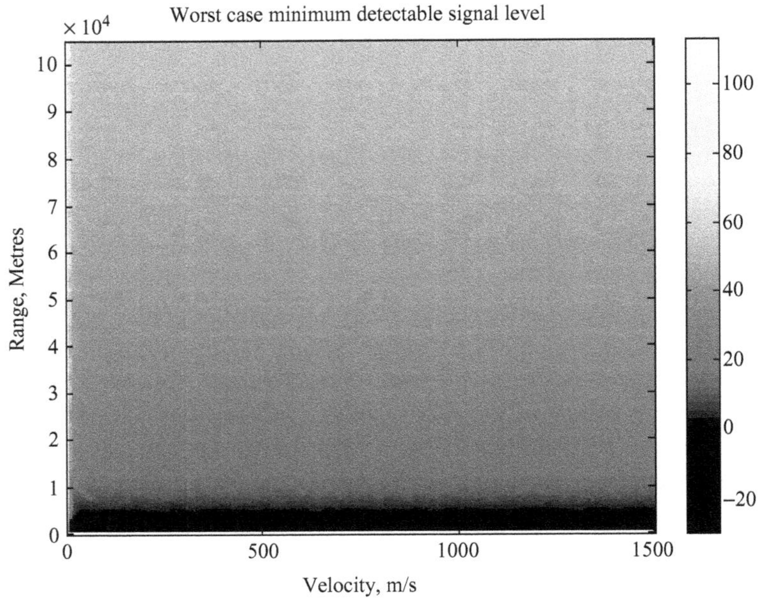

FIGURE 14-8 ■ Detectability Map

range/velocity cell, comparisons between two detectability maps can conveniently be made by comparing their matrix elements.

14.2.4 Optimization for Clutter Statistics

In Section 13.2.3 it was noted that the selection of precise values of PRF could be optimized according to the clutter scene prevalent at any time. This would suggest that it would be advantageous to perform dynamic optimization of PRF values in response to changes in the clutter scene (i.e., altitude, flight profile, platform orientation, scan angle, and terrain type). However, most medium PRF systems use a single PRF schedule designed to cope with "average" conditions. The performance of a PRF schedule designed for one set of clutter conditions but used in different clutter environment is of particular interest. Airborne radars and transportable surfaced-based systems may be used in a wide range of clutter conditions, and in many cases long-range surveillance systems may be required to cope with land clutter, sea clutter, and a mixed land/sea scene. Land and sea clutter have quite differing statistics, quite apart from the fact that clutter statistics can vary with terrain type and sea state. Boundaries between markedly different clutter types pose their own problems for CFAR detection.

An optimization procedure based on the use of an evolutionary algorithm has been used to select precise PRF values for a 3-of-8 schedule to optimize target detectability in conditions of strong surface clutter [47]. The optimization procedure was repeated for clutter statistics typical of land, sea, and a mixed land/sea scene, and different near-optimum solutions were obtained for each clutter scene. Target detectability was compared when using the PRF schedules optimized for one clutter type in scenes composed of a different clutter surface. Schedules optimized for one homogenous clutter scene (land or sea) did not perform as well in the presence of clutter from the opposite environment. However, the margin of superiority was often small and in some cases was statistically insignificant. Thus, there is little value to be gained for performing the optimization in the appropriate clutter scene. The detectability of schedules optimized for the inhomogeneous land/sea scene indicated that it was possible to design medium PRF schedules that were well suited to operations over land and sea. These schedules offered a detectability performance that was only marginally degraded with respect to those optimized for either clutter type. In summary, while medium PRF waveform diversity is advantageous to meet diverse operational conditions (e.g., change of altitude/velocity), it affords no real advantage in target detectability with diverse clutter scenes.

14.2.5 Aperture Illumination Function

Target detectability depends on the number of PRFs in which any target is visible and on the probability of detection, P_D, in each PRF. P_D in each PRF is determined by the SNCR, among other factors, and varies across the range and velocity detection space of the radar due to the ambiguous repetition of clutter across this detection space. Minimizing SLC through minimizing antenna sidelobe level is a design priority for such systems. This may be achieved by applying a tapered illumination function across the antenna aperture and can be implemented readily by appropriate amplitude and phase weightings of the elements of an active electronically scanned array (AESA) antenna. However, tapered illumination functions result in a reduction in main beam boresight gain together with a broadening of the main beam, both of which are further degraded when the beam is phase steered away from its mechanical boresight. Furthermore, phase steering tends to generate increased sidelobes. Thus, there appears to be a conflict of interests in applying tapered illumination across an array antenna as far as target detection is concerned; on the one hand the tapered illumination reduces the sidelobe level, but on the other it leads to a loss of main beam gain. Thus, both clutter and target signal strengths are reduced through the use of a tapered antenna illumination, or conversely, both are maximized for a uniformly illuminated antenna. The question arises as to whether tapered illumination actually leads to increased target detectability in scenarios in which target detection is likely to be clutter limited.

Furthermore, one advantage of an AESA antenna over a conventional design using a single high-power transmitter is the graceful degradation of the AESA performance as an increasing number of elements fail. Again, the gradual failure of more and more elements leads to a loss in main beam gain and an increase in sidelobes and SLC. A second question that arises is at what point do the number of failures lead to unacceptable degradations in target detectability when clutter limited?

Recent studies have modeled the target detectability of an airborne FCR operating in a 3-of-8 medium PRF schedule for a variety of heavy clutter scenes and for various transmitting and receiving aperture illumination functions [48, 49]. These studies concluded that target detectability was consistently at its best when tapered illumination functions giving the lowest sidelobe levels were used on both the transmitting and receiving array. In fact, the uniform illumination function used on either transmission[1] or reception always gave the worst target detectability [48, 49].

---

[1] Some references [43, p. 4.12] suggest that a uniform illumination function on transmission should be used to maximize the transmitted power. This would be advantageous in noise-limited detection conditions; however, in clutter-limited conditions increasing power affects both target signal and clutter returns equally, giving no net SNCR benefit.

Target detectability degrades as the proportion of randomly selected elements undergoing a total failure increases from 0 to 5% [48, 49]. Failed elements result in significant regions of the range/velocity detection space of degraded target detectability; however, the margins by which detectability is degraded tend to be less than the margins between detectability using the best and worst array weighting functions. A failure of 5% of the array elements results in modest, though meaningful, degradation in target detectability. Therefore, 5% would seem to be an appropriate upper limit on the proportion of failed elements.

CHAPTER 15

Methods of Pulse Repetition Frequency Selection

Chapter Outline

15.1 | A BRIEF REVIEW OF THE PULSE REPETITION FREQUENCY SELECTION REQUIREMENTS

The selection of the precise values of pulse repetition frequency (PRF) used within a medium PRF schedule needs to be made with due regard to the ability of the processing to unambiguously decode the range and velocity of targets yet to have ample resilience to ghosts. Range/velocity blindness needs to be minimized, detection performance in both noise and clutter-limited situations should be maximized, and system specifications dictating the beam dwell time, range, and velocity resolution need to be observed. The precise values of the PRFs used and other waveform design issues impact on almost every aspect of the radar's performance. Historically, many medium PRF radar modes used just one PRF schedule designed to operate in a wide range of conditions. PRI timing constraints and selection methods often led to values that were a compromise of several competing factors. As military radar specifications become more demanding, compromises become less acceptable so higher-performing PRF schedules are required.

Modern methods of PRF selection can yield PRF schedules that perform well in all areas across a wide range of clutter and platform conditions. Nevertheless, there remains some scope for selecting optimal PRF sets to maximize detection performance in clutter-limited conditions. These selection methods are capable of finding several PRF solutions, all of a similar quality. Furthermore, there is considerable merit in using multiple PRF schedules for electronic warfare (EW) considerations or to minimize the impact of ghosting. Modern systems therefore require a PRF schedule that copes well in all conditions or, more likely, the ability to select optimal PRF sets to meet clutter and jamming conditions prevalent at the time.

Very little is openly published on the methods of selecting precise PRF values, and, consequently, it is viewed as something of a black art or a game of chance. Of the open-source material available, some methods are fairly simplistic and are driven by considering just one or two aspects of the radar's performance; other more complex selection methods tend to be more thorough. The following sections of this chapter describe and review the various PRF selection methods.

15.2 PRF SELECTION FOR MAXIMUM VISIBILITY

A mean PRF may be selected to minimize blindness [13, 26], as described in Section 14.1.2. For a fixed pulse width, the optimum PRF, f_{rOPT}, is given by

$$f_{rOPT} = \sqrt{\frac{2 f_{dMBC}}{\tau}} \qquad \text{repeat of (14-9)}$$

where f_{dMBC} is the single-sided main beam clutter (MBC) rejection notch bandwidth, and τ is the transmitted pulse width. However, the mean pulse repetition interval (PRI) must also be constrained to ensure it is possible to transmit the entire schedule within the beam dwell time by

$$Beam\ dwell\ time \leq N[(n \times \overline{PRI}) + t_{SC}] \qquad \text{repeat of (12-12)}$$

where N is the number of PRFs used in the schedule, n is the fast Fourier transform (FFT) size, and t_{SC} is the space charging time.

Precise PRIs can be derived by first considering the time delay out to the maximum range of interest

$$T_A = 2\left(\frac{R_C}{c} + \tau\right) \tag{15-1}$$

where R_C is the maximum range of interest, c is the speed of light, and τ is the transmitted pulse width. Two series of PRIs are derived by dividing T_A by odd and even integers, typically in the range 9 to 17, according to

$$PRI_k = \frac{T_A}{C_1 + 2k} \tag{15-2a}$$

$$PRI_j = \frac{T_A}{C_2 + 2j + \delta_j} \tag{15-2b}$$

where

C_1 = an odd integer

C_2 = an even integer

k and j = integer indices, e.g., 0, 1, ... 4

δ_j = a small perturbation made to the even divisors, typically in the range 0.1 to 0.3 to minimize blindness

For example, if one considers a maximum range of interest, $R_C = 185$ km, and a transmitted pulse width, $\tau = 7\ \mu$s, then $T_A = 1247.33\ \mu$s. If then, $C_1 = 9$, $C_2 = 12$, and $\delta_j = 0.2$ for all j, the following PRIs are obtained:

$$PRI_k = 138.59, 113.39, 95.95, 83.16, 73.37\ \mu s$$

and

$$PRI_j = 102.24, 87.84, 77.00, 68.53, 61.75\ \mu s$$

The δ_j factors are optimized to maximize visibility, and once these have been selected this selection method is relatively straightforward. Values of C_1 and C_2 can be chosen to give a mean PRI close to the optimum mean value given by (the inverse of) equation (14-9) and such that there is adequate time within the beam dwell time to transmit the schedule, given by equation (12-12).

As it stands, this selection method does not consider decodability or the propensity to form ghosts. However, some scope to provide robust decodability margins could be possible through the optimization of δ_j.

15.3 | MAJOR–MINOR METHOD OF PRF SELECTION

The *major–minor* method of PRF selection is based on an initial selection of three major PRFs designed to provide coverage of the Doppler band of interest [28]. The major PRFs are selected to stagger their first repetitions of MBC. For each major PRF, two minor PRFs are selected to allow range ambiguities to be resolved. Range ambiguity is solved by associating detections in the minor PRFs with the respective major PRF. PRF checks are applied to ensure that the maximum duty ratio is not exceeded. All PRIs are an integer number of range cells. The selection algorithm is best illustrated by example.

Assume the following system constraints:

- MBC rejection bandwidth, $f_{dMBC} = \pm 1.67$ kHz (or a double-sided bandwidth of $2f_{dMBC} = 3.34$ kHz)
- At least 75% of the band must be clear of MBC, $P_V \geq 0.75$
- Minimum range, $R_{min} = 1050$ m
- Range resolution, $\Delta R = 75$ m
- Maximum duty ratio is 15%, $d_{max} = 0.15$

These constraints dictate the following:

The transmitted pulse width from (3.4), $\tau = 7\ \mu$s

The compressed pulse width, $\tau_C = 1/B = 0.5\ \mu$s giving a compression ratio, $CR = 14$

The maximum PRF $= d_{max}/\tau = 0.15/7\ \mu s = 21\,429$ Hz

Minimum PRF $= 4 \times 2f_{dMBC} = 13{,}360$ Hz. This is quite a high minimum value and is only marginally below the maximum PRF constraint. The high minimum value results from the 75% Doppler visibility constraint. If this is relaxed to 50% visibility then the minimum PRF is 6680 Hz, which is the value assumed through the remainder of this example.

A somewhat arbitrarily assumed value of the actual duty ratio is taken as 7.5% ($d = 0.075$), which is half the maximum allowable.

Selection of First Major PRF The nominal value of the first major PRF is selected on the basis of the duty ratio and transmitted pulse width, that is, $d/\tau = 0.075/7\ \mu s = 10{,}714$ Hz. The PRI may be composed of CR/d range cells, that is, $CR/d = 14/0.075 = 186.7$, which must be rounded down to 186 range cells. Hence, $N_{a1} = 186$. The first major PRF is adjusted to

$$PRF_{a1} = 1/(N_{a1}\tau_C) = 10{,}753 \text{ Hz.}$$

Selection of First Pair of Minor PRFs The first pair of minor PRFs is selected to ensure that any target within the range bracket of R_{mu} to $2R_{mu}$ shifts by at least 2 range cells in the minor PRFs, where

$$R_{mu} = \frac{c}{2PRF_{a1}}$$

The PRIs, in range cells, are

$$N_{a2} = N_{a1} - 2 = 186 - 2 = 184$$
$$N_{a3} = N_{a2} - 2 = 184 - 2 = 182$$

The corresponding PRFs are

$$PRF_{a2} = 1/(N_{a2}\tau_C) = 10{,}870 \text{ Hz}$$
$$PRF_{a3} = 1/(N_{a3}\tau_C) = 10{,}989 \text{ Hz.}$$

Check Decodable Range $\text{LCM}(N_{a1}, N_{a2}, N_{a3}) = \text{LCM}(186, 184, 182) = 1{,}557{,}192$

The maximum unambiguous time delay is therefore 1,557,192 range cells. This gives a maximum unambiguous range for this three PRF combination of

$$\frac{3 \times 10^8\,(1557192 \times 0.5\ \mu s)}{2} = 116.8 \times 10^6\text{m, which is more than adequate.}$$

Selection of Second Major PRF The second major PRF must be selected to shift the first repetition of MBC by its own bandwidth.

The nominal second major PRF $= PRF_{a1} + 2f_{dMBC} = 10753 + 3340 = 14{,}093$ Hz. The nominal PRI is 70.96 μs or 141.9 range cells. Rounding this down to the nearest integer gives

$$N_{b1} = 141$$

The exact second major PRF is therefore

$$PRF_{b1} = 1/(N_{b1}\tau_C) = 14{,}184 \text{ Hz.}$$

Selection of Second Pair of Minor PRFs As before, the second pair of minor PRFs must cause a second trace echo to shift by 2 range cells. Hence

$$N_{b2} = N_{b1} - 2 = 141 - 2 = 139$$
$$N_{b3} = N_{b2} - 2 = 139 - 2 = 137$$

The corresponding PRFs are

$$PRF_{b2} = 1/(N_{b2}\tau_C) = 14{,}388 \text{ Hz}$$
$$PRF_{b3} = 1/(N_{b3}\tau_C) = 14{,}599 \text{ Hz}.$$

Check Decodable Range $LCM(N_{b1}, N_{b2}, N_{b3}) = LCM(141, 139, 137) = 2{,}685{,}063$ range cells giving an maximum unambiguous range for this 3-PRF combination of 201.4×10^6 (m).

Selection of Second Major PRF As before, the third major PRF must differ by the MBC bandwidth. Taking the nominal value of the third major PRF $= PRF_{b1} + 2f_{dMBC} = 14{,}184 + 3340 = 17{,}524$ Hz, which is greater than the maximum allowable. However, taking the nominal value of the third major PRF as $= PRF_{a1} - 2f_{dMBC} = 10753 - 3340 = 7413$ Hz, which is within the allowable range. The nominal PRI is therefore 134.9 μs or 269.8 range cells. This rounds down to 269 range cells, so $N_{c1} = 269$.

The exact value of the third major PRF is

$$PRF_{c1} = 1/(N_{c1}\tau_C) = 7435 \text{ Hz}.$$

Selection of Second Pair of Minor PRFs As before, the third pair of minor PRFs must cause a second-trace echo to shift by 2 range cells. Hence

$$N_{c2} = N_{c1} - 2 = 269 - 2 = 267$$
$$N_{c3} = N_{c2} - 2 = 267 - 2 = 265$$

The corresponding PRFs are

$$PRF_{c2} = 1/(N_{c2}\tau_C) = 7491 \text{ Hz}$$
$$PRF_{c3} = 1/(N_{c3}\tau_C) = 7547 \text{ Hz}$$

Check Decodable Range $\text{LCM}(N_{c1}, N_{c2}, N_{c3}) = \text{LCM}(269, 267, 265) = 19{,}033{,}095$ range cells giving an maximum unambiguous range for this 3-PRF combination of 1.4×10^9 (m).

PRF Set The final set of PRFs is 10,753, 10,870, 10,989, 14,184, 14,388, 14,599, 7435, 7491, and 7547 Hz.

Maximum Duty Ratio Check The maximum PRF is $PRF_{a3} = 14{,}599$ Hz, which, coupled with the 7 μs transmitted pulse width, results in a maximum duty ratio of $14{,}599 \times 7\ \mu s = 0.102$ or 10.2%, which is below the maximum allowable.

Features of the Major Minor Method of PRF Selection PRFs are selected in multiples of three; hence 6 and 9 PRF schedules may be designed but not 7 and 8 PRF schedules. A schedule of 6 very carefully selected PRFs may just be adequate for some radar applications, but there may be insufficient beam dwell time to accommodate a schedule of 9 PRFs.

The major and minor PRI combinations differ by two range cells. Checks ensure that each major PRF and its associated minor PRFs are range decodable out to the maximum range of interest, but there is no guarantee that other combinations of three PRFs offer the same range decodability. It appears that a range decodability margin of two range cells is applied, but this is not necessarily the case for all combinations of three PRFs (assuming $M = 3$ decodability). Coarse range resolution results in few range cells and exacerbates the difficulty of maintaining a healthy range decodability margin. Variations on the method can easily be imagined in which major and minor PRF combinations differ by more than two range cells. This may be necessary when high compression ratios are used to ensure that the first few eclipsed zones do not overlap in too many PRFs; a larger margin would guard against blind ranges at short ranges but, conversely, might incur blindness at distant ranges and/or reduce the maximum decodable range. Further variations are possible if minor PRFs are selected on the basis of adding 2 range cells, rather than subtracting 2 range cells, from the number of range cells used by the major PRF, for example, $N_{a1} = 186$, $N_{a2} = 188$ and $N_{a3} = 190$.

The selection of the major PRFs ensures that the first repetitions of MBC do not coincide; however, this does not guarantee that higher-order repetitions of MBC do not overlap. There is no guarantee that blind velocities are avoided. If the MBC rejection bandwidth is wide, then major PRFs are widely separated and it becomes difficult to select PRFs within the allowable range. Here again, it appears as if a Doppler decodability margin equal to the MBC bandwidth is applied, but this is not necessarily the case for all combinations of 3-PRFs (assuming $M = 3$ decodability):

- No account is given to the minimization of range/velocity blindness.
- No provision is made for velocity (Doppler) decodability. The potential lack of range and velocity decodability and suspect decodability margins gives rise to fears that the schedule could be susceptible to ghosting.
- Limits on the maximum and minimum PRF and maximum duty ratio are easily applied.

In summary, the major–minor method attempts to offer PRF schedules boasting many important features such as decodability and a minimization of blindness, but these good features cannot actually be guaranteed. One is not at liberty to select any desired number of PRFs since the number of PRFs must be a multiple of three, which may be far from optimal. PRF schedules selected on this basis could have ghosting problems.

15.4 M:N METHOD OF PRF SELECTION

The M:N method of PRF selection selects N PRFs on the basis of M-of-N detection, where, typically, $M = 3$ [28, 50]. The selection of the N PRFs ensures target visibility at the first repetition of MBC in at least M PRFs, and often $(M + 1)$ PRFs. The PRIs are integer number of range cells. As with the major–minor method, the M:N method is best illustrated by an example in which the same radar parameters are assumed as those in Section 15.3 and elsewhere in this book.

As before, assume

$$N = 8$$
$$M = 3$$

Transmitted pulse width, $\tau = 7\ \mu s$ with pulse compression into a compressed pulse width, $\tau_C = 0.5\ \mu s$

$$\text{Maximum duty ratio of } 15\%,\ d_{max} = 0.15$$

MBC rejection over $f_{dMBC} = \pm 1.67$ kHz (or a double-sided bandwidth of $2f_{dMBC} =$ 3.34 kHz)

PRF Stagger The PRFs must be staggered by $\pm f_{dMBC}/(M+1) = 3340/4 = 835$ Hz.

Maximum PRF The maximum PRF allowed by the combination of the maximum allowable duty ratio and the transmitted pulse width is

$$PRF_{max} = d_{max}/\tau = 0.15/7\ \mu s = 21{,}429\ \text{Hz}$$

Choice of PRFs Starting at the maximum PRF, other nominal values of PRFs are set at the required offset frequency below the previous selection, that is,

21,429, 20,594, 19,759, 18,924, 18,089, 17,254, 16,419 and 15,584 Hz.

Rounding down the corresponding PRIs to integer range cell numbers results in

$PRF_1 = 21{,}505$ Hz, $N_1 = 93$

$PRF_2 = 20{,}619$ Hz, $N_2 = 97$

$PRF_3 = 19{,}802$ Hz, $N_3 = 101$

$PRF_4 = 19{,}048$ Hz, $N_4 = 105$

$PRF_5 = 18{,}182$ Hz, $N_5 = 110$

$PRF_6 = 17{,}391$ Hz, $N_6 = 115$

$PRF_7 = 16{,}529$ Hz, $N_7 = 121$

$PRF_8 = 15{,}625$ Hz, $N_8 = 128$

Features of the N:M Method of PRF Selection PRFs are selected to avoid overlap in the first repetition of MBC in $M + 1$ PRFs; however, there is no guarantee that higher-order repetitions of MBC will not coincide in too many PRFs and so give rise to a blind velocity.

The PRFs are quite closely grouped in frequency, which has several ramifications. The stagger in the repetitions of MBC gradually unwinds with increasing frequency, certainly more slowly than for PRFs selected using the major–minor method. This would suggest that blind velocities are less likely (but this is not *guaranteed*, as noted already) or that at least the first blind velocity occurs at such a high frequency as to be outside the band of interest. The closely grouped PRFs also suggest, but do not guarantee, that the maximum decodable Doppler (velocity) is likely to be very high.

In the time domain, the closely grouped PRIs could give rise to severely eclipsed ranges, especially at short ranges, since the first few repetitions of the blind ranges may not be sufficiently staggered in an adequate number of PRFs. This problem is exacerbated when high duty ratios are used, which may well be the case since the selection method is based on the premise of using PRFs close to the maximum allowable and hence close to the maximum duty ratio allowable.

The N:M selection method is designed to promote velocity visibility but may compromise on range visibility. The technique does not guarantee range or velocity decodability and does not consider decodability margins for minimizing ghosting. Schedules of any number of PRFs can be designed and are selected on the basis of adhering to maximum PRF and duty ratio constraints.

The methods of PRF selection described thus far apply simple rules. Their simplistic approach does not address every important consideration when selecting precise PRF values. They may yield a workable solution, but to guarantee that all factors are considered more complex methods—which often result in superior solutions—are required.

15.5 | EXHAUSTIVE SEARCH

Once the maximum and minimum PRIs have been defined, together with the PRI timing resolution, the total number of possible PRIs is fixed. If the total number of PRFs to be used within a schedule, N, is also defined, then it may be possible to exhaustively search all combinations of N PRFs. Each trial combination must be subjected to tests to screen for solutions that are not decodable and to assess their quality. The screening tests and quality assessment can be time-consuming and exhausting. The finer the PRI resolution, the greater the number of possible combinations to be trialed but the better the quality of the best solution or solutions. This leads to a trade between the time to conduct an exhaustive search and the quality of the solution it yields.

Two complications arise with this strategy. First, in some cases there are simply too many combinations to trial all possible solutions in a reasonable timescale. This will be examined further in the ensuing sections. The second complication is that the overall quality of a solution is often a combination of several factors, such as range/velocity blindness and decodability margins, which results in several optimal solutions known as a *Pareto set*, in which quality in one metric is played off against quality in another metric. Solutions in the Pareto set may be *dominated* if alternative solutions can be found that offer superior performance in all the test metrics, or *non-dominated* if this is not the case. Attempts to reduce all test metrics into a single metric by taking a weighted sum carries considerable risk since odd solutions at one extreme of the Pareto set are often declared as the favorites (depending on the weighting factors). It may be necessary to exercise some engineering judgment on the Pareto set to arrive at a preferred solution.

A trivial example of the problem may be to consider how to travel between two cities 300 miles apart. The two metrics to be considered are the *cost* and the *time* of the journey. Potential solutions under consideration may include flying, train, bus, car, cycling, and walking. The solutions are presented in Figure 15-1. The Pareto set is linked by the dashed line to form the Pareto surface. In more complex problems involving many metrics, the Pareto surface maybe multi-dimensional. However, in this simple example of only two metrics, the surface reduces to a line and any solution on the Pareto surface can be considered optimal. Train travel may be dismissed because it is a dominated solution since others beat it on both cost and time. The overall favorite may turn out to be the bus since this solution lies closest to the origin and represents the aggregated minimal cost and time. (Problems in which a function is maximized lead to preferred solutions that are farthest from the origin.) Exercising some judgment and invoking other factors not yet considered may lead to the dismissal of, or preference for, certain solutions. For example, one may readily dismiss walking, either because it is prohibitively time consuming or limits the amount of luggage that can be taken. One may have a dislike of air travel and thus dismiss that option. Alternatively, one may favor the car because there is a possibility of being joined by a second person, which would double the cost of most options but result in only a marginal increase in the cost of car travel.

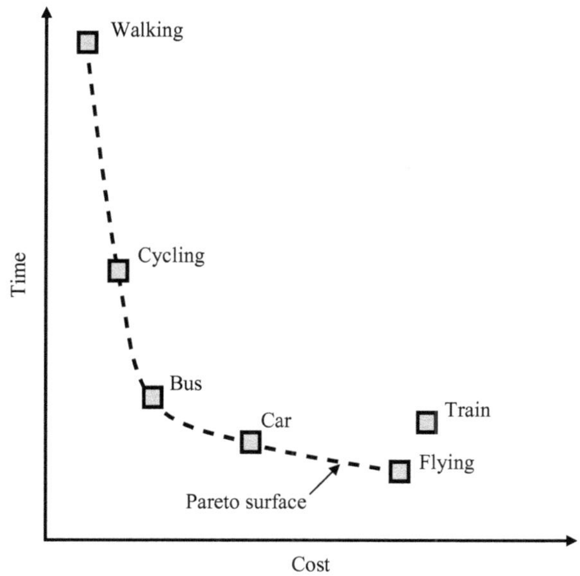

FIGURE 15-1 Optimization of Travel Options

An example of an exhaustive search for PRF values is described by Simpson [2]. PRIs in the range of 50 μs to 96 μs are considered with a PRI timing resolution of 1 μs. This results in 47 possible PRIs. A schedule of 8 PRFs is sought, which gives rise to ${}_{47}C_8 = 314{,}457{,}495$ possible solutions. Lengthy trials of this number of solutions would be very time-consuming, so a three-stage PRF selection procedure is described in which a considerable proportion of the solutions may be readily dismissed before lengthy analysis of the remaining solutions. The radar model assumes 3-of-8 range and velocity decodability. The objective is essentially to minimize blindness but also to avoid blindness-related problems.

Stage one applies a coarse screening to eliminate PRF sets that (1) fail a check for self-ghosting in the range dimension; and (2) fail a check for Doppler visibility in fewer than four PRFs due to coincidence of MBC rejection and its ambiguous repetition in the frequency domain. These checks ensure range and velocity decodability with suitable margins.

Stage two applies a PRF sorting in which all PRF sets not dismissed by the first stage are scored for their target visibility in each range/velocity cell in the space of interest. The scoring accounts for the extent and shape of blindness in range/velocity space. Three different scoring strategies are applied in differing regions of the range/velocity space that reflect the prioritization on maintaining good visibility at short ranges and the avoidance of blind velocities. Only the top-scoring PRF sets are retained and passed on to the third stage.

The third stage generates the blind zone map for each PRF set. This reveals the potential problems of (1) several small blind zones close together, which may coalesce into what is in effect a large blind zone; (2) blind zones that are confined in range but spread over a wide portion of the Doppler band; and (3) large areas where the PRF set is near-blind since near-blind areas create problems, especially at distant ranges where the probability of detection may be poor. The evaluation of the blind zone maps is a matter of expert opinion, and the quality of final solutions boils down to subjective judgment. Several PRF schedules with similar, though slightly different, qualities may be identified using this procedure.

15.6 | PRF SELECTION USING A NEURAL NETWORK

An example of PRF set selection using a neural network is described by Yi and Byun [38] (see also Section 13.3.1) in which the neural network is trained through an extensive, if not exhaustive, search. An airborne radar model is assumed, similar to that described in Table 12-1, in which the spread of allowable PRIs runs from 35 μs to 150 μs, the PRI timing resolution is 4 μs and the range cell width is 0.5 μs. This gives 29 possible PRIs. Since a 3-of-8 schedule is sought, there are ${}_{29}C_8 = 4{,}292{,}145$ combinations of 8 PRFs. A neural network is used to select an optimum 3-of-8 schedule, which maximizes target visibility in the presence of clutter in a small tracking window while also ensuring range decodability with a decodability margin of two range cells.

The neural network is trained for various combinations of engagement variables (aircraft speed, altitude, look angle, and target range and velocity). First, for each location of the target in range and velocity, the folded clutter map for each PRF is generated and the detectability of the target is assessed. If the number of PRFs in which the target may be detected is fewer than three, then the test value of the target range and velocity is not included in the training data. Otherwise, an 8-PRF candidate schedule is constructed from the 8 PRFs offering the best target detectability. Second, each candidate schedule is tested to ensure that it is range decodable with a decodability margin of at least two range cells (1 μs). These schedules form the training data for each combination of engagement variables.

The structure and parameters of the neural network established in the training phase is subsequently used by the radar to generate PRF sets. The neural network learning algorithm is able to interpolate for a wide range of clutter scenes to rapidly select an optimum PRF set. The tracking window establishes a small range/velocity space of interest in which target detectability is maximized (and hence, blindness is minimized) for the clutter scenario prevalent at the time. The neural network rapidly selects the optimum PRF set within each track update interval; a new optimum set is selected as the target being tracked moves in range and velocity and as the clutter scene varies due to changes in platform speed, altitude, and look angle.

It has also been suggested that neural networks can solve ambiguity problems in medium PRF radars [51]. Though this neural network method has been described for single targets, there is no indication of how a neural network may be trained for any arbitrary number of targets and therefore how it may handle the potential ghosting problem. In truth, modern processing algorithms work well, so there is little need for a neural network method.

15.7 | PRF SELECTION USING EVOLUTIONARY ALGORITHMS

15.7.1 Introduction to Evolutionary Algorithms

In many cases, the large number of possible PRF combinations precludes an exhaustive search. If a fine PRI resolution is used or different carrier frequencies, FFT sizes, duty ratios, or pulse widths are used in each coherent processing interval (CPI), then the number of different combinations of all these parameters rapidly escalates well beyond what can be exhaustively searched. An optimization process is required that can cope with a vast search space. Various types of evolutionary algorithms have recently been used for this task with considerable success [52].

An evolutionary algorithm (EA) is an optimization procedure that operates over a number of cycles (generations) and is designed to mimic the natural selection process through evolution and survival of the fittest [53, 54]. A population of independent individuals is maintained by the algorithm; each individual represents a potential solution to the problem. Each individual has one chromosome. This is the genetic description of the solution and may be broken into sections called genes. Each gene represents a single parameter in the problem; therefore, a problem that has eight unknowns, for example, would require a chromosome with eight genes to describe it. The genes may be binary, integer, real-value, a tree structure, an array, or a list. The chromosome may even consist of genes of different types, and the population may have chromosomes of different lengths. The chromosome level representation of potential solutions to the problem is known as the *genotypic space*.

The three simple operations found in nature—natural selection, mating, and mutation—are used to generate new chromosomes and therefore new potential solutions on each generation. The cyclical nature of the process is illustrated in Figure 15-2. In the context of PRF selection, the genes represent individual PRFs, a chromosome represents a PRF schedule, and the population represents a set of PRF schedules that are evaluated on each generation. New chromosomes may be generated by a combination of mating (otherwise known as *crossover*) and *mutation*, which makes a few random changes by applying Gaussian noise to each gene in each chromosome. Each chromosome is evaluated at every generation using an *objective function* that is able to distinguish good solutions from bad ones and to score their performance. With each new generation, some of the old individuals die to make room for the new, improved offspring. Despite being very simple to code, requiring no directional or derivative information from the objective function, and being capable of handling large numbers of parameters simultaneously, evolutionary algorithms can achieve excellent results.

One of the most awkward issues in using EAs for PRF selection is that of deriving the objective function. This is because the quality of a PRF schedule may be expressed in terms of many items of data (e.g., the detection performance at each range/velocity cell) and thus is highly dimensional or because there are several optimization goals so the quality becomes a multi-objective optimization process (e.g., minimizing blindness and maximizing decodability). Multi-objective EAs can identify solutions on the Pareto surface.

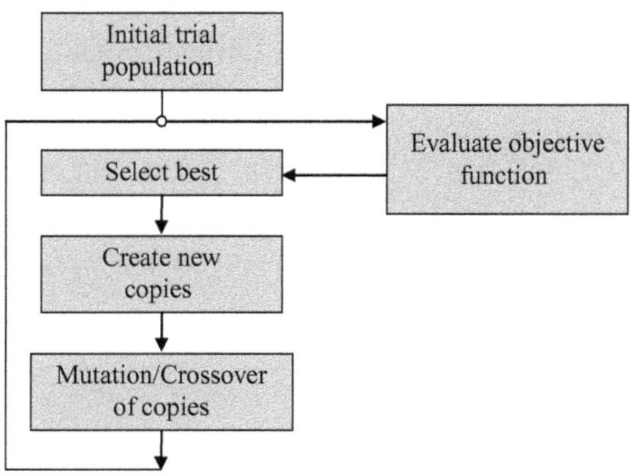

FIGURE 15-2 ■ An Evolutionary Algorithm Process

Evolutionary algorithms are powerful optimization techniques that have been successfully employed in a variety of combinatorial problems. They are particularly adept at finding near-optimum solutions very quickly when the number of possible combinations precludes an exhaustive search. It is worth noting that any EA does converge to a solution; however, the efficiency of its convergence and the quality of the solution depend upon the tuning of the algorithm (e.g., population size, crossover, and mutation rates). As of yet there is no mathematical description of the performance of an EA. In cases where an exhaustive search is not possible, one can never know for sure whether the solutions identified by an EA include the global optimum. For a well-designed EA and test for the objective function, one can be confident that the declared solutions are close to the global optimum and are quite likely to include it, but one cannot know this for sure. Hence, it is common to refer to the solutions as *near-optimum*.

Evolutionary algorithms are widely applied in academic research, industry, and the business sector. The ability of the algorithms to tackle huge, ill-defined problems with relative ease, and the availability of fast computers has led to a rapid uptake of the technology. The algorithms have been used to solve problems in electromagnetics, including the creation of novel antenna structures, allowing performance not attainable with conventional designs. Parts of satellite structures have been designed to be light, strong, and resistant to induced harmonic oscillations caused by differential heating by the sun. Large time tabling and process scheduling problems have been solved, such as the scheduling of television adverts and the scheduling of the next 30 years of planned maintenance of the UK national grid system. Evolutionary algorithms have been used for model identification in problems such as business credit scoring systems and insurance risk assessment. Techniques have been developed for identifying game playing strategies and homing guidance algorithms for missiles.

The field of evolutionary algorithms is very diverse. Some of the different algorithms are outlined in the following sections.

15.7.1.1 Genetic Algorithms

This is the type of algorithm described so far. The distinguishing characteristics are that both crossover and mutation are used to generate the new population, and that the chromosome is used to represent parameter values in the objective function. Variants of the algorithm exist that can handle multiple objective functions (multi-objective) or identify multiple optima (multi-species genetic algorithm). *Multi-species* means the ability to simultaneously output multiple local optimum solutions.

15.7.1.2 Genetic Programming

This field is concerned with evolving structures rather than parameters, such as evolving formulas rather than coefficients. Often the chromosomes represent tree structures or sequences of instructions.

15.7.1.3 Evolutionary Programming

These are some of the simplest algorithms. New populations are often generated by mutation alone, and tournament selection is used to drive the evolutionary process.

15.7.1.4 Evolutionary Strategies

The chromosome is constructed from the genes along with the standard deviations for mutating each of the genes. The mutations are therefore adaptive and evolve along with the solution.

15.7.1.5 Differential Evolution, Particle Swarm, and Population-Based Incremental Learning

New populations are generated through alternative methods of mutation and crossover. Many evolutionary algorithms use problem specific methods of creating new solution—the more domain knowledge that can be included, the simpler the optimization process.

15.7.2 PRF Selection for Minimal Blindness Using Evolutionary Algorithms

A flowchart representing an evolutionary process to select PRFs that minimize range/velocity blindness is given in Figure 15-3. The radar model accepts a chromosome from the evolutionary algorithm and decodes it into a set of PRIs. Operational parameters are passed to the clutter model, which in turn returns clutter data. A blind zone map is created, and target visibility is determined. The raw visibility data are then passed back to the evolutionary algorithm as the objective value to drive the evolutionary process. A new generation of PRFs is then produced and the process repeated.

Comparisons between evolutionary algorithms and exhaustive search techniques have been conducted for a relatively modest number of PRFs (230) to select medium PRF schedules to minimize blind zones [4]. This concluded that evolutionary algorithms offered an efficient alternative to conventional search methods and that they were capable of finding the optimum, or near-optimum, solutions in a fraction of the time taken by the exhaustive search method. The speed and flexibility of evolutionary algorithm techniques offered the potential for a radar to select PRF schedules optimally from a vast set of possible solutions, in near real time. Recent studies [3, 5-8] show that EAs may be used to minimize blindness over a range/velocity space of 185 km by ± 1500 m/s using a pool of 11,501 possible PRF values. A search for an 8-PRF schedule gives rise to ${}_{11501}C_8 = 7.57 \times 10^{27}$ combinations!

With such a vast search space available to the optimization process, it has been possible to select PRI sets for ambiguity resolution, minimization of blind zones, and the removal of blind velocities. Various PRF schedules have been optimized, some requiring $M = 3$ PRF decodability and others requiring $M = 2$. The following explanation of the PRF selection rules is based on the optimization of a 3-of-8 schedule [5] but can be readily adapted for any M-of-N criteria.

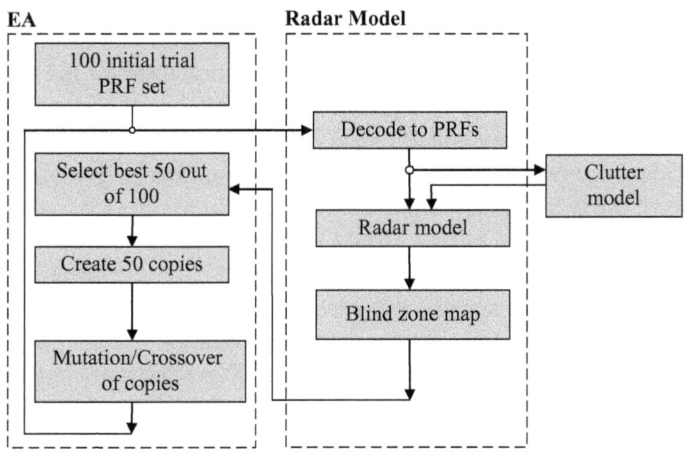

FIGURE 15-3 PRF Selection for Minimum Blindness Using an EA

Each chromosome forms a trial solution to the problem and consists of a set of eight genes that lie in the interval [0,1). These genes are then decoded into a PRI schedule, which is then used within a radar model to assess the quality of the schedule and to ensure that the schedule meets certain constraints. The chromosome is transformed into a PRI set by first generating a set containing all 11,501 possible PRIs. The first PRI is chosen at random from the set, giving a choice of 1 in 11,501, and the chosen PRI is removed from the set to prevent it from being selected again. The second PRI is also randomly selected from the remaining 11,500 and removed from the set. The remaining set is now checked. The remaining set is now checked and any PRIs that are not decodable in both range and Doppler with the first two PRIs chosen, or which may lead to severe ghosting, are removed from the set. Any desired values of range and decodability margins may be imposed; the larger these margins the greater the resilience against ghosting but the larger the number of incompatible PRIs that are pruned from the set leading to solutions of poorer blindness. Any PRIs that would also lead to a blind velocity are also pruned. The third and subsequent PRIs can now be chosen at random from those remaining and are removed from the set. For PRIs four onward, decodability must be checked between each PRI remaining in the set and each combination of PRI pairs from the PRIs already chosen. This process ensures that the PRI set is fully decodable. If there are no PRFs remaining within the set before all N PRIs are chosen, the objective is set to be totally blind. The objective function provides a measure of how well an individual performs in the problem domain [53]. In this case, the objective function is the total area of the blind zone map (in meters Hertz) with four or more ($\geq M + 1$) PRFs clear. The decoding process has already ensured that the PRF set is fully decodable with reduced ghosting and has no blind velocities.

A simple evolutionary program [54] with a base population of 50 trial solutions has been used as the evolutionary engine [5]. The evolutionary program operated by creating 50 new trial solutions at each generation and evaluating them for blind zone performance. The best 50 overall from the 100 population were then chosen for the next generation. In this particular algorithm, an initial population of 100 trial solutions was used, of which the best 50 were chosen for the next generation. To create the 50 new solutions, a typical evolutionary program cycle of crossover and mutation was applied. First, the 50 chromosomes remaining from the previous generation were copied. Each of the 50 solutions had a 70% chance of being crossed during the copy with another chromosome chosen at random from the population. If crossover was to be applied, a process of real valued intermediate crossover was used [53]. Gaussian mutation was then applied to each gene by adding a random number drawn from a zero mean Gaussian distribution with an initial standard deviation of 0.125 ($= 1/8$ of the range of the gene values). A new random number was drawn for each gene in each chromosome. The algorithm was forced to converge by reducing the standard deviation of the Gaussian distribution used for the mutation process by multiplying by a factor of 0.9 every generation. Thus, as the algorithm progresses, the size of the random numbers added to the genes reduces and forces the search to be refined to provide more repeatable results in a limited number of generations. In the first few generations of the evolutionary algorithm, the mutations are large, so a wide search is performed across the PRI search space. The reduction factor of 0.9 reduces the standard deviation of the mutations quite quickly, so after around 30 generations or so the mutation, and therefore the global search, was having little effect. The search direction was controlled more by crossover, and therefore local exploitation of the optimization surface was performed. The algorithm was terminated after 100 generations, and the best solution was selected (i.e., best blind zone performance) as the final PRI set for use. This size of population and number of generations provided a reasonable number of sample solutions from the problem domain without incurring unmanageable processing times.

The evolutionary algorithm can select near-optimal PRF sets efficiently with modest computing effort and produce a low level of range/velocity blindness [3-8, 27] (see Sections 13.2.2 and 13.2.3). The "quality" of each set is based on models of airborne fire control radar and associated clutter, so each PRF set is application/scenario specific. Repeated runs of the evolutionary algorithm identify near-optimal PRF sets that differ marginally from each other. These repeats indicate the existence of several similar local optima in the problem space and the ability of the evolutionary algorithm to find them.

15.7.3 PRF Selection for Optimal Target Detectability Using Evolutionary Algorithms

A similar evolutionary approach to that of Section 15.7.2 to select and trial PRF values is described in the flowchart of Figure 15-4 [47, 48, 49]. This is now driven to maximize target detectability and includes sophisticated antenna and clutter models [16]. In this way it has been possible to explore the extent to which the design of the medium PRF waveform may be optimized for various clutter scenes [47] and antenna radiation patterns [48, 49]. The metric used to assess the quality of each medium PRF schedule is the *detectability map* [48, 49], described in Section 14.2.3.

The optimization process is driven by an EA with an optimization goal of achieving maximum range/Doppler detectability. The EA maintains a population of trial PRF schedules whose values are refined on each iteration of the loop process (generation) along the lines of Darwinian theories of evolution and survival of the fittest. Each trial set is passed to the radar model, and the genetic description is decoded to PRF values. This decoding stage employs a variety of checks to ensure that the schedule is decodable, does not incur any blind velocities, enforces a margin to minimize the risk of ghosting, and is within the limits of maximum, minimum, and mean PRF, as dictated by the radar model. The PRFs are passed to the radar and clutter models. The clutter model returns the clutter map for each PRF, which is also passed to the radar model. The radar model is based on an airborne FCR and accepts the trial PRF schedule and clutter maps. The model then generates a detectability map. The detectability maps arising from each parent trial schedule (from the previous generation), and its offspring schedule (from the current generation)

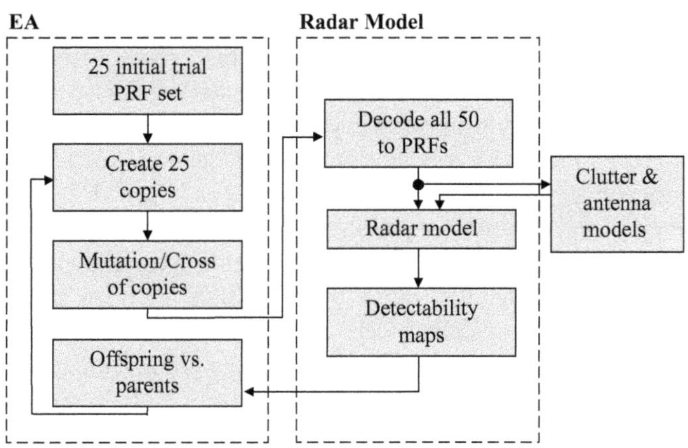

FIGURE 15-4 PRF Selection for Maximum Detectability Using an EA

are compared pixel by pixel (i.e., range/Doppler cells). The solution having the lowest radar cross section (RCS) requirement in the greater number of pixels populates the next generation of trial schedules. Over repeated generations of the process, PRF schedules of superior detectability performance evolve. In this way, no detectability map is quantified for an absolute level of performance but is merely compared with those of previous generations. This is known as *binary tournament selection*.

This process has been used to optimize PRF values for various clutter statistics [47] (Section 14.2.4) and to explore optimum antenna illumination functions for clutter-limited detection conditions [48, 49] (Section 14.2.5).

15.7.4 PRF Selection Using Multi-Objective Evolutionary Algorithms

The evolutionary algorithms described in Sections 15.7.2 and 15.7.3 are powerful techniques for PRF selection since they identify several near-optimum PRF sets from a vast number of possibilities. One may now exploit the full potential of a fine level of PRI timing resolution and combinations with other variables, such as the carrier frequency and FFT size, without fearing the complexity of the problem. These techniques yield a selection of good engineering solutions but are slowed by the lengthy process to derive the objective functions. In one case, the objective function requires the generation of the blind zone map, and in the other case a detectability map is required; both of these entail detailed radar and antenna models and a time-consuming clutter simulation. It would be desirable to rate the quality of PRF schedules purely on their own merits and without recourse to external parameters such as the factors dictating the clutter scene. This is possible since most of what can be considered "good" about a PRF schedule can be expressed by the following five optimization goals:

- Maximize the range decodability margin (Section 12.1.4)
- Maximize the velocity decodability margin (Section 12.1.4)
- Maximize the range blindness margin (Section 12.2.3)
- Maximize the velocity blindness margin (Section 12.2.3)
- Minimize the time to transmit the whole (Section 12.3.3)

Decodability is assured for range and velocity decodability margins of at least one cell in each coordinate; a greater margin offers greater resilience against self-ghosting and indeed other ghosting mechanisms. Target visibility is assured for range and velocity blindness margins of at least one cell in each coordinate; a greater margin maintains target visibility in the face of clutter patches, or other target returns, which are extended in range or velocity. The time to transmit the whole schedule should be less than the beam dwell time. Minimizing the number of coherent processing intervals reduces the necessity to incur repeated space charging times and, in turn, permits a larger FFT size to be used with its associated advantages of increased processing gain and detection performance. All of the test metrics (range and velocity decodability margins and blindness margins plus the time to transmit the entire schedule) are a function of the PRF set and a few key radar parameters alone and can be evaluated very quickly. Actually, all these metrics can be derived from a description of the *waveform* used by the radar. This now entails a multi-objective optimization procedure with the five optimization goals given in the previous bullet points.

Multi-objective EAs are very powerful tools but are far from trivial techniques. They have been used for the selection of optimum PRFs in medium PRF radar applications [37]. Since trial solutions can be evaluated very quickly, one need no longer maintain a set of PRFs and prune out incompatible values as the PRF set is constructed, which results in further time savings. These savings are offset to some extent by evaluating schedules that may not be decodable, that incur heavy blindness, or that are too long, but such schedules are discriminated against by the evolutionary process and quickly die off. Since multi-objective EAs identify the Pareto set, they are usually augmented by some degree of engineering judgment to identify a few "preferred" solutions. The solutions can be difficult to depict and interpret (one would need a five-dimensional graph).

An example of the set of solutions to the five optimization goals using a multi-objective EA is given in Figure 15-5, in which each trace represents a different solution. In this example, the PRFs for a $N = 5$ PRF schedule is sought with a constant carrier frequency of 10 GHz and with a variable FFT size designed to result in similar CPI durations and hence similar velocity resolutions. The horizontal axis indicates the median range decodability margin (MedDR), median velocity decodability margin (MedDV), median range blindness margin (MedBR), median velocity blindness margin (MedBV), minimum range decodability margin (MinDR), minimum velocity decodability margin (MinDV), minimum range blindness margin (MinBR), minimum velocity blindness margin (MinBV), and dwell time (Dwell). The vertical scale gives a normalized scaling for each of these

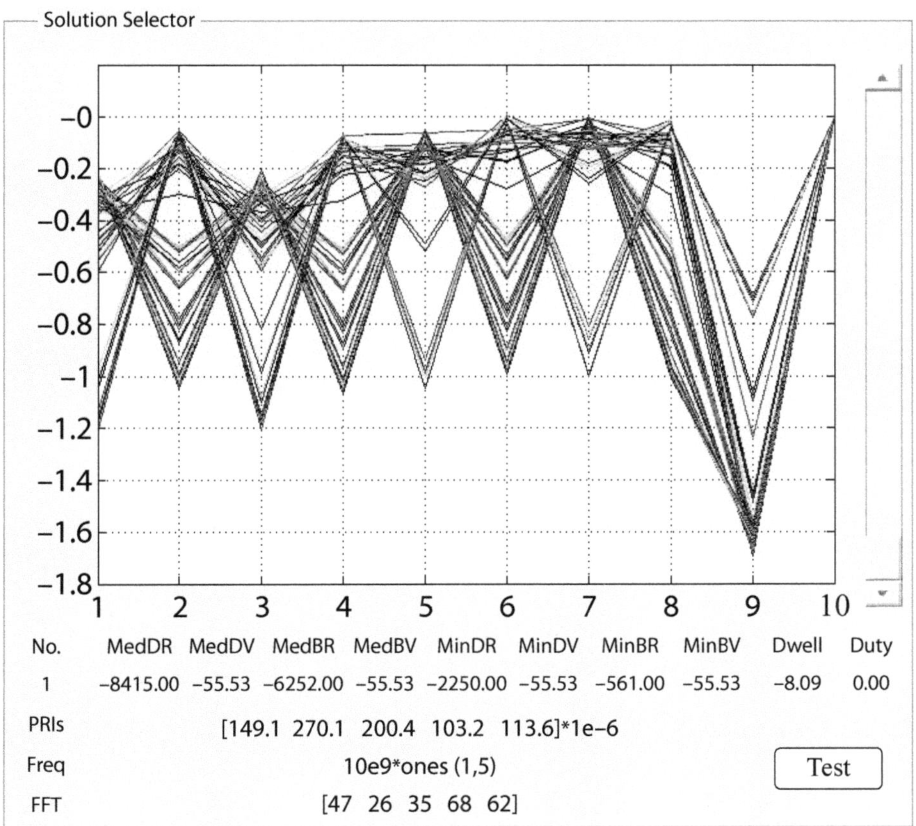

FIGURE 15-5 Solutions to Five Objective PRF Optimization Using a Multi-Objective EA

parameters. The optimization process has found several medium PRF and high PRF solutions. One of the better solutions has been highlighted:

- The-near optimum PRIs are 149.1, 270.1, 200.4, 103.2, and 113.6 μs
- A carrier frequency of 10 GHz is used in all five CPIs.
- Its associated FFT sizes are 47, 26, 35, 68, and 62.
- A read-out of the nine parameters is also given in Figure 15-5 to enable the user to judge the overall quality of the selected PRF set.

15.8 CONCLUDING REMARKS

A great deal can be done to enhance and optimize the operation of a radar through careful selection of, principally, its PRFs but also of other waveform parameters. Simplifying the method of PRF selection can give a workable solution, but far superior performance can be achieved if one is prepared to take on a more complex problem entailing fine PRF resolution and combinations of PRFs with other waveform and processing parameters. The more complex problems demand more complex methods of optimization. Modern methods described here have been used with great success in the design of radar waveforms supporting a wide range of modes such as high and medium PRF pulse Doppler and ground-moving target indication (GMTI) and applications such as airborne fire control and long-range early warning radars.

References

[1] Hovanessian, S. A., "Medium PRF Performance Analysis," IEEE Trans. Aero. & Elect. Sys., AES-18(3), May 1982, 286–296.

[2] Simpson, J., "PRF Set Selection for Pulse-Doppler Radars," in IEEE Region 5 Conference, 1988: Spanning the Peaks of Electrotechnology, 1988, 38–44.

[3] Alabaster, C. M. and Hughes, E. J., "Medium PRF Radar PRF Optimisation Using Evolutionary Algorithms," Proc. Radar 2003, Huntsville, AL, May 5–9, 2003.

[4] Davies, P. G. and Hughes, E. J., "Medium PRF Set Selection Using Evolutionary Algorithms," IEEE Trans. Aerospace and Electronic Systems, 38(3), July 2002, 933–939.

[5] Alabaster, C. M., Hughes, E. J., and Matthew, J. H., "Medium PRF radar PRF Selection Using Evolutionary Algorithms," IEEE Trans. Aerospace and Electronic Systems, 39(3), July 2003, 990–1001.

[6] Wiley, D. A., Parry, S. M., Alabaster, C. M., and Hughes, E. J., "Performance Comparison of PRF Schedules for Medium PRF Radar," IEEE Trans. Aerospace and Electronic Systems, 42(2), April 2006, 601–611.

[7] Alabaster, C. M. and Hughes, E. J., "Novel PRF Schedules for Medium PRF Radar," Paper presented at the Proceedings of Radar, Adelaide, September 3–5, 2003.

[8] Alabaster, C. M., Hughes, E. J., Parry, S. M., Wiley, D. A., Matthew, J. H., and Davies, P. G., "Medium PRF Schedules for Airborne Fire Control Radar," Paper presented at the Proceedings of the IEE Waveform Diversity & Design, Edinburgh, November 8–10, 2004.

[9] Ferrari, A., Birenguer, C., and Alengrin, G., "Doppler Ambiguity Resolution Using Multiple PRF," IEEE Transactions on Aerospace and Electronic Systems, 33(3), July 1997, 738–751.

[10] Xia, X.-G., "Doppler Ambiguity Resolution Using Optimal Multiple Pulse Repetition Frequencies," IEEE Transactions on Aerospace and Electronic Systems, 35(1), January 1999, 371–379.

[11] Long, W. H. and Harriger, K. A., "Medium PRF for the AN/APG-66 Radar," *Proceedings of the IEEE*, 73(2), February 1985, 301–311.

[12] Nevin, R. L., "Waveform Trade-Offs for Medium PRF Air-to-Air Radar," in *Proceedings of the* Radar Conference, Ann Arbor, MI, April 1998, 140–145.

[13] Moorman, R. A. and Westerkamp, J. J., "Maximizing Noise-Limited Detection Performance in Medium PRF Radars by Optimizing PRF Visibility," *Proceedings of the IEEE 1993 National Aerospace and Electronics Conference, NAECON* 93, 1, 1993, 288–293.

[14] Nooy, P. N. C. and Debuisser, J.-C., "Medium PRF Waveforms: An Optimal Solution for Various Applications in Surface Based Radar," in *Proceedings of the* 28th European Microwave Conference, Amsterdam, 1998, 572–577.

[15] Trunk, G. and Brockett, S., "Range and Velocity Ambiguity Resolution," in *Proceedings of the* Radar Conference, Record of the 1993 IEEE National, 146–149.

[16] Alabaster, C. M. and Hughes, E. J., "Clutter Mapping for Airborne Pulse Doppler Radar," Paper presented at the IET Clutter Modelling Seminar, February 19, 2008, Savoy Place, London, UK.

[17] Randig, G. W., "A Simple Method to Determine Airborne Radar Detection Performance in Clutter," in *Proceedings of the* Radar Conference, Lynnfield, MA, April 1993, 192–196.

[18] Kinghorm, A. M., Smith, M. J., and Heggie, C. A., "Pulse-Doppler Radar Detection Performance Modelling and Verification," in IEE Colloquium on Computer Modelling and Simulation of Radar Systems, London, UK, February 1993, pp. 8/1–8/4.

[19] Åström, A., Johannesson, M., and Edman, A., "An Implementation Study of Airborne Medium PRF Doppler Radar Signal Processing on a Massively Parallel SIMD processor Architecture", in *Proceedings of the* IEEE International Radar Conference, 1995, 551–557.

[20] Hovanessian, S. A., "An Algorithm for Calculation of Range in a Medium PRF Radar," IEEE Trans. Aero. & Elect. Sys., AES-12(2), March 1976, 287–290.

[21] Reddy, N. S. and Swamy, M. N. S., "Time-Domain Estimation of Unambiguous Doppler Frequency in Low and Medium PRF Radars," in *Proceedings of the* International Conference on Acoustics, Speech and Signal Processing (ICASSP) 83, Boston, MA, 1983, 687–690.

[22] Thomas, A. G. and Berg, M. C., "Medium PRF Set Selection: An Approach Through Combinatorics," *IEE Proceedings—Radar, Sonar Navigation,* **141**(6), December 1994, 307–311.

[23] Lei, W., Long, T., and Han, Y., "Resolution of Range and Velocity Ambiguity for Medium Pulse Doppler Radar," *Proceedings of the* IEEE International Radar Conference, Alexandria, VA, May 2000, 560–564.

[24] Piacentini, M., "Choice of PRF for Error-Tolerant Range Destaggering of Radar Targets," Electronics Letters, 21(24), November 21, 1996, 1128–1129.

[25] Reddy, N. S. and Swamy, M. N. S., "Resolution of Range and Doppler Ambiguities in Medium PRF Radars in Multiple-Target Environment," *Proceedings of the* International Conference on Acoustics, Speech and Signal Processing (ICASSP), 9(1), March 1984, 514–517.

[26] Lynch, Jr., D. and Kopp, C., "Multifunctional Radar Systems for Fighter Aircraft," Chapter 5 in M. I. Skolnik (Ed.), Radar Handbook, 3d ed., McGraw-Hill, 2008.

[27] Hui, C., Benxiong, H., Yaoqing, H., and Lei, R., "Research on Waveform Design of Airborne Phased Array Pulse Doppler Radar," Paper presented at the CIE'06 International Conference on Radar, Shanghai, October 2006.

[28] Morris, G. V., Airborne Pulse Doppler Radar, Artech House, 1998.

[29] Hughes, E. J. and Lewis, M. B., "Improved Detection and Ambiguity Resolution of Low Observable Targets in MPRF Radar," Paper presented at the Proceedings of the EM&RS DTC 2nd Annual Conference, Edinburgh, 2005.

[30] Hughes, E. J. and Lewis, M., "An Intelligent Agent Based Track-Before-Detect System Applied to a Range and Velocity Ambiguous Radar," EM&RS DTC 1st Annual Technical Conference, May 2004.

[31] Kinghorn, A. M. and Williams, N. K., "The Decodability of Multiple-PRF Radar Waveforms," in Proceedings of IEE Radar 97, Edinburgh, UK, Conference Publication 449, October 1997.

[32] Trunk, G. V. and Kim, M. W., "Ambiguity Resolution of Multiple Targets Using Pulse-Doppler Waveforms," *Aerospace and Electronic Systems, IEEE Transactions*, 30(4), October 1994, 1130–1137.

[33] Trunk, G. V. and Kim, M. W., "Multiple Target Ambiguity Resolution," in *Proceedings of the* Radar Conference, 1994, Record of the 1994 IEEE National, March 29–31, 1994, 91–94.

[34] Van Genderen, P. and Meijer, W. J. H., "Non Coherent Integration in a Medium PRF Radar," Paper presented at the IEEE International Radar Conference, 1995.

[35] Immediata, S., Petrucci, R., Timmonieri, L., and Vigilante, D., "New Approach to 2D Medium PRF Ambiguity Resolution with Application to Air Surveillance Radar," Paper presented at the IEEE Radar Conference, 2008.

[36] Schikorr, M., "Medium-PRF Detection by Non-coherent Integration," in *Proceedings of the 7th European Radar Conference*, September 30–October 1, 2010, Paris, France, 224–227.

[37] Hughes, E. J., "Multiple Single Objective Pareto Sampling," CEC Canberra, 2003.

[38] Yi, J. W. and Byun, Y. J., "Real-Time PRF Selection for Medium PRF Airborne Pulsed-Doppler Radars in Tracking Phase," *Proceedings of* Waveform Diversity & Design, Pisa, 2007, 116–121.

[39] Wilson, S. L. and Carlson, B. D., "Radar Detection in Multipath," IEE Proc. Radar, Sonar & Navigation, 146(1), February 1999, 45–54.

[40] Stralka, J. P. and Fedarko, W. G., "Pulse Doppler Radar," Chapter 4 in M. I. Skolnik (Ed.), Radar Handbook, 3d ed., McGraw-Hill, 2008.

[41] DiFranco, J. V. and Rubin, W. L., Radar Detection, Norwood, MA: Artech House, 1980.

[42] Shnidman, D. A., "Determination of Required SNR Values," IEEE Trans. Aerospace and Electronic Systems, 38(3), July 2002, 1059–1064.

[43] Barton, D. K., "Universal Equations for Radar Target Detection," IEEE Trans. Aerospace and Electronic Systems, 41(3), July 2005, 1049–1052.

[44] Evans, M., Hastings, N., and Peacock, B., Statistical Distributions, 3d ed., New York: John Wiley and Sons Ltd, 2000.

[45] Zhenru, Y., Zhaoda, Z., and Xinping, H., "CFAR Analysis for Medium PRF Airborne Pulse Doppler Radars," in *Proceedings of the* IEEE National Aerospace and Electronics Conference (NAECON), Dayton, OH, Vol. 1, May 1998, 259–263.

[46] Hughes, E. J. and Lewis, M., "Evaluation of Second Order Adaptive CFAR in the Littoral Environment," Paper presented at the EM&RS DTC 5th Annual Technical Conference, Edinburgh, 2008.

[47] Alabaster, C. M. and Hughes, E. J., "The Design of Medium PRF Schedules for Optimum Detectability in Diverse Clutter Scenes," Paper presented at the Proceedings of the International Waveform Diversity and Design, Lihue, HI, January 22–27, 2006.

[48] Alabaster, C. M. and Hughes, E. J., "Examination of the Effect of Array Weighting Function on Radar Target Detectability," IEEE Trans. Aerospace and Electronic Systems, 46(3), July 2010, 933–939, 1364–1375.

[49] Alabaster, C. M. and Hughes, E. J., "The Dependence of Radar Target Detectability on Array Weighting Function," Paper presented at the Proceedings of Radar 2007, Edinburgh, UK, October 15–18, 2007.

[50] Wang, Y.-L., Bao, Z., and Peng, Y.-N., "STAP with Medium PRF Mode for Non-side-looking Airborne Radar," IEEE Trans. Aerospace & Electrical Systems, 36(2), April 2000, 609–620.

[51] Wang, C.-J. and Wu, C.-H., "Analog Neural Networks Solve Ambiguity Problems in Medium PRF Radar Systems," in *Proceedings of the* IEEE International Conference on Neural Networks, San Francisco, CA, Vol. 1, March–April 1993, 120–125.

[52] Hughes, E. J., "Teaching Notes of E. J. Hughes," Defence Sensors & Data Fusion master's course, Cranfield University, Shrivenham, UK.

[53] Zalzala, A. M. S. and Flemming, P. J. (Eds.), Genetic Algorithms in Engineering Systems, London: Institution of Electrical Engineers, 1997.

[54] Goldberg, D. E., Genetic Algorithms in Search, Optimization, and Machine Learning, Reading, MA: Addison-Wesley, 1989.

PART III

Case Studies

CHAPTER 16

Airborne Fire Control Radar

Chapter Outline

16.1 INTRODUCTION

Airborne fire control radars (FCRs) on fast strike aircraft are the quintessential pulse Doppler radars. They must work in a wide variety of air-to-air and air-to-ground modes, they must be lightweight and compact, yet they have to achieve long detection ranges in the presence of extreme clutter scenes and be capable of tracking a large number of agile targets, all of which must be highly automated, especially for single seat aircraft, to minimize the workload on the aircrew.

Airborne FCR was borne out of its predecessor, airborne intercept (AI) radar. AI radar emphasizes the ability to intercept and engage enemy aircraft closing at high speeds (and often at low altitudes) from a high to medium altitude platform. AI radar had its origins in the nighttime air battles of the early 1940s over British cities during the Second World War. The limited ranges of such system required the use of ground-controlled intercept (GCI) radars to guide the night fighters toward the enemy contacts before the AI system could finish off the job. The AI Mk 24, the last of the British AI radars, was the first European airborne pulse Doppler radar to enter service in the mid-1980s and, at the time of writing (2011), is still in service onboard the Tornado F3 Air Defence Variant. The main mode of this radar is a high pulse repetition frequency (PRF) Doppler mode using PRFs in the region of 300 kHz, with a high duty ratio, incorporating frequency modulated interrupted continuous wave (FMICW) ranging. This achieves long-range detection against fast closing targets in a look-down, shoot-down attitude, which, of course, plays to the strengths of high PRF. High PRF waveforms may still be used by airborne FCRs in long-range, air-to-air velocity search, target acquisition, and tracking modes. Low PRF modes may also be used in short-range ranging applications.

More modern airborne FCR systems require all-aspect target detection, which high PRF struggles to achieve due to its sidelobe clutter (SLC) related problems. The need for this type of target detection led to the development of medium PRF pulse Doppler modes, which have become synonymous with airborne FCR [1-6]. In fact, much of Chapters 11 to

15 is based on the airborne FCR case. These chapters illustrate many of the concepts and present several figures based on a hypothetical but realistic airborne FCR model, which is then added to and grows in complexity as the description of medium PRF progresses throughout these chapters. It is therefore highly appropriate that the case study concerning airborne FCR should dwell on the medium PRF mode, and that is what this chapter does. The airborne FCR case study presented here draws on the model and data developed previously throughout Chapters 11 to 15.

16.2 SETTING THE SCENE

16.2.1 Applications

An airborne FCR is required to perform the following functions [1]:

- Long-range (100 nm) air-to-air search and targeting:
  - Look-down, shoot-down
  - Track-while-scan (TWS)
  - Raid assessment
- Short-range combat mode:
  - Weapons aiming
- Ground mapping and the acquisition of ground-based targets
- Medium-range (15–40 nm) target tracking for engagement using medium-range air-to-air missiles.
- Medium-range missile illumination (for semi-active homing systems)
- Medium- and short-range rendezvous, such as refueling tanker aircraft
- Operation in harsh electromagnetic (EM) environment of deliberate jamming, co-channel interference, and unintentional noise

Targets are agile aircraft with a radar cross section (RCS) in the 8–12 GHz band of between 2 m^2 (head-on) to 100 m^2 (maximum side-on), flying at ground velocities between 150 m/s and 750 m/s. Stealthy aircraft targets may have an RCS as low as 0.005 m^2 in the head-on aspect. Missile targets are likely to have an RCS at least one order of magnitude lower than non-stealthy aircraft and may travel at 1300 m/s (Mach 4). Ground-based targets include structures such a bridges and buildings or vehicles such as armored vehicles, which may be moving or static.

The most pernicious form of electronic countermeasures (ECM) is likely to be that from digital radio frequency memory (DRFM) based repeater jammers and transponders. This can replicate the radar's waveform with remarkable integrity and can present a plethora of false targets that may correlate within the victim receiver and processing chain. Of particular concern might be the ghosting within a medium PRF mode in response to a large number of realistic target returns.

16.2.2 Techniques

The applications outlined in the previous section require the following radar modes and waveforms:

- Pulse Doppler radar using:
  - Low, high, and medium PRF waveforms
  - FMICW waveform
  - Pulse compression
- High-resolution mapping modes:
  - Synthetic aperture radar (SAR)
  - Doppler beam sharpening (DBS)
  - Ground-moving target indicator (GMTI) radar
- Excellent detection performance:
  - High P_D
  - Low P_{FA}
  - Even against stealthy targets
- Large capacity tracker:
  - Multi-hypothesis tracker
  - Kalman filtering
  - Large computational resources
  - Eliminates false tracks quickly
- Excellent antenna design:
  - High gain
  - Low sidelobes
  - Agile scanning
- Monopulse angle locking
- Continuous wave (CW) missile illumination (semiactive homing air-air)
- Electronic protection measures (EPM):
  - Medium PRF schedule design and PRF selection
  - Target extraction algorithm
  - Adaptive waveforms
  - Steerable antenna nulls
- Low probability of intercept (LPI)
  - Low signature (stealth)
  - Spread spectrum modulations (wide signal bandwidth)
  - Transmitter power management
  - Low sidelobes

Some other important features include:

- Rugged, small, and lightweight

- Operation at high platform altitude (70,000 ft or 21,500 m):
  - Low air pressure (may require waveguide pressurization)
  - Large operational temperature range (+70°C to −50°C),
- High degree of automation:
  - Especially for single-seat aircraft
  - Man–machine interface needs to considered most carefully
- Integration with other sensors and systems:
  - Interrogation friend or foe (IFF)
  - Guided weapons
  - Inertial navigation system (INS), global positioning system (GPS), air speed
- Ease of maintenance:
  - Built-in test equipment (BITE)
  - First-line repair by line replaceable unit (LRU) replacement.

16.2.3 Some Common Features

The applications and techniques listed in the previous section tend to result in certain common features that have been widely adopted for airborne FCR applications.

16.2.3.1 Frequency

There is a trend that radar systems operating at lower frequencies (longer wavelengths) become larger and more powerful and capable of longer-range detection. This is partly due to the reduced atmospheric losses at lower frequencies and the use of a larger antenna aperture. (Recall that antenna gain varies as $1/\lambda^2$ for a fixed aperture size, which leads to the detection range being directly proportional to the diameter of the antenna, *all other factors being equal.* So a large antenna always helps.) Most strike aircraft can accommodate an antenna that is typically 60 to 80 cm in diameter, so this poses the classic compromise between wavelength, atmospheric losses, and antenna gain, beamwidth, and aperture. Almost universally, airborne FCR systems settle this compromise at around 9 GHz ($\lambda = 3.33$ cm).

16.2.3.2 Antenna

Planar phased array antennas, which are prevalent in many current systems, are now being superseded by active electronically scanned array (AESA) antennas. This facilitates the ease of beam steering to rapidly interleave various search and track modes, control the sidelobe level, and enable steerable nulls to be formed to guard against jamming from a known direction. The antenna typically has a $+/-60°$ forward field of regard about the axis of the aircraft.

The transmitter of choice in support of a planar array antenna is the traveling wave tube (TWT), which produces 2 to 5 kW of peak power. AESA antennas typically include 1500 transmit/receive modules (TRM) in which the transmitting signal source and amplification is based on gallium arsenide (GaAs) solid-state technology, whereby each TRM produces 2–5 watts (peak) of radio frequency (RF) power. Each TRM would steer

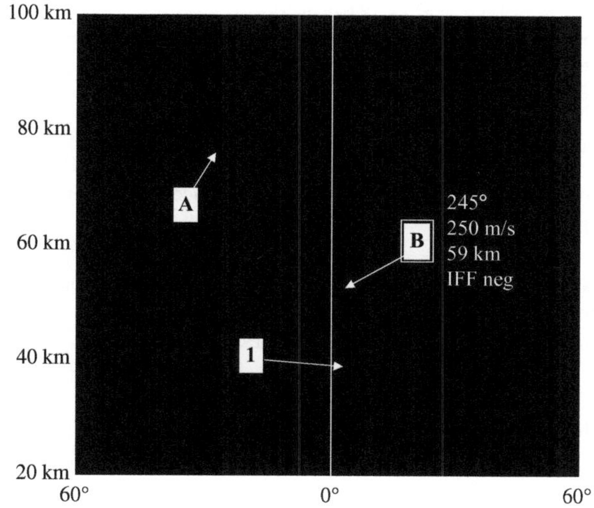

FIGURE 16-1 B-Scope Display

the beam and implement the desired aperture illumination using 6-bit phase and 6-bit amplitude control.

16.2.3.3 Display

Most FCRs present target information on computer-generated displays using the B-scope format, an example of which is given in Figure 16-1. This is a two-dimensional display of azimuth angle along the horizontal against range up the vertical. The display gives a somewhat distorted scale at short ranges. Targets are displayed using alphanumeric symbols, and their track is indicated by the arrow originating on the target symbol. The user can select a target for a readout of further information on that target. Selected target data are also often relayed to the head-up display (HUD) or helmet-mounted displays.

16.3 | MEDIUM PRF MODE

Typical parameters and performance of a medium PRF mode for an airborne FCR are presented in the following sections and are drawn from the data of previous chapters. Medium PRF is used on medium- to long-range air-to-air modes for search, target acquisition, and tracking functions. The AN/APG-66 system design was discussed by Long and Harriger [1] and Ringel, Mooney, and Long [2] and some of its details were presented by Simpson [3].

16.3.1 Typical Radar Parameters

The parameters of a typical airborne FCR are reproduced in Table 16-1 (based on Table 12-1). The pulse repetition intervals (PRIs) have been optimized for minimal range/velocity blindness. They are fully decodable over the whole range and velocity space of interest with a range decodability margin of 0.7 μs (the 0.5 μs range gate width plus an additional margin of 0.2 μs) and a velocity decodability margin of at least one velocity cell being imposed during the PRF selection process. For a space charging time of 1.85 ms (equates to 277.5 km of range), a total of 8 $\times$ 1.85 ms = 14.8 ms or 22.8% of the beam

TABLE 16-1 ■ Typical Airborne FCR Parameters

Parameter	Value
Carrier Frequency, f_{tx}	10 GHz
PRIs (μs)	63.11, 69.97, 77.07, 81.31, 90.06, 99.90, 109.75, 119.00 (mean PRI = 88.77 μs)
Ambiguity Resolution (3-of-8)	Coincidence Algorithm (range and Doppler)
Transmitted Pulse Width	7 μs
Duty Ratio	Variable: mean = 7.89%, peak = 11.09%
Pulse Compression	2 MHz linear FM, compression ratio = 14
Range Resolution	75 m (0.5 μs)
Blindness Due to Eclipsing	15 range cells
FFT Size	64
Receiver Noise Figure, F	5 dB
MBC/GMT Rejection Notch Width	±1.67 kHz (±25 m/s)
Maximum Range of Interest	185.2 km (100 nautical miles)
Maximum Velocity of Interest	±1500 m/s (±100 kHz)
Antenna Size	0.6 m (diameter), 0.28 m^2 (area)
Antenna Gain	34.2 dB_i
Antenna Beamwidth (Both Planes) with Highly Tapered Illumination	3.9°
Antenna Sidelobe Level	−30 dB
Antenna Scan Rate	60°/s
Beam Dwell Time	65 ms

dwell time is lost to space charging. A conventional planar array antenna is also assumed, although the antenna parameters could equally be achieved using an AESA antenna.

16.3.2 Blindness Performance

Since the PRF set has been optimized for the reduction of blindness in the presence of strong clutter, it is appropriate to consider its blindness performance. The blind zone map in the absence of clutter is presented in Figure 16-2. In the absence of clutter the total blindness

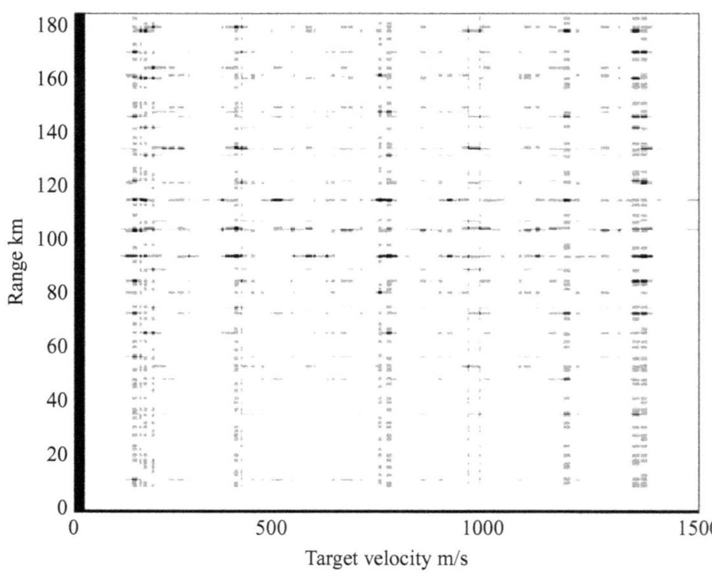

FIGURE 16-2 ■ Blind Zone Map for Optimal Eight PRF Schedule (No Clutter)

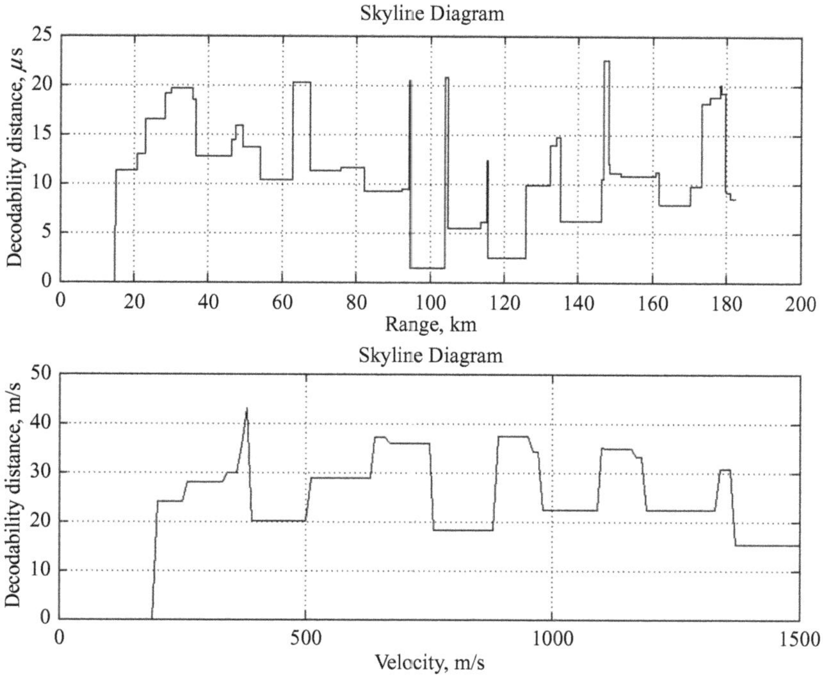

FIGURE 16-3 Decodability Margins for Optimal 3-of-8 PRF Schedule

is 6.82% (based on visibility in fewer than *four* PRFs), but this includes the blindness due to the first blind range (along the bottom of the map) and the first blind velocity (up the left-hand side of the map), which are unavoidable. These unavoidably blind regions account for 2.26% of the total blind zone map and leave 4.56% (= 6.82 − 2.26) avoidably blind, or 95.44% (= 100 − 4.56) avoidably clear in at least four PRFs. The preceding chapters discussed the viability of shorter schedules, but the 3-of-8 solution presented here is typical of the genre.

16.3.3 Decodability Margins

The PRF selection procedure imposed minimum decodability margins; however, the margins actually achieved by the optimal PRF schedule are given in Figure 16-3 and indicate that the worst-case decodability margin was 205.5 meters in range (1.37 μs) and 15.4 m/s in velocity.

CHAPTER 17

Airborne Early Warning Radar

Chapter Outline

17.1 INTRODUCTION

17.1.1 Maximum Line of Sight Range

Airborne early warning (AEW) radars provide long-range surveillance from an airborne platform [7, 8]. Their primary targets of interest are hostile aircraft.

Long-range radar surveillance coverage can be limited by the range to the horizon; however, satisfactory range can be re-established by placing the surveillance radars at high altitude. The line of sight horizon is then a respectable distance away, the range to the horizon, R_h in kilometers, from a radar at an altitude of h meters is given approximately by

$$R_h \approx 4.1\sqrt{h} \qquad \text{repeat of (6-13)}$$

For example, for a radar altitude, $h = 11,000$ m, the range to the horizon, $R_h =$ 430 km. This gives just over 25 minutes of warning against an inbound target traveling at 280 m/s. Targets beyond R_h may be detected only if they are at a suitable altitude. If h_r now represents the height of the radar and h_t is the height of the target, then the maximum range at which the radar may detect the target is given by the straight line between the two that skims the horizon, as illustrated in Figure 17-1.

17.1.2 Problems of AEW Systems

Long-range surveillance radars tend to be large powerful systems using relatively low microwave frequencies, where the atmospheric attenuation is minimal. There are certain problems associated with long-range surveillance radars in high altitude platforms:

- The antenna needs to be very large if adequate gain and beamwidth are to be achieved using a long wavelength (low frequency).

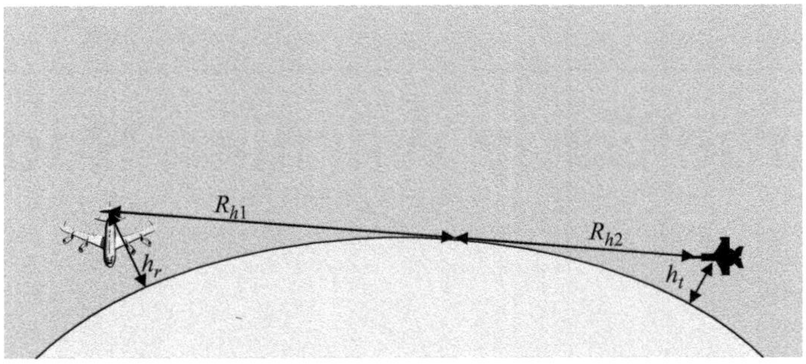

FIGURE 17-1 Maximum Line of Sight Radar Range

Line of sight maximum range, $R_h = R_{h1} + R_{h2}$

$$R_h = 4.1\sqrt{h_r} + 4.1\sqrt{h_t}$$

$$R_h = 4.1\left(\sqrt{h_r} + \sqrt{h_t}\right)$$

- Generating sufficient RF power and dissipating excess heat generated by inefficiencies on board an aircraft could be problematic.
- Ground clutter is present in any look-down attitude, and sidelobe clutter (SLC) is spread over a Doppler bandwidth corresponding to ± the platform velocity.

17.1.3 A Generic AEW Radar Specification

A generic specification for an AEW radar system is as follows [7, 8]:

- Must be able to detect airborne (and seaborne) targets while looking down against a clutter background
- Must have detection range greater than the line of sight to the horizon and have 360° coverage
- Must give an unambiguous indication of range out the specified maximum detection range
- Must be able to track large numbers of targets simultaneously
- Must be free from spurious detections caused by sidelobe clutter such as coastlines

17.1.4 Example AEW Systems

A brief survey of AEW radars indicates a wide variety of solutions to the radar requirement [7, 8]. Table 17-1 summarizes a few pertinent details concerning three AEW systems.

It can be seen from these examples that the design of AEW systems can be very different. However, it must be noted that the E2C was a small aircraft, capable of operating from a carrier; it is no longer in service with the US Navy. The small aircraft clearly limits its power generation and its antenna size. Since its role is early alert, followed by action by intercept strike aircraft, it does not need to be very precise about threat direction or about threat numbers. Once the fighters have been alerted, their own radars can sort these details out. The Searchwater Mk 2 system is in service with the UK Royal Navy and is accommodated within the Sea King Mk 7 helicopter. It, too, is used for fleet protection for surveillance over both land and sea. This radar takes an entirely different approach since

TABLE 17-1 ■ Some AEW Examples

	AEW Examples
E2C (HAWKEYE)	• Ultra high frequency (UHF) operation
	• Double Yagi antenna using monopulse beam sharpening to improve azimuth accuracy
	• Used by US Navy for fleet protection
E3 (AWACS)	• 3 GHz operation
	• Flat plate array plus radome rotates on back of aircraft fuselage
	• Aircraft acts as command and control platform as well as AEW
SEARCHWATER MK2	• 10 GHz operation
	• Parabolic antenna
	• Simultaneous over land and sea capability
	• Aircraft acts as command and control platform as well as AEW
	• Used by UK Royal Navy for fleet protection
	• Modest altitude, radar range

it uses a modest-sized antenna and modest powers and operates at 10 GHz. The AWACS aircraft, on the other hand, is much larger being based on a Boeing 707 and is required to provide the recognized air picture (RAP) for a variety of forces and to meet an exacting surveillance specification. Thus, a large antenna and a transmission frequency and power level similar to those of ground-based surveillance radars are necessary.

The following sections give some example radar parameters for two hypothetical AEW radar systems: one used for fleet protection and the other for long-range airborne surveillance.

17.2 | AEW RADAR FOR FLEET PROTECTION

A summary of the radar parameters for a hypothetical AEW radar used for fleet protection is given in Table 17-2. Its operational altitude is taken as 3000 m giving a range to the horizon of 225 km. In its long-range search above the horizon, a low pulse repetition frequency (PRF) is proposed. The absence of main beam clutter (MBC) coupled with a requirement to provide target range makes the choice of a low PRF waveform an obvious one. When searching below the horizon and therefore at ranges $\leq$ 225 km, there is strong MBC. Furthermore, at closer ranges, it becomes increasingly important to have both range and velocity data available. Low- and medium-level aircraft represent a high threat target, but surface vessels should also be detected. Therefore, it is incumbent on the radar to ensure both fast- and slow-moving targets can be detected. With these factors in mind a medium PRF solution is a reasonable choice. Here, a 3-*of*-6 schedule is assumed. Since medium PRF copes well with clutter, then it supports operations over both land and sea. If the target velocity is high, it may be assumed to be an airborne target, whereas low-speed targets can be assumed to be surface contacts.

Note that it is common to distinguish between the level of sidelobes close to the main beam (i.e., within approximately 15 degrees of the edge of the main beam) but to apply

TABLE 17-2 ■ Typical AEW (Fleet Protection) Radar Parameters

Parameter	Value
Carrier Frequency, f_{tx}	10 GHz
Transmitter Power (TWT)	80 kW (peak)
Antenna Size	1.4 (horizontal) × 0.8 (vertical) (m)
Antenna Gain	8200 (= 39.1 dB$_i$)
Antenna Beamwidth	1.5° (azimuth) ×2.6° (elevation)
Antenna Sidelobe Level	25 to 30 dB below main beam
Antenna Scan Rate	6 rpm (36°/s)
Beam Dwell Time	41 ms
Low PRF (Above Horizon) Mode	
PRF	300 Hz
Maximum Unambiguous Range	500 km
Range Bracket of Interest	200 km to 400 km
Transmitted Pulse Width	100 μs
Duty Ratio	3%
Minimum Range	15 km
Pulse Compression	15 MHz linear frequency modulation (FM)
Range Resolution	10 m
Medium PRF (Below Horizon) Mode	
Ambiguity Resolution (3-of-6)	Coincidence Algorithm
Mean PRF	8 kHz
FFT Size	32
Space Charging	2 ms (300 km)
CPI Duration	4.3 ms
Velocity Resolution	3.5 m/s (232 Hz)
Pulse Compression	As for Low PRF Mode
Range Resolution	As for Low PRF Mode
MBC/GMT Rejection Notch Width	Variable (sea/land)
Maximum Range of Interest	225 km
Maximum Velocity of Interest	±500 m/s (±33.3 kHz)

a lower sidelobe level specification to the *diffuse* sidelobes—that is, sidelobes at a far greater angular offset from the main beam [8]. The SLC contribution arising from diffuse sidelobes that are 5 dB, or better, below the close-in sidelobes, is insignificant.

17.3 | AEW RADAR FOR LONG-RANGE AIR SURVEILLANCE

A summary of the radar parameters for a hypothetical AEW radar used for long-range air surveillance is given in Table 17-3. Its operational altitude is taken as 11,000 m, giving a range to the horizon of 430 km. The long detection ranges required here favor the use of a low frequency and large antenna, but this must be tempered by the fact that the antenna has to be accommodated on an aircraft platform. A transmitter frequency of around 3 GHz is therefore proposed. A peak transmitted power in the order of 2 MW is required. In its long range (> 400 km) search above and beyond the horizon, the absence of MBC coupled with a requirement to provide target range makes the choice of a low PRF waveform

TABLE 17-3 ■ Typical AEW (Air Surveillance) Radar Parameters

Parameter	Value
Carrier Frequency, f_{tx}	3 GHz
Antenna Size	6 (horizontal) × 1 (vertical) (m)
Antenna Gain	3900 (= 36 dB_i)
Antenna Beamwidth	1.3° (azimuth) ×7.5° (elevation)
Antenna Sidelobe Level	60 dB
Antenna Scan Rate	6 rpm (36°/s)
Beam Dwell Time	36 ms
Low PRF (Above Horizon) Mode	
PRF	230 Hz
Maximum Unambiguous Range	650 km
Range Bracket of Interest	400 km to 500 km
Transmitted Pulse Width	100 μs
Duty Ratio	2.3%
Minimum Range	15 km
Pulse Compression	1.5 MHz linear FM
Range Resolution	100 m
High PRF (Below Horizon) Mode	
Mean PRF	30 kHz
Number of PRFs	4
Pulse Width	5 μs
Duty Ratio	15%
Maximum Unambiguous Range	5 km
Maximum Unambiguous Velocity	± 750 m/s
FFT Size	128
Space Charging	4.3 ms (650 km)
CPI Duration	4.3 ms
Velocity Resolution	11.7 m/s (234 Hz)
MBC/GMT Rejection Notch Width	±30 m/s (±600 Hz)
Pulse Compression	7.5 MHz linear FM
Range Resolution	20 m
Maximum Range of Interest	430 km
Maximum Velocity of Interest	±750 m/s (±33.3 kHz)

an obvious one. When searching below the horizon and therefore at ranges $\leq$430 km, there is strong MBC. Furthermore, it becomes increasingly important to have both range and velocity data available as targets appear over the horizon. The main threat comes from fast-approaching aircraft that must be detected as they break the horizon at around 430 km of range. High PRF pulse Doppler is therefore best suited to this target set but requires exceptionally low antenna sidelobes to control the sidelobe clutter. Hence, a large rotordome antenna is proposed. Generally, the use of higher PRFs drives a requirement for lower antenna sidelobes; further analysis is provided by Morchin [7]. Since the duty ratio in the high PRF mode is reasonably low, a range gated high PRF mode is proposed. Range ambiguity should be decoded and blind (eclipsed) ranges avoided by alternating the operation over four PRFs.

CHAPTER 18

Active Radar Missile Seekers

Chapter Outline

18.1 INTRODUCTION

There is a wide range of guided weapons for a whole host of differing applications, many of which are using radar-based seekers. The operating conditions are particularly harsh; an antiaircraft missile may be capable of lateral accelerations in excess of 25 g, and the constraints on seeker size and weight dictated by the missile diameter and payload capacity demand some exceedingly well-engineered radar solutions. In many cases, the radar seeker range is somewhat less than that of the missile, in which case the missile must fly out to within the range of the seeker using some other guidance mechanism before using the radar seeker for terminal phase guidance.

Fully active radar seekers are miniature fire control radars in their own right. Such seekers are increasingly being used since they offer a true fire-and-forget capability; that is, the launch platform (aircraft, ship or ground launcher) is not committed to the guidance mechanism once the missile has been launched and is free to vacate the scene for the sake of its own security.

Radar seekers are also becoming highly sophisticated systems. Not only must they acquire the target and provide terminal phase guidance, but they are also often required to classify the target to confirm that they are engaging the correct target and may even be called upon to confirm a successful engagement. These functions are especially important for fire-and-forget systems.

In any guided weapons system, the seeker requirements are dictated by the concept of operation. There is a close relationship between the seeker requirements and several other aspects of the weapon such as what information is passed to the seeker from the target acquisition asset at the firing point, the range of the missile, its diameter and payload capacity, the intended target set, the warhead and fusing mechanism, and the guidance system. The seeker is considered not in isolation but as a subsystem of the whole weapon (albeit an important one).

This chapter considers the case of a hypothetical, but realistic, medium-range air-to-air missile using a fully active pulse Doppler radar seeker for terminal phase guidance. To scope this application, this chapter starts with a short overview of guided weapons system employing radar-based seekers followed by a brief discussion on the interaction between the seeker and other aspects of the guided weapon that influence the seeker design.

18.2 | RADAR-BASED MISSILE SEEKER APPLICATIONS

18.2.1 Medium-Range Air-to-Air (Semi-active Homing)

These missiles are supersonic (Mach 4) and highly agile (lateral accelerations > 20 g) and have a missile diameter in the order of 20 cm. They are designed to engage an aircraft target beyond visual range (BVR). For air-to-air engagements in the 15 km to 40 km range, past systems such as Sparrow and its derivatives have used a semi-active guidance method over the full range of the missile.

Targets would be acquired using the fire control radar (FCR) aboard the strike aircraft. To engage a target, the FCR must achieve angle lock onto the target using amplitude or phase comparison monopulse techniques. This achieves a tracking accuracy of a fraction of a beamwidth, depending on the signal-to-noise ratio (SNR). Typically, a tracking accuracy of around 1/20 of a beamwidth is achievable for an SNR which gives decent detection performance. Tracking accuracy against long-range, low-flying targets may be degraded due to multi-pathing as a result of a strong ground reflection; this tends to degrade the accuracy in elevation more than azimuth. Angular accuracy at short ranges can also be compromised due to target glint (distortion of the wave front reflected from the target due to the multiple scattering centers distributed across the target). In both cases (multi-path and glint), the tracking errors can be minimized using frequency agility.

Having achieved angle lock, the FCR illuminates the target with a continuous wave (CW) signal, which may well be in a different band from that of the FCR. The CW missile illumination signal illuminates the target continuously, and the missile seeker acquires and tracks the illumination signal reflected from the target and homes in on the source of the reflection. The missile seeker is a passive receiver system with a gimballed antenna, employing monopulse tracking. The tracking loop is closed by directing the whole missile, not just its antenna, toward the target by actuating the missile's control fins. The FCR passes velocity (Doppler) information on the target to the missile seeker. Data are processed using a fast Fourier transform (FFT) to extract target Doppler and hence velocity information. The seeker performs a velocity search, as cued by the FCR, to locate and confirm the target in the Doppler domain and subsequently tracks the Doppler of the target and angle locks to the target. In this way, the seeker acquires the target in both angle and Doppler. The missile is then launched and continues to track the target in both domains throughout the engagement. No range tracking is possible due to the use of a CW illumination signal.

The Doppler cuing of the seeker uses a second signal known as the *rear reference*. This is a low-level sample of the missile illumination and is transmitted from an antenna directly to a rear-facing receiver on the missile. The missile seeker therefore has a sample of the missile illuminator direct from the firing platform plus the received signal reflected from the target. By comparing the frequencies of these two signals, the Doppler shift due to the platform velocity and missile velocity can be negated, leaving only the Doppler shift

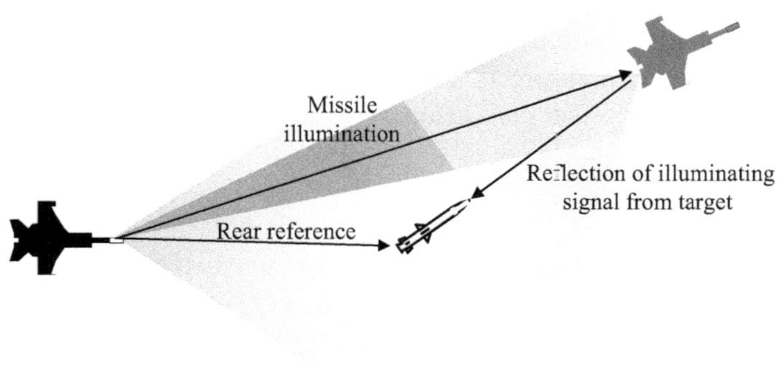

FIGURE 18-1 Semi-active Homing Air-to-Air

associated with the target velocity. The situation is depicted in Figure 18-1. Semi-active homing is an example of a CW Doppler sensing bistatic radar configuration.

The target aircraft may employ some form of electronic attack to defend itself, which may be directed at the FCR or against the seeker tracking functions. Velocity gate pull-off (VGPO) or cooperative jamming, including blink jamming, may well enjoy some success. However, many modern antiaircraft missile seekers can home in on jamming, and any active countermeasures transmitted from the target aircraft could beacon its location if they are not successful in taking control of the tracking mechanism. Dumping chaff has a temporary effect, since the chaff cloud rapidly acquires a low velocity and becomes resolvable from the target Doppler. Cross-eye jamming [9] is not likely to be successful against a bistatic (or multistatic) system but is a viable countermeasure against active (monostatic) radar seekers.

18.2.2 Medium-Range Air-to-Air (Active Homing)

A fully active radar seeker provides its own transmitting signal and receiver processing. However, due to the small antenna diameters that can be accommodated and modest transmitter powers that are available, the radar range is typically much less than that of the missile. Hence, active missile seekers used in the medium-range air-to-air role operate in the end game providing terminal phase guidance. These radars use pulse Doppler techniques and bear many similarities to a miniature fire control radar.

During the launch sequence the missile seeker is cued with the positional data (angle and range) and velocity of its target. Data on a secondary target may also be given should the seeker fail to find the first target. The missile then flies out to the approximate target zone using its inertial navigation system (INS) and autopilot. The INS may be augmented by the global positioning system (GPS) to ensure the correct flight path is taken. It is perfectly feasible that the target continues to be detected and its flight path observed (or predicted) using the track-while-scan function of the FCR or even another radar asset such as an airborne early warning radar. In this case, updates may be sent to the missile during its fly-out phase via a command guidance link. This facility improves the single-shot kill probability of the missile considerably.

As the missile nears the target location, the active radar seeker switches on and searches for the target. The seeker must search a region known as the *search basket*, which is the volume of space centered on the last known target position (or its predicted position) with dimensions encompassing all possible movements of the target given the time since the last target update and the anticipated maneuverability of the target. All being well, the seeker acquires the target and directs the missile toward an intercept. A more detailed consideration of the radar seeker parameters is given in Section 18.4.

18.2.3 Anti-ship Missile

An anti-ship missile (ASM) may be air launched or launched by another ship often at a considerable range. It flies out toward the target at very low altitudes and high subsonic speeds using its onboard INS/GPS and autopilot systems. At a range of about 25 km to 30 km, the target ship becomes visible on the horizon and the radar seeker goes active, acquires the target, and provides homing guidance down to the point of impact.

Since the missile must carry a significant warhead, typically flies subsonically, and must achieve long ranges, it is somewhat larger than air-to-air missiles and may have a diameter of 40 cm or so. Thus, a larger radar seeker can be accommodated that supports a longer detection range. Typically, a 10 GHz radar is used since this provides a good compromise between atmospheric attenuation (notoriously bad at low altitudes over the sea surface due to rain, fog, mist, and spray) and antenna gain. Strong sea-surface clutter can be expected all the way, so pulse Doppler techniques are the order of the day. Targets may have very low radial velocities, so radar sensitivity must be maintained well inside the bandwidth of main beam clutter. However, the targets of interest are huge; the radar cross section (RCS) of a destroyer seen at broadside is in the order of 10^5 to 10^6 m^2. It is worth noting that at broadside the radial velocity is lowest, the target RCS is greatest, and the probability of a hit and kill is greatest. The range resolution may therefore be matched to the beam of the ship. In a head-on (or stern) attack, range profiling for target recognition and confirmation can be achieved using a resolution in the order of a few meters (as opposed to a fraction of a meter required for armored targets or airborne targets).

18.2.4 Air-Launched Anti-armor Missiles

In recent years several short-range guided weapons systems have emerged that are designed to defeat ground-based armored targets such as a main battle tanks (MBTs) or lightly armored targets such as air defense systems, personnel carriers, and trucks. These may be launched from a variety of helicopter and fixed-wing platforms. The missiles are supersonic (Mach 1.2 to 2), short-range (5 km to 20 km), and small (diameters < 18 cm) and must deliver a sophisticated warhead to a preferred impact point. Targets may be stationary or moving, may be partially concealed, and are detected in the presence of strong ground clutter and potential multi-pathing problems. Political sensitivities also demand that the seeker or targeting system should be able to resolve legitimate military targets from civilian vehicles.

The seeker solutions operate in the millimeter wave band (30 to 300 GHz), with the atmospheric window frequencies of 35 and 94 GHz often being the preferred choices [10, 11] The exploitation of the millimeter wave band is driven by the need to achieve decent gain and narrow beamwidth from an antenna of only 14 cm in diameter. Operation in the millimeter wave band also offers fine velocity resolution (for a given coherent processing interval [CPI]) and wide bandwidth, which supports fine range resolution. The fine resolution in angle, range, and velocity help minimize the impact of the clutter while also supporting target recognition techniques such as range profiling, Doppler signature analysis, and polarization agility. However, the millimeter wave band is subject to high atmospheric losses, even in clear air, which worsens in conditions of rain, mist, fog, and dust obscuration. Furthermore, existing solid-state transmitters output just a few watts of radio frequency (RF) power, and millimeter wave receivers suffer from high front-end losses. The limited powers, high receiver noise figures, large atmospheric losses, and small antennas all contribute to modest detection ranges, often considerably shorter than the missile range. In many cases, the active radar seekers are used in a lock-on after launch (LOAL) mode to provide terminal phase guidance. Modern advances in RF generation and low noise amplifier/receiver designs are, however, facilitating longer detection ranges, and in the near future the next generation of these weapons may push into the medium-range bracket.

Current radar seekers used in ground attack typically use frequency modulated continuous wave (FMCW) techniques (Section 10.1). The radar alternates between modes for the detection of moving targets (and hence the rejection of main beam clutter [MBC]) and stationary target detection, which can be achieved only at shorter ranges when strong MBC is present. Estimated detection ranges for a 35 GHz system against a moving target are in the order of 5 km, while at 94 GHz this may reduce to 2 km. The radar seeker attempts to perform target recognition as it closes and the signal-to-noise-plus-clutter ratio (SNCR) improves. (In some cases it may be sufficient to simply recognize that the target conforms to the profile of an MBT, and hence the recognition reduces to a binary decision: *is it a tank or not?*) Seekers use monopulse techniques to guide the weapon toward the target. At very short ranges, the seeker then scans the surface of the target, sequentially sampling its RCS, to provide aim point selection. This ensures that the warhead is delivered to the optimum point to defeat the armor.

18.2.5 Surface-to-Air Missile Systems

Surface-to-air missiles (SAMs) using millimeter wave active radar seekers are being deployed against fast strike aircraft. These seekers bear certain similarities to those of the air launched anti-armor system considered in Section 18.2.4 and the active air-to-air systems considered in Section 18.2.2, although the rest of the missile, particularly its warhead, is quite different. Short-range pulse Doppler radar seekers are used to provide terminal phase guidance. Again, use of the millimeter wave band is common, and the requirement for target recognition is important.

18.2.6 Gun-Launched Antia-Armor Munitions

Several countries have developed gun-launched terminally guided submunitions (TGSM) to prosecute MBT targets. These weapons fire a 155 mm round over a formation of enemy tanks. The round dispenses, typically, two submunitions at the apogee of its trajectory, each of which deploys a drogue chute to arrest its rate of descent. A fully solid-state 94 GHz FMCW radar becomes active and looks vertically down to target a tank. The design of the drogue chute imparts a precessional movement to the round, which causes the radar to scan a spiral pattern as it drifts over potential targets. At vertical incidence, the ground clutter is confined to a strong altitude line only and is not distributed in range or Doppler. The top of a MBT is therefore seen at about 3 meters ahead of the clutter return (the height of the vehicle). A signal bandwidth of 100 MHz achieves a range resolution, $\Delta R = 1.5$ m and would be sufficient to resolve the top of the MBT from the ground below. The estimated detection range against an MBT is about 400 to 200 meters. These munitions detonate an explosively formed projectile (EFP) warhead at a stand-off range of about 50 m. The radar seeker provides the necessary cuing information at the appropriate range and orientation to kill the tank in an overhead top attack. An infrared sensor may also be employed to determine whether the target has a significant heat signature, indicating that it has already been hit and that the munition can seek another target.

These radar seekers are not especially sophisticated, although they have one rather interesting requirement: they must be engineered to survive the 25,000 g shock of the launch! This demands that there can be no moving parts (no scanning antenna) and that the whole thing must be potted into one solid lump.

18.3 | RADAR SEEKER AS A SUBSYSTEM

The radar seeker cannot be considered in isolation to the rest of the missile system. Other aspects of the guided weapon dictate the requirements on the seeker and heavily influence its design. A brief summary of these considerations is given in the following sections.

18.3.1 Guidance Mechanism

The examples considered in Section 18.2 are all active or semi-active homing systems. Many radar seekers provide only terminal phase guidance and are combined with INS/GPS guided fly-out phases. Some may also be combined with command guidance to provide midcourse corrections.

The missile guidance mechanism requires an input on the target parameters at regular intervals. The guidance update rate may be as low as 10 Hz for relatively sluggish or fixed targets or in the order of 100 Hz for more agile targets. For a guidance update rate of 100 Hz, the seeker is therefore limited to a processing interval of 10 ms with minimal processing delay.

To target an object, the guidance may require only angular information (in two planes), which can be provided by monopulse angle tracking. However, monopulse tracking is often supported by tracking in range since this isolates the tracking function from other targets in the beam, countermeasures deployed by the target, and noise residing at ranges other than that of the target. Furthermore, target range data may also be required for target confirmation. High-range resolution profiling also offers great potential for target recognition of both moving and stationary targets. Velocity data can also be used by the seeker for target confirmation (as with semi-active homing) and supports Doppler signature analysis for target recognition. Furthermore, Doppler discrimination is useful in combating clutter and countermeasures. The guidance mechanism may also require target range and radial velocity information to control an appropriate flight path.

Two of the most popular guidance laws used are (1) pursuit guidance, in which the missile aims directly at the current target location, and (2) proportional navigation (PN), in which the missile is directed toward an anticipated impact point.

18.3.1.1 Pursuit Guidance

The missile flies directly toward the last known position of the target and requires only angular information. This is typically used against slow-moving targets but is also an inevitable result of semi-active homing. In an air-to-air head-on aspect, the missile closes

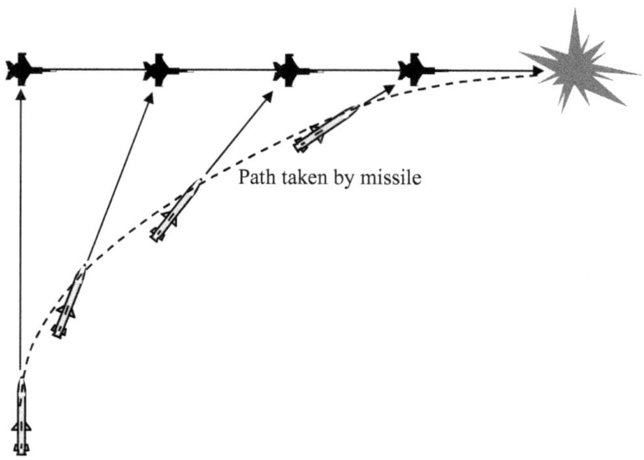

FIGURE 18-2 Side-On Pursuit Guidance

at high speed along a straight-line path, presuming the target holds a constant track. If the missile misses and overshoots the target, then the target falls outside the seeker's field of view, and it is unlikely that the missile could turn itself around to reacquire the target. In an air-to-air tail-on aspect, the closing velocity is somewhat lower, but otherwise the situation is much like the head-on aspect. In an air-to-air side-on aspect, the missile loops around and engages from the rear; the tightness of the loop and the exact approach angle are dependent on the speed advantage that the missile has over the target. This is depicted in Figure 18-2. For a fast-moving target this is a very inefficient flight path. In addition, since in general the majority of engagement scenarios are to an extent side-on, proportional navigation might be considered because it flies a more direct route to an anticipated impact point.

18.3.1.2 Proportional Navigation (PN)

PN is used in some form or another by most homing air target missiles. It is based on the fact that two vehicles are on a collision course when their direct line of sight (LOS) does not change direction. PN dictates that the missile velocity vector should rotate at a rate proportional to the LOS rotation rate and in the same direction. A typical engagement scenario is shown in Figure 18-3.

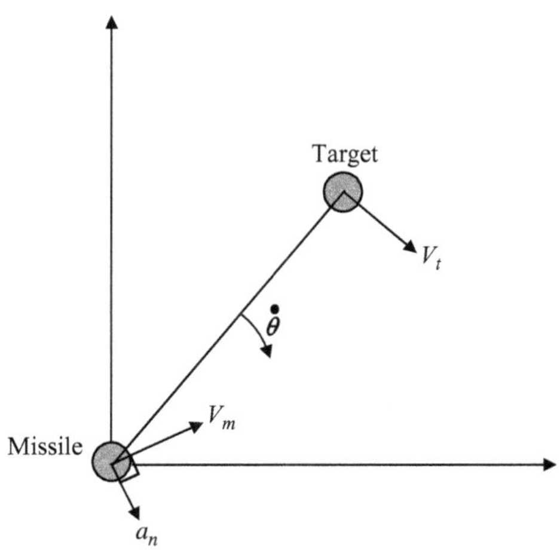

FIGURE 18-3 Proportional Navigation

The missile produces an acceleration, a_n, perpendicular to its velocity (hence it is known as a *lateral acceleration*), which is proportion to the rate of change of the line of sight between the missile and target. The guidance law is summarized as

$$a_n = \frac{N\dot{\theta}V_r}{\cos\theta} \tag{18-1}$$

in which N is the navigation constant and determines the flight path described by the missile, and V_r is the radial velocity of the missile with respect to the target. V_r is, of course, readily obtainable from the Doppler and provides a useful input to the guidance mechanism to fly an efficient PN trajectory.

18.3.2 Warhead

The whole point of any guided weapon is to deliver the warhead to the correct place at the correct time; it is the precious payload of the missile. However, since the seeker represents the "eyes" of the weapon, it enjoys the privileged location right at the front. One must be aware of the effects that the seeker and other electronics may have on the effectiveness of the warhead placed behind them.

Different targets require different warheads, and the kill mechanism varies with each type. To defeat armor, a shaped charge warhead may be used, which must be delivered to a point on the target where the armor is comparatively thin. Furthermore, the warhead must be detonated at a stand-off distance of just a few cone diameters. This can be ensured by using a crush fuse on the front of the nose cone and positioning the warhead at a suitable distance behind this point. When detonated, the plasma jet formed by the warhead cuts through the seeker (and armor) like a hot knife through butter. The seeker must therefore guide the missile down to the impact point. One must be mindful that pulsed radars have a minimum range within which the radar is blind, which could be a severe limitation for seeker applications. Short-range anti-armor weapons therefore tend to use FMCW waveforms, which do not suffer from blind (minimum) range problems. An EFP warhead operates over a much longer stand-off range, so seeker data are required to indicate the target range and determine the point in time when the warhead is directed toward the target.

Anti-ship missiles need to penetrate the hull of the ship, ideally close to the waterline, and, using a delayed fusing mechanism, to explode within the ship. ASM seekers do tend to use pulsed radars, and therefore there are potential minimum range problems. As the ASM approaches the target ship, the waveform is switched to narrower and narrower transmitted pulse widths to reduce the minimum range. The narrower pulse width contains less energy (unless the pulse repetition interval [PRI] is also changed), but then less energy is required to detect the target at decreasing ranges. At very close ranges the ASM seeker goes blind, but by this time the fate of the engagement is already sealed. For example, if the minimum range is 15 meters and the missile is on course to hit the ship at the point where the seeker goes blind, then there is nothing the ship can do to evade the hit within the last 15 meters of the missile flight. Conversely, if the missile is not on course to hit its target, then there is nothing it can do over the last 15 meters of its flight to bring it on course. Incurring a short minimum range is acceptable when the target is both large and not particularly agile.

Antiaircraft missiles might typically use a blast fragmentation warhead, which is detonated a short distance from the target aircraft. The destruction is brought about by a hail of high-energy fragments and, possibly, the shock wave. This requires a proximity fuse to detonate the warhead a few meters from the target, ideally ahead of the target so that it flies into the shower of fragments. In these circumstances there is a simple choice: to invest in an accurate seeker to deliver the warhead exactly where it needs to be so that only a small warhead is necessary; or to use a less accurate seeker but require a bigger warhead. The size and weight of the warhead have obvious implications for the propulsion requirements and the range and agility of the missile. In surface attack applications, a large blast fragmentation warhead also has implications on the collateral damage inflicted on nearby structures, vehicles, and personnel.

18.3.3 Missile Dimensions and Shape

The diameter of a missile is a function of its velocity, range, and the size of the warhead it must deliver. ASMs can accommodate an antenna of 30 to 40 cm and must operate in wet-weather conditions. This favors the use of radar seekers operating at around 10 GHz. Anti-armor weapons are much faster and smaller and can accommodate an antenna of about 14 cm in diameter. This drives the operating frequency into the millimeter wave band. Medium-range air-to-air missiles and surface-to-air missiles are very fast and agile. Their seekers may operate toward the upper end of the centimetric wave band (10 GHz to 15 GHz) or the lower part of the millimeter wave band (35 to 45 GHz). The nose cones of supersonic missiles are typically tangent-ogive, which is not very sympathetic to the radar seeker requirements within, whereas subsonic missiles have a more rounded nose cone, which tends to help preserve lower sidelobe levels.

18.4 | HYPOTHETICAL AIR-TO-AIR ACTIVE RADAR SEEKER PARAMETERS

18.4.1 Summary of Requirement

Consider a hypothetical medium-range air-to-air missile having a diameter of 18 cm and traveling at a velocity of 1300 m/s (Mach 4) that is designed to engage strike aircraft at a maximum range of 50 km. An active radar seeker is required for terminal phase guidance over the last 10 km with an update rate of 100 Hz (period = 10 ms). Detections at high altitude or in a look-up scenario are noise limited, but significant MBC and sidelobe clutter (SLC) can be expected in low-altitude engagements or look-down scenarios. It can therefore be assumed that the detection space of interest is from 0 to 10 km in range and between ±750 m/s ground velocity (or +550 m/s to +2050 m/s with respect to the missile).

18.4.2 Seeker Parameters

A medium pulse repetition frequency (PRF) pulse Doppler approach is taken using just five PRFs. This is justified to provide all aspect target detection in the presence of strong surface clutter. Full range and velocity decodability may not be necessary since ambiguities could be resolved from the target parameters provided by the FCR [12, 13]. Since the detection space of interest is great in the velocity domain but is very low in the range domain, a high mean PRF is anticipated. The update period of 10 ms, coupled with the requirement for

TABLE 18-1 ■ Hypothetical Medium Range Air-to-Air Missile Seeker Parameters

Parameter	Value
Carrier Frequency, f_{tx}	12 GHz
Mean PRI/PRF	59.375 μs/16.842 kHz
Medium PRF Schedule	3-of-5
Transmitted Pulse Width	4.16 μs
Duty Ratio (Mean)	7%
Pulse Compression	7.5 MHz linear frequency modulation (FM), compression ratio = 31.2
Range Resolution	20 m
FFT Size	32
Receiver Noise Figure, F	3.16 (5 dB)
Peak Transmitted Power	50 watts
Total Losses	5 (7 dB)
SNR (per Pulse)	31.6 (15 dB)
CPI Duration (Mean)	1.9 ms
Velocity Resolution	6.6 m/s
Antenna Size	16 cm (diameter), 0.02m^2 (area)
Antenna Gain	270 (24.3 dB$_i$)
Antenna Beamwidth (Both Planes) with Highly Tapered Illumination	12°
Maximum Range of Interest	10 km
Maximum Velocity of Interest	±750 m/s (±60 kHz)

five CPIs and a space charging time of 100 μs (15 km of range) gives a CPI duration of 1.9 ms. The range resolution is matched to the target dimensions (20 m). The mean RCS of the target is assumed to be 10 m^2. The parameters of the hypothetical radar seeker for this requirement are summarized in Table 18-1.

18.4.3 Ambiguities

The mean PRF of 16.842 kHz achieves a maximum unambiguous range of 8.9 km, only marginally less than the detection range of the radar. Therefore, all targets are either first or second trace, and resolution of the range ambiguity needs to resolve only these two cases. However, there could be up to 8 orders of velocity ambiguity for this PRF over the full velocity space of interest. Once the target has been acquired, the tracking window could reduce the range/velocity detection space of interest to a much smaller region centered on the target coordinates [12, 13] (see Section 13.3.1).

18.4.4 Range Walk

The issue of range walk is also worthy of consideration. Moving targets may range walk during the time it takes to transmit the entire schedule, which causes the following:

- Range gate straddling losses
- Decorrelation in range across successive PRFs

Range walk can be problematic under any (or all) of the following conditions:

- Long schedules
- High velocities
- Fine range resolution

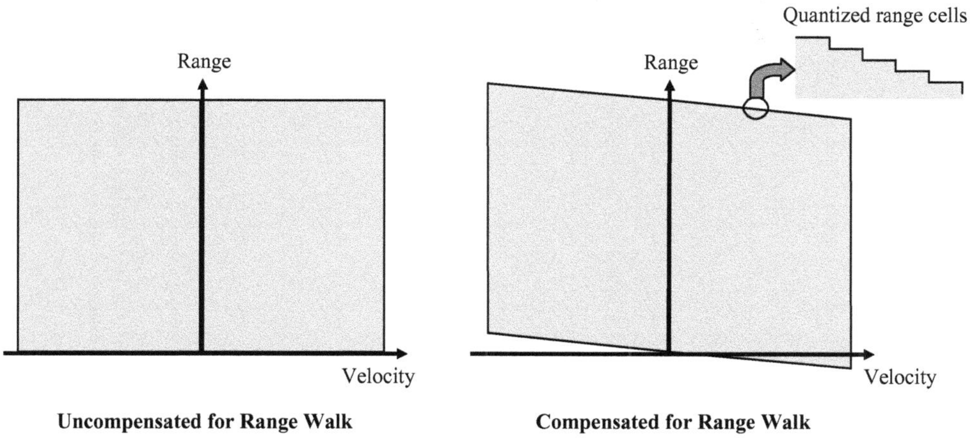

FIGURE 18-4 Distortion of Range/Velocity Space Due to Compensation for Range Walk

The maximum closing rate between the missile and its target could be as great as 1300 + 750 = 2050 m/s. Over the update period of 10 ms the missile and target could close by 20.5 meters, which just exceeds the range resolution of the seeker. Thus, over the period of one update, a target could be subject to range walk of about one range cell. A range decodability margin of at least two range cells is therefore required if the ambiguous range is to be decoded robustly. To avoid range ambiguity, some form of platform motion compensation (PMC) may be applied so that the missile velocity is negated. With PMC the maximum range walk over 10 ms is now 7.5 meters.

Range walk can be compensated for when correlating returns and constructing detectability maps by offsetting the data for each PRF in range as a function of velocity and the time from the start of the schedule. Note that range walk can still occur within a CPI but not from one CPI to the next within a schedule. The correction can be computed for each velocity cell as Range walk = $V_r T$ (normally quantized into range cells), where V_r is the radial velocity of the test cell, and T is the time from start of schedule. A correction of 7.5 m (or 20.5 m without PMC) would be required at $T = 10$ ms for the seeker and target parameters previously considered. This correction has the effect of distorting the range/velocity detection space for each PRF as shown in Figure 18-4.

CHAPTER 19

Ground-Based Air Defense Radar

Chapter Outline

19.1 INTRODUCTION

Chapter 17 considered the case of airborne early warning (AEW) radar for long-range air surveillance. The point was made that a radar of this type has certain similarities with ground-based systems for long-range air surveillance. Broadly, these similarities may be summarized as follows:

- A distinction can be made between the clutter scenarios seen at high elevation angles and those seen at low elevations.
- The high- and low-angle cases can operate in two different modes, or at least possess differing clutter cancellation requirements.
- A low frequency is likely to be used to minimize atmospheric attenuation and hence preserve long-range detection.
- A large antenna is required to achieve high gain and narrow beams at such low frequencies.
- The tracking of a large number of targets in a track-while-scan mode is necessary.

However, there are also some fundamental differences between the airborne and ground based cases:

- The spread of clutter in the range and velocity domains
- Antenna pattern and scan pattern for the determination of target elevation
- The range to the horizon
- Differing constraints on size, weight, and cooling capacity

The first two points on the previous list are considered further in the next two sections. This chapter closes with the description of a hypothetical short-range ground-based air surveillance radar, typical of the type used for target acquisition as part of a short-range air defense system. This type of radar is chosen because it differs quite considerably from the longer-range applications and is illustrative of the wide variety of pulse Doppler radars in military service.

19.2 | DIFFERENCES IN THE CLUTTER SITUATION BETWEEN AIRBORNE AND GROUND-BASED PULSE DOPPLER RADARS

Many of the preceding chapters of this book have presented airborne applications of pulse Doppler radar. This is partly due to the historical context but also because the airborne case represents the *general case*. The situation for a ground-based pulse Doppler radar is a *specific case* that can be derived from the airborne situation by setting the platform altitude to zero and the platform speed to zero. (For ship-borne radars, the altitude may be set at the height of the radar above the waterline, and the platform velocity may be set to the ship's speed.) The key differences in the spread of clutter in the range and velocity dimensions for a ground-based radar with respect to an airborne one are as follows:

- A look-down aspect is not normally required for a ground-based radar. Main beam clutter (MBC) from the surface is incurred only at very low elevation angles corresponding to low altitude search and track functions or perhaps when the main beam illuminates high ground such as distant mountains or cliffs.
- Clutter at high elevations is dominated by volume clutter sources such as cloud, whereas at lower elevations the clutter is dominated by the ground return. The clutter backscatter coefficient, statistics, and Doppler characteristics are different in each case.
- Sidelobe clutter (SLC) and MBC are confined to a narrow Doppler band determined by the internal clutter motion. There is no spread in MBC or SLC due to variation in radial velocity over a wide range of look angles because the platform velocity is zero. Therefore, clutter is confined to a narrow band about zero Doppler and occupies a very small proportion of the Doppler band. Moving target indicator (MTI) filtering can be used to reject all clutter (both MBC and SLC). Moving targets are therefore detected in a noise-limited region (never clutter limited). This can affect the selection of PRFs for medium and high PRF pulse Doppler processing.
- SLC is incurred from zero range (zero altitude) out to the range to the horizon or the maximum detection range, whichever is shorter. For an antenna height of 10 m, the range to the horizon over the sea surface is 13 km. The range to the horizon on land maybe more or less depending on the local topography. Ground-based radars are influenced by clutter over shorter range brackets than their airborne counterparts, so fewer ambiguous repetitions of clutter over the receiving period for a given PRF are likely.
- The formation of a clutter map is a distinctly more viable proposition for ground-based radars than for airborne ones because the clutter scene varies very slowly for radars at a fixed location.

19.3 THREE-DIMENSIONAL RADARS

19.3.1 The Problem of Providing 3-D Target Data

A ground-based radar is typically required to give 360° azimuth coverage and elevation coverage from 5° below the horizon to 30° above the horizon. A modern active electronically scanned array (AESA) antenna would direct a pencil beam to a series of positions in rapid succession to cover the required volume of the sky in an efficient and timely manner. If an AESA antenna beam can be steered through ±60°, then three such antenna faces would be required for continuous 360° azimuth coverage. Some hardware (and cost) saving can be made by mechanically rotating one or two AESA antennas, and this solution has been adopted for naval multi-function radars.

Prior to the advent of AESA antennas, it was not viable to mechanically scan a narrow pencil beam over a large solid angular range since the time to complete one scan would be prohibitively long to support short-range applications. For example, a beam of 6° × 6° would have to adopt 360 beam positions to cover 360° (azimuth) by 36° (elevation). If the beam dwell time required is 50 ms, then an entire scan takes 18 seconds. For a target closing at 250 m/s, the target range decreases by 4.5 km between updates. In short-range air defense, an engagement range of around 10 km is typical, so the radar detection range may be set at around twice or three times this distance (i.e., 20 to 30 km). If one presumes that a target detection is confirmed at 25 km (confirmation usually requires consistent target data seen on two successive scans), then the radar would provide only three updates by the time the target closes on the desired engagement range and time has almost run out. Short-range air defense radars require a revisit rate of around once per second.

Conventional antennas transmit a beam that is narrow in azimuth but sufficiently broad in elevation as to provide the total elevation coverage required. A scan rate of 60 rpm (360°/s) ensures a 1 Hz frame time. If the same beam pattern is used on reception, then the azimuth resolution is fine but the elevation resolution is coarse. Pulsed radar operation yields the target range (and azimuth data are also available), but the system lacks target data in the third coordinate—elevation. The problem now becomes one of determining the target elevation. A variety of methods can achieve this and are collectively known as 3-D radars. One of the most popular 3-D radar techniques is known as *stacked beams* and may be readily implemented on a mechanically steered planar array antenna.

19.3.2 Stacked Beams

Stacked beam radars produce a series of *receiving* beams that are stacked vertically, one above another, as shown in Figure 19-1. Signals received from each row of elements are phase weighted to produce a narrow receiving beam that is steered to an appropriate elevation. Several parallel channels of phase weighting can be implemented in a beam former to produce a series of receiving beams stacked vertically. Each beam has a narrow elevation beamwidth, but the collection of the whole gives the total elevation coverage required. Thus, 6 beams of elevation beamwidth = 6° can cover 36° in elevation, assuming each beam is offset by its own beamwidth. Each receiving beam feeds its own receiver, and all receivers should be identical. Therefore, the number of receiving channels required must equal the number of stacked beams. The beam former is a hardware implementation of a series of spatial filters providing a discrete Fourier transform of data received in the elevation domain into the appropriate receiver.

FIGURE 19-1 Stacked Beams

The elevation resolution of the stacked beam approach is governed by the elevation beamwidth of the individual stacked beams, which, in turn, is proportional to the ratio of the wavelength to the vertical dimension of the antenna, λ/D (see equation (8-26), Section 8.5). For resolved targets, the elevation accuracy may be improved by comparing the strength of the target return in neighboring beam positions (i.e., neighboring receiver channels). An accuracy of around 1/20 the beamwidth is possible, depending on the signal-to-noise ratio (SNR) of the target return (see Section 3.8 for a discussion on accuracy and resolution issues).

The stacked beam approach also offers the advantage that the processing in each channel may differ. The clutter cancellation and constant false alarm rate (CFAR) detection in the lower beams can therefore be optimized to cope with the ground clutter scene, whereas in the upper beams they may be optimized to combat the weather clutter.

19.3.3 Cosecant Squared Radiation Pattern

Consider an aircraft target closing toward a ground-based air surveillance radar at a constant altitude. The target is initially seen at a low elevation, close to the horizon. As it closes, the radar sees the target at an increasing elevation angle, even though it continues to fly at a constant altitude. The situation is portrayed in Figure 19-2, in which the target is detected in two positions: long-range/low angle and short-range/high angle.

In general, the slant range to the target, R_S, is related to the elevation angle, θ_{el} by

$$R_S = \frac{h_t}{\sin\theta_{el}} = h_t \text{cosec}\,(\theta_{el}) \tag{19-1}$$

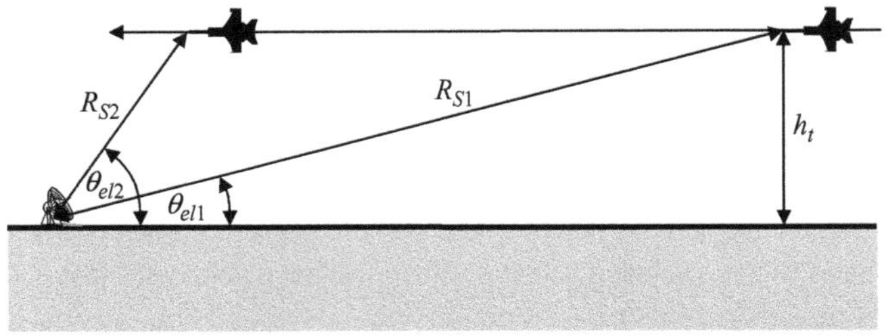

FIGURE 19-2 Target Closing at Constant Altitude

Recalling the radar range equation (2-3c)

$$R^4 = \frac{P_{pk} G^2 \lambda^2 \sigma}{(4\pi)^3 S}$$

Rearranging this for the received signal strength and setting the range to the slant range in the previous scenario results in

$$S = \frac{P_{pk} G^2 \lambda^2 \sigma}{(4\pi)^3 R_S^4}$$

Substituting in (19-1) for R_S gives

$$S = \frac{P_{pk} G^2 \lambda^2 \sigma}{(4\pi)^3 (h_t \mathrm{cosec}(\theta_{el}))^4} \tag{19-2}$$

One can see that if the gain of the antenna is shaped so that it is a function of $\mathrm{cosec}^2(\theta_{el})$, then the signal strength passed to the receiver remains constant for all angles of elevation, and hence slant range, as the target closes at a constant altitude. This gives rise to a cosec^2 radiation diagram. Therefore

$$G = G_0 \mathrm{cosec}^2(\theta_{el}) \tag{19-3}$$

and equation (19-2) now becomes

$$S = \frac{P_{pk} G_0^2 \mathrm{cosec}^4(\theta_{el}) \lambda^2 \sigma}{(4\pi)^3 h_t^4 \mathrm{cosec}^4(\theta_{el})} = \frac{P_{pk} G_0^2 \lambda^2 \sigma}{(4\pi)^3 h_t^4} \tag{19-4}$$

which is independent of θ_{el} and R_S. The cosecant-squared radiation pattern ensures that a constant signal strength is received from a target flying at a constant altitude. The SNR and detection performance of the radar therefore remain fixed throughout the flight path of the target.

In practice, a cosec^2 radiation diagram is implemented over a restricted range of elevation angles since it would demand infinite gain at zero elevation and zero gain at the vertical. The final radiation diagram would end up looking like that shown in Figure 19-3.

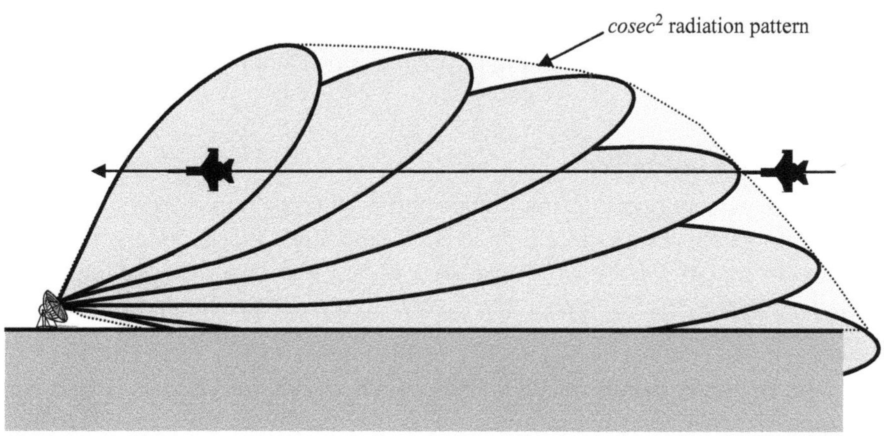

FIGURE 19-3 cosec² Radiation Diagram

The cosec^2 radiation diagram can easily be combined with the stacked beam approach by weighting the gain of each beam appropriately. Since both the transmit and receive beams must have the cosec^2 gain pattern, transmission can be made by feeding the transmitted power through all the channels so that the transmitted beam pattern is taken as the sum of all the stacked (receive) beams.

19.4 | PARAMETERS OF A HYPOTHETICAL SHORT-RANGE GROUND-BASED AIR SURVEILLANCE RADAR

19.4.1 Introduction

The pulse Doppler radar technique has an important application in short- and medium-range ground-based air defense systems. It is used as a surveillance system for initial target detection and threat evaluation, both of which are enabled significantly by the use of pulse Doppler. Having detected a target, the surveillance radar cues a high-resolution radar to track the target in support of a missile engagement. The surveillance and tracking functions may be combined into a single multi-function radar using an AESA antenna.

19.4.2 Target Detection

A short- to medium-range missile has a typical range of about 10 km, so the surveillance radar system used for initial target detection requires a range of around 30 km against low-flying fixed-wing aircraft, helicopters, and possibly surface-directed missiles. This extra range gives the operator a greater situational awareness and enables the radar to preempt engagements against distant targets before they encroach within lethal range. Targets can be handed over to the tracker at 20 km of range, which still leaves time for the tracking radar to acquire the target and guide a missile to an interception at a range of 10 km. Low altitude targets have to be detected against a background of considerable surface clutter; furthermore, the Doppler shift from slow-moving or stationary targets merges with the clutter. High altitude targets, which may also have small radar cross section (e.g., missiles and agile fighter aircraft seen head-on), must be detected against weather clutter but are more likely to be fast moving, which separates their echoes both spatially and in Doppler from surface clutter. Clearly, there is a requirement for clutter rejection through MTI or pulse Doppler processing, which should ideally be dependent on elevation angle.

19.4.3 Threat Evaluation

It has already been mentioned that long-range initial detection improves situational awareness, which obviously allows more time to assess the threat that each target poses to the point or area being defended and for decisions to be taken regarding the target priority. However, the use of a pulse Doppler radar assists target prioritization in a number of other areas:

- Movement of target toward or away from point or area of defense. Pulse Doppler processing presents real-time information on the speed and direction of targets from the air defense system. Aggressive maneuvers by targets such as rapid closing speeds at low levels are immediately recognized and attract the higher priorities.

- Coverage of target by weapons system. Knowledge of a target's position, speed, and direction of travel and the rates of changes of these parameters, together with the direction of any changes reveal all aspects of the target's current flight pattern. This knowledge can be used to compute whether the target is within the range of the weapon system or whether it will soon enter or leave the range of the weapon system given its current flight pattern.

19.4.4 Radar Parameters

The parameters of a hypothetical ground-based short-range air surveillance radar are summarized in Table 19-1. The proposed solution is a medium pulse repetition frequency (PRF) pulse Doppler approach [14] based on a 3-*of*-5 detection scheme. This is a very short schedule but is necessary given the short illumination time of a rapidly scanned narrow beam. The relatively high value for the mean PRF (20 kHz) results only two orders of velocity ambiguity and only four orders of range ambiguity. It should be perfectly feasible to select precise values of the PRFs to support range and velocity decodability, especially as the decodable range/velocity space of interest is quite modest [14]. Furthermore, sidelobe clutter (SLC) does not obscure the detection of moving targets, so there is no blindness due to overwhelming SLC. The coarse range and velocity resolution is adequate to determine the threat level of targets and provides data of sufficient resolution and accuracy to cue a tracking radar to take over and support a missile engagement.

TABLE 19-1 ■ Typical Ground Based Short Range Air Surveillance Radar Parameters

Parameter	Value
Carrier Frequency, f_{tx}	12 GHz
Antenna Size	1m (horizontal) $\times$ 0.5 m (vertical)
Antenna Gain	6 700 (38.3 dB$_i$)
Antenna Beamwidth	2° (azimuth) $\times$ 4° (elevation)
Stacked Beams	8 beams giving coverage from $-2°$ to $+30°$
Antenna Sidelobe Level	$<$ 35 dB lower than main beam
Antenna Scan Rate	60 rpm (360°/s)
Beam Dwell Time	5.6 ms
Ambiguity Resolution	3-of-5 coincidence algorithm
PRFs	20 kHz (mean)
FFT Size	16
Space Charging	300 μs (45 km)
CPI Duration	800 μs
Velocity Resolution	15.6 m/s
MBC/GMT Rejection Notch Width	$\pm$16 m/s ($\pm$1280 Hz)
Transmitted Pulse Width	6.5 μs (fixed)
Duty Ratio	Variable, 13% (mean)
Minimum Range	1 km
Pulse Compression	13-bit biphase Barker coding
Range Resolution	75 m
Maximum Range of Interest	30 km
Maximum Velocity of Interest	$\pm$500 m/s ($\pm$40 kHz)

References

[1] Long, W. H. and Harriger, K. A., "Medium PRF for the AN/APG-66 Radar," *Proceedings of the IEEE*, 73(2), February 1985, 301–311.

[2] Ringel, M. B., Mooney, D. H., and Long, W. H., "F-16 Pulse Doppler Radar (AN/APG-66) Performance," IEEE Transactions on Aerospace & Electronic Systems, AES-19(1), January 1983, 147–158.

[3] Simpson, J., "PRF Set Selection for Pulse-Doppler Radars," in IEEE Region 5 Conference, 1988: Spanning the Peaks of Electrotechnology, 1988, 38–44.

[4] Kinghorn, A. M., Smith, M. J., and Heggie, M. J., "Pulse-Doppler Radar Detection Performance Modelling and Verification," in IEE Colloquium on Computer Modelling and Simulation of Radar Systems, London, UK, February 1983, 8/1–8/4.

[5] Morris, G. V., Airborne Pulse Doppler Radar, Artech House, 1998.

[6] Lynch, Jr., D. and Kopp, C., "Multifunctional Radar Systems for Fighter Aircraft," Chapter 5 in M. I. Skolnik, Radar Handbook, 3d ed., McGraw-Hill, 2008.

[7] Morchin, W. C., Airborne Early Warning Radar, Artech House, 1990.

[8] Long, M. W., Ed., Airborne Early Warning Systems Concepts, Artech House, 1992.

[9] Schleher, D. C., Introduction to Electronic Warfare, Artech House, 1994.

[10] Currie, N. C. and Brown, C. E., Eds., Principles and Applications of Millimeter-Wave Radar, Artech House, 1987.

[11] Klein, L. A., Millimeter-Wave and Infrared Multisensor Design and Signal Processing, Artech House, 1997.

[12] Blanton, J. L., "Cued Medium-PRF Air-to-Air Radar Using Stretch Range Compression," in *Proceedings of the Radar Conference*, Ann Arbor, MI, May 1996, 208–213.

[13] Yi, J. W. and Byun, Y. J., "Real-Time PRF Selection for Medium PRF Airborne Pulsed-Doppler Radars in Tracking Phase," in *Proceedings of* Waveform Diversity & Design, Pisa, 2007, 116–121.

[14] Nooy, P. N. C. And Debuisser, J.-C., "Medium PRF Waveforms: An Optimal Solution for Various Applications in Surface Based Radar," in *Proceedings of the 28th* European Microwave Conference, Amsterdam, 1998, 572–577.

Closing Remarks

Part I of this book (Chapters 1 to 8) covered basic principles of radar, much of which is covered in similar text books. It laid the foundations for the material specifically covering pulse Doppler radar that was to come in later chapters. Special emphasis was given to continuous wave Doppler sensing radars and fast Fourier transform processing techniques as they apply to Doppler sensing radars. Three explanations of fast Fourier transform processing were offered: an intuitive approach; the matrix transformation; and some shortcuts to simplify the processing and their depiction using butterfly diagrams.

Part II dealt with the techniques of pulse Doppler radar. This was subdivided into Part IIA (Chapters 9 and 10), covering high pulse repetition frequency (PRF) pulse Doppler and its associated techniques of frequency modulated interrupted continuous wave ranging, and Part IIB (Chapters 11 to 15), which were devoted to medium PRF pulse Doppler. It drew on standard material, the author's own experience, and recent research work (at least what is published in open-source material up until the time of writing, July 2011). Part II gave a thorough treatment of some quite difficult subjects to impart a deep and up-to-date understanding of the subject for engineers, users, those involved in the specification and procurement of radar equipment, and researchers, academics, and postgraduate students with an interest in radar. Of course, it would be perfectly feasible to apply this material to reverse-engineer a radar, in which case the readership may be extended to those in the radar electronic warfare community.

Part III (Chapters 16 to19) presented four case studies featuring different military applications of pulse Doppler radar. These case studies illustrate practical implementations of the techniques covered in the earlier chapters; they are in many respects the culmination of all the previous material. Each case study also tended to highlight specific radar techniques that are peculiar to the application. All the case studies presented parameters of hypothetical, albeit realistic, examples. These examples are best understood in the context of a much more significant weapon system of which they form a key component. While Part III presents four types of pulse Doppler radars, there are many more military applications of this technology. Additionally, any of the following applications might be equally considered: weapons locating radars, battlefield surveillance, long-range surveillance, naval multi-function radars, and high-frequency over-the-horizon radars. Then again, there are many other specialist radars using pulse Doppler or pulse Doppler-like techniques, such as synthetic aperture radars for ground mapping and inverse synthetic aperture imaging radars, ground-moving target indication radars for the location of moving ground-based and low altitude targets, radar altimeters (frequency modulated continuous wave radars), and fuses. These lists are not exhaustive, but the four case studies considered here suffice to illustrate the relevance of the technology.

In spite of the prevalence of pulse Doppler radar, the open press is not full of academic papers on the subject. However, it would be incorrect to assume that the subject does not attract research interest or that the technology has stood still. Much of the work conducted in industry or private consultancy is, by its nature, not openly available. The apparent lack of publications should not be taken for a lack of interest. Many of the recent advances are in related subjects such as semiconductor materials, microwave devices, computational techniques and algorithms, and, of course, processing power. Nevertheless, a trickle of papers has been making an appearance in the open literature. The subject of pulse Doppler radar is indeed a mature one but is not entirely honed to perfection. Advances continue to be made, and there remains plenty of work to be done. As the technology continues to evolve and the costs reduce, radars are increasingly being used in many new ways in both the military and civilian sectors. Pulse Doppler radar has been the mainstay of military radar sensors for many years and looks set to remain so for many years to come.

Index

E

F

www.ingramcontent.com/pod-product-compliance
Ingram Content Group UK Ltd.
Pitfield, Milton Keynes, MK11 3LW, UK
UKHW050950290926
15470UKWH00005B/1744

9 781891 121982